Analytical Profiles of Drug Substances

Analytical Profiles of Drug Substances

Volume 20

Edited by

Klaus Florey

Bristol-Myers Squibb
New Brunswick, New Jersey

Contributing Editors

Abdullah A. Al-Badr
Gerald S. Brenner
Harry Brittain
George A. Forcier
David J. Mazzo
Timothy J. Wozniak

ACADEMIC PRESS, INC.
Harcourt Brace Jovanovich, Publishers
San Diego New York Boston London Sydney Tokyo Toronto

Academic Press Rapid Manuscript Reproduction

This book is printed on acid-free paper. ♾

Academic Press, Inc.
San Diego, California 92101

United Kingdom Edition published by
Academic Press Limited
24–28 Oval Road, London NW1 7DX

Library of Congress Catalog Number: 89-659072

International Standard Serial Number: 0099-5428

International Standard Book Number: 0-12-260820-8

PRINTED IN THE UNITED STATES OF AMERICA
91 92 93 94 9 8 7 6 5 4 3 2 1

CONTENTS

AFFILIATIONS OF EDITORS AND CONTRIBUTORS

H. Abdel-Aliem, Faculty of Pharmacy, University of Mansoura, Mansoura 35516, Egypt

Mohammad A. Abounassif, College of Pharmacy, King Saud University, Riyadh 11451, Saudi Arabia

Iqbal Ahmad, Faculty of Pharmacy, University of Karachi, Karachi-75270, Pakistan

Tauqir Ahmad, Faculty of Pharmacy, University of Karachi, Karachi-75270, Pakistan

Abdullah A. Al-Badr, College of Pharmacy, King Saud University, Riyadh 11451, Saudi Arabia

Abdulrahman M. Al-Obaid, College of Pharmacy, King Saud University, Riyadh 11451, Saudi Arabia

Fahad J. Al-Shammary, College of Applied Medical Sciences, King Saud University, Riyadh 11451, Saudi Arabia

John F. Bauer, Abbott Laboratories, North Chicago, Illinois 60064

F. Belal, Faculty of Pharmacy, University of Mansoura, Mansoura 35516, Egypt

Susan J. Bogdanowich, Bristol-Myers Squibb, New Brunswick, New Jersey 08903

Gerald S. Brenner, Merck Sharp and Dohme Research Laboratories, West Point, Pennsylvania 19486

Glenn A. Brewer, Bristol-Myers Squibb, New Brunswick, New Jersey 08903

Harry G. Brittain, Bristol-Myers Squibb, New Brunswick, New Jersey 08903

David E. Bugay, Bristol-Myers Squibb, New Brunswick, New Jersey 08903

Zui L. Chang, Abbott Laboratories, North Chicago, Illinois 60064

Alexandra Davies, Bracco Industria Chimica Spa, Milan, Italy

Joseph DeVincentis, Bristol-Myers Squibb, New Brunswick, New Jersey 08903

Maher M. El-Domiaty, College of Pharmacy, King Saud University, Riyadh 11451, Saudi Arabia

Klaus Florey, Bristol-Myers Squibb, New Brunswick, New Jersey 08903

George A. Forcier, Pfizer Inc. Central Research, Groton, Connecticut 06340

Lee T. Grady, The United States Pharmacopeia, Rockville, Maryland 20852

Maurizio Grandi, Bracco Industria Chimica Spa, Milan, Italy

John R. Hicks, Lilly Research Laboratories, Eli Lilly and Company, Indianapolis, Indiana 46285

Mohamed S. Hifnawy, Faculty of Pharmacy, Cairo University, Cairo, Egypt

Leonard J. Kostek, Pfizer Inc. Central Research, Groton, Connecticut 06340

Geoffrey Lewen, Bristol-Myers Squibb, New Brunswick, New Jersey 08903

G. William Martin, Burroughs Wellcome Co., Research Triangle Park, North Carolina 27709

Claude Mazuel, Pharmaceutical Research & Development, Merck Sharp & Dohme-Chibret, Riom, France

David J. Mazzo, Rhone-Poulenc Rorer Central Research, Fort Washington, Pennsylvania 19034

Mohammad Saleem Mian, College of Pharmacy, King Saud University, Riyadh 11451, Saudi Arabia

Neelofur Abdul Aziz Mian, College of Applied Medical Sciences, King Saud University, Riyadh 11451, Saudi Arabia

Ezzat M. Abdel Moety, College of Pharmacy, King Saud University, Riyadh 11451, Saudi Arabia

Farid J. Muhtadi, College of Pharmacy, King Saud University, Riyadh 11451, Saudi Arabia

Ann W. Newman, Bristol-Myers Squibb, New Brunswick, New Jersey 08903

Cynthia L. Obetz, Rhone-Poulenc Rorer Central Research, Fort Washington, Pennsylvania 19034

Davide Pitré, Faculty of Pharmacy, University of Milan, Milan, Italy

T. A. Plomp, The Netherlands Institute for Drugs and Doping Research, University of Utrecht, 3521 GE Utrecht, The Netherlands

T. Negendra Vara Prasad, TTK Pharma, Madras 600-043, India

E. Venkata Rao, Department of Pharmaceutical Sciences, Andhra University, Waltair-530 003, India

B. Sarveswara Sastry, Department of Pharmaceutical Sciences, Andhra University, Waltair-530 003, India

C. Surya Prakasa Sastry, School of Chemistry, Andhra University, Waltair-530 003, India

Manohar L. Sethi, College of Pharmacy, Howard University, Washington, D.C. 20059

Judith E. Shuster, Rhone-Poulenc Rorer Central Research, Fort Washington, Pennsylvania 19034

Edward Smith, Food and Drug Administration, Rockville, Maryland 20853

Mohammad Tariq, College of Pharmacy, King Saud University, Riyadh 11451, Saudi Arabia

K. Usmanghani, Faculty of Pharmacy, University of Karachi, Karachi-75270, Pakistan

Timothy J. Wozniak, Lilly Corporate Center, Indianapolis, Indiana 46285

PREFACE

Although the official compendia define a drug substance as to identity, purity, strength, and quality, they normally do not provide other physical or chemical data, nor do they list methods of synthesis or pathways of physical or biological degradation and metabolism. Such information is scattered through the scientific literature and the files of pharmaceutical laboratories.

I perceived a need to supplement the official compendial standards of drug substances with a comprehensive review of such information, and nineteen years ago the first volume of *Analytical Profiles of Drug Substances* was published. That we have been able to publish one volume per year is a tribute to the diligence of the editors to solicit articles and even more so to the enthusiastic response of our authors, an international group associated with pharmaceutical firms, academic institutions, and compendial authorities. I would like to express my sincere gratitude to them for making this venture possible.

Over the years, we have had queries concerning our publication policy. Our goal is to cover all drug substances of medical value, and therefore, we have welcomed any related articles of interest.

With the increasing sophistication of formulating drug products of reproducible quality, the precise physico-chemical characterization of pharmaceutical excipients has become of great importance. On the initiative of Dr. Harry Brittain, the Editorial Board has decided to include profiles of well-defined excipients in our series. The profile on anhydrous lactose in this volume is the first such effort and should serve as a model for future contributions in this area.

Klaus Florey

ANALYTICAL PROFILE OF AMIODARONE

T.A. Plomp

The Netherlands Institute for Drugs and Doping
Research, University of Utrecht,
Vondellaan 14, 3521 GE Utrecht, The Netherlands

CONTENTS

Table 1

Molecular weight of amiodarone analogues

Compound*)	M.W.	Compound	M.W.
1. LB 33530	589.22	4. L3373	545.91
2. L 6355	519.42	5. L 6424	420.02
3. L 3937	393.51	6. L 3372	294.13

M.W. = molecular weight

*) structural formulae are shown in Fig. 2

1.4. Elemental composition

The elemental composition of amiodarone and desethylamiodarone and their hydrochlorides are summarized in Table 2.

Table 2

Elemental composition of amiodarone and desethylamiodarone

Compound	Element (%)					
	C	H	Cl	I	N	O
Amiodarone	46.53	4.53	-	39.33	2.17	7.44
Amiodarone hcl	44.04	4.44	5.20	37.25	2.06	7.04
Desethylamiodarone	44.80	4.08	-	41.10	2.27	7.78
Desethylamiodarone hcl	42.30	4.00	5.42	38.80	2.14	7.34

1.5. Appearance, Color, Odor and Taste

Amiodarone hydrochloride is a white to pale beige, fine powder. It is odorless or almost odorless with almost no taste.
Desethylamiodarone hydrochloride is a fine, white, odorless powder.

1.6. Dosage forms

White to off-white scored round biconvex tablets. containing 100 mg and 200 mg of amiodarone hydrochloride (Cordarone ® 100 and 200, Sanofi Labaz).
Injection, contains per ml water 50 mg of amiodarone hydrochloride, 20 mg of benzylalcohol and 100 mg of polysorbate 80 in ampoules of 3 ml (Cordarone ®i.v., Sanofi Labaz).

2. PHYSICAL PROPERTIES

2.1. Melting point and melting range

The melting points for amiodarone hydrochloride reported in the literature are 156°C (2), 159 ± 2°C (4) and 161°C (3).
The melting range for amiodarone hydrochloride reported by the manufacturer (Sanofi-Labaz, Brussels, Belgium) is 159°C - 163°C and 177°C - 178°C respectively (5). The results of own melting point and melting range analyses of 4 different batches of reference substance of amiodarone hydrochloride and 2 batches of desethylamiodarone hydrochloride supplied by Sanofi-Labaz are summarized in Table 3. In this table also the melting point data of the amiodarone analogues are included. The melting point analyses were carried out on a Mettler melting apparatus

Table 3

Melting points and melting ranges of amiodarone, desethylamiodarone and amiodarone analogues

Compound*	Melting Point °C Heating Rate 2°C/min	Melting Point °C Heating Rate 10°C/min	Melting Range °C Heating Rate 10°C/min
Amiodarone hcl	156.5	162.6 ± 0.8	162.0-162.9
Desethylamiodarone hcl	178.0	181.7 ± 1.1	180.7-181.6
LB 33530	-	210	-
L 6355**	n.r.	106.7	94.2-96.0
L 3937***	130.3	133.1	131.9-132.0
L 3373	144.9	147.1	144.7-146.5
L 6424	146.4	149.2	148.0-149.0
L 3372	119.5	121.1	121.0-122.7

- = not performed

n.r. = not reproducible

* = structural formula of compounds, see Figs. 1 and 2

** = oxalate salt

*** = hydrochloride salt

model FP1 using heating rates of 2°C/min and 10°C/min respectively. Amiodarone hydrochloride and desethylamiodarone hydrochloride and most of the amiodarone analogues melt with decomposition.

2.2. Solubility

Amiodarone hydrochloride was found to be freely soluble in chloroform and methanol, soluble in ethanol, sparingly soluble in isopropanol, slightly soluble in acetone, dioxane, ether and carbon tetrachloride and very slightly soluble in water (5). However, a 5% (w/v) aqueous solution of the drug may be prepared due to autosolubilisation. Amiodarone hydrochloride is practically insoluble in alkali hydroxides. The solubility of desethylamiodarone hydrochloride in various solvents is comparable to that of amiodarone hydrochloride. It is freely soluble in methanol and chloroform and very slightly soluble in water. Bonati et al. (4) studied the solubility of amiodarone hydrochloride in various solvents at 25°C. The data are summarized in Table 4.

Table 4

Solubility of amiodarone hydrochloride in various solvents at 25°C

Solvent	Solubility (g/100 ml)
Chloroform	44.51
Methylene chloride	19.20
Methanol	9.98
Ethanol	1.28
Benzene	0.65
Tetrahydrofuran	0.60
Acetonitrile	0.32
1-Octanol	0.30
Ether	0.17
1-Propanol	0.13
Water	0.07
Hexane	0.03
Petroleum ether	0.001

2.3. pH-value

A 5% solution of amiodarone hydrochloride was reported to have a pH of 3.4 to 3.9 (5) or 3 to 4 (6). The solution used for the determination of the pH value was prepared by dissolving 1.25 g of the compound in 25 ml of bidistilled water of 80°C. After cooling down and a waiting period of 15 min the pH of the solution was measured with a pH meter equipped with a glass electrode (5).

2.4. Loss on Drying

Not more than 0.5%, determined with 2.00 g by drying for 4 hours at 50°C over magnesium perchlorate at a pressure not exceeding 5 mm Hg (5).

2.5. Dissociation constant

The apparent dissociation constant (pKa) was determined spectrophotometrically by Bonati et al. (4) and found to be 6.56 ± 0.06 at 25°C. This value was substantially higher than the previous reported pKa of 5.6 (5,7). Own pKa measurements of solutions of 4 different batches of amiodarone hydrochloride reference substance using the spectrophotometric method of Clarke et al. (8) showed a mean pKa value of 6.64 ± 0.28. Assay of the pKa value of solutions of 3 different reference batches of desethylamiodarone hydrochloride showed a mean value of 5.58 ± 0.35. The pKa value of both compounds was assessed by UV titration at λ max of aqueous solutions containing 10 μg/mlof the drugs respectively.

A range of pH 1.4 - 3.5 was obtained by addition of various volumes of 0.1 M. hydrochloric acid, pH 5.5 - 8.8 by various volumes of phosphate/sodium hydroxide and borate/sodium hydroxide buffers, while values to pH 11.4

were reached by addition of various volumes of 0.1 M. sodium hydroxide. All solutions were prepared immediately before use by diluting methanolic stock solution of the drugs (1 mg/ml) with the various buffers to concentrations of 10 μg/ml. The pH values were checked at 20°C using a standard pH-meter.

The dissociation constant was determined according to the equation

$$pKa = pH + {}^{10}\log \frac{E\ base - E\ buffer}{E\ buffer - E\ acid}$$

where E acid and Ebase are the specific extinctions ($E^{1\%}$1 cm) of the drug in acidic and in alkaline solution and where E buffer is specific extinction of the drug in buffer of known pH which approximate the pKa value. In the dissociation constant experiments we also found that the UV absorption curves of amiodarone and desethylamiodarone recorded at the differend pHs showed characteristic bathochromic shifts of the maximum absorbance (λ_{max}) from pH 5.5 to 8.8. A shift of λ_{max} from 242 nm to 252 nm and from 242 nm to 250 nm were observed respectively for amiodarone and desethylamiodarone with in both cases a decrease of the specific extinction. These findings were in good agreement with previous reported observations (4,7).

2.6. Partition coefficient

Riva et al (9) found the apparent partition coefficient (P_{app}) of amiodarone hydrochloride between n-octanol and pH 7.4 phosphate buffer to have a value of 80 at room temperature where P_{app} = Coctanol/Cbuffer. The concentrations of amiodarone in n-octanol (C octanol) and buffer pH 7.4 (C buffer) were measured with a reversed phase liquid chromatographic method. The apparent (P_{app}) and true (P_{corr}) partition coefficients at 25°C of amiodarone hydrochloride between n-octanol and buffers with pH values varying from 2.5 to 12.0 were reported by Bonati et al (4). The P_{app} was calculated from the concentrations of amiodarone, determined spectrophotometrically, present in the various buffer solutions before (C_1) and after (C_2) shaking for 2 hours.

P_{app} = C octanol / C buffer = $(C_1 - C_2)/C_2$.

The P_{corr} was calculated as: $P_{corr} = P_{app}/(1-\alpha)$, where α is the degree of ionization at a given pH.

The apparent partition coefficient of amiodarone was found to proceed from ca 5 at pH 2.5 to 0.2 at pH 12 with a maximum of 7 between pH 3.5 and 5.5. The corrected partition coefficients at pH 7.5 and 8.5 were approximately 70 and 300 respectively. Own measurements of apparent partition coefficient between n-octanol and 0.01 M phosphate buffer pH 7.4 showed a mean value of 82 ± 34 (n=16) . The assay was carried out by gently shaking for 1 h of a mixture of 5 ml phosphate buffer pH 7.4 containing 10 µg/ml of amiodarone hydrochloride with 5 ml of n-octanol. After separation of the phases by centrifugation the concentrations of the drug in the aqueous and organic layer were determined using an own developed reversed phase high-performance liquid chromatographic method (10). Our P_{app} findings were similar to those of Riva et al. (9), but they were substantially higher than the P_{app} reported by Bonati et al. (4). This discrepancy may be explained by the more specific assay method used in our procedure.

2.7. Stability

2.7.1. Dry State

Amiodarone hydrochloride in solid state, stored in a dark, closed container at room temperature showed no decomposition after 3 years, when examined by TLC, HPLC and UV-spectrophotometry. Desethylamiodarone hydrochloride in dry state, stored at room temperature was also stable for at least 3 years.

2.7.2. Aqueous solutions

Andreasen et al. (11) reported a decrease in concentration of 20% after 12 min standing at 20°C and of 90% after 4 days storage at 5°C respectively of a solution of 5 μg/ml of amiodarone hydrochloride in phosphate buffer of pH 7.4. The results of own experiments with solutions of amiodarone hydrochloride and desethylamiodarone hydrochloride reference substance in distilled water, phosphate buffer pH 7.4, TRIS buffer pH 7.4 and 50% methanol in disposable polypropylene-capped glass centrifuge tubes (16 x 100 mm) are summarized in Table 5.

Table 5

Stability of amiodarone (A) and desethylamiodarone (DA) in various aqueous solutions

Solution	Loss in potency of 50 μg/ml solutions in %			
	Standing *) for 1 h at 20°C		Shaking for 1 h at 20°C	
	A	DA	A	DA
distilled water	5	10	40	60
50% methanol	0	0	1	4
phosphate buffer pH 7.4	5	28	57	73
TRIS buffer pH 7.4	0	18	80	90

*) Loss of potency of amiodarone after 4 h: 14% and after 3 days: 69%

The stability of the compounds in the various aqueous solutions was tested in the following way: one series of samples (5.00 ml) was kept at 20°C for

1 h and the other were horizontally, gently shaken for 1 h with a mechanic shaker (Marius type 67SL, Utrecht, The Netherlands). Before and after the testing period volumes of 100 μl were withdrawn from the tubes and after mixing with 900 μl of internal standard solution in methanol (5 μg/ml) subjected to HPLC analysis for determination of the concentrations (10). The data in Table 5 show that amiodarone hydrochloride and desethylamiodarone hydrochloride are not stable in aqueous solutions and their decomposition is enhanced by shaking.

2.7.3. Organic solvents

Methanolic solutions of amiodarone hydrochloride and desethylamiodarone hydrochloride in concentrations varying from 0.1 μg/ml to 1000 μg/ml were stable for at least 3 months if stored at 4°C in the absence of light in glass tubes (10,12).

Solutions of both compounds are also stable in 50% methanol and in 50% ethanol when stored at 4°C in the dark (11).

The solutions of amiodarone hydrochloride and desethylamiodarone hydrochloride in acetonitrile at concentrations of 2 μg/ml to 2000 μg/ml were stable for at least 9 months if stored at 0-4°C in glass (13).

In chlorofrom amiodarone hydrochloride undergoes a rapid dehalogenation process and this solvent is therefore unsuitable for preparation of solutions of the drug or for usage as extraction solvent or elution solvent.

2.7.4. Biological fluids

In plasma, serum and in urine constant values for the concentration of amiodarone and desethylamiodarone were found for 1 month if the samples were kept in glass tubes at 4-5°C (10-12). No loss in potency of either drug was observed over a 25 days period, when plasma, urine and bile samples spiked with known concentrations were stored in disposable polypropylene tubes at -10°C (13).

2.8. Incompatibility

The visual compatibility of amiodarone hydrochloride injection (containing amiodarone hydrochloride 50 mg and polysorbate 80 100 mg per ml) with 13 other injectable drugs were investigated by Hasegawa and Eder (14). After mixing of 5 ml of amiodarone hydrochloride in 5% dextrose or 0.9% sodium chloride at a concentration of 4 mg/ml with 5 ml of dilutions of other injectable drugs, they observed no visible alterations after 24 hours in any of the drug mixtures, except in those containing aminophylline. These latter mixtures in both 5% dextrose and 0.9% sodium chloride were hazy at 15 minutes and a white precipitate was evident at six hours. Ravin et al. (15) reported that chloride, acetate, citrate, tartrate and sulfate ions may disrupt the micellar state and depress the equilibrium solubility of amiodarone hydrochloride. This finding is the basis of their recommendation to dilute amiodarone hydrochloride injection only with electrolyte free solutions. Cairns (16) reported the incompatibility of amiodarone hydrochloride with heparin. Precipitation was observed when amiodarone injection and heparin injection were concomitantly administered through a common central venous line (16). Weir et al. (13) reported a decrease in amiodarone and desethylamiodarone concentration of 50% to 60% when 5% dextrose solutions containing 10 μg/ml of both drugs had come in contact with laboratory film or rubber stopper tops for 60 min at ambient temperature. Substantial decrease in amiodarone hydrochloride concentration of about 60% was observed after 5 days storage in polyvinyl chloride bags at room temperature (17). Comparable drop in concentration was found by these authors when amiodarone solutions were perfused through polyvinyl chloride giving sets (17). The losses in potency were atrributed to the presence of the plasticizer di-2-ethylhexylphtalate.

2.9. Spectral Properties

2.9.1. Ultraviolet Spectrum

Amiodarone was first characterized by Bonati et al. (4) as having ultraviolet absorbance maxima at 208 nm and 241 nm with 2 shoulders near 270 nm and 280 nm in methanol. The molar extinction coefficients (e) at these wavelength were 47000, 44000, 18000 and 16000 respectively.

The ultraviolet spectrum of amiodarone hydrochloride in ethanol at a concentration of 20 μg/ml showed according to the manufacturer (Sanofi Labaz, Brussels, Belgium) a maximum at 242 ± 1 nm, a minimum at 222

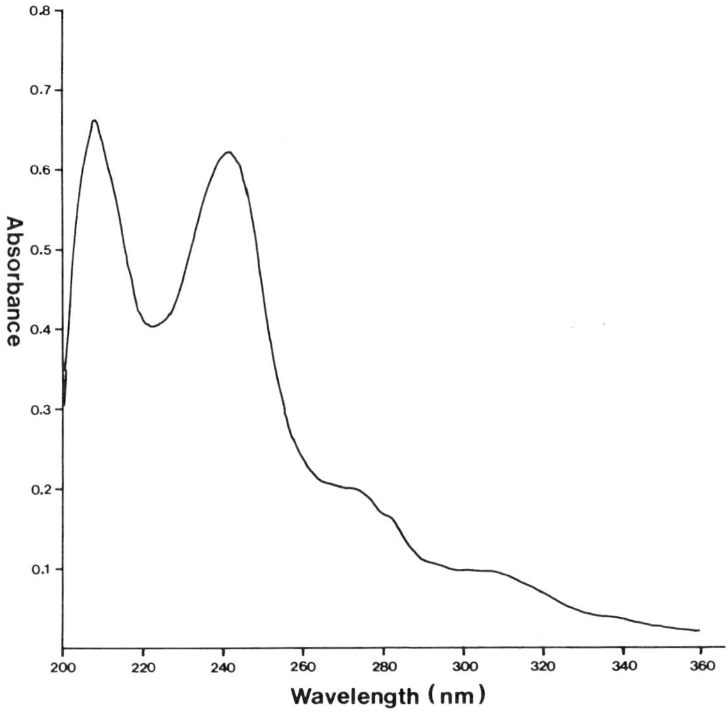

Fig. 3 Ultraviolet spectrum of amiodarone in methanol at concentration of 10 μg/ml

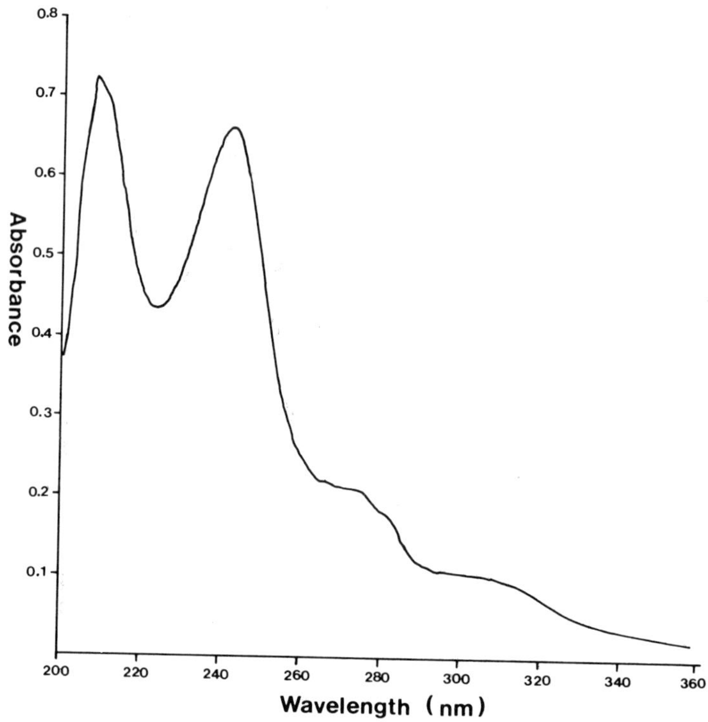

Fig. 4 Ultraviolet spectrum of desethylamiodarone in methanol at concentration of 10 μg/ml

± 1 nm and a flat part between 269 nm and 272 nm (5). Clarke lists a U.V. absorbance maximum at 241 nm in aqueous acid and a λ max at 251 nm in aqueous alkali respectively (7). Own ultraviolet measurements for the assessment of the absorbance maxima (λ max) and minima (λ min), the specific extinction (E 1%, 1 cm) and the molar extinction coefficient (e) of amiodarone, desethylamiodarone and amiodarone analogues in methanol using 1 cm quartz cells were performed on a Pye Unicam SP1800 spectrophotometer. Methanolic solutions of 4 reference batches of amiodarone hydrochloride, 3 batches of desethylamiodarone hydrochloride and 1 batch of the amiodarone analogues all containing 10 μg/ml of the bases were used for the experiments. The mean values of the U.V.

Table 6

The absorbance maxima (λmax) and minima (λmin), specific extinction (E 1%, 1 cm), molar extinction coefficient (e) and specific extinction ratio of amiodarone (A), desethylamiodarone (DA) and desdiethylamiodarone (DDA) in methanol

UV parameters	Compounds *)		
	A	DA	DDA
λmax (nm)	208	208 ± 1	209
	242 ± 1	242 ± 1	242
shoulder (nm)	274 ± 1	275 ± 1	276
	283 ± 1	282 ± 1	282
λmin (nm)	223 ± 1	223 ± 1	224
E(1%, 1 cm)			
208 nm	662 ± 8	718 ± 2	710 ± 6
242 nm	623 ± 10	664 ± 6	669 ± 1
274 nm	195 ± 4	204 ± 6	201 ± 2
283 nm	154 ± 5	168 ± 6	173 ± 2
223 nm	405 ± 6	445 ± 9	454 ± 4
e			
208 nm	43000	44000	42000
242 nm	40000	41000	39000
274 nm	13000	13000	12000
283 nm	10000	10000	10000
223 nm	26000	27000	27000
E 242 nm/223 nm	1.54 ± 0.02	1.49 ± 0.02	1.47

*) For chemical structures see Figs. 1 and 2

parameters of amiodarone, desethylamiodarone, desdiethylamiodarone and amiodarone analogues are summarized in Tables 6 and 7 respectively. Figures 3 and 4 show an ultraviolet absorption spectrum of amiodarone and desethylamiodarone in methanol at a concentration of 10 μg/ml. The spectra

Table 7

The absorbance maxima (λmax) and minima (λmin) specific extinction (E 1%, 1 cm), molar extinction coefficient (e) and specific extinction ratio of amiodarone analogues.

UV parameters	Compounds *)				
	L6355	L3937	L3373	L6424	L3372
λ max	208	208	208	208	208
(nm)	240	277	244	240	233
	275	283	276	283	283
			283		
Shoulder	282	242		276	257
(nm)					276
					290
λ min (nm)	221	254	224	223	222
	262		268	264	262
E(1%,1cm)					
208 nm	724 ± 13	781 ± 8	759 ± 8	860 ± 9	1081 ± 17
± 242 nm	638 ± 2	385 ± 3	636 ± 3	706 ± 2	684 ± 5**)
± 275 nm	314 ± 2	420 ± 2	252 ± 2	343 ± 2	543 ± 2
± 283 nm	286 ± 1	414 ± 4	247 ± 2	356 ± 2	594 ± 2
± 223 nm	376 ± 4	-	419 ± 5	464 ± 2	624 ± 5
± 264 nm	261 ± 1	-	237 ± 1	257 ± 1	365 ± 3
e					
208	38000	31000	41000	36000	32000
± 242 nm	33000	15000	35000	30000	20000**)
± 275 nm	16000	17000	14000	14000	16000
± 283 nm	15000	16000	13000	15000	18000
± 223 nm	20000	-	23000	20000	18000
± 264 nm	14000	-	13000	11000	11000
E242nm/223nm	1.70	-	1.52	1.52	-

*) For chemical structures see Fig. 2
**) E (1%, 1 cm) or e at 233 nm

were recorded with a Pye Unicam SP1800 spectrophotometer. From the comparison of the ultraviolet parameters of amiodarone and its analogues can be seen that the absorbance maxima and minima and the shoulders found in the UV spectra of amiodarone, desethylamiodarone and desdiethylamiodarone are similar. This could be expected from the minor structural differences of these compounds from each other (Table 6). According to our findings identification of these compounds by UV spectrometry can be obtained from measuring the specific extinction (E 1%, 1 cm) at 242 nm and 223 nm and calculation of the extinction ratio at these wavelengths. The extinction ratio observed in this study for amiodarone of 1.54 ± 0.02 is in good agreement with the ratio of 1.54 ± 0.05 reported by Crispin et al (5).

Amiodarone and its N-deethylated derivatives can be differentiated from the other amiodarone analogues (Table 7) by the presence in the latter compounds of additional maxima at ±275 nm and/or ± 283 nm and an additional minimum at ± 264 nm. The absence of λmax at 242 nm in the UV spectrum of L3937 and L3372 is typical for amiodarone analogues not containing iodine (Table 7).

Table 8

Infrared band assignments for amiodarone hydrochloride in KBr reported in the literature

Bonati et al. (4)		Crispin et al. (5)	
Wavenumber(cm^{-1})	Assignment	Wavenumber(cm^{-1})	Assignment
3000-3070	ArCH	3060	aromatic C-H stretch
2860-2960	aliphatic CH	2965,2940,2870	aliphatic C-H stretch
2200-2700	tert-amine NH^+	2300-2700	tert-amine-NH^+ stretch
1635	diAr C=O	1635	diarylketon stretch
1560,1015	benzofuran C=C	1575,1565,1535,1480	aromatic C-C stretch
1380	aliphatic CH_3	1250,1025,1015	C-O-C stretch
1285	ketonic C-C	900,750	aromatic C-H
1250,1075	ether C-O-C		out of plane deformation
1225,1025	tert-amine C-N		

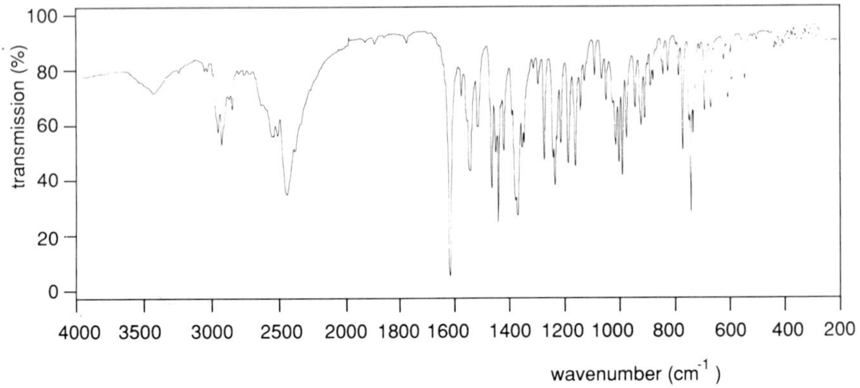

Fig. 5 Infrared spectrum of amiodarone hydrochloride

Sample: 1 mg in 300 mg KBr

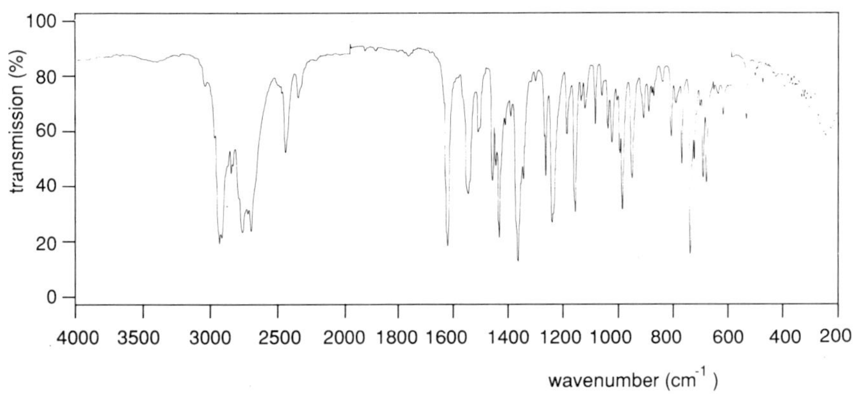

Fig. 6 Infrared spectrum of desethylamiodarone hydrochloride

Sample: 1 mg in 300 mg KBr

2.9.2. Infrared Spectrum

The infrared absorption spectrum of amiodarone hydrochloride in KBr in the range of 4000-600 cm^{-1} has been reported by Bonati et al. (4) and Crispin et al. (5). The major band assignments given by these authors are summarized in Table 8. Clarke (7) reported that the principal peaks for amiodarone hydrochloride in KBr disc in the range of 2000-650 cm^{-1} are at wavenumbers 1630, 1558, 1245, 1170, 998 and 748.

Own infrared spectra of amiodarone hydrochloride and desethylamiodarone hydrochloride are presented in Figures 5 and 6.

The spectra were obtained in KBr at a concentration of 1 mg/300 mg on a Perkin Elmer 283 infrared spectrophotometer in the range of 4000-200 cm^{-1}. The significant absorption bands in the spectra of amiodarone and desethylamiodarone are summarized in Tables 9 and 10 respectively.

Table 9

Infrared spectral assignments of amiodarone hydrochloride in KBr

Band (cm^{-1})	Assignment
3040-3060	aromatic = C-H stretch
2860-2960	aliphatic C-H asymmetric and symmetric stretch
2200-2700	tert-amine N^{+}-H stretch
	(plus overtone and combination bands)
1630	diaromatic C=O stretch
1555 and 1528	aromatic C=C ring-quadrant stretch
1477 and 1454	aromatic C=C ring-semicircle stretch
1380	aliphatic CH_3 symmetric scissoring
	deformation mode
1284	ketonic C=O binding
1250	aromatic ether = C-O-C stretch
750	aromatic C-H out of plane binding of the four
	adjacent aromatic hydrogen atoms

Our IR absorption bands of amiodarone are in good agreement with previously reported ones (4,5). Major differences between the infrared spectrum of amiodarone and desethylamiodarone are only observed in the region between 3000-2000 cm^{-1}, where the bands characteristic for the tertiary amine group of amiodarone (2200-2700 cm^{-1}) and the secundary amine group of desethylamiodarone (2840-2660 cm^{-1}, 2470 cm^{-1}, 2380 cm^{-1}) are located (Figs 5,6).

Table 10

Infrared spectral assignments of desethylamiodarone hydrochloride in KBr

Band (cm^{-1})	Assignment
3060	aromatic = C-H stretch
2860-2960	aliphatic C-H asymmetric and symmetric stretch
2840-2660 and 2470	$-^{+}NH_2$ stretch
and 2380	(plus overtone and combination bands)
1640	diaromatic C=O Stretch
1565 and 1525	aromatic C=C ring-quadrant stretch
1475 and 1450	aromatic C=C ring-semicircle stretch
1380	aliphatic CH_3 symmetric scissoring
	deformation mode
1280	ketonic C=O binding
1255	aromatic ether = C-O-C stretch
750	aromatic C-H out of plane binding of the four adjacent
	aromatic hydrogen atoms

Amiodarone HCL Proton NMR Spectrum in $CDCL_3$ + TMS

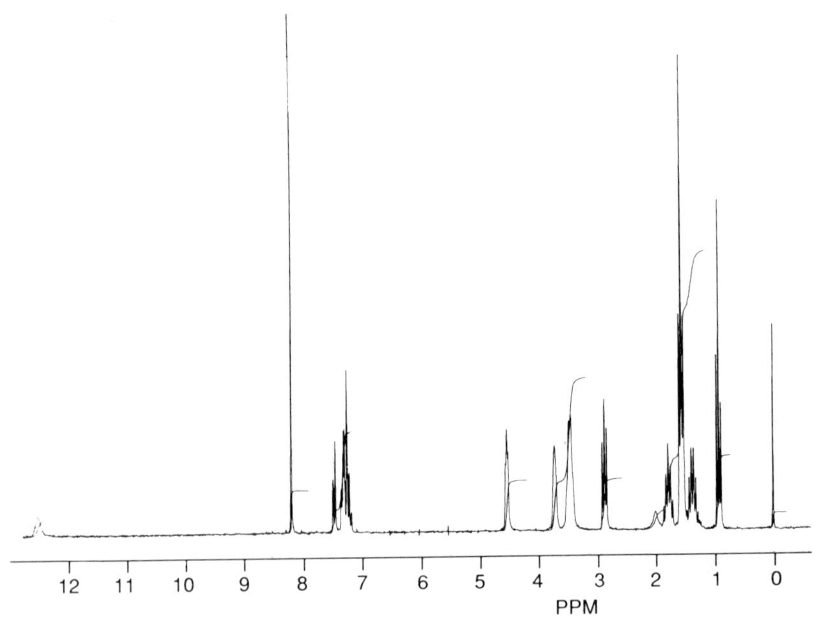

Fig. 7 Proton magnetic resonance (PMR) spectrum of amiodarone hydrochloride

2.9.3. Nuclear Magnetic Resonance Spectra

2.9.3.1. Proton Magnetic Resonance Spectra

The proton magnetic resonance (PMR) spectrum of amiodarone hydrochloride in $CDCl_3$ (20 mg/0.5 ml) is presented in Fig. 7. The spectrum was run on a Bruker WP 200 NMR spectrometer at 200 MHz. Shifts are reported in ppm downfield from tetramethylsilane (TMS) which was used as an internal reference.

The spectral assignments are shown in Fig. 8.

The PMR spectrum of desethylamiodarone hydrochloride in $CDCl_3$ (20 mg/0,5 ml) recorded on a Bruker NMR spectrometer at 200 MHz using

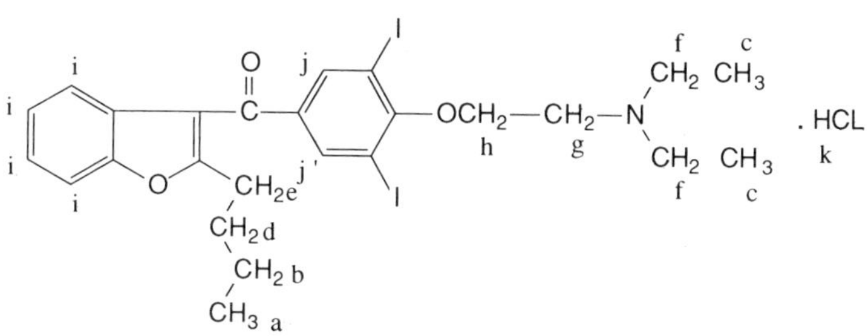

Band (ppm)	Splitting pattern	Assignment
0.93	triplet	a
1.38	multiplet	b
1.56	triplet	c
1.78	quintet	d
2.89	triplet	e
3.48	line broaden quartet	f
3.74	line broaden triplet	g
4.56	line broaden triplet	h
7.21 - 7.50	overlapping multiplets	i
8.22	double doublet	j, j'
12.55	line broaden singlet	k

Fig. 8 The PMR spectral assignments of amiodarone hydrochloride

TMS as an internal reference is shown in Fig. 9. Characteristic peaks of the proton MR spectrum of the compound are summarized in Fig. 10.

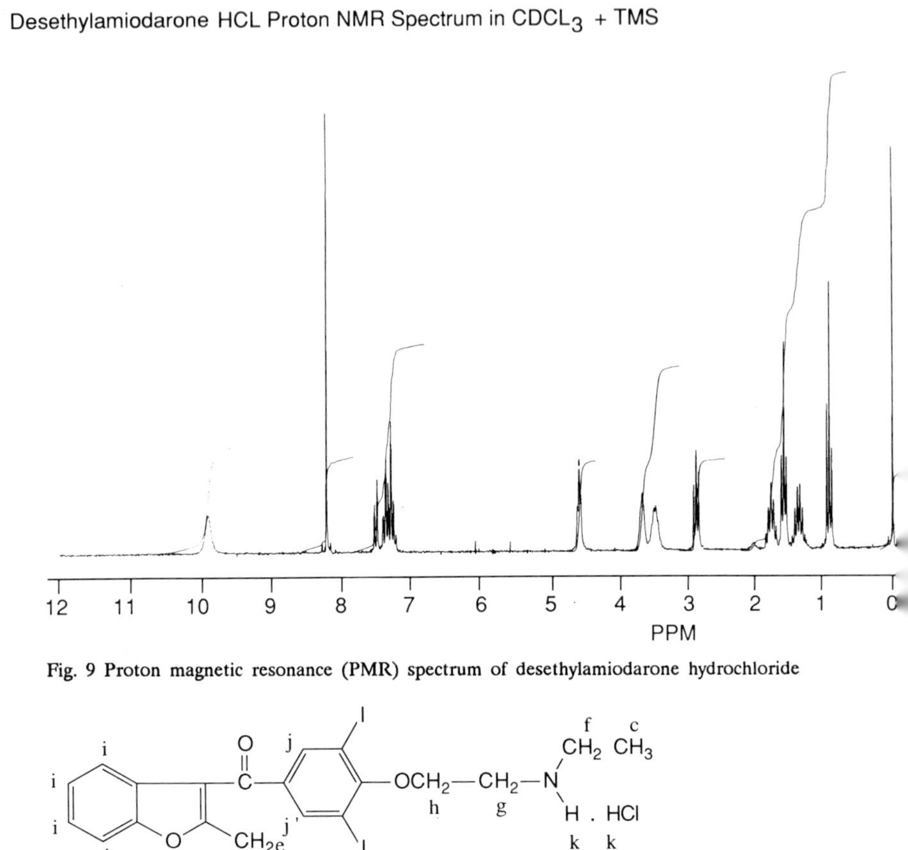

Fig. 9 Proton magnetic resonance (PMR) spectrum of desethylamiodarone hydrochloride

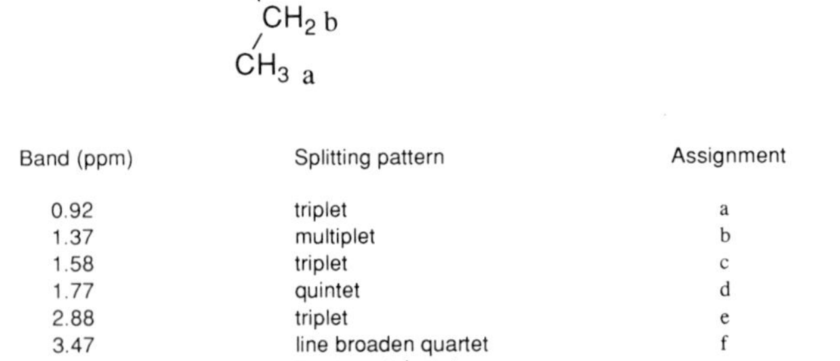

Band (ppm)	Splitting pattern	Assignment
0.92	triplet	a
1.37	multiplet	b
1.58	triplet	c
1.77	quintet	d
2.88	triplet	e
3.47	line broaden quartet	f
3.66	line broaden triplet	g
4.56	line broaden triplet	h
7.20 - 7.51	overlapping multiplets	i
8.20	double doublet	j, j'
9.91	line broaden singlet	k

Fig. 10 The PMR spectral assignments of desethylamiodarone hydrochloride

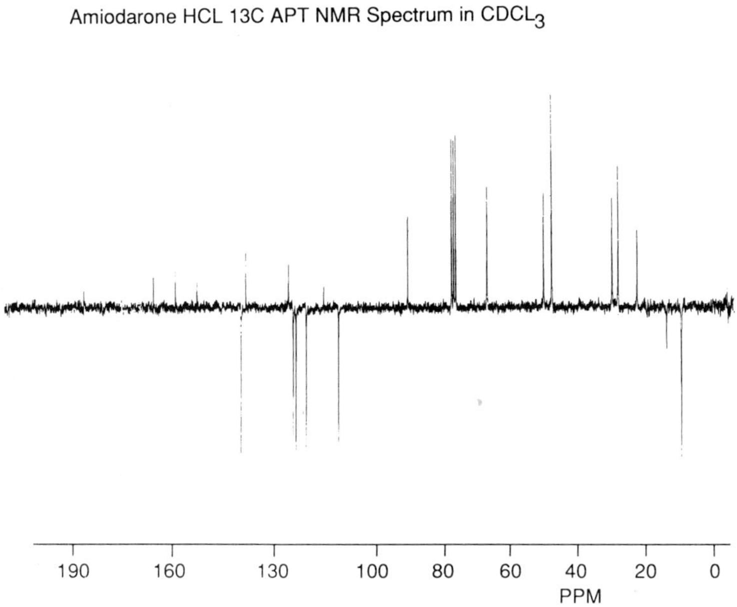

Fig. 11 Carbon magnetic resonance (^{13}C NMR) spectrum of amiodarone hydrochloride

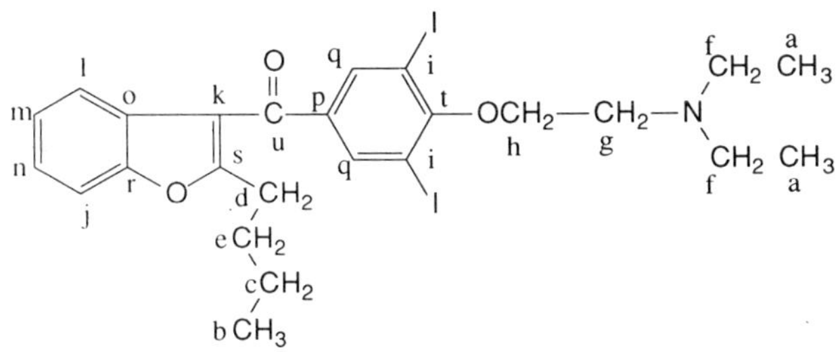

Band (ppm)	Assignment	Band (ppm)	Assignment
9.2	a	120.9	l
13.7	b	123.9	m
22.5	c	124.8	n
28.1	d	126.1	o
30.0	e	139.1	p
47.9	f	140.7	q
50.3	g	153.6	r
67.1	h	160.0	s
90.6	i	166.5	t
111.2	j	187.3	u
115.6	k		

Fig. 12 The ^{13}C NMR spectral assignments of amiodarone hydrochloride

2.9.3.2. Carbon Magnetic Resonance Spectra

The carbon magnetic resonance (^{13}CNMR) of amiodarone hydrochloride was run on a Bruker WP 200 NMR spectrometer operating at 50 MHz. The ^{13}C NMR spectrum of amiodarone hydrochloride in $CDCl_3$(112 mg/0,5 ml) using the APT method (22) is shown in Fig. 11. Shifts are reported in ppm downfield from the shift of tetramethylsilane as referenced to a shift of 77.0 ppm for $CDCl_3$, which was used as an internal reference. The chemical shift assignments are listed in Fig. 12. The ^{13}C NMR spectrum of desethylamiodarone in $CDCl_3$ (100 mg/0.5 ml) using the APT method (22) was obtained using the same NMR instrument and conditions as described for amiodarone and is presented in Fig. 13. Shifts are reported in ppm downfield from the shift of TMS as referenced to a shift at 77.0 ppm for $CDCl_3$. The carbon chemical shift values are derived from the spectrum and are outlined in Fig. 14.

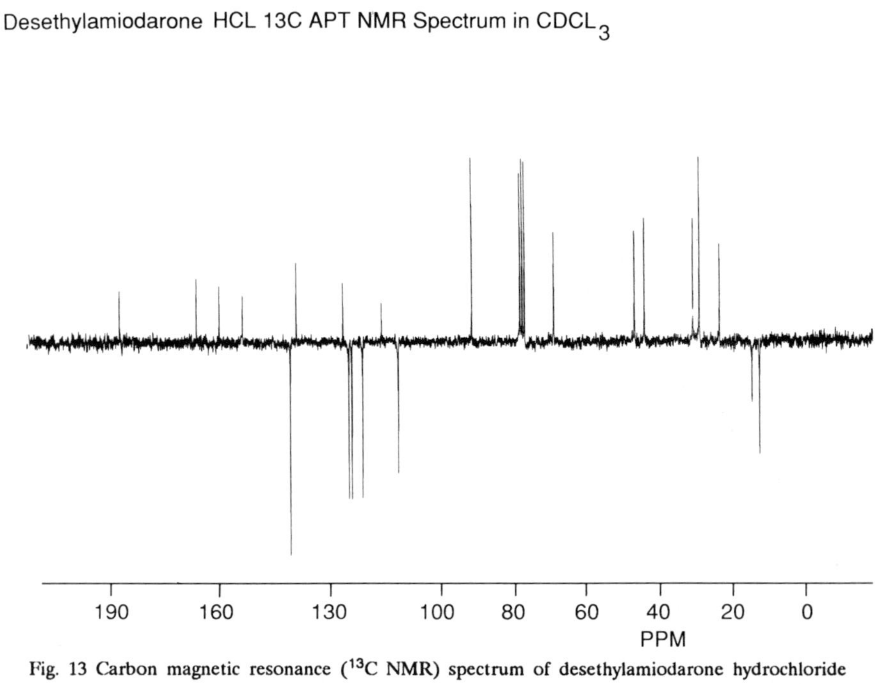

Fig. 13 Carbon magnetic resonance (^{13}C NMR) spectrum of desethylamiodarone hydrochloride

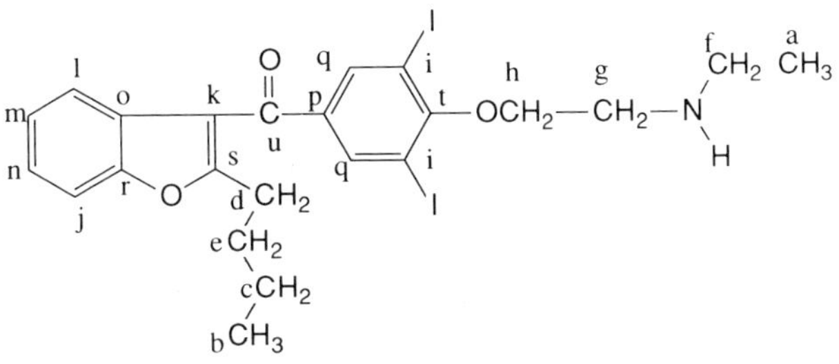

Band (ppm)	Assignment	Band (ppm)	Assignment
11.6	a	120.9	l
13.7	b	123.9	m
22.5	c	124.7	n
28.1	d	126.2	o
30.0	e	138.9	p
43.4	f	140.7	q
46.2	g	153.6	r
68.4	h	160.1	s
90.9	i	166.5	t
111.1	j	187.4	u
115.7	k		

Fig. 14 The ^{13}C NMR spectral assignments of desethylamiodarone hydrochloride

2.9.4. Mass Spectrum

2.9.4.1. Electron-Ionization Impact Spectrum

Direct inlet low-resolution electron impact mass spectra of amiodarone and fourteen analogues have first been described by Flanagan et al. (18) For the electron impact (EI) operation they used an electron energy of 70 eV, a trap current of 200 μA and a source temperature of 190°C - 200°C. Significant ions in the EI spectra were observed for amiodarone at m/z 645 (M^+), 630 $(M\text{-}CH_3)^+$, 546 $(M\text{-}CH_2CH_2N(C_2H_5)_2)^+$ and 86 $(CH_2{=}N(C_2H_5)_2)^+$; for desethylamiodarone at m/z 617 (M^+), 602 $(M\text{-}CH_3)^+$, 546 $(M\text{-}CH_2CH_2NHC_2H_5)^+$ and 58 $(CH_2{=}NHC_2H_5)^+$ and for desdiethylamiodarone at m/z 589 (M^+), 574 $(M\text{-}CH_3)^+$, 546 (M-

$CH_2CH_2NH_2)^+$ and 44 $(CH_2CH_2NH_2)^+$ respectively. The base peak of these compounds was found at m/z 86, 58 and 44 respectively. The molecular ions (M^+) showed a very low intensity and the high mass fragments were also weak and not very abundant although the ion at m/z 546 was observed in all three compounds. Low resolution EI mass spectrometry of amiodarone, desethylamiodarone and desdiethylamiodarone has also been reported by Latini et al. (19). Electron impact spectra of these compounds obtained by direct probe introduction were found to have similar characteristics to those reported by Flanagan et al. (18). The molecular ions were also quite weak and the high mass fragments were also not abundant. This was shown in their study for desethylamiodarone. The principal peaks observed in the mass spectrum of this compound were at m/z 617, 602, 560, 546, 517, 490, 420, 373, 201 and 159 (19). In addition these authors (19) reported one characteristic and relatively intense peak at the high masses (m/z 518, 490, 462) for amiodarone, desethylamiodarone and desdiethylamiodarone respectively due to the loss of one iodine from the molecular ion of these compounds. The characteristic peak at m/z 490 in the EI spectrum of desethylamiodarone was also found by Marchiset et al. (20) together with a characteristic ion at m/z 446 corresponding to the desiodinated and desaminated fragment (M-I $NHC_2H_5)^+$. The low resolution EI spectrum of amiodarone hydrochloride, obtained via direct probe insertion and using an electron energy of 20-60 eV and ion source temperature of 180°C, were reported by Bonati et al (4). They found a molecular peak at m/z 645 and characteristic fragments of the drug with peaks at m/z 630 $(M\text{-}CH_3)^+$, 546, 517 $(M\text{-}HI)^+$, 420, 391, 373, 201 and 159. The ion observed at m/z 546 corresponds to cleavage of the aromatic ether with elimination of the aliphatic aminic chain. The other peaks found below m/z 500 resulted from the fragmentation of the aromatic moiety of the molecule. According to Clarke (7) the m/z values of the 8 most intense ions in the electron impact spectrum of amiodarone were in descending order of intensity 86, 36, 87, 84, 58, 56, 44 and 38. Own mass spectrometric investigations were carried out on a Finnigan-Mat 45 Triple Stage Quadrupole (TSQ) Mass Spectrometer. For EI mass spectrometry an electron energy of 70 eV, an emission current of 0.20 mA and a source and manifold temperature of 120°C and 135°C respectively was used. The probe of the mass spectrometer was programmed from 30°C to 280°C with a rate of 50°C/min. The direct inlet low resolution positive EI spectrum of amiodarone hydrochloride and desethylamiodarone hydrochloride are depicted in Figs. 15, 16. The major fragmentation ions were found for amiodarone at m/z 545, 492, 420, 391, 377, 295, 264, 250, 249, 247, 228, 221, 165, 159, 127, 121, 115, 102, 100, 92, 86, 74, 72, 63, and 58 and for desethylamiodarone at m/z 490, 420, 391, 294, 263, 250, 249, 247, 221, 200,

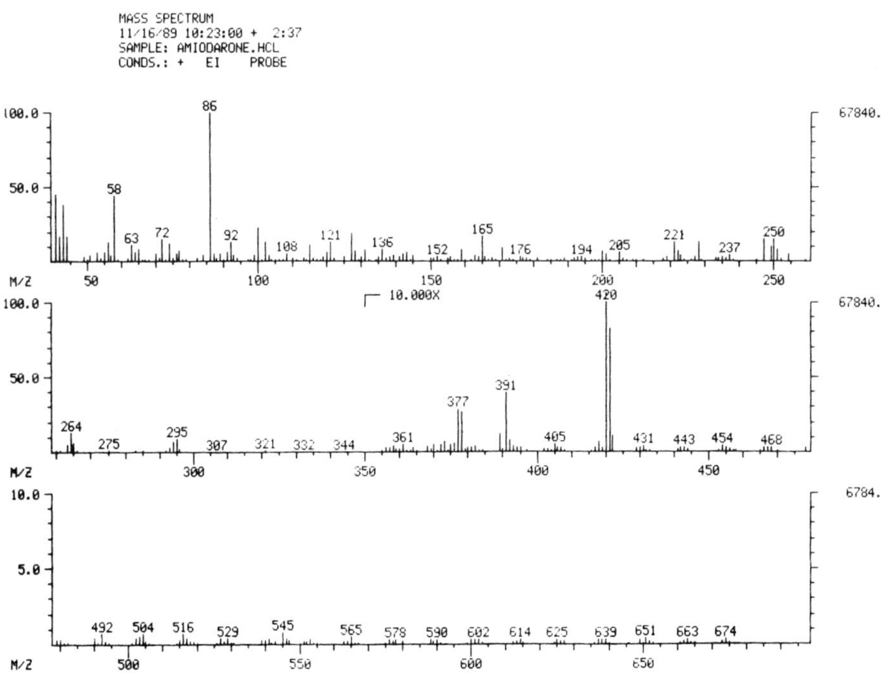

Fig. 15 Positive electron-ionization mass spectrum of amiodarone hydrochloride

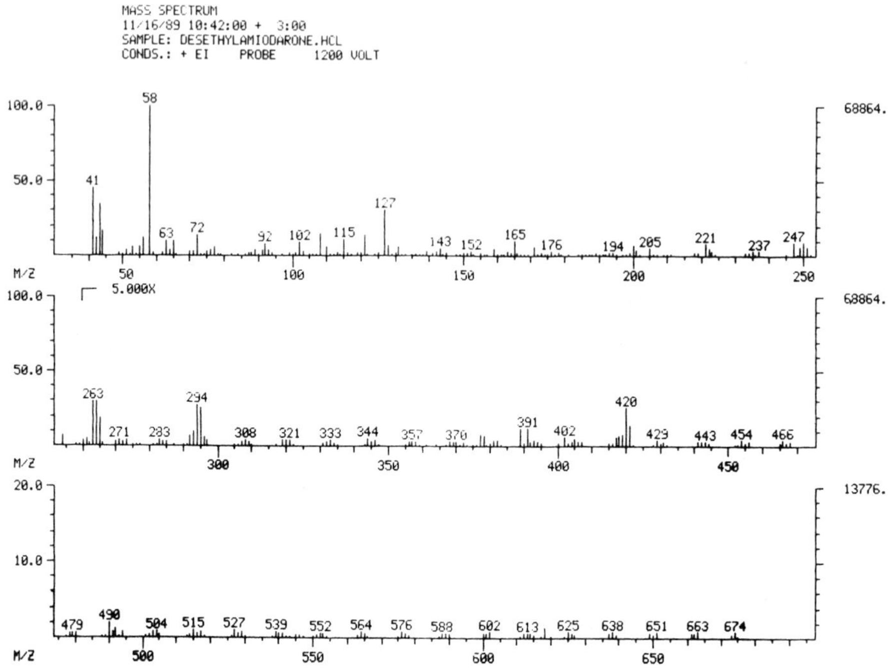

Fig. 16 Positive electron-ionization mass spectrum of desethylamiodarone hydrochloride

165, 127, 121, 115, 108, 72, 65, 63, and 58 respectively. The base peak of amiodarone and desethylamiodarone was observed at m/z 86 and 58 and a molecular ion with a very low intensity was found for the respective compounds at m/z 646 and 618. Since the mass fragments obtained for the 2 compounds are almost identical it can be concluded that positive EI mass spectrometry is not a suitable method for the differentiation between the two drugs. Our positive EI data were comparable with literature values (7,18,19) except for the peaks with masses higher than m/z 500, which showed in our study very little intensity. This was probably due to differences in probe temperatures used in the various investigations. Negative electron impact mass spectrometry using direct probe insertion showed in contrast to positive EI less fragmentation and only a few

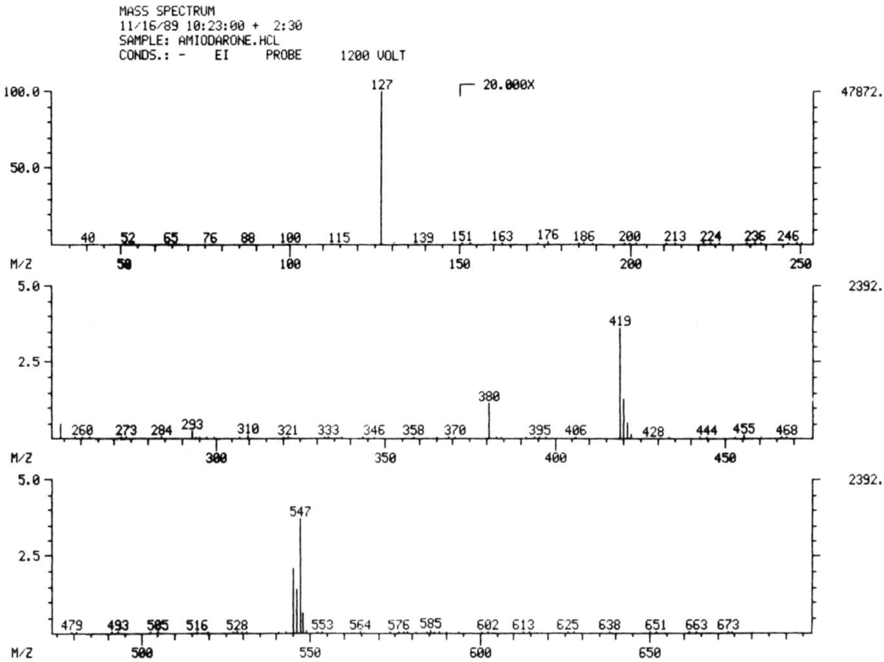

Fig. 17 Negative electron-ionization mass spectrum of amiodarone hydrochloride

relatively intense mass fragments were found for amiodarone at m/z 547, 419, 380, 293, 127 (base peak) and for desethylamiodarone at m/z 546, 419, 380, 293 and 127 (base peak) respectively (Figs. 17,18). These peaks of amiodarone and desethylamiodarone, however, show in analogy to positive EI no marked differences and suggest that negative EI mass spectrometry is unsuitable as a identification method for the two compounds.

2.9.4.2. Fast Atomic Bombardment Spectrum

Fast atomic bombardment (FAB) mass spectrometry of amiodarone and its two N-deethylated metabolites have been reported by Latini et al (19). FAB mass spectrometry was performed on a Kratos MS-50S mass spectrometer equipped with a 23 kg magnet. A standard Kratos 8 keV Xe FAB gun and

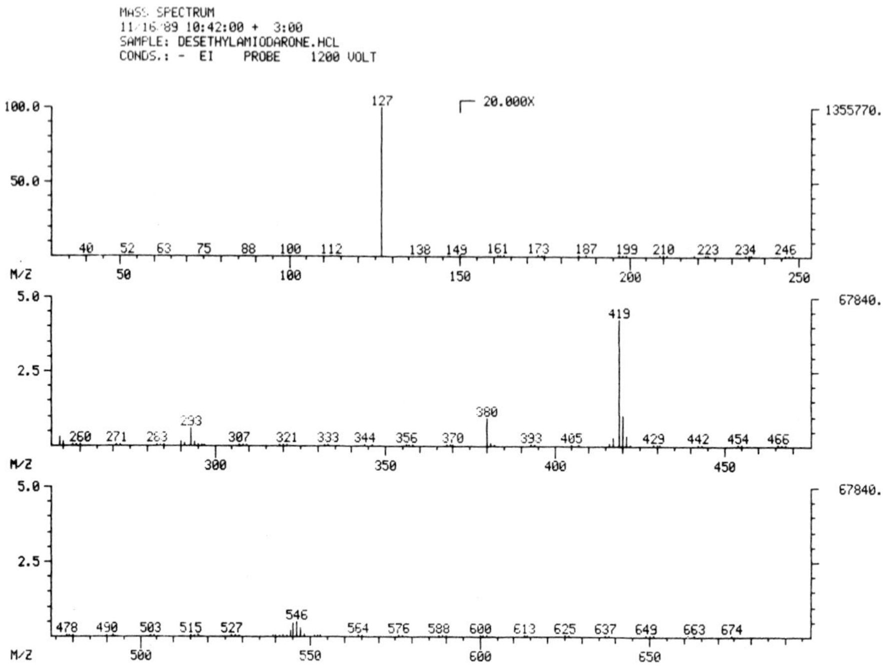

Fig. 18 Negative electron-ionization mass spectrum of desethylamiodarone hydrochloride

ion chamber were used to obtain a neutral xenon primary FAB beam. Samples of the drugs were dissolved onto a glycerol/oxalic acid matrix and deposited on a copper target surface. The FAB mass spectral fragment ions of amiodarone, desethylamiodarone and desdiethylamiodarone are listed in Table 11. The FAB mass spectra of the 3 compounds showed rather abundant pseudomolecular ions and some interesting high mass fragments. Two fragmentation patterns were observed: one due to the sequential loss of the two iodines, giving rise to specific ions for each of the three substances and the other due to the loss of the amine group with the aliphatic chain. These last fragments are non-specific. Two other non-specific peaks were found at m/z 373 and 421.

Table 11

The masses of the significant ions with their relative abundance in parentheses of amiodarone (A), desethylamiodarone (DA) and desdiethylamiodarone (DDA) by FAB mass spectrometry according to Latini et al (19)

	A	DA	DDA
	m/z	m/z	m/z
1*	646 (17.5)	618 (25.1)	590 (36.3)
2	547 (1.7)	547 (21.1)	547 (43.7)
3	520 (8.3)	492 (16.7)	464 (14.2)
4	421 (1.7)	421 (17.2)	421 (15.2)
5	394 (2.0)	366 (6.1)	338 (1.1)
6	373 (1.3)	373 (5.0)	373 (18.4)

1* pseudomolecular ion $[MH]^+$

2.9.4.3. Chemical Ionization Spectrum

The chemical ionization (CI) mass spectrum of amiodarone, desethylamiodarone and desdiethylamiodarone have been reported by Flanagan (18). The direct probe spectra of the compounds were obtained with an emission current of 500 μA, ionizing voltage of 50 eV, ion source temperature of 140°-160°C using isobutane as a reagent gas.

Chemical ionization with isobutane yield intense MH^+ ions for amiodarone, desethylamiodarone and desdiethylamiodarone at m/z of 646, 618 and 590 respectively. An abundant ion at m/z 547 was observed in the C1 mass spectra of these compounds assigned to $[(MH)\text{-}(CH_2 = CHN\ (C_2H_5)_2]^+$, $[(MH) - (CH_2 = CH\ NH\ C_2H_5)]^+$ and $[(MH) - (CH_2 = CH\ NH_2)]^+$ respectively. The base peak of amiodarone and desethylamiodarone was found at m/z 100 $[(CH_2\ CH_2\ N\ (C_2H_5)_2)^+]$ and at m/z 72 $[(CH_2CH_2NHC_2H_5)^+]$ respectively. In addition an other intense peak was

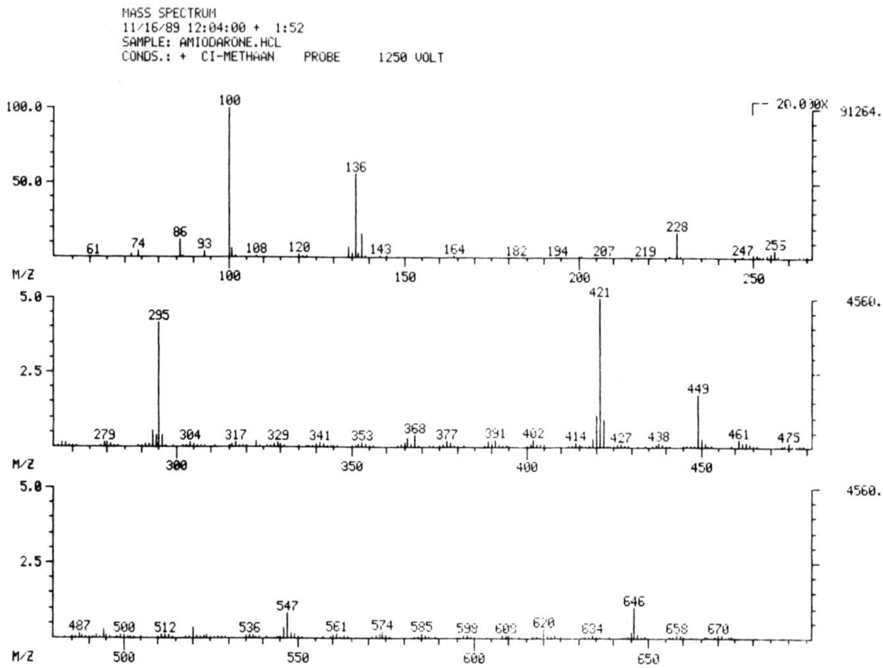

Fig. 19 Positive methane chemical ionization mass spectrum of amiodarone hydrochloride

observed for amiodarone at m/z 86 [$(CH_2 = N(C_2H_5)_2)^+$] and for desethylamiodarone at m/z 58 [$(CH_2 = NHC_2H_5)^+$] respectively. With respect to desdiethylamiodarone the situation was difficult to interpret as the mass fragments at m/z 44 and 30 were obscured by the CI reagent background. The direct inlet CI spectrum of desethylamiodarone using ammonia as the reactant gas yielded predominantly the MH^+ at m/z 618 and minor peaks at m/z 547, 492, 421, 405, 366, 338, 295, 279, 200, 149, 136, 123, 108, 90, 89 and 72 respectively (21).

Own chemical ionization (CI) mass spectra of amiodarone and desethylamiodarone were recorded on a Finnigan Mat 45 TSQ mass spectrometer with ion source pressures of 0.3 and 0.2 Torr for methane and ammonia, ion source temperature of 120°C, manifold temperature of

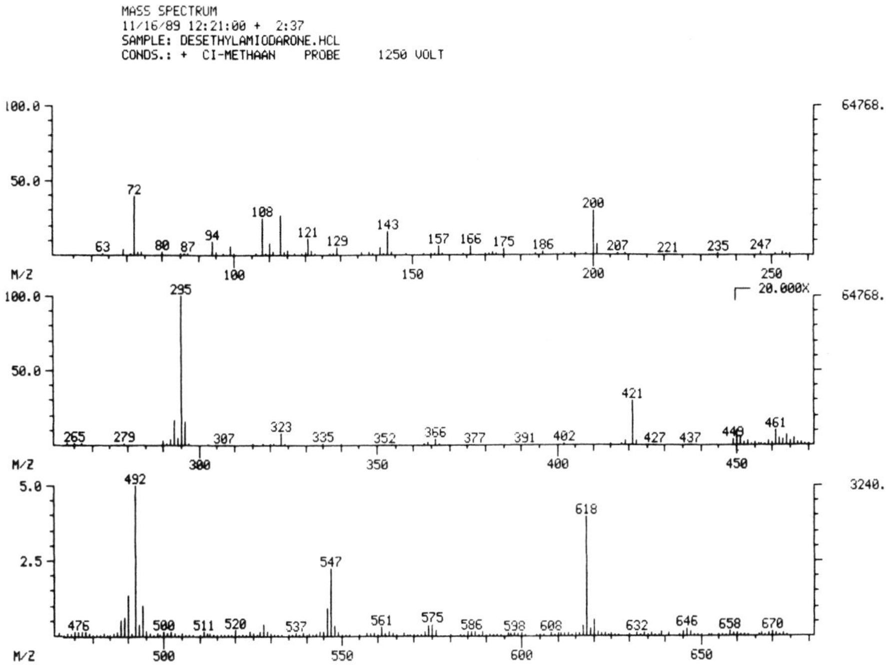

Fig. 20 Positive methane chemical ionization mass spectrum of desethylamiodarone hydrochloride

135°C, emission current of 200 μA, electron energy of 58.2 eV using methane and ammonia as reactant gasses. The direct inlet probe was heated from 30°C to 280°C with a rate of 50°C/min. The positive CI spectra of amiodarone and desethylamiodarone using methane as a reagent gas are shown in Figs. 19,20. The mass spectrum of amiodarone shows a molecular ion (MH^+) peak at m/z 646 and other characteristic peaks at m/z 547, 449, 421, 295, 228, 138, 100 (base peak), and 86 respectively. A molecular ion peak is observed in the spectrum of desethylamiodarone at m/z 618. Other significant peaks in the CI spectrum of this compound are found at m/z 547, 492, 421, 295 (base peak), 200, 143, 108, and 72. Our data are consistent with the values reported by Flanagan et al (18).

In addition our findings indicate that positive CI with methane is a suitable

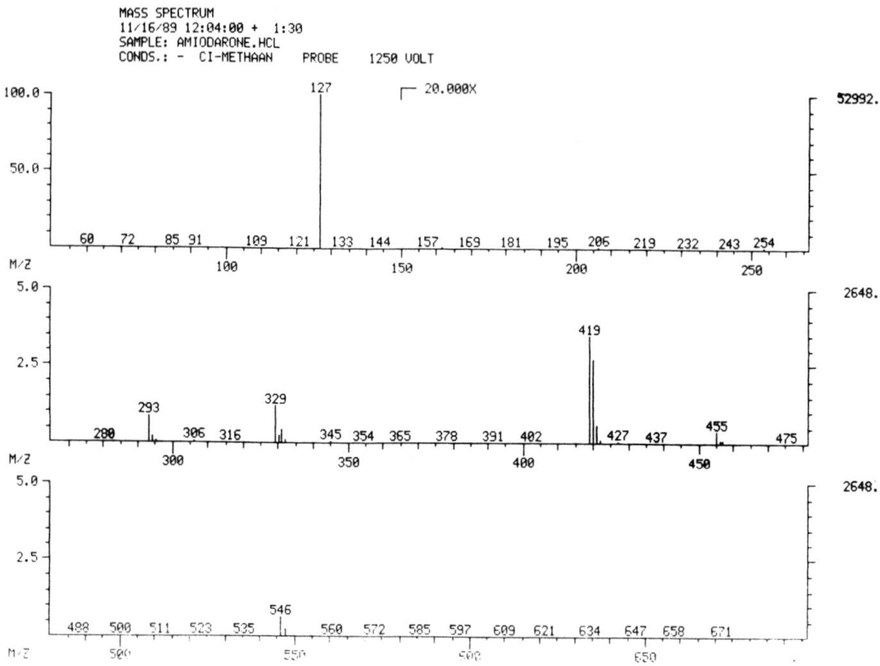

Fig. 21 Negative methane chemical ionization mass spectrum of amiodarone hydrochloride

mass spectrometric method for identification of these two compounds. Negative chemical ionization with methane yields predominantly the iodine ion at m/z 127 for both compounds and minor peaks at m/z 546, 455, 419, 329 and 293 for amiodarone and at m/z 419 and 293 for desethylamiodarone respectively (Figs. 21,22). Compared to positive CI, negative CI with methane results for both amiodarone and desethylamiodarone in a lesser degree of fragmentation, however, with the formation of similar ions from the two compounds. Therefore it can be concluded that this mass spectrometric method is inefficient for the identification of these substances. The positive chemical ionization spectra of amiodarone and desethylamiodarone using ammonia as a reactant gas are depicted in Figs. 23,24. The mass spectrum of amiodarone shows a molecular ion (MH^+) at m/z 646 and other prominent fragments at m/z

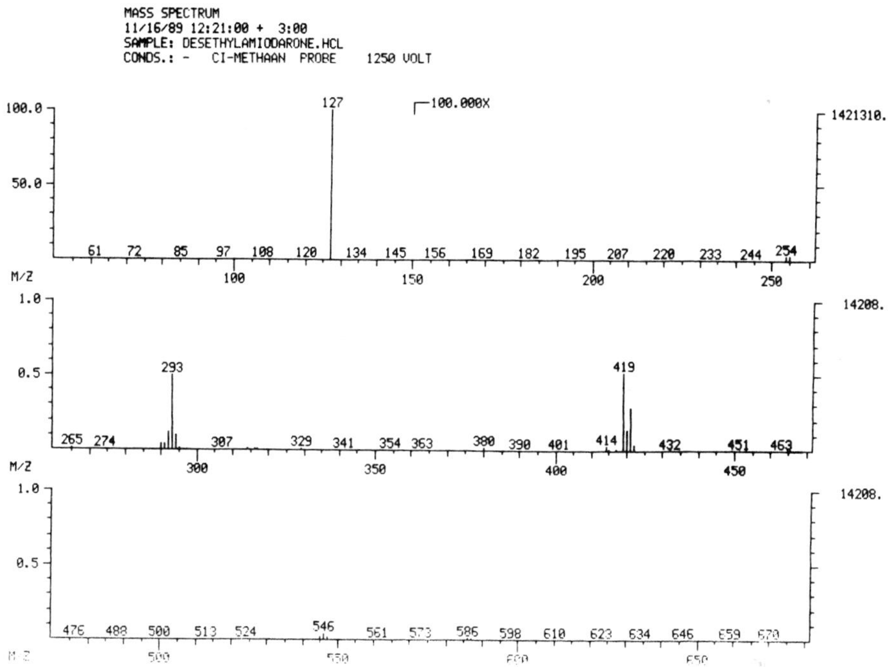

Fig. 22 Negative methane chemical ionization mass spectrum of desethylamiodarone hydrochloride

547, 520, 421, 394, 366, 295, 228, 173, 138, 136 (base peak), 100, 86 and 74. The mass spectrum of desethylamiodarone shows a molecular peak at m/z 618 and other characteristic peaks at m/z 547, 492, 438, 421, 366, 295, 200, 143, 110, 108 (base peak), 89, and 72 respectively. The peaks observed in our CI spectrum are in good agreement with the peaks reported in the CI spectrum of desethylamiodarone by Berger (21). The negative chemical ionization spectra of amiodarone and desethylamiodarone using ammonia as a reagent gas are comparable with the spectra obtained with methane (Fig. 25,26). Both spectra mainly show the iodine ion at m/z 127 (base peak) and some minor peaks at m/z 420, 329, 293 for amiodarone and at m/z 546, 419, 293 for desethylamiodarone respectively. Finally our mass spectrometric studies indicate that positive chemical ionization using ammonia as a

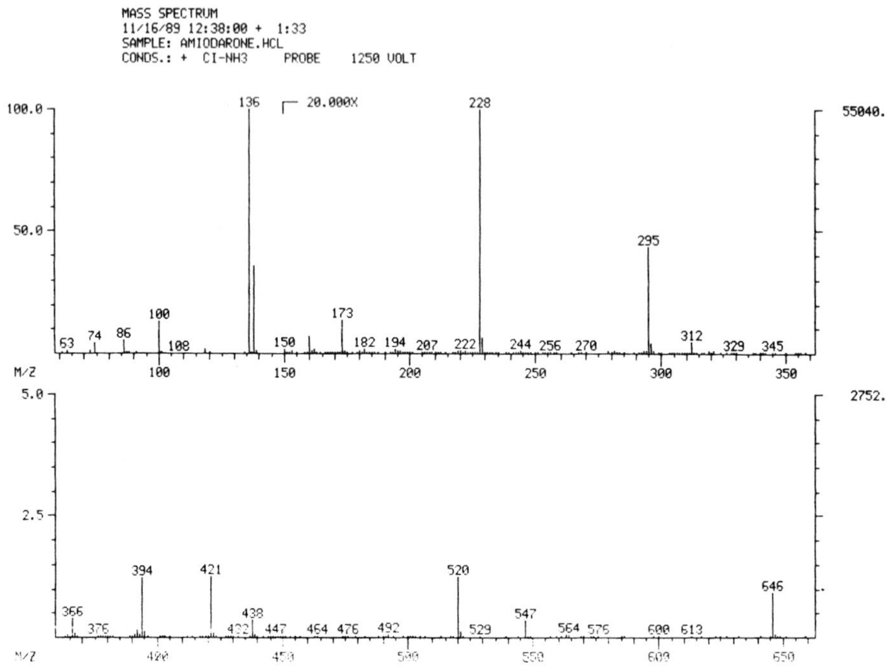

Fig. 23 Positive ammonia chemical ionization mass spectrum of amiodarone hydrochloride

reagent gas is the method of choice for the identification of amiodarone and desethylamiodarone because this ionization mode yields first an relatively intense molecular ion peak at m/z 646 and 618 respectively and second the most specific fragments at m/z 646, 520, 394, 228, 173, 138, 136, 100, 86, 74 for amiodarone and at m/z 618, 492, 200, 143, 110, 108, 89, 72 for desethylamiodarone respectively. As a good alternative positive C1 using methane can be used as identity test, which yields a little less number of specific fragments for amiodarone at m/z 646, 228, 138, 100, 86 and for desethylamiodarone at m/z 618, 492, 200, 143, 108, 72 respectively.

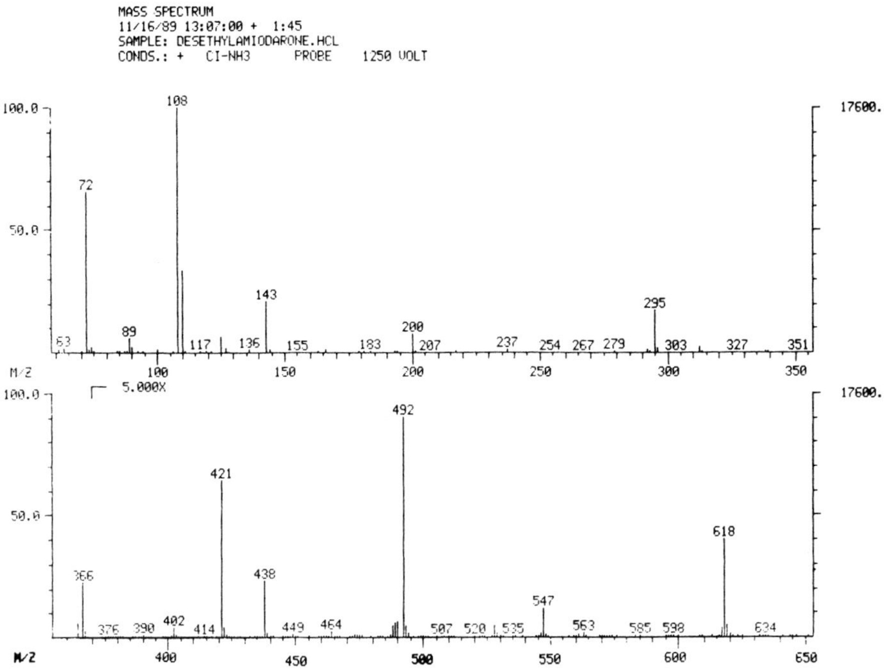

Fig. 24 Positive ammonia chemical ionization mass spectrum of desethylamiodarone hydrochloride

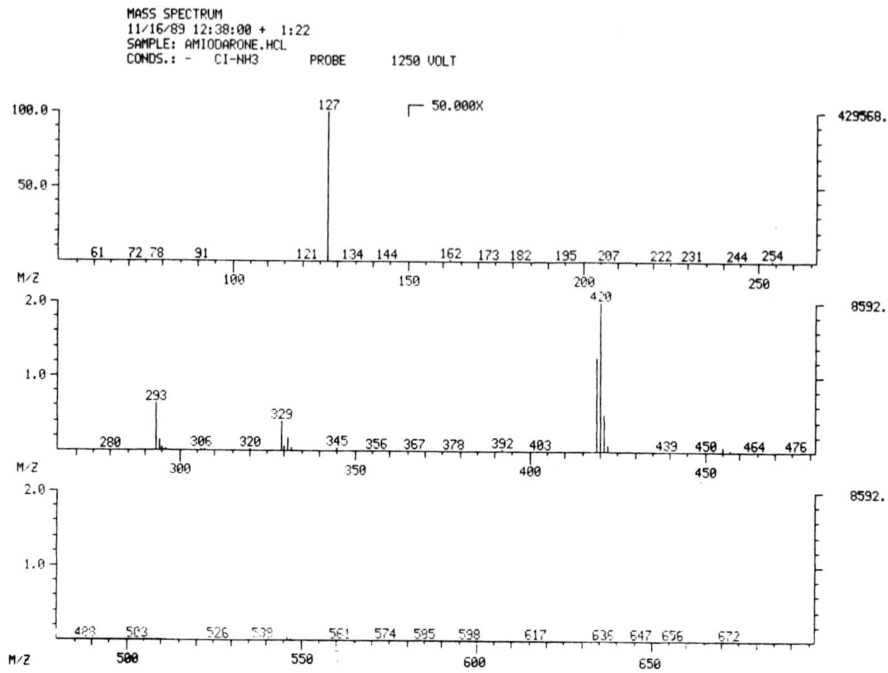

Fig. 25 Negative ammonia chemical ionization mass spectrum of amiodarone hydrochloride

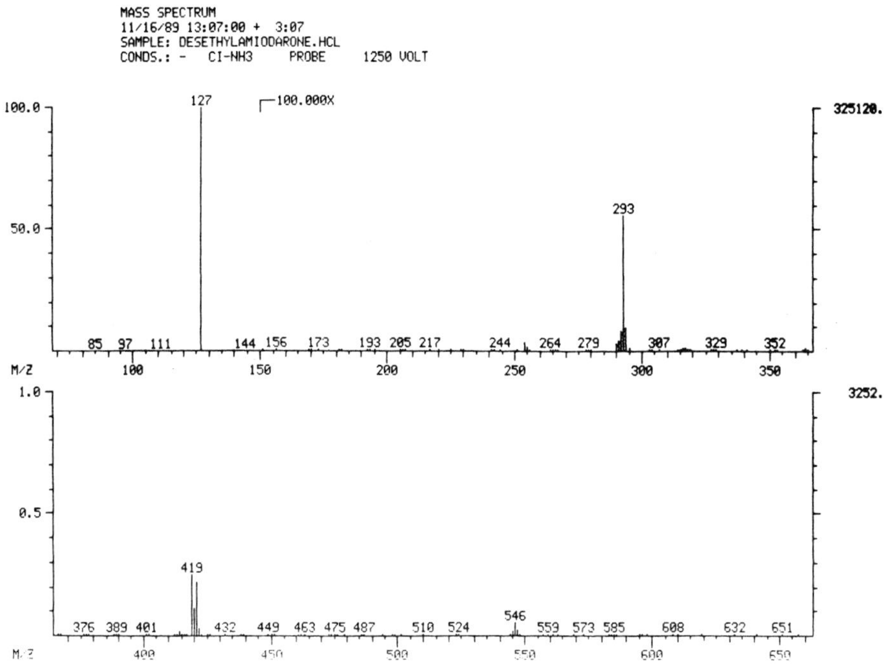

Fig. 26 Negative ammonia chemical ionization mass spectrum of desethylamiodarone hydrochloride

3. SYNTHESIS

The synthesis of amiodarone hydrochloride was first accomplished in 1961 by R. Tondeur and F.Binon. The preparation of the drug according to US Patent 3, 248, 401 (1966) is described here (Fig. 27).

2-n-Butyl-3-(3,5-diiodo-4-hydroxy-benzoyl) benzofuran (1) dissolved in ethyl carbonate was treated with a solution of sodium methoxide (2) in methanol. All of the methanol introduced was distilled off by heating until the vapours attained a temperature of approximately 100°C. Then the suspension of the sodium salt of the phenol (3) was cooled to a temperature of about 50°C and the chlorinated amine: β-

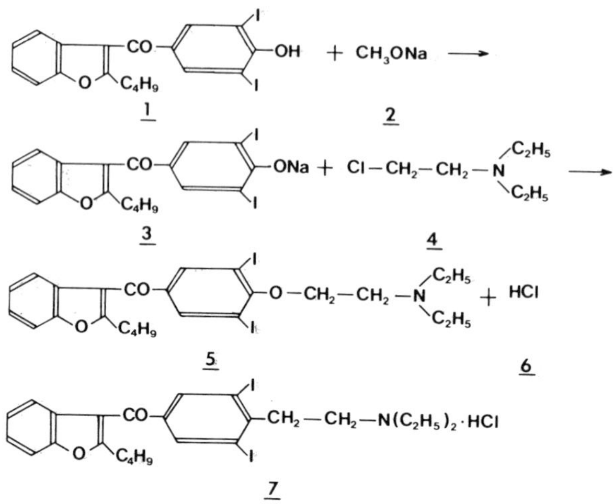

Fig. 27 Route of synthesis of amiodarone hydrochloride

diethylaminoethylchloride (4), which had been obtained from the hydrochloride in ethylcarbonate, was introduced drop by drop into the suspension under continually stirring. The mixture was heated to a temperature of approximately 90°C, which was maintained for approximately 2 hours. The mixture was cooled and allowed to stand overnight during which time the sodium chloride settled down. The ethyl carbonate solution containing diethylaminoethylether (5) was extracted with increasingly diluted aqueous hydrochloride acid solutions while stirring. Extraction was continued until the alkalized solution produced no further precipitate. The combined aqueous solutions were washed with ether and then made strongly alkaline with aqueous sodium hydroxide. Extraction with ether was carried out three times. The organic layers were washed with water and then dried over anhydrous potassium carbonate. In order to produce the hydrochloride of amiodarone the carbonate was filtered off and then the hydrochloride was precipitated from the ether solution with an ethereal hydrochloric acid solution (6). After the solution had been allowed to stand for a few hours, decantation was performed and the syrupy hydrochloride residue was taken up in boiling acetone. Amiodarone hydrochloride [2-n-butyl-3-(3,5-diiodo-4-β-N-diethylaminoethoxy-benzoyl) benzofuran hydrochloride] in the form of a crystalline powder which melts at 156°C was obtained (7). If, after alkalization of the aqueous solutions containing the basic ether (amiodarone base), extraction is carried out with benzene instead of with ether, the hydrochloride in crystalline form can be obtained directly after drying the benzene solution of the basic ether by treating this benzene solution with gaseous hydrogen chloride or with a benzene solution of the latter. In the same way, if the etherification is carried out in benzene or toluene instead of in ethyl carbonate, cooling can be effected after the reaction with the β-diethylaminoethyl-chloride, the sodium chloride formed during the reaction can be filtered off in the presence of activated charcoal and the hydrochloride can be directly precipitated in crystalline form by introducing gaseous hydrogen chloride just until an acid reaction is obtained.

4. METHODS OF ANALYSIS IN RAW MATERIALS AND PHARMACEUTICAL FORMULATIONS

4.1. Identification Tests

4.1.1.

A 1% aqueous solution of amiodarone hydrochloride responds positively to the test for chloride.

4.1.2.

Upon heating on a flame a 0.1 g sample of amiodarone hydrochloride liberated violet vapours of iodine.

4.1.3.

Upon the addition of a few drops of concentrated sulfuric acid to 10 mg of amiodarone hydrochloride; a yellow orange colour is produced (reaction on bezofuran nucleus).

4.1.4.

Upon the addition of 20 ml of a 0.12% solution of sodium dioctyl sulphosuccinate to 5 ml of a 1% solution of amiodarone hydrochloride; a white turbidity is produced. Upon the addition of 10 ml of chloroform and 10 ml of 2M sulfuric acid followed by shaking; the white turbidity disappeared from the aqueous phase (anionic-cationic reaction on the tertiary amine group).

4.1.5.

Upon the addition of a few drops of a 10% solution of sodium nitrite in sulphuric acid (Liebermann's reagent) to 10 mg of amiodarone hydrochloride; a brown yellow color is produced.

4.1.6.

The ultraviolet absorption spectrum of a solution of amiodarone hydrochloride in ethanol (2mg/100 ml) exhibits a maximum at 242 ± 1 nm, a minimum at 222 ± 1 nm and a flat part between 269 nm and 272 nm (5). Refer to Section 2.9.1.

4.1.7.

The infrared spectrum of amiodarone hydrochloride taken in KBr disk shows absorption maxima at 3060, 2965, 2940, 2870, 2300-2700, 1635, 1575, 1565, 1535, 1480, 1250, 1025, 1015, 900, and 750 cm^{-1} (5). Refer to Section 2.9.2.

4.1.8.

Characteristic thin-layer chromatographic Rf-values and high-performance liquid chromatographic retention times are described for amiodarone in the text and may be used as confirmatory tests for the drug (Section 4.4)

4.2. Titrimetric Analyses

4.2.1. Aqueous

Dissolve about 500 mg of a sample of amiodarone hydrochloride, accurately weighed, in 5 ml of methanol and dilute with distilled water to a volume of 50 ml. Pipet 5 ml of the solution into a titration vessel of 100 ml with stopper and add 20 ml of distilled water, 5 ml of 2 M sulphuric acid, 1 ml of indicator solution (0.01% dimethyl yellow in ethanol) and 20 ml of chloroform. With a buret 25 ml of a 0.12% solution of sodium dioctyl sulphosuccinate (D.O.S.S.) was added and the mixture vigorously shaken: a yellow chloroform layer is observed. Subsequently titration is performed with volumes of 0.1 ml to 0.2 ml of D.O.S.S. with vigorous shaking after each addition. The end-point of the titration is indicated by the alteration of the color of chloroform layer from yellow to red-orange. Each ml of 0.12% (0.027 mol) D.O.S.S. is equivalent to 18.41 mg of amiodarone hydrochloride.

4.2.2. Non-aqueous Titration

Dissolve about 660 mg of amiodarone hydrochloride, accurately weighed, in a mixture of 50 ml of glacial acetic acid and 10 ml of a 3% solution of mercuric acetate in acetic acid. Add 5 drops of a 0.5% solution of Oracet Blue B in acetic acid and titrate with standardized 0.1 N perchloric acid in acetic acid until the color of the solution had changed from blue to lilac. Perform a blank determination and make any necessary correction. Each ml of 0.1 N perchloric acid is equivalent to 68.18 mg of amiodarone hydrochloride. Desethylamiodarone hydrochloride can be assayed in the same way using the above described titration with perchloric acid. Each ml of 0.1 N perchloric acid is equivalent to 65.38 mg of desethylamiodarone hydrochloride (21).

4.3. Spectrophotometric Analysis

Transfer about 100 mg of a predried sample of amiodarone hydrochloride (W_1), accurately weighed, to a 100 ml volumetric flask. Dissolve in and dilute to volume with 95% ethanol (Solution A). Transfer 5.0 ml of solution A to a 50 ml volumetric flask and dilute to volume with the same solvent (Solution B). Transfer 10 ml of solution B to a 100 ml volumetric flask and dilute to volume with 95% ethanol (Solution C). Prepare in the same way a solution of a predried reference standard of amiodarone hydrochloride containing 1.0 mg per 100 ml 95% ethanol (W_2). Record the ultraviolet absorption spectrum of sample solution C in 1 cm cells on a suitable spectrophotometer from 200 nm to 320 nm using 95% ethanol as a blank. An absorbance maximum (λ max) and minimum (λ min) should be observed at 242 $\pm$ 1 nm and 222 $\pm$ 1 nm respectively. Measure the extinction of the solution at λ max (E_1) and at λ min and calculate the extinction ratio E 242 nm / E 222 nm. The ratio should be between 1.49 and 1.59. Determine in the same way the extinction at 242 nm of reference standard solution of 1 mg/100 ml (E_2). The amiodarone hydrochloride content in the sample is calculated by using the formula

$$W_2 \times E_1 \times 100\% \ / \ E_2 \times W_1$$

where W_1 and W_2 are the weight of sample and reference standard in mg; E_1 and E_2 are the extinction of sample and reference standard.

4.4. Chromatographic Analyses

4.4.1. Thin-Layer Chromatography

A thin-layer chromatographic (TLC) assay of amiodarone and some of its analogues in raw material have been reported by Crispin et al (5) using the following system.

Absorbent	: Silica Gel GF, 250 microns
Activation	: drying of the plate at 110°C during 30 min
Eluting Solvent	: chloroform-methanol-formic acid (90:7:3,v/v)
Equilibration	: 1 h in a closed paper lined tank
Concentration	: 50 mg/ml chloroform
Detection	: Ultraviolet at 254 nm and iodine absorption
Spotting	: 250-500 μgof amiodarone hcl
	1.25 μgof amiodarone analogues

The solvent is allowed to rise to the 15 cm line, the plate is removed from the chromatographic chamber and dried in a current of air until no solvent odor is detected. The plate is examined under UV light at 254 nm and thereafter placed in a chamber containing iodine crystals for about thirty minutes. All products observed under UV irradiation as darkening blue spots formed yellow brown spots after iodine exposure. The Rf of amiodarone is 0.28 and for its analogues L3937, L6355, L3372, L6424 and L3373 0.17, 0.23, 0.53, 0.62 and 0.73 respectively. (For structures of analogues refer to Fig. 2). Clarke (7) described 3 solvent systems for thin-layer chromatography of amiodarone on Silica Gel G plates. Solvent system 1 consisted of methanol - 25% ammonia (100:1.5), system 2 of cyclohexane - toluene - diethylamine (75:15:10) and system 3 of chloroform - methanol (90:10) respectively. The Rf- values of amiodarone in these systems were 0.72, 0.62 and 0.68 respectively.

As visualizing agents were used: Dragendorff spray (positive reaction), acidified iodoplatinate solution (positive reaction) and Marquis reagent (yellow spot). A TLC assay for desethylamiodarone on Silica Gel GF 254 plates, which uses hexane - isopropanol (85:15) saturated with 25% ammonia as solvent system and UV as detection method has been described by Berger (21). The RF value for desethylamiodarone in this sytem is 0.45.

Own thin-layer chromatographic experiments on pre-coated 20 cm x 20 cm Silica Gel 60 F254 plates (Merck) using 3 solvent systems for the identification of amiodarone, desethylamiodarone, desdiethylamiodarone and five amiodarone analogues are described below.

Absorbent	: Silica Gel 60 F254, 250 microns
Activation	: drying of the plate at 110°C during 30 min
Eluting Solvents	: A. methanol-25% ammonia (99:1)
	: B. chloroform-methanol-formic acid (90:7:3)
	: C. hexane-isopropanol-25% ammonia (84:15:1)
Equilibration	: 30 min in a closed paper lined tank
Concentration	: 2 mg/ml methanol of amiodarone and its analogues
Detection	: 1) UV 254 nm
	: 2) Dragendorff spray
	: 3) acidified iodoplatinate solution
	: 4) iodine vapour
Spotting	: 2.4 μg of all compounds using a Camag nano-applicator

The solvents are allowed to rise to the 15 cm line, the plates are removed from the chromatographic chamber and dried in a current of air until no solvent odor is detected. The plates are examined under UV light at 254 nm and the spots are visualized using one of the detection agents. The mean Rf-values of amiodarone and its analogues obtained from five determinations in solvent system A, B, and C are presented in Table 12.

Under UV irradiation at 254 nm all compounds were observed as purple spots on a light green background. After spraying with Dragendorff reagent amiodarone and its deethylated and its deiodinated derivates (Figs 1 and 2) give orange spots on a yellow background. With acidified iodoplatinate solution amiodarone and its deethylated and deiodinated derivatives were observed as purple spots while the deaminated amiodarone analogues (Fig. 2) were seen as white spots on a brown background. After iodine exposure amiodarone and its analogues give brown spots on a light brown background. The minimal detectable amount of amiodarone and its

Table 12

The mean Rf-values of amiodarone, amiodarone analogues and internal standard in 3 solvent systems on Silica Gel 60 F254 plates

Compound*	Rf-value		
	System A	System B	System C
amiodarone	0.64 ± 0.01	0.26 ± 0.01	0.58 ± 0.03
desethylamiodarone	0.53 ± 0.01	0.23 ± 0.01	0.30 ± 0.04
desdiethylamiodarone	0.49 ± 0.01	0.11 ± 0.01	0.14 ± 0.02
L 6355	0.52 ± 0.02	0.21 ± 0.02	0.28 ± 0.04
L 3937	0.52 ± 0.02	0.15 ± 0.01	0.29 ± 0.04
L 3373	0.89 ± 0.01	0.73 ± 0.01	0.03 ± 0.01
L 6424	0.85 ± 0.01	0.57 ± 0.01	0.10 ± 0.01
L 3372	0.81 ± 0.01	0.48 ± 0.01	0.28 ± 0.01
L 8040 (IS)	0.64 ± 0.01	0.32 ± 0.02	0.69 ± 0.03

A: methanol:25% ammonia (99:1)

B: chloroform-methanol-formic acid (90:7:3)

C: hexane-isopropanol-25% ammonia (84:15:1)

* for structures of compounds see Figs. 1 and 2

deethylated and its deiodinated derivatives in the 3 TLC systems using UV detection was 80-100 ng. With the detection agents a sensitivity of about 200 ng was attained for these compounds.

4.4.2. High-Performance Thin-Layer Chromatography

High-performance thin-layer chromatographic (HPTLC) experiments for the identification of amiodarone and its analogues were performed on pre-coated 10 cm x 10 cm HPTLC Silica Gel 60 F254 plates using the conditions, solvent systems, and detection methods as decribed by us under thin layer chromatography (4.4.1). For the assessment of the Rf-values of amiodarone and it analogues 1 μg of each compound was spotted on the plate with a Camag nano-applicator and the plates were developed over a distance of 8 cm. The mean Rf-values (n=5) of amiodarone and its analogues in the solvent systems A, B, and C are listed in Table 13. The minimal detectable quantity of amiodarone and desethylamiodarone on HPTLC Silica Gel plates using UV detection and I_2 exposure was 20 ng and 40 ng respectively for both compounds. Preliminary experiments on reversed phase HPTLC RP18 plates using methanol - 25% ammonia (99:1) and chloroform - methanol - formic acid (90:7:3) as developing solvents showed in the first system a poor separation of amiodarone, desethylamiodarone and desdiethylamiodarone with Rf of 0.14, 0.14, and 0.17 and in the second system elution of all compounds in the solvent front with Rf of 0.89, 0.85, and 0.77 for the forementioned substances.

4.4.3. Gas Chromatography

Clarke (7) has described a gas chromatographic method for the identification of amiodarone using a glass column packed with 2.5% SE 30 on 80-100 mesh Chromosorb G and a flame ionization detector. The column temperature was set at 300°C and nitrogen was used as carrier gas at a flow of 45 ml/min. Under these conditions for amiodarone and 2 major on column decomposition products of the drug retention indices of 3335, 2590, and 2780 were reported respectively. Own gas chromatographic analyses of amiodarone, desethylamiodarone, and six amiodarone analogues (Figs 1 and 2) were carried out on 2 glass columns packed with 3% OV1 on 80-100 mesh Gas Chrom Q and 3% OV17 on 80-100 mesh Gas Chrom Q using for both columns a flame ionization detector (FID) and a ^{63}Ni electron capture detector. For both columns an injector temperature of 250°C, a detector temperature for FID and ^{63}Ni detector of 280°C and a carrier gas flow of 30 ml/min of nitrogen was employed. Column temperatures of 100°C, 120°C, 140°C, 180°C, 210°C, 230°C, and 250°C were used for the gas chromatographic screening of amiodarone and its analogues. Following injection of 1 μl of methanolic standard solutions containing 1 mg/ml of amiodarone and its analogues, no peaks were observed within 1 hour in the chromatograms of all compounds at all selected column temperatures. This lack of response may be explained by the fact that amiodarone as well as its analogues undergoes extended thermal decomposition with evolution of iodine at temperatures between 150°C and 200°C as observed in the melting point determination of these substances (Section 2.1)

4.4.4. High-Performance Liquid Chromatography

A method for the chromatographic identification and determination of amiodarone in raw material by high-performance liquid chromatography (HPLC) was described by Bonati et al (4). Separation is carried out through a reversed phase 25 cm x 4 mm ID Hibar Lichrosorb RP-8 7 microns column using a 2 ml/min flow and an isocratic phase consisting of acetonitrile - water - acetic acid (80:19:1) adjusted to pH 5.8 with ammonia. The column effuent was monitored at 254 nm with a fixed wavelength UV detector. No retention time of amiodarone was given. The sensitivity of the method was 0.05 μg/ml. The calibration curve was linear over the concentration range of 0.5 to 5.0 μg/ml.

A normal phase HPLC system for the analysis of desethylamiodarone in raw material was reported by Berger (21). He used a 12.5 cm x 4 mm ID Silica Si 100 5 microns column with a mobile phase of iso-octane - tetrahydrofuran - methanol - diethylamine (100:40:40:0.5 v/v) at a flow rate of 1 ml per min and UV detection of the effluent at 254 nm. No data on the retention time of the drug were given in his procedure.

Clarke (7) reported a normal phase HPLC method for the identification of amiodarone and desethylamiodarone in raw materials and biological fluids. He employed a Spherisorb S 5 W 5 μm column, 12.5 cm x 4.9 mm ID; the mobile phase was a solution containing 1.175 g (0.01 M) of ammonium perchlorate in 1000 ml of methanol; adjusted to pH 6.7 by the addition of 1 ml of 0.1 M sodium hydroxide in methanol; the UV detector was set at 241 nm. Under these conditions, desethylamiodarone had a column capacity ratio (k^1) of 1.8 and amiodarone a k^1 of 2.4. An own reversed phase HPLC method for the identification and determination of amiodarone in raw material and tablets is presented below. Mobile phase: methanol - 25% ammonia (99.7:0.3). Assay preparation: weigh and finely powder twenty tablets. Weigh accurately a portion of powder (W_s) equivalent to 50 mg of amiodarone hydrochloride into a 100 ml volumetric flask, add 50 ml of methanol, sonicate for 5 min and bring to the mark with methanol. Pipet 10 ml of this solution into a 50 ml volumetric flask, add 10 ml of internal standard solution (1 mg/1 ml of L 8040 in methanol, Sanofi-Labaz), mix and bring up to volume. Filter the solution over a 0.45 μmPTFE membrane filter and inject 10 μl of the obtained filtrate into the liquid chromatograph. Standard preparation: Transfer about 50 mg of amiodarone hydrochloride reference standard (W_R) accurately weighed into a 100 ml volumetric flask and dissolve in 100 ml of methanol. Pipet 10 ml of this solution into a 50 ml volumetric flask and proceed as described under assay preparation. The liquid chromatograph was adjusted to the following operating conditions.

Column: 10 cm x 8 mm ID Radial-PAK C_{18} cartridge in combination with a Model RCM 100 module for compressing of the cartridge (Waters Ass.)

Flow rate : 2.5 ml/min

Detector : variable UV set at 242 nm

Retention times : amiodarone 5.1 min. Internal standard 7.8 min.

Calculation : mg amiodarone hcl/tablet =

$R_s \times W_R \times P \times T_a / R_R \times W_s$

where: R_s = peak area ratio of amiodarone in sample to internal standard

: R_R = peak area ratio of amiodarone in standard to internal standard

: W_R = weight of reference standard (mg)

: W_s = weight of sample of tablet powder (mg)

: P = potency of reference standard in pct

: T_a = mean tablet weight (mg)

5. METHODS OF ANALYSIS IN BIOLOGICAL FLUIDS AND TISSUES

5.1. High-Performance Thin-Layer Chromatography

High-performance thin-layer chromatography (HPTLC) has been employed for the identification of amiodarone and its major metabolite desethylamiodarone in various rat tissues and human tisses. The rat tissues used for HPTLC analysis were lung, liver, renal and adipose tissue and were obtained from rats which had been treated orally with 100 mg/kg of amiodarone per day for 13 days. The human tissues, used for the assay were lung, liver, heart and adipose tissue and were collected from 2 post-mortem cases, who had been on long-term treatment with 200 mg per day of amiodarone. Methanol extracts of the rat and human tissue samples were prepared as follows. Tissue samples were finely minced, dried between Kleenex[R] tissues and portions of about 100 mg of lung, liver, renal and heart tissue and 50 mg of adipose tissue were homogenized with 1 ml and 5 ml of methanol respectively, using a Potter apparatus. Subsequently the tissue homogenates were vortexed for 30 min and after centrifugation 10 μl to 20 μl of the clear supernatants were applied on the plates as spots or bands. Chromatography of the methanolic extracts and methanolic reference solutions of amiodarone and desethylamiodarone was carried out on activated precoated 10 cm x 10 cm HPTLC Silica Gel F254 plates in a saturated tank using chloroform-methanol-formic acid (90:7:3) and hexane-isopropanol-25% ammonia (84:15:1) as solvent systems (See section 4.4.2). Visualization of the spots was done by UV 254 nm and exposure to iodine vapors. Comparison of the Rf values of the spots observed in the methanolic extracts to the references similarily chromatographed in the 2 solvent systems serves to identify amiodarone and desethylamiodarone in the various tissues. Using chloroform-methanol-formic acid (90:7:3) as solvent system (B), HPTLC assay of the methanolic extracts of lung, liver, renal and adipose tissue of the rat on 4 plates showed in all tissues examined 4 to 6 spots including 2 spots with mean Rf values of 0.25 ± 0.01 and 0.21 ± 0.01 corresponding to Rf values of amiodarone and desethylamiodarone (Table 13). Chromatographic analysis of the same rat tissue extracts with eluting solvent C (hexane-isopropanol-25% ammonia = 84:15:1) on 4 HPTLC plates revealed in all tissues 4 to 7 spots including 2 spots with average Rf-values of 0.52 ± 0.04 and 0.28 ± 0.04 corresponding to the Rf of amiodarone and its metabolite. The HPTLC assay of the methanolic extracts of liver, lung, heart and adipose tissue samples obtained at autopsy

Table 13

The mean Rf-values of amiodarone, amiodarone analogues and internal standard in 3 solvent systems on HPTLC Silica Gel 60 F254 plates

Compound*	Rf-value		
	System A	System B	System C
amiodarone	0.66 ± 0.02	0.23 ± 0.02	0.54 ± 0.01
desethylamiodarone	0.54 ± 0.03	0.20 ± 0.01	0.29 ± 0.01
desdiethylamiodarone	0.48 ± 0.03	0.10 ± 0.01	0.11 ± 0.01
L 6355	0.53 ± 0.03	0.17 ± 0.02	0.25 ± 0.01
L 3937	0.53 ± 0.03	0.14 ± 0.01	0.26 ± 0.01
L 3373	0.86 ± 0.01	0.76 ± 0.01	0.03 ± 0.01
L 6424	0.83 ± 0.02	0.59 ± 0.02	0.15 ± 0.01
L 3372	0.80 ± 0.05	0.47 ± 0.03	0.33 ± 0.01
L 8040 (IS)	0.63 ± 0.03	0.31 ± 0.03	0.64 ± 0.02

A: methanol:25% ammonia (99:1)

B: chloroform-methanol-formic acid (90:7:3)

C: hexane-isopropanol-25% ammonia (84:15:1)

* for structures of compounds see Figs. 1 and 2

showed in system B and C on 3 plates in all extracts 4 to 8 spots of which 2 spots with Rf values of 0.26 ± 0.03 and 0.23 ± 0.02 (B) and 0.51 ± 0.02 and 0.26 ± 0.01 (C) correspond to Rf of amiodarone and desethylamiodarone in the respective systems (Table 13). Finally , our HPTLC analyses of various tissues on Silica Gel 60 F254 plates indicate that system C is the mobile phase of first choice for the identification of amiodarone and desethylamiodarone in biological material, because it gives by far the best separation between the two compounds.

5.2. High-Performance Liquid Chromatography/Mass Spectrometry

High-performance liquid chromatography (HPLC) in combination with probe-injection mass spectrometry of column effluent fractions have been used to identify amiodarone and desethylamiodarone in human plasma (18,20) and desdiethylamiodarone in dog plasma and myocardial tissue (19).

a. Flanagan et al (18) reported a method for the identification and measurement of desethylamiodarone in plasma from amiodarone-treated patients by HPLC with mass spectrometric detection. Plasma is extracted with methyl t-butyl ether. The ether phase is evaporated to dryness under a stream of air and the residue is redissolved in methanol before HPLC assay. The methanolic solution is then applied either to a straight phase Spherisorb S5W Silica column or a reversed phase Spherisorb S5ODS 1 column operating at ambient temperature, with methanol containing 0.04% (v/v) perchloric acid and methanol containing 0.02% (v/v) perchloric acid as mobile phases for the respective columns. For both columns a flow rate of 4.0 ml/min was used and the UV detector was set at 240 nm. Under these conditions the retention times are 10.2 min for amiodarone and 6.9 min for desethylamiodarone in the Silica system and 11.6 min and 5.7 min for the compounds in the ODS system. For mass spectrometric confirmation the column effluent fractions corresponding to the chromatographic peaks of amiodarone and desethylamiodarone were collected manually from the detector outlet into conical glass test-tubes. Each effluent fraction was treated to remove perchloric acid by the addition of distilled water, evaporation of the organic component of the eluent (methanol) under a stream of compressed air, and extraction of the remaining contents of the tubes with hexane. The hexane extracts were subsequently evaporated to dryness and reconstituted in a small volume of methanol before deposition onto the direct insertion probe of the mass spectrometer. The methanolic solutions were analyzed by chemical ionization (CI) mass spectrometry using isobutane as reactant gas. The isobutane CI spectrum of the peak in the HPLC chromatogram of the plasma extract with retention time identical with amiodarone showed fragments with m/z at 646 (MH^+), 547, 100 and 86 which were very similar to the reference CI spectrum of amiodarone (See Section 2.9.43). The peak in the HPLC chromatogram of the plasma extract with retention time corresponding with that of desethylamiodarone gave a CI mass spectrum with MH^+ ion at m/z 618 and prominent fragments at m/z 547, 72 and 58, which showed a high similarity with CI reference spectrum of desethylamiodarone (See Section 2.9.43). In view of the spectral similarity and retention time similarity with reference standards it can be concluded that amiodarone and desethylamiodarone are present in plasma specimens from amiodarone treated patients.

b. Marchiset et al (20) described a method for the identification of desethylamiodarone in the plasma of patients receiving a single intravenous or oral dose of amiodarone using mass spectrometry following extraction and HPLC separation. The method involves a single solvent extraction of amiodarone, desethylamiodarone and the internal standard (L8040) from plasma at pH 5.4 with diethyl ether. The organic layer is evaporated to dryness under a stream of nitrogen at 40°C, and a solution of the residue in chloroform is submitted to HPLC on a 25 cm x 4 mm Lichrosorb SI 100 (7μm) column at 30°C with chloroform-ethanol-ammonia (99.7:0.25:0.04 v/v/v) as the mobile phase and a fixed wavelength (254 nm) UV detector. A flow rate of 2 ml/min was used. Under these chromatographic conditions the retention times of amiodarone and the internal standard were approximately 6.5 min and 8.1 min respectively; a supplementary peak, which was identified as desethylamiodarone appeared at the retention time of 9.2 min. The identification of desethylamiodarone in plasma extract was performed by direct probe electron impact (EI) mass spectrometry of the column effluent fraction. The EI mass spectrum of the peak in the HPLC chromatogram with a retention time of 9.2 min showed 2 characteristic ions at m/z 490 (M-I) and at m/z 446 corresponding to the desiodinated and desaminated fragment, which were similar to those fragments found in EI spectrum of desethylamiodarone reference (See Section 2.9.4.1).

c. Latini et al (19) reported a sensitive method for the qualitative analysis of amiodarone and its metabolites desethylamiodarone and desdiethylamiodarone in plasma and myocardium from dogs and of amiodarone and desethylamiodarone in human plasma using HPLC isolation and fast atomic bombardment (FAB) mass spectrometric identification. Amiodarone and its metabolites are extracted from plasma or myocardial tissue at pH 5.4 with hexane. The organic phase is evaporated to dryness at 35°C under nitrogen. The residue is then dissolved in methanol and an aliquot injected onto the chromatographic column. Myocardial tissue samples are homogenized in methanol. The methanol extract is evaporated and the residue redissolved in phosphate buffer pH 5.4 and then extracted with hexane and further treated as described for plasma. They used an Alltech Lichrosorb RP-8 (25 cm x 3.2 mm) reversed phase column; the mobile phase was a mixture of 80% acetonitrile and 20% acetate buffer pH 5.4; the flow rate was 2-3 ml/min; the UV detector was set at 240 nm. Under these conditions desdiethylamiodarone eluted at approximately 7 min, desethylamiodarone at 9 min and amiodarone at 13 min respectively. Amiodarone and its metabolites were isolated from the biological samples by fractional collection of the HPLC effluent, appropriate fractions were identified by UV monitoring at 240 nm. The fractions were dried at 60°C under a stream of nitrogen and then redissolved in phosphate buffer solution of pH 7.4. The obtained solution is extracted with benzene and the benzene extract is evaporated to dryness under nitrogen. The residue is then redissolved in small volume of methanol prior to mass spectrometric analysis. Positive FAB mass spectrometric analyses of the methanolic extracts of plasma obtained from patients on long-term oral treatment with amiodarone revealed the presence of amiodarone with characteristic ions at m/z 646, 547, 520, and 421 and desethylamiodarone with prominent fragments at m/z 618, 547, 492, and 421 respectively (See Section 2.9.4.2). Positive FAB mass spectrometric assay of plasma and heart tissue of dogs chronically treated with oral dosages of amiodarone showed besides amiodarone and desethylamiodarone also the presence of desdiethylamiodarone with characteristic ions at m/z 590, 547, 464, 421, and 373 respectively (See Section 2.9.4.2).

5.3. High-Performance Liquid Chromatography

Several methods have been reported which are suitable for the determination of amiodarone alone or in combination with its major metabolite desethylamiodarone in biological fluids and tissues. The assay procedures which are assessing only amiodarone levels in biological material are nowadays obsolete because the main metabolite of amiodarone, desethylamiodarone is present in human plasma and tissues of patients treated with the drug in such relatively large amounts and possess pharmacological activity that it is obvious that the compound should be included in assay methods used for pharmacokinetic studies or clinical monitoring of the anti-arrhythmic agent amiodarone. For these reasons the HPLC methods used for amiodarone will only be reviewed briefly (Section 5.3.1) and the methods adequate for simultaneous analysis of amiodarone and desethylamiodarone in biological material will be outlined extensively in Section 5.3.2.

5.3.1. HPLC Methods for the Measurement of Amiodarone

Method I (23)

Type of sample : plasma

Column : 7 μm Hibar-Lichrosorb Si 100 (250 x 4 mm)

*Mobile phase : chloroform-ethanol-25% ammonia (99.48:0.5:0.02)

Detection : UV at 250 nm

Internal standard : brominated benzothiophene L8040, Labaz

Principle : Amiodarone is extracted from plasma after addition of acetate buffer pH 3.6 with diethyl ether

Recovery : 91%

Limit of sensitivity: 0.02 μg/ml

Method II (24)

Type of sample : plasma

Column : Hibar RP-8 (250 x 2.6 mm)

Mobile phase : acetonitrile-1.5% acetic acid (75:25) pH 4.7 with ammonia

Detection : UV at 250 nm

Internal standard : Crysene

Principle : Amiodarone is extracted from plasma after addition of 0.5 N hydrochloric acid with chloroform

Recovery : 87%

Limit of sensitivity : 1 μg/ml

Method III (25)

Type of sample : plasma

Column : 7 μm Hibar RP-8 (250 x 4 mm)

Mobile phase : acetonitrile-5% acetic acid (80:20) pH 5.9 with ammonia

Detection : UV at 254 nm

Internal standard : Benzopyrene

Principle : Amiodarone is extracted from plasma after deproteinization with trichloroacetic acid (0.2 N) with benzene

Method IV (26)

Type of sample : plasma

Column : 7 μm Hibar Si 100 (250 x 4 mm)

*Mobile phase : chloroform-ethanol-25% ammonia (99.84:0.5:0.02)

Detection : UV at 250 nm

Internal standard : L8040, Labaz

Principle : Amiodarone is extracted from plasma after addition of acetate buffer pH 3.6 with diethyl ether

Limit of sensitivity : 0.05 μg/ml

Method V (12)

Type of sample : plasma or serum

Column : 5 μm Spherisorb Si (125 x 5 mm)

Mobile phase : methanol-diethyl ether-perchloric acid (85:15:0.02)

Temperature : ambient (22°C)

Flow rate : 2 ml/min

Retention time : 2.7 min

Detection : UV at 240 nm

Internal standard : Fenethazine

Principle : Amiodarone is extracted from plasma or serum after addition of phosphate buffer pH 4.5 with diisopropyl ether

Recovery : 99%

Limit of sensitivity : 0.05 μg/ml

Method VI (27)

Type of sample : plasma

Column : 7 μm Hibar Si 100 (250 x 4 mm)

*Mobile phase : chloroform-ethanol-25% ammonia (99.71:0.25:0.04)

Detection : UV at 250 nm

Internal standard : L8040, Labaz

Principle : Amiodarone is extracted from plasma after addition of phosphate buffer pH 5.4 with diethyl ether

Recovery : 79%

Limit of sensitivity : 0.025 μg/ml

Method VII (11)

Type of sample : serum

Column : 10 μm μBondapak C_{18} (300 x 3.9 mm)

Mobile phase : methanol-0.01 M phosphoric acid-acetic acid (84:14:2)

Flow rate : 2 ml/min

Retention time : 2.2 min

Detection : UV at 254 nm

Principle : Amiodarone is extracted from serum by deproteinization with ethanol

Limit of sensitivity : 0.1 μg/ml

Method VIII (28)

Type of sample : serum

Column : 10 μm Micropak MCH-10 C_{18} (300 x 4 mm)

Mobile phase : methanol-water-25% ammonia (98.99:1:0.01)

Flow rate : 1 ml/min to 2 ml/min with 0.1 ml/min

Retention time : 8.1 min

Detection : UV at 242 nm

Internal standard : L8040, brominated benzothiophene (Labaz)

Principle : Amiodarone is extracted from serum after addition of acetate buffer pH 3.6 with diethyl ether

Recovery : 79%

Limit of sensitivity : 0.1 μg/ml

Method IX (29)

Type of sample : serum, urine or tissue

Column : 10 μm Porasil (300 x 4 mm)

*Mobile phase : chloroform-methanol-25% ammonia (98.95:1.0:0.05)

Flow rate : 0.8 ml/min

Retention time : 8.2 min

Detection : UV at 254 nm or 280 nm

Internal standard : L8040, Labaz

Principle : Amiodarone is extracted from serum or urine after addition of acetate buffer pH 3.8 with hexane. Amiodarone is extracted from tissues, after homogenizing with distilled water and followed by mixing of tissue homogenate with acetate buffer, with hexane.

Recovery : 80% from serum

Limit of sensitivity : 0.05 μg/ml

Method X (30)

Type of Sample : plasma

Column : 13 μm Amino-Sil X1 (250 x 2.6 mm)

*Mobile phase : n-heptane sat. with water-chloroform-ethanol (85:14.8:0.2)

Detection : UV at 250 nm

Internal standard : 2-ethyl N-dimethyl analogue of amiodarone

Principle : Amiodarone is extracted from plasma as follows: after addition of acidified acetonitrile and heptane and discarding of heptane layer; sodium hydroxide 1 M is added to the acetonitrile layer and the drug is extracted with chloroform.

Limit of sensitivity : 0.025 μg/ml

Method XI (31)

Type of sample : serum

Column : 10 μm μ Porasil (300 x 4 mm)

*Mobile phase : chloroform-methanol-25% ammonia (98.5:1.0:0.5)

Flow rate : 0.8 ml/min

Retention time : 9.6 min

Detection : UV at 280 nm

Internal standard : L8040, Labaz

Principle : Amiodarone is extracted from serum after addition of acetate buffer pH 4.0 with hexane

Limit of sensitivity : 0.05 μg/ml

Method XII (32)

Type of sample : blood or plasma

Column : 7 μm Hibar RP-8 (250 x 4 mm)

Mobile Phase : acetonitrile: 5% acetic acid (80:20), acetic acid brought up to pH 5.9 with ammonia

Flow rate : 2 ml/min

Retention time : 9.6 min

Detection : UV at 254 nm

Internal standard : Benzopyrene

Principle : Amiodarone is extracted from blood or plasma, after acidification with trichloroacetic acid, with benzene

Recovery : 87% from blood

Limit of sensitivity : 0.02 μg/ml

Method XIII (33)

Type of sample : plasma

Column : 5 μm Lichrosorb SI-100 (250 x 4 mm)

Mobile phase : methanol-diethyl ether with 0.02% perchloric acid (70:30)

Flow rate : 2.0 ml/min

Retention time : 2.9 min

Detection : UV at 242 nm

Internal standard : benziodarone derivative

Principle : Amiodarone is extracted from plasma after addition of acetate buffer pH 3.8 with diethyl ether

Recovery : 98%

Limit of sensitivity : 0.02 μg/ml

*

It should be noted that all solvent systems containing chloroform should be used very cautiously since amiodarone as well as the frequently used internal standard L8040 (brominated benzothiophene) undergo a rapid dehalogenation in this organic solvent, which may result in incorrect amiodarone levels.

5.3.2. HPLC Methods for Determination of Amiodarone and Desethylamiodarone

a. Berger et al (34) have studied 4 different chromatographic systems for amiodarone analyses in order to retain a reference method for amiodarone and desethylamiodarone pharmacokinetic studies. The criteria for the selection of the reference method were: a low detection limit, a high reproducibility, lack of interferences with potential metabolites, an easy use and unversatile chromatographic conditions. In the first system they used a straight-phase 10 μmLiChrosorb Si 100 column (250 x 4 mm) which was eluted with a mixture of chloroform-ethanol-25% ammonia (99.69:0.30:0.01) at a flow rate of 0.8 ml/min and UV detector set at 250 nm. In the second system a reversed-phase 7 μm LiChrosorb RP-18 (250 x 4 mm) column was employed; the mobile phase was a mixture of methanol-water-25% ammonia (98.99:1:0.01); the flow rate was maintained at 1 ml/min and the UV detector was set at 242 nm. In the third system the authors used a 5 μm Nucleosil CN (150 x 4.6 mm) column with an eluting solvent consisting of n-hexane-isopropyl alcohol-concentrated sulfuric acid (49.98:49.98:0.04) at a flow rate of 0.8 ml/min. The column effluent was monitored at 242 nm. In the fourth system a 5 μm LiChrosorb NH_2 (250 x 4 mm) column was used. The elute was a mixture of n-hexane-water-glacial acetic acid-isopropyl alcohol (85.4:0.6:1.5:12.5) with a flow rate of 1 ml/min. The absorbance was measured at 242 nm using UV detector. This last system has been selected by the investigators as the most suitable chromatographic system for pharmacokinetic studies of amiodarone and desethylamiodarone, because of its lack of interferences by potential metabolites, especially for the desethylamiodarone assay and absence of degradation of amiodarone and internal standard L8040, which had been observed with LiChrosorb RP-18 and Nucleosil CN phases and LiChrosorb Si 100 phase respectively (See section 5.3.1). The extraction procedure used, involves a double solvent extraction of amiodarone, desethylamiodarone and internal standard L8040 from acidified plasma, followed by evaporation and HPLC analysis. Concentrations of the compound between 0.05 and 2.0 μg/ml can be measured with a coefficient of variation varying from 5 to 10%. Calibration graphs (peak-height ratio versus concentration) is rectilinear in the range of 0.05 to 2.0 μg/ml. The limit of detection of this method was 20 ng/ml of plasma for both drugs. Sample is prepared by taking 1 ml plasma in a tube containing 1 μg of L8040 in methanol as internal standard. To this 4 ml of a 0.067 M phosphate buffer pH 5.4 was added and the tube shaken for a few seconds. The sample is then extracted twice with 5 ml of hexane by shaking for 5 minutes each time, followed by centrifugation at 3500 rpm for 5 minutes. The organic phases are separated and dried at 30°C under nitrogen. The residue is dissolved in 200 μl of mobile phase and 100 μl aliquot is injected in the chromatograph.

b. Storey and Holt (35) developed a sensitive and specific HPLC procedure for the simultaneous determination of amiodarone and desethylamiodarone in plasma or serum at levels attained after a single oral dose. The method involves a single solvent extraction of the compounds and internal standard from acidified plasma, followed by evaporation of organic phase and HPLC assay. Plasma or serum (1 ml) is extracted after the addition of 100 μl of a 0.2 mg/l solution of internal standard L8040 [2-ethyl-3-(3,5-dibromo-4-γ-di-n-propylaminopropoxybenzoyl)benzothiophene] in phosphate buffer pH 4.5, by vortex mixing for 10 min with 1 ml diisopropyl ether. After centrifugation at 3000g for 5 min the organic phase is separated and evaporated to dryness using a stream of compressed air. The residue is dissolved in 120 μl of methanol and 100 μl of this solution is applied to a column (125 x 5 mm) of 5 μm Spherisorb 5 Silica operated at 22°C, with a mixture of methanol-diethyl ether (85:15) containing 0.03% (v/v) perchloric acid as mobile phase (2 ml/min) and detection at 240 nm.

Retention times are 2.8 min for desethylamiodarone, 3.2 min for internal standard and 4 min for amiodarone respectively. The calibration graphs for both compounds (peak-height ratio versus concentration) is linear for 2 to 100 μg/l. The intra-assay coefficients of variation of plasma sample with concentrations ranging from 5 to 30 μg/l for both compounds varied from about 7 to 2%. The inter-assay coefficient of variation of plasma containing 20 μg/l of amiodarone and desethylamiodarone were 3% and 4% respectively. The lower limit of detection was 2 μg/l for both compounds and there was no interference from endogenous plasma material.

c. Brien et al (36) reported a rapid and reliable assay for the measurement of amiodarone and desethylamiodarone in serum using high-performance liquid chromatography. The procedure involves addition of absolute ethanol to serum previously adjusted to pH 6, followed by centrifugation and analysis of supernatant. Serum (1 ml) is adjusted to pH 6 with 30 μl of 1.2 M hydrochloric acid and then 2.0 ml of absolute ethanol was added followed by shaking on a Vortex mixer for 30 sec and after 15 min standing, centrifugation at 1350 g for 5 min. A 50 μl aliquot of the supernatant is submitted to HPLC on a reversed phase column (300 x 3.9 mm) 10 μm micro Bondapak C_{18} with methanol-water-25% ammonia (89.7:10:0.3) as the mobile phase at a flow rate of 1.3 ml/min. The column effluent is monitored at 254 nm. The retention times of desethylamiodarone and amiodarone were 9.2 and 10.9 min respectively. There is no interference from endogenous substances or other commonly used antiarrhythmic drugs. The mean overall recovery of the HPLC assay is 96.9 ± 9.1% for amiodarone over the concentration range 0.10-10.0 μg/ml and 99.8 ± 4.8% for desethylamiodarone over the concentration range 0.10 - 7.5 μg/ml. The within-day coefficient of variation is less than 7% for both compounds and the lower limit of quantitative sensitivity is 0.10 μg/ml for both drugs.

d. Mostow et al (37) described a high-performance liquid chromatographic procedure for the simultaneous analysis of amiodarone and its N-deethylated metabolite in serum. They used a (300 x 4 mm) Varian NH_2-10-alkylamine column operated at 30°C; UV detection at 248 nm, the mobile phase is a mixture of methylene chloride-hexane-methanol-glacial acetic acid-0.117 M perchloric acid in methanol (50:39:6:1:1 v/v). At a flow rate of 1.5 ml/min the internal standard L8040 (Sanofi) eluted at 4 min, desethylamiodarone at 5.5 min and amiodarone at 6.5 min respectively. For serum extraction 10 μl of internal standard solution of 100 μg/ml in methanol, 200 μl of 3M sodium acetate buffer pH 4.8 and 0.50 ml serum were subsequently pipetted in a disposable glass tube. This was mixed briefly followed by the addition of 5 ml of hexane. The tube was then vortexed vigorously for 2 min and centrifuged at 7000 g for 1 min. The hexane phase is evaporated to dryness under reduced pressure. The residue was redissolved in 65 μl of mobile phase and a 25 μl aliquot of this solution was injected in the chromatograph. The recovery of amiodarone and internal standard from serum containing levels ranging from 0.2 to 20 μg/ml varied from 90 to 99%. The recovery of desethylamiodarone was not stated because of the relatively low value obtained by this procedure. The method is linear between 0.05 and 20.0 μg/ml of amiodarone. The within-day coefficient of variation for amiodarone serum concentrations of 2.0 μg/ml was 3.2%. The method is sufficiently sensitive to measure low amiodarone levels of 50 ng/ml and is free from interferences by cardiac and non-cardiac drugs and endogenous serum substances.

e. Storey at al (38) reported a high-performance liquid chromatographic method for the determination of amiodarone and desethylamiodarone in plasma and small tissue samples after enzymatic digestion. The method involves tissue digestion with a proteolytic enzyme, with exception of adipose tissue for which lipase enzyme is used and after the addition of internal standard, extraction of the analytes from the homogeneous digest at acidic pH into an organic solvent and analysis with HPLC. The tissue digestion is carried out by adding to a preweighed tube 100 mg of finely minced tissue and 1 ml of 2 mg/ml of Subtilisin A solution in phosphate buffer pH 7.4 or for fat 1 ml of 10 mg/ml of Lipase type II solution in phosphate buffer pH 7.4 followed by incubation of the stoppered tube at 50°C for 16 h. Subsequently, the contents of the tube were vortex mixed to yield a homogeneous suspension. To the fat digest approximately 4 mg of sodium lauryl sulphate was added before vortex mixing to give a more stable homogeneous mixture. The extraction of the compounds was performed by adding to a small glass test tube: 100 μl of tissue digest, 100 μl of analyte-free human plasma, 20 μl of 0.05 mmol fenethazine (I.S.) in 2M sodium dihydrogen orthophosphate and 200 μl methyl tert-butyl ether. The contents of the test tube were vortex mixed for 30 sec, centrifuged at 9950 g for 1 min and 100 μl of the organic phase is submitted to HPLC on a column (125 x 5 mm) of Spherisorb 5 Silica operating at ambient temperature (22°C). The column effluent was monitored at 240 nm. The mobile phase consisted of methanol:2.2.4-trimethyl pentane:methyl tert-butyl ether (80:10:10) containing 6 mmol/l potassium bromide and was delivered at a flow-rate of 2.0 ml/min. Under these conditions desethylamiodarone eluted at 1.7 min, amiodarone at 2.4 min and fenethazine at 4.0 min. The intra-assay precision for human heart tissue samples containing 35 μg/g of amiodarone and 200 μg/g of desethylamiodarone and for adipose tissue samples containing 170 μg/g of amiodarone and 40 μg/g of desethylamiodarone varied for both compounds between 7 and 9% (n=10). The limit of sensitivity for the determination of both compounds was 0.1 μg/g net weight using a 100 mg tissue sample. The recovery of amiodarone and desethylamiodarone from tissues was higher than 95% for both drugs. No interferences have been observed from endogeneous sources or from concomitantly used cardio-active drugs.

f. Gupta and Connolly (39) described a method for the simultaneous determination of amiodarone and desethylamiodarone in plasma by HPLC. In a glass tube 0.5 ml of plasma sample, 1 ml of 250 μg/l solution of prazepam (I.S.) in 0.2 mol/l acetate buffer pH 4.8 and 4 ml of pentane is placed. The tube is gently rotary mixed for 10 min, centrifuged, and the upper organic phase is collected, after which the aqueous phase is extracted again with 4 ml of pentane. The pooled pentane extracts are evaporated in a water bath at 45-50°C, and a solution of the residue in mobile phase is applied to a column (150 x 4.6 mm) of 5 μm Ultrasphere C_8 with a mixture of acetonitrile - water - 70% perchloric acid - 23 % tetramethylamonium hydroxide in methanol (750 : 250 : 0.5 : 0.5) as mobile phase (1.2 ml/min) and detection at 240 nm. Retention times are approximately 2 min for prazepam, 5 min for desethylamiodarone and 7 min for amiodarone. The calibration curve (peak area ratios versus concentration) is linear for both compounds over the range of 0.1 to 10 μg/ml. The recovery from plasma of amiodarone and desethylamiodarone after double pentane extraction varied from 85% to 90%

Analysis of plasma supplemented with a 1 μg/ml of each of amiodarone and desethylamiodarone showed both within-batch and between-batch coefficients of variations to be less than 5% for each compound.

g. A high-performance liquid chromatographic method for the measurement of serum concentrations of amiodarone and desethylamiodarone was reported by Rotmensch et al. (40). Serum samples (0.5 ml) were adjusted to pH 5.4 with phosphate buffer and extracted with hexane (5 ml). The hexane layer was transferred and evaporated to dryness. The residue was redissolved in mobile phase and injected onto a Spherisorb 5 μm silica column. The mobile phase consisted of chloroform - methanol - 25% ammonia (98.95 : 1.0 : 0.05) at a flow of 1.5 ml/min and UV detection was done at 254 nm. Intra-assay variations for five replicates at concentrations of 10, 1.0 and 0.1 μg/ml were ± 1.7%, 4.8%, and 11.8% respectively. The lower limit of reliable quantitation was 0.5 μg/ml in serum.

h. Weir and Ueda (13) developed a rapid high-performance liquid chromatographic assay for the assessment of amiodarone and desethylamiodarone in plasma, urine and bile. Analysis was carried out on a reversed phase 10 μm Waters C_{18} column (300 x 3.9 mm) using a mobile phase consisting of methanol - water - 58% ammonia (94 : 4 : 2) delivered at a flow rate of 1.5 ml/mm. The eluant was monitored at 244 nm. Under these conditions, desethylamiodarone, amiodarone and the internal standard L8040 eluted with retention times of 4.6, 5.5, and 6.8 min, respectively.
Samples (100 μl) of plasma were prepared by precipitating the plasma proteins with 200 μl of acetonitrile containing the internal standard L8040 (1 μg/ml) and after mixing for 30 sec and centrigugation for 4 min at 9500 g, injection of a 200 μl aliquot of the supernatant directly onto the column. Samples (100 μl) of urine and bile were prepared for injection by acidifying the sample with 20 μl of concentrated hydrochloric acid and then extracting the mixture with 600 μl of 2.2 dimethoxypropane. Following centrifugation at 9500 g for 4 min, the supernatant was transferred and evaporated to dryness under nitrogen. The residue was redissolved in 300 μl of internal standard solution, and a 200 μl aliquot was injected into the chromatograph. The mean recovery from plasma over the concentration range of 0.1 - 10 μg/ml varied for amiodarone from 91 - 101 % and for desethylamiodarone from 89 - 102%. The recovery from urine and bile was 80 - 90% for amiodarone and 60 - 65% for desethylamiodarone. The limit of sensitivity of both compounds in plasma was 0.1 μg/ml. For urine and bile, the detection limits were 1 and 5 μg/ml respectively. Over the plasma concentration range of 0.1 - 10.0 μg/ml, the within-day coefficient of variation ranged from 1 to 10% for amiodarone and from 1 to 8% for

desethylamiodarone. The day-to-day coefficient of variation varied from 2 to 12% and from 1 to 17% for the respective compounds. Forty-four drugs were tested for their potential to interfere with the assay. It appeared that chloroquine, nortriptyline, thioridazine and thiothixene interferred with the desethylamiodarone assay and desipramine with the analysis of amiodarone. No interferences were observed from endogenous plasma, urine or bile substances. Standard graphs were constructed by plotting the peak height ratios of amiodarone or desethylamiodarone to the internal standard versus amiodarone or N-deethyl metabolite concentraction and were linear over the range 0.1 - 10 μg/ml for plasma and 1 - 100 μg/ml for urine and bile respectively.

i. Somani et al (41) developed a high-performance liquid chromatographic procedure for measurement of both amiodarone and desethylamiodarone in small samples of plasma and tissue without enzymatic breakdown. Briefly, to 0.50 ml aliquots of plasma, 100 μl of the internal standard L8040 (0.01 μg/ml in methanol) and 25 μl of 1.5 M potassium hydrogen phosphate buffer pH 4.5 were added and the mixture was vortexed. The drugs were extracted for 30 min in the isoamyl alcohol (10 μl)/n-pentane (5 ml) layer, to which 100 μl of 5 M sodium hydroxide was added. After centrifugation of the pentane/alkali mixture, the aqueous layer was frozen in dry ice methanol and the organic layer was decanted and evaporated to dryness in air at 85°C. The residue was dissolved in 100 μl of mobile phase and a 50 μl aliquot was injected into the HPLC system.

For tissue samples a homogenate in water (0.3 g/ml) was prepared with a polytron and protein concentration was adjusted with buffer. After addition of the internal standard L8040 to the homogenate, extraction was performed on 2 ml of homogenate buffered with 70 μl of phosphate and the procedure was similar to that for plasma except that no isoamyl alcohol was added.

A straight phase (100 x 8 mm) Waters Silica 10 μm cartridge column was used and eluted with methanol-perchloric acid (100 : 0.03). The UV detector was set at 254 nm. At a flow rate of 1.5 ml/min the retention times for desethylamiodarone, L8040 and amiodarone were 6.3, 7.9 and 10.1 min respectively.

Other cardiac drugs did not interfere with the peaks of the abovementioned compounds.

The percent extraction of amiodarone and desethylamiodarone from plasma and tissues ranged between 78 and 90% and a linear concentration-peak height relationship was observed over the concentration range of 0.1 to 4 μl/ml or g.

j. Lam (42) developed a reliable and efficient HPLC procedure for amiodarone and desethylamiodarone in serum by the micro-deproteinization technique using zinc sulfate in combination with acetonitrile as the precipitants. Samples (100 μl) of serum were prepared by addding 10 μl of 10% (w/v) zinc sulfate solution and mixing of the content of the tube. Then, 100 μl of acetonitrile were added with mixing. The tube was centrifuged at 3000 g for 1 min and an aliquot of 100 μl of the supernatant was injected into the chromatograph with a variable wavelength detector, using (150 x 4.2 mm) Spherosil C_{18} 5 μm column, absorbance was measured at 254 nm. The mobile phase is a mixture of acetonitrile - water - phosphoric acid - diethylamine (40 : 10 : 0.05 : 0.025) with a flow rate of 3 ml/min. Under these conditions desethylamiodarone and amiodarone eluted with retention times of 3 and 5 min respectively. The sera spiked with 1.0 μg/ml and 5.0 μg/ml of amiodarone were recovered 100% and 108% respectively from the supernatant. The standard curves of the compounds over the concentration range of 0.5 to 5.0 μg/ml exhibit good linearity. The detection limits are 0.05 μg/ml and 0.025 μg/ml for amiodarone and desethylamiodarone, respectively. The within-day precision for the compounds at concentrations of 1.0 μg/ml was 2.2 and 2.3% and the day-to-day precision at concentrations of 1 μg/ml and 0.5 μg/ml was 6.7% and 6.5% respectively.

k. Bliss et al (43) developed an HPLC assay for the simultaneous measurement of amiodarone and desethylamiodarone in human serum samples which has the precision and sensitivity to allow long term measurement of these 2 compounds after a single intravenous or oral dose. Sample is prepared by taking 1 ml serum in a glass tube containing 20 μl of the internal standard, triflupromazine (1.0 μg/l). To this 8 ml of hexane is added and the samples are extracted by horizontally shaking at a low speed for 30 min, followed by centrifugation at 1500 g for 5 min. The organic layer is separated and evaporated to dryness at 40°C under a stream of nitrogen. The resulting residue is reconstituted in 40 μl of methanol and a 20 μl aliquot is applied to a column (250 x 4.6 mm) of Altex Ultrasphere-Si 5 μm operated at room temperature with a mixture of methanol-diethyl ether (80 : 20) to which 12-15 μl of triethylamine per l was added, as mobile phase and dectection at 254 nm. At a flow rate of 2.2. ml/min the retention times for amiodarone, internal standard and desethylamiodarone were approximately 3.0, 4.5 and 6.0 min. The calibration graph (peak-height ratio versus concentration of both compounds is linear for 2 to 200 μg/l of the drugs. The limit of detection of this method was 1 μg/l of serum for amiodarone and 3 μg/l for desethylamiodarone.

Coefficients of variation at concentrations for amiodarone between 25 - 100 μg/l and for desethylamiodarone between 10 - 180 μg/l were within-day and between-day varying for both drugs from 6 to 16% and 5 to 9% respectively. The recovery from serum at the abovementioned concentrations ranged for amiodarone from 99 to 113% and for desethylamiodarone from 100 to 101% respectively. No interferences were observed from several cardio-active drugs.

l. De Smet and Massart (44) reported a standardised extraction and chromatographic optimisation procedure for the determination of amiodarone and desethylamiodarone in plasma.

The standardised extraction procedure is briefly as follows: To 1 ml of plasma sample in a glass centrifuge tube 100 μl of internal standard solution L8040 (5 μg/ml) in acetonitrile were added.

Then 2 ml of acetonitrile was added dropwise with continuous vortexing. After centrifugation for 15 min, the supernatant was transferred to a clean tube and evaporated under nitrogen at 60°C. To the residue in the tube 10 ml of phosphate buffer pH 3.0 containing 0.05 M sodium-N-octyl sulphate and 5 ml chloroform-hexane (3 : 2 v/v) was added and the tube was shaken for 30 min. After centrifugation 4 ml of the organic phase was evaporated at 30°C under nitrogen. The residue was reconstituted in 200 μl of acetonitrile and 50 μl was injected onto the chromatograph. A reversed phase Lichrosorb CN-column (250 x 4 mm) with particle size of 5 μm was used and eluted at a flow rate of 1 ml/min with acetonitrile-water-propylamine (56 : 44: 0.01). Ultraviolet detection was done at 254 nm. Retention times are approximately 6 min for amiodarone, 8 min for desethylamiodarone and 15 min for internal standard respectively. The calibration graph for both compounds (peak area ratio of the analytes to the internal standard versus the concentration) is linear for 0.5 to 5 μg/ml.

The within-day precision determined at a concentration level of 1 μg/ml and 5 μg/ml ranged for amiodarone from 3.2 - 3.5% and for desethylamiodarone from 4.1 to 6.2% respectively. The recovery from plasma for both drugs using ion-pair extraction was approximately 80%. The detection limit was estimated to be 0.02 μg/ml. No interference was observed from plasma compounds or from many other drugs.

m. An own developed simple and sensitive high-performance liquid chromatographic method for the simultaneous assay of amiodarone and desethylamiodarone in plasma, urine and tissues is described below (10).

Our HPLC method should be considered as a reference method for the determination of amiodarone and desethylamiodarone in biological fluids and tissues. It was the first reversed phase HPLC method described for the drugs and has several times been used as basis for the development of other later published HPLC procedures (13, 36, 42, 44). Samples of plasma and serum are prepared by introducing 1.00 ml of plasma/serum and 2.00 ml of internal standard solution in acetonitrile (2.5 μg/ml of L8040, Sanofi Labaz) in a disposable polypropylene-capped glass centrifuge tube. The tube was vortex mixed for 30 sec, allowed to stand at room temperature for 15 min and centrifuged at 2000 g for 15 min. Subsequently the supernatant was transferred to a clean tube and 20 μl of this solution were injected into the liquid chromatograph.

Samples of urine are prepared by successively adding to the top of a Clin-Elut 1003 disposable extraction column 2.00 ml phosphate buffer pH 7.0, 40 μl of the stock internal standard solution in acetonitrile (125 μg/ml) and 1.00 ml of urine. After waiting for 3 min, the column was extracted two times with 6 ml of a mixture of diisopropyl ether-acetonitrile (95 : 5 v/v) and the collected combined eluates were evaporated to dryness at room temperature under reduced pressure. The residue then was redissolved in 3 ml of a mixture of methanol-acetonitrile (1 : 2 v/v) and a 20 μl aliquot was injected into the chromatograph. Tissue samples were finely minced, dried between Kleenex[R] tissues and then portions of about 20 mg of thyroid tissue, 40 mg of adipose tissue and 100 mg of the other tissues were homogenized with 1 ml of 50% ethanol using a Potter apparatus. Subsequently the tissue homogenate was vortexed for 30 min with 2 ml of internal standard solution (2.5 μg/ml) in acetonitrile and after centrifugation 20 μl of the clear supernatant were injected into the HPLC.

In the original procedure (10) a reversed phase (100 x 8 mm) Radial-PAK C_{18} 10μm radially compressed column (Waters Ass.) was used at ambient temperature with a mixture of methanol - 25% ammonia (99.3 : 0.7 v/v) as mobile phase and UV detection at 242 nm.

At a flow rate of 2 ml/min the retention times for desethylamiodarone, amiodarone and internal standard were 5.3, 6.3 and 9.4 min, respectively.

In the modified procedure of 1985, a new reversed phase (100 x 8 mm) Nova-Pak C_{18} 5μm radially compressed column (Waters Ass.) was used with a flow rate of the mobile phase of 1 ml/min (45). Compared to the previously employed Radial-Pak C_{18} column this new column showed a higher efficiency and resolution at similar time of chromatographic analysis. Under these chromatographic conditions the retention times of

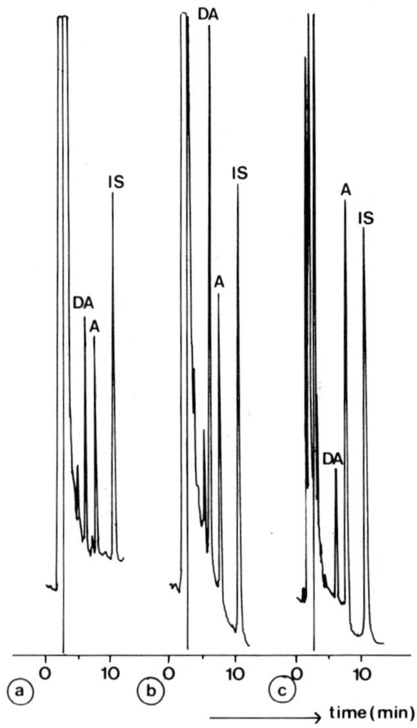

Fig. 28 Chromatogram of (a) plasma containing 1.41 μg/ml A and 1.14 μg/ml of DA, (b) heart tissue containing 19.1 μg/g A and 59.6 μg/g DA, and (c) adipose tissue containing 226 μg/g A and 60 μg/g DA from a heart-transplantation patient who had received 300 mg of amiodarone for 2 years

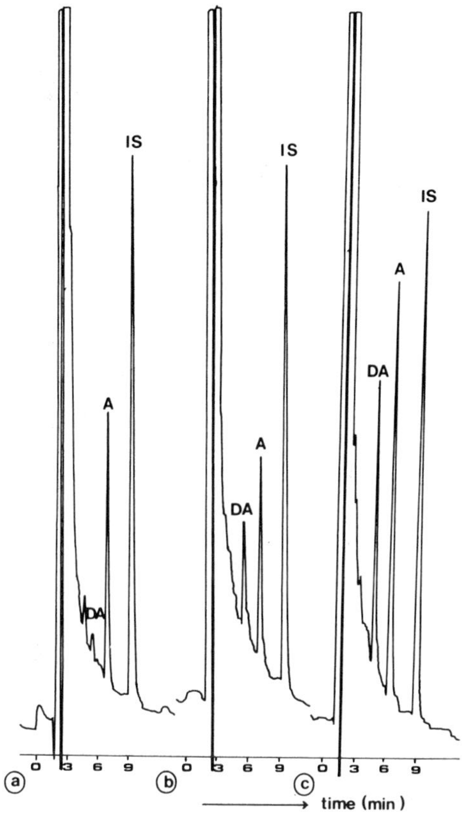

Fig. 29 Chromatogram of (a) serum sample containing 1.35 μg/ml A and 0.20 μg/ml DA, (b) heart tissue sample containing 12.4 μg/g A and 5.8 μg/g DA, and (c) renal tissue sample containing 70.7 μg/g A and 45.0 μg/g DA obtained from a rat orally treated with 100 mg/kg of amiodarone per day for desethylamiodarone (DA), amiodarone (A) and internal standard (IS) were 13 days

desethylamiodarone (DA), amiodarone (A) and internal standard (IS) were 5.9, 7.6 and 10.3 min, respectively. Typical chromatograms using this modified chromatographic conditions are depicted in Fig. 28 and 29. Fig. 28 shows the chromatograms of a plasma sample (a) and a heart (b) and adipose (c) tissue sample obtained from a patient who underwent a heart transplantation and was treated with 300 mg amiodarone per day for more than 2 years.

Fig. 29 shows the chromatograms of an extract of a serum sample (a) and a renal (b) and heart (c) tissue sample obtained from a rat orally treated with 100 mg/kg of amiodarone per day for 13 days.

Standard calibration curves were constructed by plotting the peak area ratios of amiodarone and desethylamiodarone to the internal standard versus analytes concentration and were linear over the range 0.1 - 5.0 μg/ml for plasma, urine and tissues.

The mean recovery from plasma over the concentration range of 0.1 - 5.0 μg/ml varied for amiodarone from 98.9 - 104.8% (n = 8) and for desethylamiodarone from 100.8 - 105.6%. The mean recovery from urine was 81 - 91% for amidarone and 72 - 86% for desethylamiodarone. The mean recovery from tissues was 94.2 ± 7.2% for amiodarone and 94.1 ± 4.5% for desethylamiodarone respectively.

Over the plasma concentration range of 0.75 - 4.0 μg/ml the intra-assay coefficient of variation ranged from 1 to 3% for amiodarone and from 1 to 5% for desethylamiodarone. The inter-assay coefficient of variation varied from 2 - 6% and from 3 - 6% for the respective compounds.

The detection limit of both compounds in plasma, urine or tissue was 0.01 μg/ml or g with a precision of approximately 20%.

Endogenous substances or a variety of drugs concomitantly used in amiodarone therapy did not interfere with the assay. Finally own experiments with activated Bondelut C_{18}, C_8 and CN disposable solid phase extraction columns at pH 7, 6, 5 and 1 using methanol as eluting solvent showed in all cases low and variable recoveries for amiodarone and desethylamiodarone from plasma due to no retention of the analytes on these sorbents under these pH conditions. Our preliminary results suggest that reversed-phase solid phase extraction is not suitable as extraction procedure for amiodarone and its metabolite from plasma.

6. PHARMACOKINETICS AND METABOLISM

6.1. Pharmacokinetic studies in man

The pharmacokinetics of amiodarone in blood or plasma has been investigated in healthy volunteers and patients with cardiac arrhythmias.

6.1.1. Single-dose intravenous administration

Andreasen et al (11) studied the intravenous pharmacokinetics of amiodarone in seven patients with cardiac arrhythmias. The serum concentration of amiodarone was assessed by HPLC. The sensitivity of the assay used was 0.1 μg/ml. Following intravenous bolus of 400 mg of the drug over 2 min, the mean serum level decayed rapidly from about 22 μg/ml at 5 min to 0.3 μg/ml at 24 h after administration.

The individual serum curves did not show a uniform pattern and often had secondary peaks between 4 h and 12 h after dosing, which did not allow a reliable estimate of the kinetic parameters. Amiodarone concentrations were in all patients below 0.1 μg/ml after 48 h or 96 h. An estimated serum elimination half-life of 11 h and a distribution volume of 400 l, derived from the average serum curve, were reported.

A relatively short serum elimination half-life of 5.8 h was observed in one patient after a single intravenous dose of 5 mg/kg of amiodarone by Kannan et al (50). Intravenous pharmacokinetic studies of amiodarone had been conducted by Anastasiou-Nana et al (51) in 8 patients who received a 150 mg intravenous bolus of the drug. Amiodarone plasma concentrations were determined by HPLC. Amiodarone plasma levels after intravenous administration, with one exception, declined biphasically. About seven hours after administration drug levels were less than 0.1 mg/l.

A mean plasma clearance of 197 ± 53 ml/min and a mean apparent volume of distribution of 1.26 ± 0.43 l/kg were found. After intravenous injection the drug concentration declined during the second phase with a mean elimination half-life of 4.3 ± 1.0 h.

Amiodarone was administered intravenously in a dose of 150 mg to 3 healthy volunteers and in a dose of 5 mg/kg of the drug to 3 patients with supraventricular tachycardia by Riva et al (32). Blood was collected at intervals for 24 h in patients and up to 32 h in healthy volunteers.

The amiodarone blood concentration was measured by a high performance liquid chromatographic method. Following intravenous bolus of 150 mg in 3 volunteers mean blood levels declined from 7 μg/ml at 15 min to circa 0.07 μg/ml at 8 h and then to 0.05 μg/ml at 24 h.

The average amiodarone blood concentration in 3 patients after 5 mg/kg injection decreased from 3.3 μg/ml at ½ h to 0.4 μg/ml at 4 h and then stayed at the same level to 8 h and thereafter decreased slowly to 0.2 μg/ml at 24 h.

No difference in mean pharmacokinetic parameters was found between the two groups. The kinetic parameters, calculated on the intravenous data using a two-compartment model, indicate a very large volume of distribution of 9.3 to 17.2 l/kg (mean 12.9 l/kg) in healthy volunteers and 6.9 to 21.1 l/kg (mean 11.7 l/kg) in patients.

The elimination half-life varied from 12.1 to 20.7 h (mean 17.4 h) in normal subjects and from 11.6 to 19.6 h (mean 16.2 h) in patients. The mean total blood clearance observed in this study was quite high and were for the respective groups 35.8 l/h and 27.2 l/h.

The single-dose pharmacokinetics of amiodarone have been studied by Holt et al (52) in six healthy male volunteer subjects given 400 mg doses by 10 minute intravenous infusion. Serial blood samples were collected for up to 84 days (means 56 days) following intravenous dosage for the assessment of plasma amiodarone and desethylamiodarone.

The limit of accurate measurements for these compounds was 5 μg/l. Plasma concentration-time data following intravenous dosage were best described by a polyexponential function with four terms. The kinetic parameters calculated on the intravenous data show the compound to have a very large volume of distribution ranging from 18.3 to 147.8 l/kg (mean 65.8 l/kg), a low total clearance from 6.5 to 11.1 l/h (mean 8.6 l/h) and a long and variable terminal elimination half-life ranging from 9.3 to 44.1 days (means 24.8 days).

The single-dose kinetics of amiodarone were investigated by Mostow et al (37) in five healthy male volunteers.

Amiodarone, 5 mg/kg, was administered as an intravenous infusion over 15 min and serum sampled at designated intervals for 24 to 48 h. The amiodarone serum concentrations were measured by an HPLC assay. The serum concentration time data were best described by a biexponential function.

The kinetic parameters, calculated from the intravenous fit parameters, revealed an apparent volume of distribution varying from 4.8 to 10.8 l/kg (mean 6.4 l/kg), a body clearance from 15.8 to 43.8 l/h (mean 28.0 l/h) and a terminal elimination half-life ranging from 8.7 to 14.9 h (mean 11.3 h) respectively.

The kinetics of the plasma concentrations of amiodarone were studied by Marchiset et al (20) in 3 patients after a single administration of amiodarone by intravenous route, using a sensitive and specific HPLC method.

Following infusion of a dose varying from 5.6 to 6.7 mg/kg (mean 6.1 mg/kg) over 20 min period, a plasma clearance rate for the drug ranging from 0.2 to 0.5 l/h/kg of body weight (mean 0.36 l/h/kg) was observed and the mean elimination half-life of amiodarone was found to be 18.5 h (range 15-23 h). Mostow et al (53) reported the intravenous kinetics of amiodarone in 11 patients with complex ventricular arrhythmias. Each patient in the study received a single 5 mg/kg infusion of the drug for approximately 15 min. At the end of the infusion blood was sampled over a 72 h period. Serum was analyzed for amiodarone utilizing an HPLC method. A terminal elimination half-life ranging from 20.4 h to 53.5 h (mean 33.6 ± 8.8 h) and a clearance rate varying from 203 to 438 ml/min (mean 315 ± 81 ml/min) was found by these authors.

The time course and the mode of antiarrhythmic action of amiodarone was studied after acute intravenous injection of the drug (5 mg/kg in 5 min) to four female patients with junctional reciprocating tachycardia by Giani et al (54).

Blood was collected during 24 h after amiodarone application for analysis by an HPLC method. Plasma concentrations of amiodarone varied from 0.83 to 1.97 μg/ml after 1 h and from 0.41 to 1.53 μg/ml after 2 h. The plasma decay of the drug from 0 to 24 h followed a two-compartment open model. The pharmacokinetic parameters calculated from this model showed values for elimination half-life ranging from 6.4 to 21.7 h (mean 13.5 h), for apparent volume of distribution from 5.2 to 21.3 l/kg (mean 9.9 l/kg) and for plasma clearance varying from 3.6 to 14.3 ml/min/kg (mean 9.1 ml/min/kg) respectively.

Nolan et al (55) investigated intravenous single-dose pharmacokinetcs of amiodarone in 3 healthy adult males, which received 10 mg/kg of the drug as injection. Serial blood sampling followed each single i.v dose for 3 months. Serum amiodarone levels were assayed by a sensitive and specific HPLC procedure.

Disposition parameters of amiodarone were determined by model-independent area analysis. The mean values and ranges for apparent distribution volume, elimination half-life and clearance were respectively: 173.7 l/kg (140-195 l/kg), 30.2 days (23-52 days) and 2.76 ml/min/kg (1.70-

4.10 ml/min/kg).

Own single dose intravenous pharmacokinetic studies were performed in seven healthy volunteers (56). Following a single intravenous dose of 400 mg amiodarone blood samples were collected from each subject over a 96 h period and amiodarone and desethylamiodarone concentrations in plasma were measured by an own developed sensitive HPLC method (45).

The mean plasma concentration time data after single intravenous and oral administration of the drug are outlined in Table 14.

The intravenous plasma concentration time data were best fitted by a triexponential equation. The kinetic parameters derived from the

Table 14

Mean amiodarone plasma concentrations after a single intravenous and oral dose of 400 mg of amiodarone to 7 healthy volunteers

Time after administration in hours	Amiodarone plasma concentration (μg/ml)	
	Intravenous study	Oral study
0.125	16.2 ± 6.4	0.0
0.25	10.8 ± 12.2	0.004 ± 0.008
0.50	3.20 ± 2.88	0.037 ± 0.045
1.0	1.69 ± 1.04	0.116 ± 0.131
2.0	1.04 ± 0.24	0.270 ± 0.184
4.0	0.81 ± 0.12	0.326 ± 0.222
6.0	0.64 ± 0.17	0.371 ± 0.203
8.0	0.46 ± 0.13	0.267 ± 0.169
12.0	0.38 ± 0.12	0.210 ± 0.050
24.0	0.20 ± 0.06	0.076 ± 0.050
48.0	0.09 ± 0.07	0.041 ± 0.039
96.0	0.03 ± 0.05	0.006 ± 0.008

Mean body weight : 66 ± 8 kg

Mean age : 28 ± 5 yr

Table 15

Pharmacokinetic parameters of amiodarone after intravenous bolus administration of various doses of the drug

Dose (mg)	No. of subjects	Cl (l/h)	Vd (l/kg)	t½e	Sampling period	Ref
150 mg	8 (pts)	11.8	1.3	4.3 h	12 h	51
150 mg	3 (vol)	35.8	12.9	17.4 h	32 h	32
270 mg*	3 (pts)	27.2	11.7	16.2 h	24 h	32
300 mg*	4 (pts)	32.9	9.9	13.5 h	24 h	54
350 mg*	5 (vol)	28.0	6.4	11.3 h	48 h	37
365 mg*	11 (pts)	18.9	-	33.6 h	72 h	53
400 mg	7 (pts)	-	± 5.7	± 11.0 h	96 h	11
400 mg	6 (vol)	8.6	65.8	24.8 d	56 d	52
400 mg	7 (vol)	14.7	5.7	34.5 h	96 h	56
420 mg**	3 (pts)	25.0	-	18.5 h	48 h	20
700 mg***	3 (vol)	11.6	173.7	30.2 d	90 d	55

pts = patient, vol = healthy volunteer, h = hour, d = day

* = dose derived from: 5 mg/kg dose x mean body weight

** = dose derived from 6 mg/kg dose x mean body weight

*** = dose derived from 10 mg/kg dose x mean body weight

Cl = clearance, Vd = apparent volume of distribution, t½e = elimination half-life

triexponential functions showed the following mean values and ranges: terminal elimination half-life 34.5 h (8.7-116.1 h), apparent volume of distribution 376 l (43-1151 l) and plasma clearance 14.7 l/h (7.1 - 22.2 l/h) respectively.

A summary of the reported pharmacokinetic parameters of amiodarone assessed after interavenous administration is presented in Table 15.

6.1.2. Single dose oral administration

Most of the groups which have reported the intravenous pharmacokinetics of amiodarone, have also extensively studied the single oral kinetics of the drug (11, 20, 32, 37, 50-52). Several of these studies have indicated that amiodarone is absorbed variably and erratically from the gastro-intestinal tract. The observed plasma concentration time curves are irregular and there is a substantial variability in maximum plasma concentration, the time to reach the peak concentration, the elimination half-life and in absolute bioavailability (11, 32, 50).

The slow and variable absorption of amiodarone was first extensively described by Andreasen et al (11) in a study carried out in 7 fasting patients with cardiac arrhythmias. After the patients had received an oral dose of 400 mg, the concentration of the drug in serum was followed for 24 h using a specific HPLC procedure. The individual serum curves of amiodarone showed a variable profile with maximum concentrations ranging approximately from 0.2 to 1.0 μg/ml (mean 0.6 μg/ml) and the time to attain these concentrations (T_{max}) varying from 4.2 to 12.0 h (mean 7.3 ± 2.9 h). The amount of amiodarone reaching the general circulation in 24 h after oral intake was also variable and incomplete indicated by an absolute bioavailability ranging from 22 to 80% (mean 42%).

This slow and incomplete absorption is possibly due to first-pass metabolism in the gut wall or the liver.

Amiodarone kinetics after high single oral dosages varrying from 1400 to 1800 mg were investigated in 6 patients with ventricular tachyarrhythmias by Kannan et al (50).

Following a single oral dose, serum levels of amiodarone and desethylamiodarone, were determined over a 24 h period using an HPLC assay. Mean peak concentrations for amiodarone of 6.9 ± 4.2 μg/ml (range

3 - 14 µg/ml) and for the metabolite of 0.7 µg/ml were attained in 4.9 ± 1.2 h. The mean absorption and elimination half-life after single oral dose was 2.5 ± 0.9 h and 7.5 ± 5.0 h respectively. The mean kinetic data obtained after the 3 high doses are outlined in Table 16.

Riva et al (32) studied the kinetics of amiodarone in 3 healthy volunteers and in 3 patients with supraventricular tachycardia after a single oral 400 mg dose.

Table 16

Pharmacokinetic parameters of amiodarone after single oral administration

Dose (mg)	No. of subjects	C_{max} (µg/ml)	t_{max} (h)	F (%)	t½e (h)	Sampling period	Ref
200	5 (vol)	-	3.0	-	17.1	30 h	51
200-400	8 (pts)	-	4.5	34	17.1	30 h	51
400	7 (pts)	0.6	7.3	43	11.0	24 h	11
400	3 (pts)	0.4	2.3	58	35.9	24 h	32
400	3 (vol)	0.3	8.0	-	9.2	50 h	32
400	6 (vol)	-	-	35	-	38 d	52
400	12 (vol)	0.6	4.6	65	-	72 h	58
400	7 (vol)	0.4	4.8	31	32.0	96 h	56
800	8 (pts)	1.7	5.2	-	4.6	12 h	57
1400	2 (pts)	8.0	5.0	-	5.1	24 h	50
1400	1 (vol)	2.2	2.5	48	11.3	24 h	37
1600	2 (pts)	4.1	4.5	-	5.5	24 h	50
1800	2 (pts)	8.7	5.1	-	11.0	24 h	50

pts = patient, vol = healthy volunteer, h = hour, d = day,

C_{max} = maximum concentration, t_{max} = time to reach C_{max},

F = absolute bioavailability, t½e = elimination half-life

After oral intake of the drug, blood was drawn at intervals for 24 h in patients and up to 50 h from volunteers.
Amiodarone blood concentrations were measured by HPLC. The oral absorption showed to be slow and erratic, with in patients a variation in systemic bioavailability of 22 to 86%. The mean peak blood level in patients and volunteers was 0.37 μg/ml (range 0.15 - 0.60 μg/ml) and 0.32 μg/ml (range 0.25 - 0.40 μg/ml) respectively. These maximum levels were attained at 2.3 h and 8 h. The elimination half-life ranged from 12.1 to 79.7 h in patients and from 6.9 to 12.8 h in volunteers.
Oral pharmacokinetic studies were conducted by Anastasiou-Nana et al (51) in five volunteers who received 200 mg of amiodarone per os and in 8 patients who received 200 mg (6) or 400 mg (2) of the drug. For the oral administration blood samples were drawn before and at intervals up to 30 h after administration. Plasma concentrations of amiodarone were determined by HPLC. The decline of the plasma curves after oral administration was monophasic in all subjects. The elimination half-life in volunteers varied from 10.2 to 25 h (mean 17.1 h) and in patients from 11.7 to 24.0 h (mean 17.1 h).
After oral administration of 200 mg maximum levels were reached after approximately 3 h. The time to reach maximum levels for the two cases reveiving 400 mg was prolonged to 5-6 h. The systemic availability of amiodarone calculated in 4 patients who received the drug orally as well as intravenously varied from 23 to 50%. Absorption and disposition kinetics of a single oral 800 mg dose of amiodarone were studied in eight patients by Haffajee et al (57). Serum amiodarone concentrations were measured by HPLC for up to 12 h after administration of the drug. The serum concentration time data of amiodarone showed a wide interpatient variability, a delayed appearance of the drug in serum with a mean lag time

of 1.44 h, a mean absorption half-life of 1.62 h and a mean time to reach peak serum concentration (T_{max}) of 5.2 ± 0.6 h. Maximal concentrations of serum amiodarone (C_{max}) ranged from 0.58 to 3.12 μg/ml (mean 1.70 ± 0.35 μg/ml). The disappearance of amiodarone from the serum in all patients was relatively fast as indicated by a mean elimination half-life of 4.6 h. The single-dose pharmacokinetics of amiodarone have been investigated by Holt et al (52) in six volunteer subjects given 400 mg doses by the oral route.

Serial blood samples were collected for up to 63 days (median 38 days) following oral dosage for the measurement of plasma amiodarone and desethylamiodarone by HPLC. The data of this study showed that oral absorption proceeded for up to 15 h after a lag time in the range of 0.4 h to 3 h. The rates of absorption were irregular and showed when followed in time multiple irregular peaks. Mean aborption times of 6.5 ± 0.6 h and mean oral bioavailability of 35 ± 9% (range 22 - 46%) were observed in this study.

Single dose oral pharmacokinetic study in a normal male volunteer was reported by Mostow at al (37). Following a single oral dose of 1400 mg, a maximum serum level of 2.2 μg/ml was attained in about 2.5 h and thereafter amiodarone disappeared from serum with a half-life of 11.3 h. The absorption of the drug was incomplete as shown by an availability of the oral preparation of amiodarone of 48%.

The kinetics of amiodarone were studied by Marchiset et al (20) in four patients after a single oral administration of doses of amiodarone varying from 6.7 to 13.0 mg/kg, using a sensitive and specific HPLC method. The observed mean elimination half-life of amiodarone was found to be 23.6 ± 1.5 h and the oral bioavailability, as assessed in one patient, was abouty 60%.

The relative and absolute bioavailability of different oral forms of amiodarone was examined in 12 normal subjects by Pourbaix et al (58). In this study fasting volunteers received an 400 mg oral dose of 2 tablet formulations and a drinkable solution of the drug. Blood samples were collected before and during a 72 h period after dosing. Plasma levels of amiodarone and desethylamiodarone were assayed by HPLC. All amiodarone formulations showed a slow absorption after oral dosing with peak levels being reached between 3 h and 7 h with a mean value of 4.6 ± 1.2 h.

Amiodarone peak concentrations showed a substantial intersubject variation. Mean concentrations after the oral solution, the commercial tablets and the 'new' tablets were 0.56 μg/ml (range 0.39 - 0.85 μg/ml), 0.54 μg/ml (range 0.22 - 1.13 μg/ml) and 0.55 μg/ml (range 0.28 - 0.86 μg/ml).

Also the observed absolute bioavailability of the oral forms varied from subject to subject with mean values of 68% (range 43 - 105%), 65% (range 22 - 104%) and 60% (range 29 - 105%) for the forementioned formulations.

The similarity of the kinetic parameters obtained from the tablets and the drinkable solution indicate bioequivalence of the oral forms studied and permits exclusion of poor or erratic in vivo dissolution characteristics of the tablets to explain the incomplete bioavailability of amiodarone. Therefore it can be suggested that the variability in absorption is in part a result of fluctuations in the disposition of the drug during first-pass metabolism in the intestinal wall, portal vein, and liver (particularly liver blood flow), or in the absorption process through the lymphatic system.

Own single dose oral pharmacokinetic studies were conducted in the same seven volunteer subjects used in the intravenous study (6.1.1.).

Following a single oral 400 mg dose of amiodarone blood was sampled from each volunteer over a 96 h period and amiodarone and desethylamiodarone concentrations were determined by HPLC method.

The observed mean plasma levels of the drug and its metabolite are summarized in Table 14.

The oral plasma concentration time data were most satisfactorily fitted by a triexponentail equation. The kinetic parameters obtained from analysis of the individual plasma concentration time data revealed the following mean values and ranges: peak level (C_{max}) 0.37 μg/ml (0.15 - 0.70 μg/ml), time to reach peak concentration (t_{max}) 4.8 h (2.7 - 6.5 h), elimination half-life 32 h (5 - 68 h) and absolute bioavailability (F) 31% (3 - 69%).

A summary of oral kinetic parameters of amiodarone reported in the literature is outlined in Table 16.

6.1.3. Steady-state plasma concentrations after chronic oral therapy

Mean steady-state plasma or serum amiodarone and desethylamiodarone concentrations in patients with various arrhythmias on long-term amiodarone therapy with various maintenance doses are summarized in Table 17.

The first data on average trough concentrations in serum of amiodarone during prolonged oral therapy with 200 mg three times daily have been reported by Andreasen et al (11).

The mean serum concentration before the morning dose observed in 7 patients averaged 0.61 μg/ml after 24 h, 0.76 after 48 h, 1.18 after 1 week and 1.56 μg/ml after 1 month. The data indicate that during long-term administration amiodarone accumulates in plasma and steady-state levels are not reached for at least 30 days, with a two- to threefold increase occuring from day 1 to day 30.

Amiodarone serum kinetics after long-term therapy were investigated by Kannan et al (50) in patients with ventricular tachyarrhythmias.

They found in 12 patients given a mean dose of 1327 ± 338 mg per day of amiodarone for 4.1 ± 2.3 weeks, mean serum amiodarone concentration of 3.8 ± 2.9 μg/ml (range 0.9 - 12.0 μg/ml); in 3 patients simultaneous

Table 17

Plasma amiodarone (A) and desethylamiodarone (DA) concentrations in patients with various arrhythmias on chronic oral therapy

Maintenance dose per day (mg)	Duration of therapy (month)	n	Mean plasma level (μg/ml) A	DA	Ref
200	1	101	1.06 ± 0.43	1.04 ± 0.34	52
200	3	14	1.45 ± 0.50	1.38 ± 0.74	60
200	11.5	41	1.20 ± 0.50	0.70 ± 0.40	40
200	24	1	1.46	1.28	61
200	9	32	0.99 ± 0.40	0.85 ± 0.31	56
300	3	14	2.40 ± 0.80	1.52 ± 1.06	60
300	9	12	1.47 ± 0.57	1.29 ± 0.54	56
320	5	14	1.60	-	59
390	9.3	122	1.70	-	57
400	1	51	1.93 ± 0.80	1.79 ± 0.65	52
400	3	14	3.05 ± 1.36	2.33 ± 1.22	60
400	6.4	75	2.50 ± 1.10	1.50 ± 0.50	40
400	7.2	9	1.53 ± 0.56	1.08 ± 0.23	61
400	9	30	1.50 ± 0.61	1.25 ± 0.59	56
600	1	6	1.56	-	11
600	1	18	3.46 ± 1.50	2.79 ± 1.17	52
600	3	14	3.78 ± 1.61	2.92 ± 1.40	60
600	4.3	36	3.70 ± 1.00	1.90 ± 0.70	40
600	8.5	2	4.59 ± 0.04	1.81 ± 1.94	61
600	9	13	2.90 ± 1.00	2.10 ± 0.70	56
800	24	1	7.64	6.07	61
800	9	15	3.34 ± 1.22	1.76 ± 0.60	56
1330	1	12	3.80 ± 2.90	-	50
1400	1.3	1	11.9	5.4	50

n = number of patients, - = not determined

measurement of levels of amiodarone and desethylamiodarone showed that the concentration of the metabolite was about 50% of the parent drug during long-term therapy. In addition a study of the course of amiodarone and desethylamiodarone serum concentration over 41 days period in 1 patient using 1400 mg/day showed a considerable increase of amiodarone and desethylamiodarone levels from 0.7 and 0.3 μg/ml at day 1 to 11.9 and 5.4 μg/ml respectively at day 41 and demonstrate clearly the plasma accumulation of both compounds.

Using an HPLC assay, Haffajee et al (57), measured serum amiodarone concentrations serially in 122 patients treated with amiodarone for 1.5 to 53 months (mean 9.3 months) for control of atrial and ventricular tachyarrhythmias.

The maintenance dosage used in the group of 58 patients with atrial tachyarrhythmias varied from 200 to 600 mg per day (mean 380 ± 125 mg/day) and in the group of 68 patients with ventricular tachycardia from 200 to 800 mg per day (mean 394 ± 140 mg/day). They reported maintenance serum amiodarone concentrations in the 122 patients varying from 0.4 to 3.3. μg/ml with a mean level of 1.7 μg/ml. In 14 patients they observed additionally a reasonably good correlation between steady-state serum concentration and daily maintenance dose of amiodarone in mg/kg.

A good correlation between mean steady-state serum concentration of amiodarone and the prescribed daily dose of amiodarone in mg/kg was also reported by Stäubli et al (59) in a study of 12 patients.

Using maintenance doses varying from 200 to 600 mg per day steady-state levels ranging from 0.4 to 3.7 μg/ml (mean 1.6 μg/ml) were found in 14 patients.

The relationship between dose and plasma concentration of amiodarone and desethylamiodarone in a much larger group of patients (103 men and 67 women) was investigated by Holt et al (52). Using maintenance dosages of

200 mg, 400 mg and 600 mg daily for at least 1 month, mean plasma levels of amiodarone and desethylamiodarone for the respective doses of 1.06 and 1.04 μg/ml, 1.93 and 1.79 μg/ml, and 3.46 and 2.79 μg/ml were observed. In addition a linear relationship between dose and mean amiodarone concentration at these doses was reported in this group of patients.

In another study by Heger et al (60) the relationship between amiodarone dosage and drug concentration during long-term management was examined in 55 cardiac patients, who were given amiodarone for at least 3 months and had received a constant daily dosage ranging from 200 to 800 mg per day. Plasma concentrations of amiodarone and desethylamiodarone reported for each dosage group are outlined in Table 17.

During maintenance therapy, they observed a correlation between amiodarone concentration and dosage, however, with a wide interpatient variability for any given dosage level. No significant difference in drug concentration was detected between the 600 and 400 mg dosage groups. A significant difference in plasma levels was only observed between 200 mg/day dosage compared with the 600 mg/day dosage.

The relationship between amiodarone dosage and apparent steady-state serum concentrations of amiodarone and desethylamiodarone in patients after at least 2 months (mean 9.8) of treatment with daily maintenance doses of 200 to 600 mg was studied by Rotmensch et al (40).

The mean daily maintenance dose of amiodarone given in this study of 127 patients was 392 ± 143 mg and the mean serum concentrations of amiodarone and its metabolite at apparent steady-state were 2.4 ± 1.4 μg/ml and 1.4 ± 0.6 μg/ml, respectively. The mean serum levels of amiodarone and desethylamiodarone obtained after the various dosages of the drug are for 200 mg dose, 1.2 and 0.7 μg/ml, for 400 mg, 2.5 and 1.5 μg/ml and for 600 mg 3.7 and 1.9 μg/ml respectively. Although the follow-up periods were different for patients on different doses, the obtained results indicate a linear increase of steady-state serum concentrations with the prescribed daily dose. Despite this linear relationship, also these authors stated that serum concentrations achieved from a given dose showed a substantial interpatient variability.

Steady-state plasma amiodarone and desethylamiodarone concentrations in 14 patients treated with 200 mg to 800 mg of oral amiodarone for 2 to 24 months (mean 10 ± 2) were reported by Somani et al (61).

Using a mean oral dosage of 446 ± 40 mg per day, an average plasma level of amiodarone and desethylamiodarone of 2.4 ± 0.6 and 1.6 ± 0.4 μg/ml were found.

The mean plasma concentrations observed in this study in each dosage group are listed in Table 17.

No significant correlation was seen in this study between the reported plasma levels of the drugs and the maintenance dosage applied.

An own steady-state plasma concentration study, performed in 106 patients using a mean oral daily maintenance dosage of 440 ± 253 mg during mean period of therapy of 9.1 months, showed mean levels of 1.85 ± 1.17 μg/ml for amiodarone and 1.35 ± 0.71 μg/ml for desethylamiodarone respectively (56). The mean steady-state plasma amiodarone and desethylamiodarone concentrations obtained in our study with maintenance doses varying from 200 mg to 800 per day are presented in Table 17. A reasonably good correlation between concentration and maintenance dose was found in our study.

6.1.4. Elimination kinetics of amiodarone and desethylamiodarone after long-term oral dosing

Several groups which have reported steady-state plasma concentrations of amiodarone and desethylamiodarone following long-term oral therapy, have also determined terminal elimination half-lives of the drugs after cessation of treatment (11, 50, 52, 57, 59, 61). A review of the terminal elimination half-lives of amiodarone and desethylamiodarone is outlined in Table 18. Andreasen et al (11) reported an elimination half-life (t½e) of amiodarone of 13.7 days in one patient who had been on 300 mg per day of amiodarone for 8 months. In this patient after discontinuation of therapy a fall in plasma drug level from circa 1 μg/ml to 0.1 μg/ml was observed in 1 month.

Table 18

Elimination kinetics of amiodarone (A) and desethylamiodarone (DA) after long-term oral amiodarone therapy

Maintenance dose per day (mg)	Duration of therapy (month)	n	Mean elimination half-life (day) A	DA	Ref
200	13.3	3	51.0	62.0	52
200	5.0	7	37.0	-	59
300	8.0	1	13.7	-	11
300	5.0	2	37.0	-	59
390	5.3	3	19.3	-	57
400	29.3	3	59.0	63.7	52
400	5.0	2	50.5	-	59
400	16.5	2	40.2	53.1	56
600	1.5	2	39.0	-	50
600	15.0	2	45.5	56.5	52
600	5.0	1	59.0	-	59
600	25.0	2	39.2	61.7	56
800	1.4	1	22.0	-	50
800	24.0	1	34.0	34.0	61
1000	0.7	1	17.0	-	50

n = number of patients, - = not determined

A much longer mean t½e of amiodarone of 29 ± 19 days was reported by Kannan et al (50) in 4 patients which had been treated for a mean period of 39 days with a mean daily dose of 750 mg. Mean serum levels after drug withdrawal fell in these patients over approximately 2 month time from 1.1 μg/ml to 0.18 μg/ml.

Relatively very long mean t½e of 52.6 ± 23.7 days for amiodarone and 61.2 ± 31.1 days for desethylamiodarone were found after cessation of therapy by Holt et al (52) in 8 patients who had received maintenance doses varying from 200 to 600 mg for 2 to 52 months. In one patient who had been using 400 mg for 25 months these authors reported a decrease in plasma level of amiodarone and desethylamiodarone over a 5 months period from 4.0 to 1.0 μg/ml and 3.0 to 1.3 μg/ml respectively.

A comparable mean t½e of amiodarone of 41 ± 19 days (range 21 - 78 days) was reported by Staubli et al (59) in 12 patients who had received mean daily oral doses of 287 ± 124 mg for at least 5 months.

The terminal elimination half-life obtained per dosage level are listed in Table 18.

Elimination kinetics of amiodarone were also investigated by Haffajee et al (57) in 3 patients in whom amiodarone was discontinued after long-term oral treatment with a mean dose of 390 mg/day for a period of 4 to 7 months. The t½e of amiodarone found in these patients was 13, 15 and 30 days, respectively. The drug was discernable in serum for up to 6 months after cessation in the patient who took amiodarone for 7 months.

A terminal elimination half-life of 34 days for both amiodarone and desethylamiodarone was reported by Somani et al (61) after discontinuation of therapy in one patient who took 800 mg/day of amiodarone for 2 years.

Own examinations of elimination kinetics of amiodarone and desethylamiodarone after cessation of amiodarone therapy were performed in 4 patients who had used maintenance doses of 400-600 mg for a period varying from 2 to 48 months (56). We found t½e for amiodarone varying from 27 to 52 days (mean 40 days) and for desethylamiodarone from 30 to 94 days (mean 57 days).

The compounds were still detectable in plasma for about 10 months after withdrawal in one patient who was treated with amiodarone for 48 months.

The average terminal elimination half-lives of amiodarone reported in the literature (Table 18) varied from 14 to 59 days. This is definitely longer than that reported after single-dose oral administrations (Table 16) and are more in agreement with the slow disappearance rate of the therapeutic effects described in many clinical studies (46, 47). The observed differences can be ascribed to time-dependent kinetics of amiodarone and/or to analytical restrictions of amiodarone assay methods.

6.1.5. Tissue concentrations of amiodarone and desethylamiodarone

6.1.5.1. Localisation of iodine in human tissues

The first data on body distribution in man were already reported in 1969 by Broekhuysen et al (49) using single intravenous and oral doses of ^{131}I-labeled amiodarone.

These data suggest that amiodarone accumulates extensively in the skin, subcutaneous fat and/or muscle.

Determination of iodine content of various tissues obtained at autopsy from 3 patients who had been treated long term with 600 mg/day of amiodarone revealed that adipose tissue and skeletal muscle are the most important reservoirs of the drug (49).

6.1.5.2. Tissue concentrations obtained at post-mortem

Since the introduction of several specific and sensitive HPLC methods for the determination of amiodarone and desethylamiodarone levels in biological fluids and tissues, data had been published on the concentration of amiodarone and its metabolite in various tissues obtained from patients at autopsy and at surgery. A review of post-mortem human tissue concentrations of amiodarone and desethylamiodarone reported in the literature is presented in Table 19.

Table 19

Post-mortem human tissue concentrations of amiodarone (A) and desethylamiodarone (DA)

Tissues	Concentration (µg/g or µg/ml) Case 1 A	Case 2 A	Case 3 A	Case 4 A/DA	Case 5 A/DA	Case 6 A/DA	Case 7 A/DA	Case 8 A/DA	Case 9 A/DA
Ad.Tissue	-	385	117	316/76	245/87	210/45	500/112	-	-
Liver	111	187	7	391/2354	890/6500	430/1520	188/570	7/64	112/307
Lung	243	-	47	198/952	-	191/810	183/620	28/238	178/541
Bone Mar.	157	152	-	-	-	-	-	-	-
Adrenal	121	-	-	137/437	-	-	-	-	-
Pancreas	109	145	9	-	-	-	-	-	-
Lymphnode	-	-	-	83/316	-	-	-	-	-
Testes	35	-	-	89/470	-	-	-	-	-
Spleen	32	98	-	-	-	-	-	-	-
Heart	22	34	6	40/169	35/161	21/72	-	4/19	18/52
Salivary	23	-	-	-	-	-	-	-	-
Kidney	-	-	5	57/262	-	-	-	3/19	16/53
Thyroid	17	-	3	14/64	15/92	12/40	12/47	3/13	-
Muscle	-	-	2	22/51	8/3	8/19	10/30	-	-
Brain	6	-	-	8/54	4/49	8/44	3/25	-	-
Aorta	-	-	4	-	-	-	-	-	10/13
Plasma	0.6	1.0	0.9	-	2.1/2.5	2.3/2.1	-	-	0.8/0.4

Case 1 and 2 : Ref 57
Case 3 : Ref 62
Case 4 (n=9) : Ref 52
Case 5,6 and 7 : Ref 38
Case 8 and 9 : Ref 56

(-) : no sample taken
Ad. tissue = Adipose tissue
Bone Mar. = Bone Marrow

The first data on tissue concentration of amiodarone in 2 patients who died during long-term amiodarone therapy had been reported by Haffajee et al (57).

Using a specific HPLC assay, they found immediately after death a high content of amiodarone in adipose tissue, fat-laden organs (bone marrow, liver, adrenals and pancreas) and lung tissue and a relatively moderate concentration in heart tissue (Table 19).

Maggioni et al (62) reported amiodarone distribution in human tissues after sudden death of a patient who had been treated with 200 mg of amiodarone per day for about 2 months. HPLC analysis of samples of various organs collected at necropsy showed relatively high concentrations in adipose tissue (110 μg/g) and lung (50 μg/g) and relatively low levels (< 10 μg/g) in organs like liver, heart, kidneys and thyroid (Table 19).

Tissue distribution of amiodarone and desethylamiodarone was investigated by Holt et al (52) in a variety of postmortem samples collected from nine patients who had received amiodarone before death in doses ranging from 200 to 800 mg daily for a period varying from 2 weeks up to 18 months. Amiodarone tissue concentration, assessed by HPLC, were the highest in liver and fat followed by lung, lymph node, heart, skeletal muscle, thyroid gland and brain in decreasing order. Interestingly, the desethylamiodarone levels encountered in all tissues, except in adipose tissue, were consistently higher than those observed for amiodarone (Table 19).

Storey et al (38) reported the concentration of amiodarone as well as desethylamiodarone in post-mortem samples collected from three cases in which the patients died whilst receiving amiodarone therapy. These patients had received daily doses of amiodarone ranging from 200 to 600 mg for a period of 1 to 10 months. An appreciable accumulation of both compounds was found in some tissues like liver and lung, with fat forming a potentially large tissue reservoir of the drug. Of note is the fact that desthylamiodarone levels were in all tissues examined, with exception of fat, substantially higher than those of the parent compound. A finding consistent with the observations reported by Holt et al (52).

Relatively high concentrations of amiodarone and desethylamiodarone of 562 μg/g and 2925 μg/g were found in lung tissue of one patient without signs of pulmonary toxicity by Heger et al (60).

Adams et al (69) reported very high concentrations of amiodarone and desethyl-amiodarone respectively ranging from 25 to 1067 μg/g and 72 to 3691 μg/g in necropsy samples of testis from 5 patients who had received therapy with amiodarone with daily doses of 250 to 800 mg for a period varying from 2 weeks to 21 months.

Own investigations of tissue distribution of amiodarone and desethylamiodarone were performed in 2 autopsy cases which before death had chronically received 200 mg of amiodarone daily for 2 and 12 months respectively (56). An extensive tissue accumulation of amiodarone and desethylamiodarone was observed with particularly high levels in lung, liver and adipose tissue and moderate concentrations in kidney, heart and thyroid gland (Table 19). With exception of adipose tissue, in all tissues under investigation the levels of desthylamiodarone were found to be 2 to 9 times higher than the parent compound.

6.1.5.3. Tissue concentrations obtained at biopsy or at surgery

Harris et al (63) reported very high liver concentrations of 1020 μg/g for amiodarone and 5050 μg/g for desethylamiodarone in a liver biopsy sample of a patient on long-term amiodarone treatment. Also very high amounts of the drug and its metabolite of 734 μg/g and 2551 μg/g were found by Heger et al (60) in a lung biopsy specimen from a patient who showed signs of pulmonary toxicity after chronic amiodarone therapy.

Data on the concentration of amiodarone and desethylamiodarone in tissue specimens and plasma samples collected during cardiac surgery from patients on amiodarone were reported in a few studies (38, 56, 64, 65) and are summarized in Table 20.

Debbas et al (64) reported plasma and cardiac amiodarone concentrations in a group of 8 patients who had received before surgery doses varying from 200 to 400 mg per day for more than 3 months.

They found mean amiodarone plasma, right atrium and left ventricle concentrations of 0.2 μg/ml (range 0.1 - 0.5 μg/ml), 4.4 ± 1.2 μg/g (range 2.8 - 6.3 μg/g) and 4.4 ± 1.3 μg/g (range 2.3 - 6.3 μg/g) respectively. No significant difference was seen between atrial and ventricular concentrations of the drug. Mean plasma and myocardial levels of amiodarone and desethylamiodarone in 61 patients following a single oral dose of 30 mg/kg (1200 - 3000 mg) given 7 to 96 h before open heart surgery were published by Escoubet et al (65). Maximum myocardial concentrations of amiodarone of 13 ± 8 μg/g were found after 7 h in the atrium and in the ventricle after 24 h (17 ± 11 μg/g). Maximum levels of the metabolite of 1.5 μg/g were achieved at 24 h in atrium as well as in the ventricle. The amiodarone myocardial to plasma concentration ratio was similar in the atrium and in the ventricle and averaged 22 and 10 for amiodarone and desethylamiodarone respectively.

Myocardial, skeletal, muscle, skin and adipose tissue levels of the 2 compunds were reported by Storey et al (38) in 2 patients who received 800 mg daily for 5 months and 600 mg for 1 month respectively. High levels of amiodarone and desethylamiodarone were found by these authors in adipose tissue, relatively moderate concentrations in heart and relatively low levels in skin and muscle (Table 20). Own examinations in 2 patients who received amiodarone 200 mg/day for 2 months and 600 mg/day for 8 months respectively showed a distribution pattern comparable with that reported by Storey et al (38) (Table 20).

Table 20

Tissue concentration of amiodarone (A) and desethylamiodarone (DA) in surgical patients

Tissues	Concentration (μg/g or μg/ml) Patient 1 A	Patient 2 A/DA	Patient 3 A/DA	Patient 4 A/DA	Patient 5 A/DA	Patient 6 A/DA
Left ventricle	4.4	17/1.5	47/95	12/47	-	-
Right atrium	4.4	13/1.5	-	-	-	-
Myocardium	-	-	-	-	12/64	49/114
Endocardium	-	-	-	-	-	29/48
Muscle	-	-	23/46	8/12	-	-
Skin	-	-	8/24	5/9	-	-
Adipose tissue	-	-	470/135	120/30	112/65	364/135
Plasma	0.2	2.2/0.2	2.1/2.4	1.4/1.3	0.8/0.9	1.5/1.7

Patient 1 (n=8) : Ref 64
Patient 2 (n=61) : Ref 65
Patient 3 and 4 : Ref 38
Patient 5 and 6 : Ref 56

(-) : no sample taken

6.2. Metabolism and urinary excretion

The metabolism of amiodarone has not been fully unraveled. However, since the drug is a tertiary amine it can be anticipated, that N-dealkylation must be a significant route of metabolism. In addition it has been shown that tertiary amines usually undergo a two-step N-dealkylation giving rise first to a secondary and subsequently to a primary amine (66). This metabolic pathway was indeed demonstrated for amiodarone where the secondary amine desethylamiodarone and the primary amine desdiethylamiodarone have been positively identified (19). The latter has been detected in dogs but was not present in measurable quantities in humans (19). The former has been found in high concentrations in humans (20, 38, 40, 56, 60-63).

Desethylamiodarone appears in the plasma within hours after an single intravenous or oral dose of amiodarone in very low levels of about 10 to 100 ng/ml and was still present at concentrations of less than 10 ng/ml for at least 1 month after administration (20, 52, 56).

Desethylamiodarone level was substantially lower than that of amiodarone for at least 2 days and thereafter showed values comparable to that of amiodarone (20, 52).

During long-term therapy, its plasma level is usually comparable to that of the parent drug (20, 38, 40, 56, 60-63). The two drugs accumulate in various organs in similar amounts, although desethylamiodarone levels are consistently higher than those of the parent compound with exception of adipose tissue (38, 52, 56, 60, 63).

The elimination half-life of the metabolite is longer than that of the parent drug. The significance of the presence of desethylamiodarone in plasma and tissues in relation to amiodarone therapeutic and adverse effects is still not completely elucidated, however, a recent study of Yabek et al (48) suggests a pharmacological activity comparable to that of amiodarone.

Besides the abovementioned dealkylated metabolites of amiodarone, the presence of deiodinated amiodarone metabolites might be expected from several studies which have reported the excretion of free iodine during amiodarone therapy (11, 49, 51, 59).

In an early study of Broekhuysen et al (49), using a single intravenous dose of ^{131}I labeled amiodarone, about 0.7% of the administered radioactivity was found in 24 h urine fraction. Following intravenous administration of 400 mg of amiodarone to 2 subjects, Andreasen et al (11) found in 2 h urine fraction, iodine (in the form of iodide) in an amount corresponding to about 5% of the iodide given as amiodarone. Anastasiou-Nana et al (51) collected urine for 24 h after application of single intravenous doses of 150 mg to healthy volunteers and reported an amount of iodine equivalent to only 0.5% of the content of the administered dose. Staubli et al (59) determined the total urinary output of iodine at steady-state in 3 patients on daily amiodarone doses of 4 to 9 mg/kg and found that 8 to 15% of the iodine ingested as amiodarone was excreted per 24 h. Due to the reported presence of various amounts of iodine in urine of amiodarone treated patients, it can be suggested that amiodarone undergoes deiodination. Up to now, however, only a small peak was found by Flanagan et al (18) in pooled plasma from patients on amiodarone which may possibly correspond to the deiodinated metabolite deiododesethylamiodarone.

Neither amiodarone nor desethylamiodarone was detected in 24 h urine samples of volunteers and patients after single doses or during chronic therapy with amiodarone (51, 56, 57). Very low levels of amiodarone of less than 100 ng/ml were reported by Andreasen et al (11) in 2 h urine fractions of 2 volunteers following a single 400 mg intravenous dose. Harris et al (67) was able to measure very low concentrations of amiodarone and desethylamiodarone in the urine of 10 patients on chronic amiodarone therapy with normal renal function. They found mean urinary concentration for amiodarone of 0.029 ± 0.027 μg/ml and for desethylamiodarone of 0.149 ± 0.132 μg/ml, with mean corresponding plasma concentrations of 2.08 ± 1.47 μg/ml and 1.48 ± 0.90 μg/ml for the respective drugs. No correlation was reported between urinary excretion of the compounds and serum creatinine or amiodarone dosage.

6.3. Excretion in other biological fluids

Broekhuysen et al (49) examined the recovery of total iodine in the faeces of 2 subjects to whom amiodarone was administered in a dose of 300 mg per day for 1 month. They found a substantial mean faecal excretion of iodine over the period of observation ranging from 36 to 64% of the daily oral dose of iodine ingested as amiodarone. No faecal levels of amiodarone and desethylamiodarone were given in this study. Biliary concentrations of amiodarone fifty times higher than the corresponding serum amiodarone level were reported by Andreasen et al (11) in a preliminary study in one patient 2 h after a single intravenous dose of 400 mg.

Post-mortem bile and blood levels of amiodarone of 20 μg/ml and 0.94 μg/ml respectively were reported by Maggioni et al (62) in a case who received before death 200 mg daily for 2 months.

Serial measurements on seven patients receiving 600 mg of amiodarone per day showed less than 0.1 μg/ml of amiodarone in tears, when the corresponding amiodarone level in serum were less than 1.2 μg/ml, but at higher serum concentrations a rapid increase in the concentration in tears was observed (68). Preliminary measurements of amiodarone in saliva and plasma suggest that salivary levels of amiodarone are about 2% of the corresponding plasma levels and indicate a high binding of amiodarone to plasma proteins (67).

Neither amiodarone nor desethylamiodarone were detected in dialysis fluid from a patient on intermittent peritoneal dialysis receiving 200 mg of the drug daily or from another patient on heamodialysis receiving 400 mg amiodarone per day. The plasma concentrations in these patients showed no difference between pre-dialysis and post-dialysis and were in the normal range for their applied doses (67).

Seminal concentrations of amiodarone and desethylamiodarone were reported by Adams et al (69) and Ward et al (70) in 2 patients on long-term amiodarone (200-400 mg/day).

In the first case the concentration of amiodarone and desethylamiodarone reported in plasma and semen were 0.8/0.9 μg/ml and 0.2/0.8 μg/ml respectively (69). In the second case, in which an epididymitis associated with amiodarone treatment was observed, concentration values in serum for amiodarone of 1.13 μg/ml and for desethylamiodarone of 0.85 μg/ml and in semen of 0.37 and 4.21 μg/ml for the respective drugs were found (70). Both amiodarone and desethylamiodarone are partially transferred across the placenta and can be detected in cord blood, amniotic fluid and neonatal plasma of patients treated with amiodarone during the last part or throughout the whole of pregnancy (71-77).

A review of the data available in the literature about levels in maternal serum, infant serum, and breast milk are outlined in Table 21.

Candelpergher et al (71) reported neonatal levels of amiodarone and desethylamiodarone of 27% and 54% of the maternal levels at birth in 1 patient who received 200 mg of amiodarone during the last 3 months of pregnancy. The neonatal/maternal serum concentration ratio indicate clearly a limited placental transfer of both compounds.

A limited placental transfer of amiodarone and desethylamiodarone in the order of 10% and 25% respectively was also reported by Pitcher et al (72) in a woman on 200 mg amiodarone for the last 3 weeks of gestation.

A similar maternal-fetal transfer of amiodarone and its desethyl metabolite was described by McKenna et al (73) in another case treated with amiodarone during the last 7 weeks of pregnancy.

Maternal and infant plasma concentrations of amiodarone and desethylamiodarone and infant/maternal plasma concentration ratios

Table 21

Concentration of amiodarone (A) and desethylamiodarone (DA) in serum, amniotic fluid and milk after administration of amiodarone to mother before and after delivery

Treatment period	Case 1 last 3 months	Case 2 last 3 weeks	Case 3 last 25 weeks	Case 4 last 7 weeks		Case 5 last 24 weeks	Case 6 last 17 weeks	Case 7 entire pregnancy	Case 8 entire pregnancy	
dose (mg/day)	200	200	200	400		200	400/600	200	400	
sampling time	At birth	At birth	After 5 ds	At birth	After 14 ds	At birth	At birth	At birth	At birth	After 14 ds
Reference	(71)	(72)	(74)	(73)	(73)	(75)	(76)	(76)	(77)	(77)
Maternal plasma										
A (μg/ml)	0.55	0.70	0.65	1.5	-	0.84	1.1	0.5	0.50	0.20
DA (μg/ml)	0.64	0.72	0.46	1.4	-	-	1.2	0.8	0.36	0.33
Cord plasma										
A (μg/ml)	-	0.07	-	0.2	-	0.08	0.16	0.05	0.30	-
DA (μg/ml)	-	0.18	-	0.4	-	-	0.26	0.15	0.05	-
Amniotic fluid										
A (μg/ml)	-	-	-	-	-	-	-	0.02	-	-
DA (μg/ml)	-	-	-	-	-	-	-	0.11	-	-
Breast milk										
A (μg/ml)	-	1.1	-	-	13.3	-	-	-	2.40	2.60
DA (μg/ml)	-	0.6	-	-	5.7	-	-	-	0.74	1.24
Infant plasma										
A (μg/ml)	0.15	-	0.14	0.4	0.4	0.06	-	-	-	<0.10
DA (μg/ml)	0.35	-	0.26	0.35	0.25	-	-	-	-	<0.10
Placenta										
A (μg/g)	-	-	-	-	-	7.2	72	34	-	-
DA (μg/g)	-	-	-	-	-	-	492	300	-	-

comparable to those reported by Candelpergher et al (71) were observed by De Wolf et al (74) 5 days after delivery of a patient who was treated with 200 mg daily during the last 25 weeks of gestation (Table 21).

A relatively low transplacental passage of amiodarone, with umbilical cord serum levels of 10% of the maternal serum levels and placental tissue concentration of 7 μg/g were reported by Penn et al (75) in a woman who was treated with 200 mg/day during the last 24 weeks of gestation.

Placental transfer of amiodarone and desethylamiodarone varying from 10-15% and 20-22% respectively were reported by Robson et al (76) in 2 cases who were receiving amiodarone throughout pregnancy and for the final 17 weeks of pregnancy. This degree of transplacental passage of the 2 compounds was similar to that previously reported (72, 73, 75).

In addition the relatively high concentrations of both drugs, particularly of desethylamiodarone found in placental tissue should be noted.

In disagreement with the transplacental findings of Robson et al (76) and others (72, 73, 75), Strunge et al (77) reported a transplacental transfer of amiodarone and desethylamiodarone of 60% and 14% respectively at birth in 1 patient who received 400 mg of amiodarone during whole pregnancy and during breast feeding period. No explanation for these discrepant results was given. Amiodarone as well as its desethyl metabolite enter human breast milk (Table 21). They are present in the milk at higher concentrations than in the mother's plasma. Levels of the parent compound in breast milk are higher than the metabolite and both levels showed a wide variation from sample to sample (72, 73, 77).

The accumulation in breast milk and the within-day and day-to-day variation in breast milk concentration was first reported by McKenna et al (73). In the first 2 weeks post-partum they found milk concentration, in a woman using 400 mg amiodarone during breast feeding, varying for amiodarone from 3.6 to 13.3 μg/ml and for desethylamiodarone from 1.3 to 5.7 μg/ml respectively. Sample to sample variation in milk levels over a 24 h period in this mother showed values for amiodarone of 3.7 to 14.6 μg/ml and for desethylamiodarone of 1.2 to 5.7 μg/ml, whereas at the same time for maternal and infant plasma, concentrations of amiodarone and desethylamiodarone of 1.6/1.5 μg/ml and 0.4/0.15 μg/ml were found respectively. The estimated daily ingestion of amiodarone and its metabolite by the breastfed infant was around 1.5 and 0.6 mg/kg body weight respectively, which was equivalent to a low maintenance dose of an adult. Substantial lower breast milk levels were reported by Pitcher et al (72) in one patient on amiodarone for the last 3 weeks of gestation. Samples expressed on the 2nd and 3rd days post-partum showed concentrations ranging from 0.5 to 1.8 μg/ml for amiodarone and from 0.4 to 0.8 for desethylamiodarone. Comparable breast milk levels of a patient who was treated with 400 mg per day during lactation were reported by Strunge et al (77). They found in samples taken over a 1 month period concentrations of amiodarone and desethylamiodarone varying from 1.06 to 2.88 μg/ml and from 0.50 to 1.24 μg/ml respectively.

REFERENCES

1. Chem. Abstr. 61 (1965) 2375 g.
2. Windholz, M. The Merck Index 10th Ed., Merck and Co., Inc. Rahway, N.J., U.S.A. (1983) 486.
3. Reynolds, J.E.F. and Prasad, A.B. The Extra Pharmacopoeia, 29th Ed., The Pharmaceutical Press, London, U.K. (1989) 71-73.
4. Bonati, M., Gaspari, F., d'Aranno, V., Benfenati, E., Neyroz, P., Galletti and Tognoni, G. Physicochemical and analytical characteristics of amiodarone. J.Pharm.Sci. 73 (1984) 829-830.
5. Crispin, S., Alexis, J. and Mouffe, P. Methode de Controle pour le Chlorhydrate d'Amiodarone - L3428 Qualité P. Report D.T. 8.029. S.A.Labaz 1980.
6. Reynolds, J.E.F. and Prasad, A.B. The Extra Pharmacopoeia, 28th Ed., The Pharmaceutical Press, London, U.K. (1983) 1374-1375.
7. Moffat, A.C., Jackson, J.V., Moss, M.S., Widdop, B. Clarke's Isolation and Identification of Drugs. 2nd Ed., The Pharmaceutical Press, London, U.K. (1986) 344-345.
8. Clarke, E.G.C. Isolation and identification of drugs, 1st Ed. The Pharmaceutical Press, London, U.K. (1971) 90-91.
9. Riva, E., Gerna, M., Neyroz, P., Urso, R., Bartosek, I. and Guaitani, A. Pharmacokinetics of amiodarone in rats. J. Cardiovasv. Pharmacol. 4 (1982) 270-275.
10. Plomp, T.A., Engels, M., Robles de Medina, E.O. and Maes, R.A.A. Simultaneous determination of amiodarone and its major metabolite desethylamiodarone in plasma, urine and tissues by high-performance liquid chromatography. J. Chromatog. 273 (1983) 379-392.
11. Andreasen, F., Agerbaek, H., Bjerregaard, P. and Gotzsche, H. Pharmacokinetics of amiodarone after intravenous and oral administration. Eur. J. Clin. Pharmacol. 19 (1981) 293-299.
12. Flanagan, R.J., Storey, G.C.A. and Holt, D.W. Rapid high-performance liquid chromatographic method for the measurement of amiodarone in blood, plasma or serum at the concentrations attained during therapy. J. Chromatogr. 187 (1980) 391-398.

13. Weir, S.J. and Ueda, C.T. Rapid liquid chromatographic assay for the determination of amiodarone and its N-deethyl metabolite in plasma, urine and bile. J. Pharm. Sci. 74 (1985) 460-465.

14. Hasegawa, G.R. and Eder, J.F. Visual compatibility of amiodarone hydrochloride injection with other injectable drugs. Am. J. Hosp. Pharm. 41 (1984) 1379-1380.

15. Ravin, L.J., Shami, E.G. and Rattie, E.S. Micelle formation and its relationship to solubility behavior of 2 butyl-3-benzofuranyl 4-(2-(diethyl(amino)ethoxy)-3,5-diiodophenyl ketone hydrochloride. J. Pharm. Sci. 64 (1975) 1830-1833.

16. Cairns, C.J. Pharm. J. 1 (1986) 68.

17. Weir, S.J., Szucs Myers, V.A., Bengtson, K.D. and Ueda, C.T. Sorption of amiodarone to polyvinyl chloride infusion bags and administration sets. Am. J. Hosp. Pharm. 42 (1985) 2679-2683.

18. Flanagan, R.J., Storey, G.C.A., Holt, D.W. and Farmer, P.B. Identification and measurements of desethylamiodarone in blood plasma specimens from amiodarone-treated patients. J.Pharm.Pharmacol. 34 (1982) 638-643.

19. Latini, R., Reginato, R., Burlingame, A.L. and Kates, R.E. High-performance liquid chromatographic isolation and fast atom bombardment mass spectrometric identification of di-n-desethylamiodarone, a new metabolite of amiodarone in the dog. Biom. Mass. Spectrom. 11 (1984) 466-471.

20. Marchiset, D., Aubert, C., Sumirtapura, Y.C., Egre, A. and Cano, J.P. Mass spectrometric identification of amiodarone N-monodesethyl metabolite and application of an HPLC method to a pharmacokinetic study. Eur.J.Drug Metab. and Pharmacokinet. 9 (1984) 123-128.

21. Berger, Y. Internal report of Sanofi Labaz 1983.

22. Shoolery, J.N. in Interface 8 (1981) Varian NMR Newsletter.

23. Pacco, M. and Matteazzi, J.R. Quantitative determination of amiodarone in plasma by high-performance liquid chromatography. Internal report Labaz (Brussels) 1979.

24. Garattini, S. Personal communication. January 1979.

25. Garattini, S. Personal Communication. April 1980.

26. Cano, J.P. Personal communication. December 1980.

27. Cano, J.P. Personal communication. April 1981.

28. Cervelli, J.A., Kerkay, J., and Pearson, K.H. Quantitative serum amiodarone determination by high-performance liquid chromatography. Analytical Letters 14 (1981) 137-153.

29. Lesko, L.J., Marion, A., Canada, A.T., and Haffajee, C. High-pressure liquid chromatography of amiodarone in biological fluids. J. Pharm. Sci. 70 (1981) 1366-1368.

30. Stathaki, S., Anastasiou, M., Levis, G.M., and Moulopoulos, S.D. Determination of amiodarone and detection of one possible metabolite in human plasma by high- pressure

liquid chromatography. Hell. Biochem. Biophys. Soc. Newsletter 16 (1981) 6-8.

31. Canada, A.T., Lesko, L.J., and Haffajee, C.I. Disposition of amiodarone in patients with tachyarrhythmias. Current Therap. Res. 30 (1981) 968-974.

32. Riva, E., Gerna, M., Latini, R., Giani, P., Volpi, A., and Maggioni, A. Pharmacokinetics of amiodarone in man. J. Cardiovasc. Pharmac. 4 (1982) 264-269.

33. Duranti, L., Caracciolo, M., and Orlani, G. New, accurate semi-automatic high-performance liquid chromatographic method for routine monitoring of amiodarone plasma levels. J. Chromatogr. 277 (1983) 401-407.

34. Berger, Y., Mateazzi, J.R. and Paco, M. Quantitative analysis of amiodarone (L3428) and one of its metabolites (LB32812) in plasma by-high performance liquid chromatography. Scientific Report, Sanofi, Brussels. 1981.

35. Storey, G.C.A., and Holt, D.W. High-performance liquid chromatographic measurement of amiodarone and desethylamiodarone in plasma or serum at the concentrations attained following a single 400 mg dose. J. Chromatrogr. 245 (1982) 377-380.

36. Brien, J.F., Jimmo, S., and Armstrong, P.W. Rapid high-performance liquid chromatographic analysis of amiodarone and N-desethylamiodarone in serum. Can. J. Physiol. Pharmacol. 61 (1983) 245-248.

37. Mostow, N.D., Noon, D.L., Myers, C.M., Rakita, L., and Blumer, J.L. Determination of amiodarone and its N-deethylated metabolite in serum by high-performance liquid chromatography. J. Chromatogr. 227 (1983) 229-237.

38. Storey, G.C.A., Adams, P.C., Campbell, R.W.F., and Holt, P.W. High-performance liquid chromatographic measurement of amiodarone and desethylamiodarone in small tissue samples after enzymatic digestion. J. Clin. Pathol. 36 (1983) 785-789.

39. Gupta, R.N., and Connolly, S. Liquid-chromatographic determination of amiodarone and its N-desethyl metabolite in plasma. Clin. Chem. 30 (1984) 1423-1424.

40. Rotmensch, H.H., Belhassen, B., Swanson, B.N., Shoshani, D., Spielman, S.R., Greenspan, A.M., Vlasses, P.H., and Horowitz, L.N. Steady-state serum amiodarone concentrations: relationships with antiarrhythmic effects and toxicity. An. of Int. Med. 101 (1984) 462-469.

41. Somani, P., Simon, V.A. and McManus, K. Pharmacokinetic, antiarrhythmic effects, and tissue concentrations of amiodarone and desethylamiodarone in dogs with acute coronary artery occlusion. Can. J. Cardiol. 1 (1985) 56-65.

42. Lam, S. High-performance liquid chromatography of amiodarone and desethylamiodarone in serum after microscale protein precipitation. J. Chromatogr. 381 (1986) 175-178.

43. Bliss, M., Mayersohn, M., and Nolan, P. High-performance liquid chromatographic analysis of amiodarone and desethylamiodarone in serum. J. Chromatogr. 381 (1986) 179-184.

44. De Smet, M., and Massart, D.L. Determination of amiodarone and desethylamiodarone with a standarized extraction and chromatographic optimisation procedure. J. Pharm.

Biomed. Analysis 6 (1988) 277-284.

45. Plomp, T.A., Wiersinga, W.M., and Maes, R.A.A. Tissue distribution of amiodarone and desethylamiodarone in rats after repeated oral administration of various amiodarone dosages. Arzneim. Forsch. Drug Res. 35 (1985) 1805-1810.

46. Rosenbaum, M.B., Chiale, P.A., Halpern, M.S., Nau, G.J., Przybylski, J., Levi, R.J., Lazzari, J.O. and Elizari, M.V. Clinical efficacy of amiodarone as an antiarrhythmic agent. Am. J. Cardiol. 38 (1976) 934-944.

47. Rosenbaum, M.B., Chiale, P.A., Ryba, D., Elizari, U.M. Control of tachyarrhythmias associated with W.P.W. syndrome by amiodarone hydrochloride. Am. J. Cardiol. 34 (1974) 215-223.

48. Yabek, S.M., Kato, R., Singh, B.N. Effects of amiodarone and its metabolite desethylamiodarone, on the electrophysiologic properties of isolated cardiac muscle. J. Cardiovasc. Pharmacol. 8 (1986) 197-207.

49. Broekhuysen, J., Laruel, R. and Sion, R. Recherches dans la série des benzofurannes XXXVII. Etude comparée du transit et du metabolisme de l'amiodarone chez diverses especes animales et chez l'homme. Arch. Int. Pharmacodyn 177 (1969) 340-359.

50. Kannan, R., Nademanee, K., Hendrickson, J.A., Rostami, H.J. and Singh, B.N. Amiodarone kinetics after oral doses. Clin. Pharmacol. Ther. 31 (1982) 439-444.

51. Anastasiou-Nana, M., Levis, G.M. and Moulopoulos, S. Pharmacokinetics of amiodarone after intravenous and oral administration. Int. J. Clin. Pharmacol. Ther. and Toxicol. 20 (1982) 524-529.

52. Holt, D.W., Tucker, G.T., Jackson, P.R. and Storey, G.C.A. Amiodarone pharmacokinetics. Am. Heart J. 108 (1983) 840-846.

53. Mostow, N.D., Rakita, L., Vrobel, T.R., Noon, D., and Blumer, J. Amiodarone: Intravenous loading for rapid suppression of complex ventricular arrhythmias. J. Am. Coll. Cardiol. 4 (1984) 97-104.

54. Giani, P., Maggioni, A.P., Volpi, A., Cavalli, A., Latini, R., Neyroz, P., Riva, E. and D'Aranno, V. Blood levels and electrophysiological effects of intravenous amiodarone in patients with junctional reciprocating tachycardia. Acta Cardiol. 39 (1984) 9-18.

55. Nolan, P.E., Mayersohn, M., Fenster, P.E., Bliss, M. Single-dose pharmacokinetics of amiodarone. Drug Intel. Clin. Pharm. 19 (1985) 463.

56. Plomp, T.A., van Rossum, J.M., Robles de Medina, E.O., van Lier, T. and Maes, R.A.A. Pharmacokinetics and body distribution of amiodarone in man. Arzneim. Forschung/Drug Res. 34 (1984) 513-520.

57. Haffajee, C.I., Love, J.C., Canada, A.T., Lesko, L,J., Asdourian, G. and Alpert, J.S. Clinical pharmacokinetics and efficacy of amiodarone for refractory tachyarrhythmias. Circulation 67 (1983) 1347-1355.

58. Pourbaix, S., Berger, Y., Desager, J.P., Pacco, M., and Harvengt, C. Absolute bioavailability of amiodarone in normal subjects. Clin. Pharmacol. Ther. 37 (1985) 118-123.

59. Stäubli, M., Bircher, J., Galeazzi, R.L., Remund, H. and Studer, H. Serum concentrations of amiodarone during long-term therapy. Relation to dose, efficacy and toxicity. Eur. J. Clin. Pharmacol. 24 (1983) 485-494.

60. Heger, J.J., Prystowsky, E.N. and Zypes, D.P. Relationships between amiodarone dosage, drug concentrations and adverse side effects. Am. Heart J. 106 (1983) 931-934.

61. Somani, P., Simon, V.A., Temesy-Armos, P.N., Gross, S.A. and Didido, L.J.A. Amiodarone associated changes in human neutrophils. Am. J. Cardiol. 57 (1986) 666-672.

62. Maggioni, A.P., Maggi, A., Volpi, A., D'Aranno, V., Tognoni, G. and Giani, P. Amiodarone distribution in human tissues after sudden death during Holter recording. Am. J. Cardiol. 52 (1983) 217-218.

63. Harris, L., McKenna, W.J., Rowland, E., Holt, D.W., Storey, G.C.A. and Krikler, D.M. Side-effects of long-term amiodarone therapy. Circulation 67 (1983) 45-51.

64. Debbas, N.M., Du Cailer, C., Sassine, A., Derancourt, J., Demaille, J.G., Puech, P. Determination of cardiac and plasma drug levels during long-term amiodarone therapy. Eur. J. Clin. Invest. 13 (1983) 123-127.

65. Escoubet, B., Coumel, P. Poirier, J.M., Maison-Blanche, P., Jaillon, P., Leclercq, J.F., Menasche, P., Cheymol, G. Piwnica, A., Lagier, G. and Slama, R. Suppression of arrhythmias within hours after a single oral dose of amiodarone and relation to plasma and myocardial concentrations. Am. J. Cardiol. 55 (1985) 696-702.

66. Mandel, H.G. (1971) in Fundamentals of Drug Metabolism and Drug Disposition. Eds. La Du, B.N., Mandel, H.G. and Way, E.L. Williams and Wilkins, Baltimore, 149-186.

67. Harris, L., McKenna, W.J., Krikler, S.J., Hind, C.R.K., Savage, C., Storey, G.C.A. Renal elimination of amiodarone and its desethylmetabolite. Postgrad. Med. J. 59 (1983) 440-442.

68. Bjerregaards, P., Neilson, C.E., Andreasen, P. and Gotzsche, H. Dose, concentrations and effects of amiodarone. Br. J. Clin. Pharmacol. 14 (1982) 577.

69. Adams, P.C., Holt, P. and Holt, D.W. Amiodarone in testis and semen. Lancet I (1985) 341.

70. Ward, M.J., Routledge, P.A., Hutchings, A., Morris, J.M. Association of seminal desethylamiodarone concentrations and epididymitis with amiodarone treatment. Br. Med. J. 296 (1988) 19-20.

71. Candelpergher, G., Buchberger, R., Suzzi, G.L. and Padrini, R. Passagio trans-placentare dell'Amiodarone: evidenza elettrocardiografica e documentazione farmacologica in un neonato. G. Ital. Cardiol. 12 (1982) 79-82.

72. Pitcher, D., Leather, H.M., Storey, G.C.A., Holt, D.W. Amiodarone in pregnancy. Lancet 1 (1983) 597-598.

73. McKenna, W.J., Harris, L., Rowland, E., Whitelaw, A., Storey, G. and Holt, D. Amiodarone therapy during pregnancy. Am. J. Cardiol. 51 (1983) 1231-1233.

74. De Wolf, D., De Schepper, J., Verhaaren, H., Deneyer, M., Smitz, J. and Sacre- Smits, L. Congenital hypothyroid goiter and amiodarone. Acta Paediatr. Scand. 77 (1988) 616-618.

75. Penn, I.M., Barrett, P.A., Pannikote, V., Barnaby, P.F., Campbell, J.B., and Lyons, N.R. Amiodarone in Pregnancy. Am. J. Cardiol. 56 (1985) 196-197.

76. Robson, D.J., Jeeva Raj, M.V., Storey, G.C.A. and Holt, D.W. Use of amiodarone during pregnancy. Postgrad. Med. J. 61 (1985) 75-77.

77. Strunge, P. Frandsen, J. and Andreasen, F. Amiodarone during pregnancy. Eur. Heart J. 9 (1988) 106-109.

ACKNOWLEDGEMENT

The author would like to thank Mrs M. Sterman for typing the manuscript.

ANALYTICAL PROFILE OF

APOMORPHINE HYDROCHLORIDE

Farid J. Muhtadi and Mohamed S. Hifnawy*

Department of Pharmacognosy
College of Pharmacy, King Saud University,
Riyadh, Saudi Arabia

* Present Address : Pharmacognosy Department,
Faculty of Pharmacy, Cairo University,
Cairo, Egypt.

Apomorphine Hydrochlride

Contents

9. Methods of Analysis.

9.1 Identification
9.2 Microcrystal Tests
9.3 Titrimetric Determinations
9.4 Electrochemical Determinations
9.5 Spectrophotometric Methods

9.5.1 Colorimetric Determinations
9.5.2 Ultraviolet Determinations
9.5.3 Spectrofluorimetry

9.6 Chromatographic Methods

9.6.1 Paper Chromatography
9.6.2 Thin Layer Chromatography
9.6.3 Gas Liquid Chromatography
9.6.4 High Performance Liquid Chromatography

9.7 Other Methods of Determination

9.7.1 Radiometric Assay
9.7.2 Enzymatic Radioisotopic Assay
9.7.3 Turbidimetric Determination.

Acknowledgement.

References.

1. Description

1.1 Nomenclature

1.1.1 Chemical Names

For apomorphine:

(R)-5,6,6a,7-Tetrahydro-6-methyl-4H-dibenzo *[de, g]* quinoline-10, 11-diol.
- 4H-Dibenzo *[de, g]* quinoline-10,11-diol, 5,6,6a,7-tetrahydro-6-methyl-*(R)*.
- 6a β-Aporphine-10,11-diol.
- *(R)* - 10,11-dihydroxy-6a-aporphine.

For apomorphine hydrochloride:

- 4H-Dibenzo *[de, g]* quinoline-10,11-diol, 5,6,6a,7-tetrahydro-6-methyl hydrochloride, hemihydrate *(R)*.
- 6a β-Aporphine-10,11-diol hydrochloride, hemihydrate.
- *(R)*- 10,11-dihydroxy-6a-aporphinium chloride hemihydrate.
- (6a *R*)-5,6,6a,7-Tetrahydro-6-methyl-4H-dibenzo *[de, g]* quinoline-10,11-diol hydrochloride hemihydrate.

1.1.2 Generic Names

Apomorphine hydrochloride; Dihydroxy aporphinium chloride.

1.2 Formulae

1.2.1 Empirical

$C_{17}H_{17}NO_2$ (Apomorphine)

$C_{17}H_{18}Cl\ NO_2$ (Apomorphine HCl)

$C_{17}H_{18}Cl\ NO_2,\ \frac{1}{2}\ H_2O$ (The hemihydrate)

1.2.2 Structural

Apomorphine is produced upon dehydration of morphine with a strong acid under pressure (1) *(morphine loses one molecule of water and undergoes structural rearrangement to form apomorphine)*.

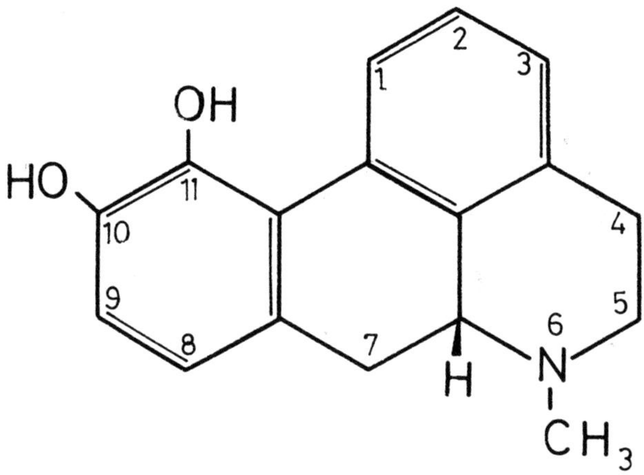

The structure was established by the work of Pschorr (2-4) and others (5). Its absolute configuration as *(R)*-(-)-apomorphine was also determined (6).

1.2.3 CAS Registry Number

[58-00-4] Apomorphine
[314-19-2] Apomorphine hydrochloride (anhydrous)
[41372-20-7] Apomorphine hydrochloride (hemihydrate)
[602-81-3] Apomorphine methylbromide.

1.2.4 Wiswesser Line Notation (7)

T C6666 1A Q KN & TT & J DQ
EQK (Apomorphine)

T C6666 1A Q KN & TT & J DQ
EQ K & GH & QH (Apomorphine hydrochloride hemihydrate).

1.3 Molecular Weight

267.32 (Apomorphine)
303.79 (Apomorphine hydrochloride)
312.80 (Apomorphine HCl hemihydrate)

1.4 Elemental Composition

C, 76.38%; H, 6.41%; N, 5.24%; O, 11.97%. (Apomorphine)

C, 67.22%; H, 5.97%; Cl, 11.67%; N, 4.61%; O, 10.53%. (Apomorphine hydrochloride)

C, 65.28%; H, 6.12%; Cl, 11.33%; N, 4.48%; O, 12.79%. (Apomorphine HCl hemihydrate)

1.5 Appearance, Color and Odor

- Hexagonal plates (from chloroform and petroleum ether), oxidizes rapidly in air and becomes green (8,9); odorless (Apomorphine).
- White or greyish white crystals or microcrystalline powder, but rapidly assuming a greenish tint on exposure to air and light (10); odorless (Apomorphine hydrochloride).

2. Physical Properties

2.1 Melting Range

About 195° (dec.) (8,9) apomorphine.
Apomorphine hydrochloride melts with decomposition:- between 225° and 236° (8); between 200° - 210° (7); between 220° - 260° (by microscope hot stage technique) (11).

2.2 Eutectic Temperature

The eutectic temperature of apomorphine hydrochloride was determined by microscope hot stage technique (11): sal. 178°; Dic. 168°; phenolphth. 217°

---

sal. = Acetaminosalol; Dic. = Dicyandiamide; phenolphth. = phenolphthalein.

2.3 Solubility Data

Soluble in alcohol, acetone, chloroform, slightly soluble in water, benzene and ether; solutions darken rapidly (9).

One gram of the hydrochloride dissolves in 50 ml water, in about 20 ml water at 80°, in 50 ml ethanol (96%), practically insoluble in chloroform and very slightly soluble in ether (10).

2.4 Acidity

pH of a 1% w/v solution of the hydrochloride is 4.5 to 5.5 (10).

2.5 Specific Optical Rotation

The following data have been reported for apomorphine hydrochloride:-

$[\alpha]D^{23}$	- 48° (c = 1.2) (7,9)
$[\alpha]D$	- 48.5° to - 52° (in a solution prepared by dissolving 0.30 g in sufficient 0.02M HCl to produce 30 ml) (10).
$[\alpha]D^{25}$	- 49° to - 51° (12)
$[\alpha]D$	Between - 60.5° and - 63.0° (calculated on the dried basis, determined in a solution in dimethylsulfoxide containing the equivalent of 150 mg of apomorphine HCl in 10 mL) (13).

2.6 Loss on Drying

When 0.5 g of the hydrochloride salt is dried to constant weight at 100° to 105°, it loses 2.5 to 4.0% of its weight (10).

When apomorphine hydrochloride is dried at 105° for 2 hours; it loses between 2.0% and 3.5% of its weight (13).

2.7 Dissociation Constant

pKa of apomorphine = 7.0 (14)
pKa of apomorphine hydrochloride 7.2, 8.9 (15)

2.8 X-Ray Powder Diffraction

The X-ray diffraction pattern of apomorphine hydrochloride was determined using a Philips X-ray diffraction spectrogoniometer equipped with PW 1730 generator. Radiation was provided by a copper target (Cu anode W, γ= 1.5480A°).
High intensity X-ray tube operated at 40 KV and 30 MV was used as generator slitting. The monochromator was a curved single crystal (PW 1752). Divergance

slit and the receiving slit were 1° and 0.1° respectively. The scanning speed of the goniometer used was 0.02 2θ per second.
The X-ray pattern of apomorphine hydrochloride is presented in Fig. 1. Interplanner distance and relative intensity are shown in table 1.

Table 1 : X-Ray Powder Diffraction Pattern of Apomorphine Hydrochloride

d(A°)	I/Io	d(A°)	I/Io	d(A°)	I/Io	d(A°)	I/Io
11.23	34.8	4.19	22.6	2.93	15.8	2.33	15.2
10.15	29.4	4.06	93.6	2.90	18.4	2.30	13.1
8.77	34.3	4.02	72.9	2.86	18.7	2.19	20.1
8.13	79.4	3.83	52.5	2.79	18.3	2.18	23.6
7.69	26.1	3.68	55.2	2.72	15.4	2.14	15.7
6.67	52.4	3.53	47.3	2.67	25.3	2.08	23.1
6.08	20.1	3.47	65.7	2.63	23.0	2.03	25.0
5.89	18.1	3.33	81.2	2.60	17.0	1.97	12.4
5.58	42.7	3.29	50.6	2.57	14.7	1.92	15.5
5.39	42.3	3.25	40.0	2.51	26.8	1.91	17.0
5.04	24.5	3.16	52.7	2.47	12.3	1.90	14.8
4.77	65.4	3.13	25.4	2.44	14.2	1.86	15.0
4.41	100	3.02	18.4	2.38	16.6	1.84	16.6
4.28	45.8	3.00	22.7	2.35	22.4		

d = interplannar distance, I/Io = relative intensity (based on the highest intensity of 100).

2.9 Spectral Properties

2.9.1 Ultraviolet Spectrum (UV)

The UV absorbance spectrum of apomorphine hydrochloride in methanol was scanned from 200 to 400 nm using a Pye-Unicum SP 8-100 spectrophotometer (Fig. 2).
Apomorphine hydrochloride exhibited the following absorptivity values (Table 2).

Table 2: UV absorptivity Values

max. nm	ε	A(1%, 1 Cm)
205	-	-
268	9592.17	315.75
305	2012.60	66.25

Other reported UV data for apomorphine

Solvent	Max. nm	Ref.
98% alcohol	336, 399	(9)
0.1 NH_2SO_4	273 (E 1%, 1 cm 741) 305 (E 1%, 1 cm 144)	(8)
Aqueous acid	272 (A_1^1 = 626a)	(15)
Aqueous alkali	253	

2.9.2 Infrared Spectrum (IR)

The IR spectrum of apomorphine hydrochloride as KBr-disc (1:100) was recorded on a Perkin Elmer 580 B Infrared Spectrophotometer (Fig. 3). Assignment of the functional groups have been correlated with the following frequencies (Table 3).

Table 3: IR Characteristics of Apomorphine

Frequency Cm^{-1}	Function Group
3470	OH (Free)
3170	OH (Intermolecular H bonding)

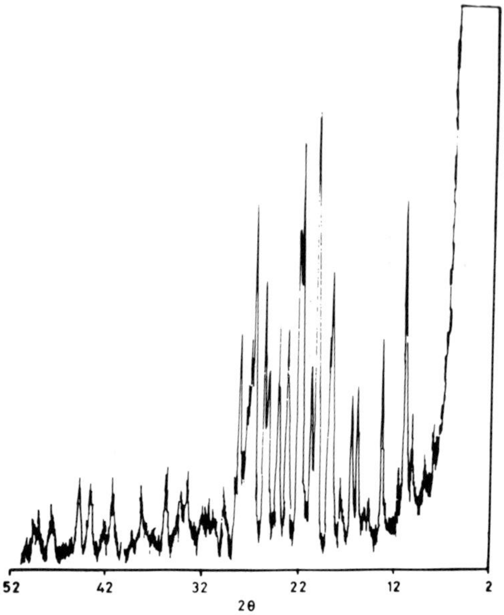

Fig. 1: X-Ray Diffraction Pattern of Apomorphine Hydrochloride.

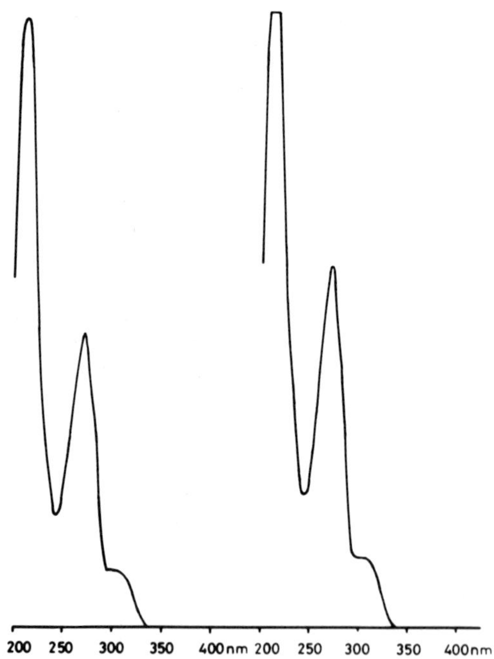

Fig. 2: UV Spectrum of Apomorphine HCl (MeOH).

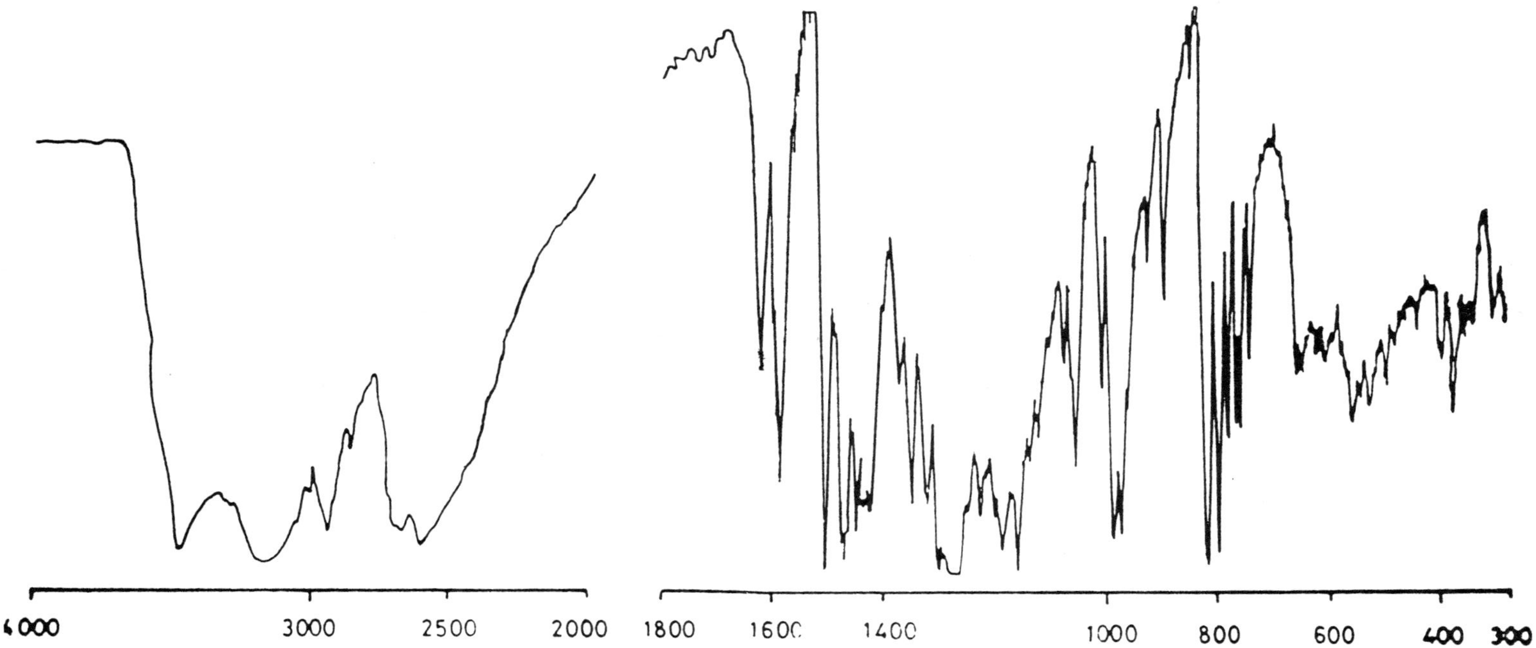

Fig. 3: IR Spectrum of Apomorphine HCl (KBr-Pellet).

Frequency Cm^{-1}	Function Group
2935	CH (stretch)
2700	$N-CH_2$
1590, 1622	C=C (aromatic)

The IR exhibited the following other absorption bands: 2595, 1510, 1470, 1450, 1438, 1425, 1375, 1352, 1328, 1305, 1275, 1230, 1190, 1160, 1075, 1055, 1008, 985, 970, 922, 895, 850, 820, 795, 775, 755, 738, 655, 620, 605, 558, 528, 495, 395 and 375 cm^{-1}.
Clarke (8) reported the following principal peaks for apomorphine base in KBr disc: 1265, 1460, 752, 790, 985 cm^{-1}.
The following principal peaks at wavenumbers: 1265, 1298, 1204, 752, 985 and 790 cm^{-1} were reported for apomorphine hydrochloride using a KBr disc (15).

2.9.3 Nuclear Magnetic Resonance

2.9.3.1 ^{1}H-NMR Spectrum

The proton spectrum of apomorphine hydrochloride in DMSO was done on a Varian FT80A (80 MHz) NMR Spectrophotometer (Fig. 4) using Tetramethylsilane (TMS) as an internal reference. The proton chemical shifts are assigned and presented in table 4.

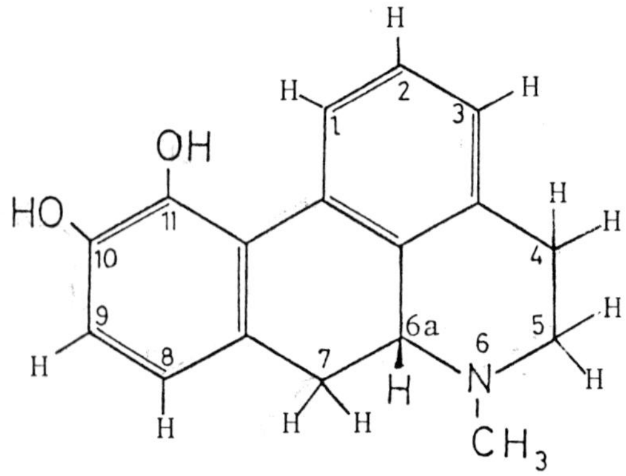

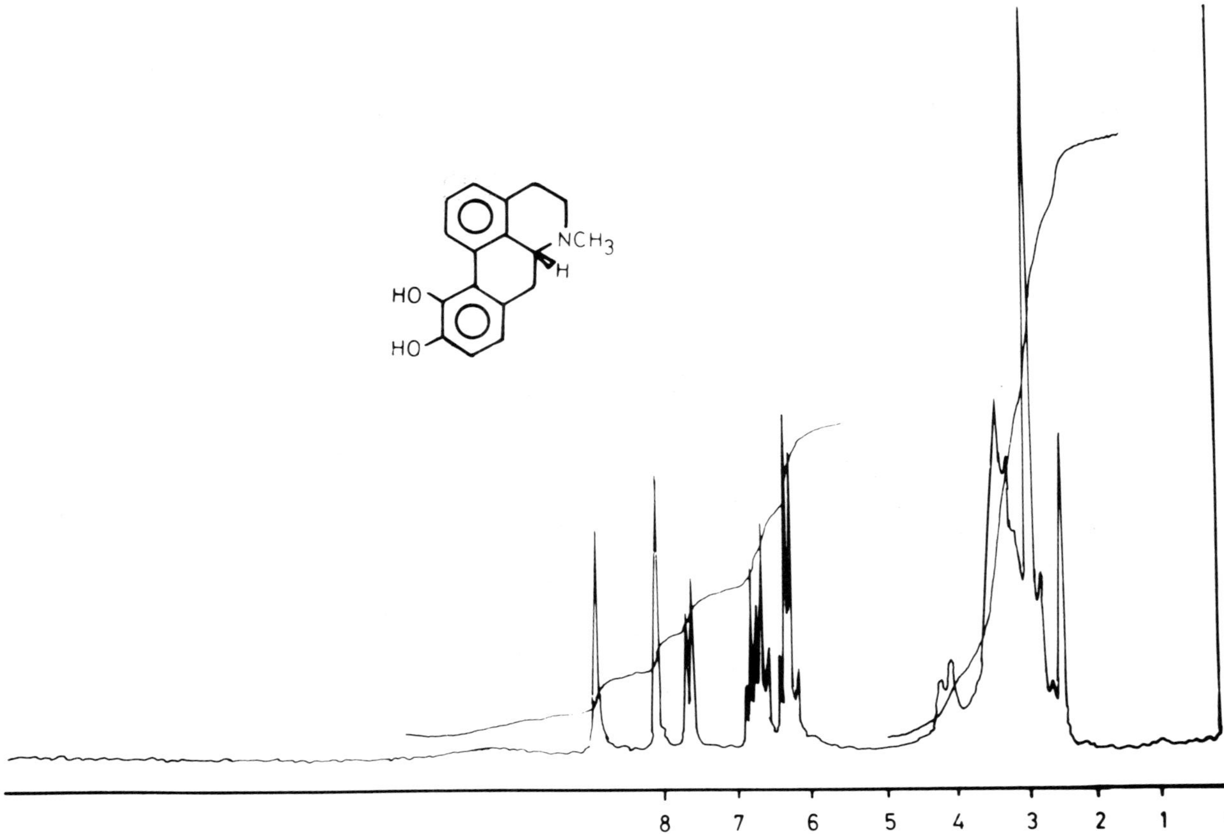

Fig. 4: ^{1}H-NMR Spectrum of Apomorphine (DMSO-d6).

Table 4: ^{1}H-NMR Characteristics of Apomorphine

Proton Assignment	Chemical Shifts δ
$C_{1,2,3,8,9}$	Aromatic protons absorb as a group of signals between δ 8.36 and 6.53 integrating for 5 protons.
H at C_6 a	4.11 (d)
2H at C_5	3.28 (s)
$N\text{-}CH_3$	2.46 (s)

s= singlet, d = doublet.

^{1}H-NMR data of some aporphinoid alkaloids has been reported (16).

2.9.3.2 Carbon-13 NMR Spectra

The ^{13}C-NMR spectra of apomorphine hydrochloride have been recorded once in DMSO-d_6 (to which 2 drops $CDCl_3$ were added), using tetramethylsilane (TMS) as an internal reference (Fig. 5 and 6), and another in D_2O and using 4,4-dimethyl-4-silapentane sulfonic acid (DSS) as an internal reference (Fig.7). Both spectra were recorded over 6024 Hz using a Joel FX-100 NMR Spectrophotometer. The carbon chemical shifts were assigned and listed in table 5.

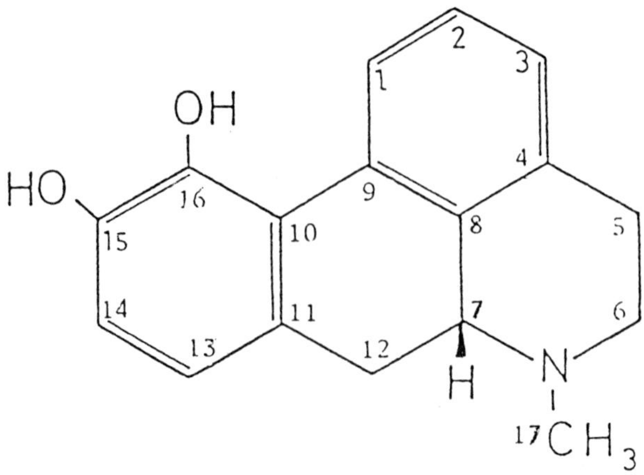

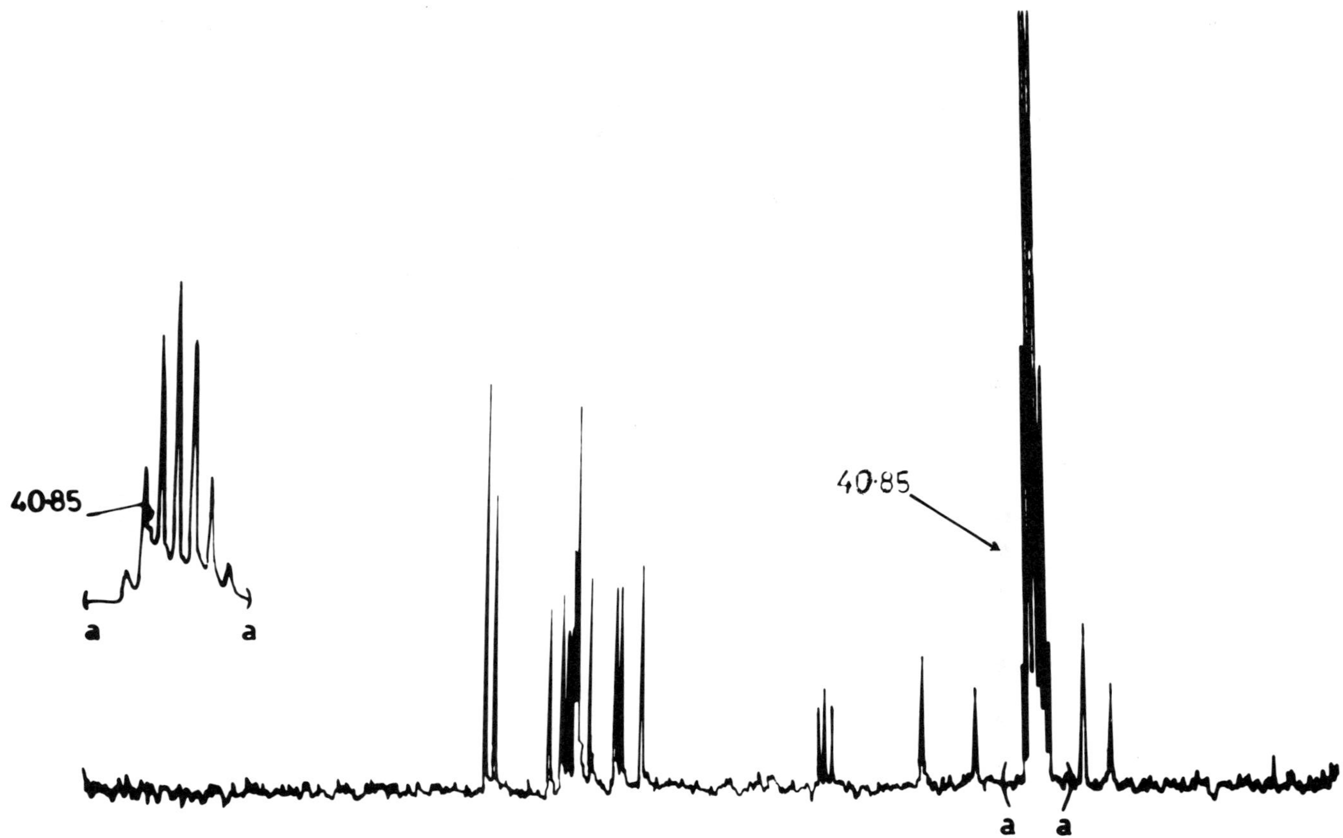

Fig. 5: Carbon—13 NMR Decoupled Spectrum of Apomorphine (DMSO-d6).

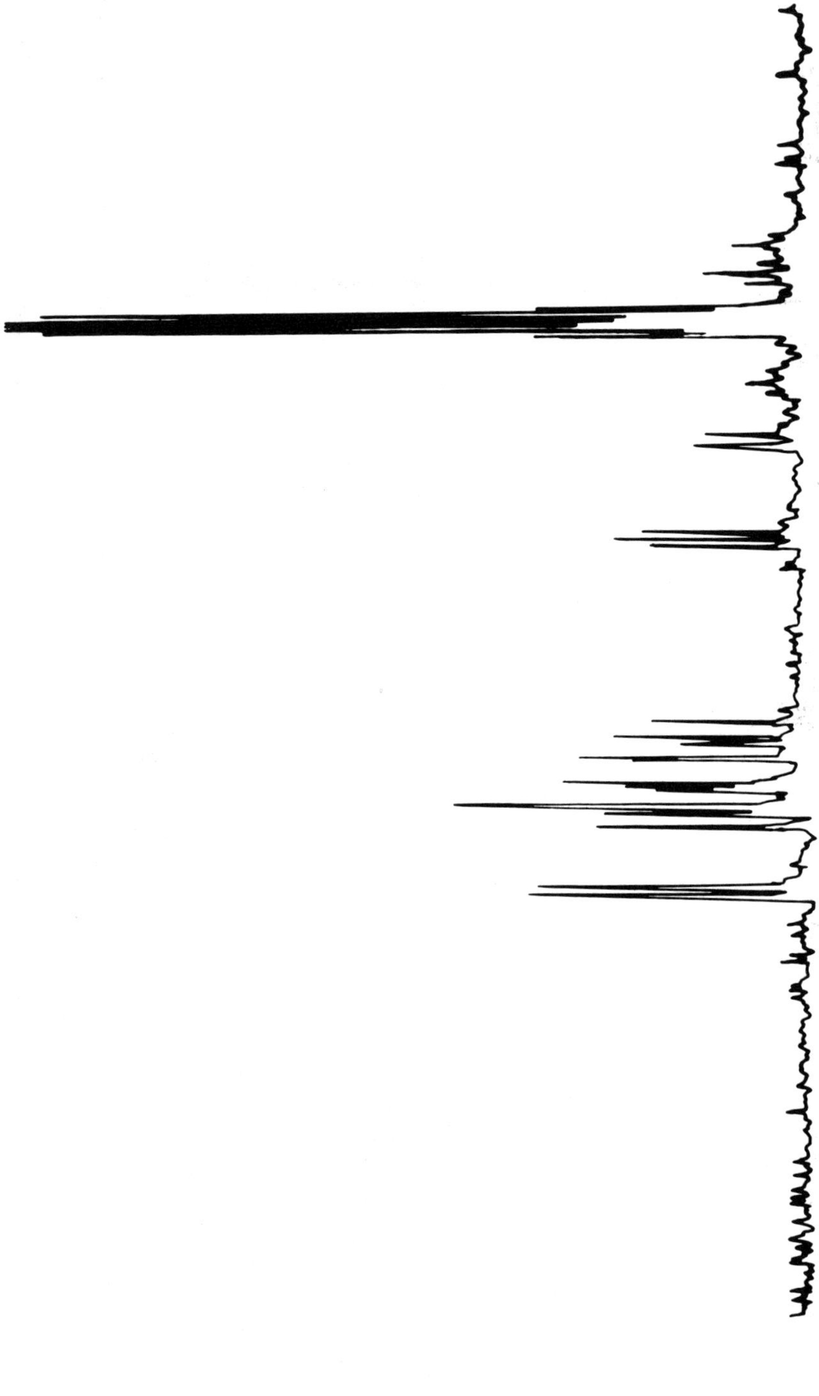

Fig. 6: Carbon-13 NMR OFF-Resonance Spectrum of Apomorphine (DMSO-d6).

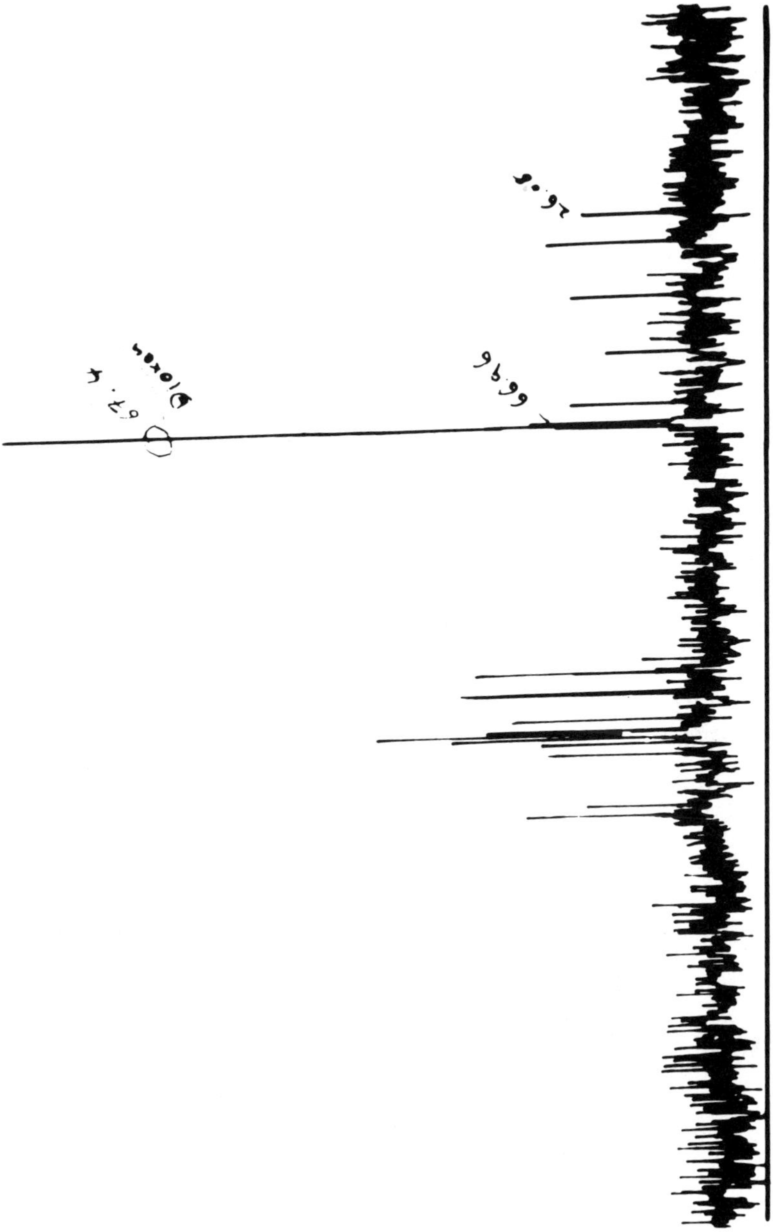

Fig. 7: Carbon-13 NMR Spectrum of Apomorphine HCl(D_2O).

Table 5 : Carbon-13 Chemical Shifts of Apomorphine.

Carbon Assignment	Chemical Shifts [δ]		Multiplicity
	DMSO-d_6	D_2O	
C_{15}	144.86	144.60	Singlet
C_{16}	143.13	142.72	Singlet
C_{11}	132.20	132.32	Singlet
C_{10}	129.69	130.27	Singlet
C_8	128.44	129.21	Singlet
C_4	127.21	128.42	Singlet
C_3	126.68	127.68	Doublet
C_1, C_2	124.30	125.60	Doublet
C_9	119.61	120.52	Singlet
C_{13}	118.46	120.28	Doublet
C_{14}	114.41	116.14	Doublet
C_7	64.88	66.96	Doublet
C_6	60.17	62.87	Triplet
$C_{17}(CH_3)$	42.02	42.14	Quartet
C_{12}	40.35	31.48	Triplet
C_5	30.51	26.08	Triplet

Carbon-13 NMR data for aporphine alkaloids have been reported (16,17).

2.9.4 Mass Spectrum

The electron impact ionization (EI) mass spectrum of apomorphine hydrochloride is presented in Fig. 8. The spectrum was obtained using Finnigan MAT 5100 series GC/MS spectrophotometer operating with an ionization potential of 70 eV.

The spectrum shows a base peak at mass/charge (m/z) ratio of 40.2.

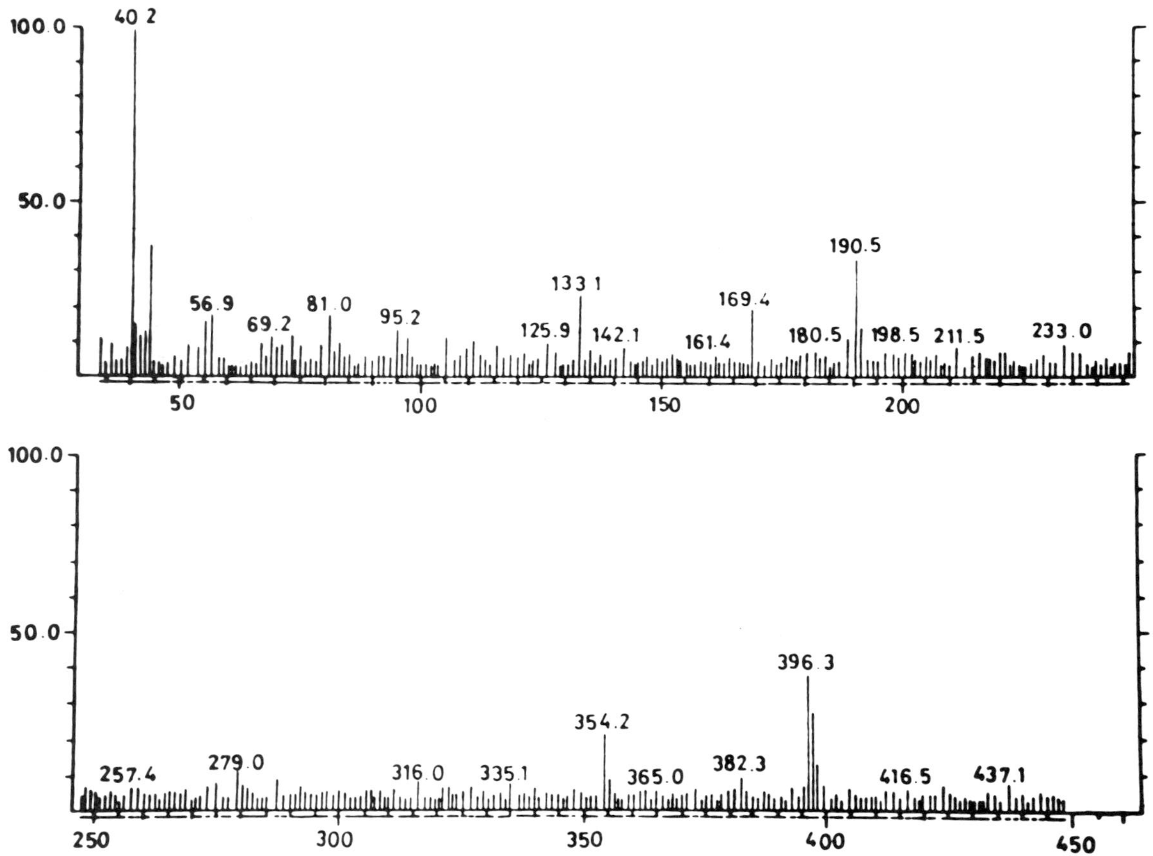

Fig. 8: Mass Spectrum of Lobeline.

The molecular ion peak of apomorphine (267) is absent or undetectable.

The most prominent ions and their.relative intensities are listed in Table 6.

Table 6: The Mass Fragments of Apomorphine.

m/z	Relative Intensity %	m/z	Relative Intensity %
40	100	190.5	35.6
44	38.4	198.5	6.8
56	17.8	211.5	8.2
69	11.0	233	9.6
81	17.8	257.5	6.8
95	15.1	279	11.0
126	9.6	288	8.2
133	24.7	316	8.2
142	8.2	354	21.9
169	20.5	396	44.0
180.5	6.8	-	-

Principal peaks of mass spectrum at m/z 266, 267, 224, 220, 268, 44, 250, 248 have been reported (15).

The mass spectra of the aporphines shows no characteristic fragmentation pathways. Loses are mainly of small fragments, $-\dot{H}$, $-\dot{C}H_3$, $-\dot{O}H$, $-\dot{O}CH_3$ (18).

3. Extraction of Apomorphine

Extraction of apomorphine and metabolites from biological fluids and tissues has been discussed by several authors.

Clarke (8) described the following general procedure for amphoteric drugs including apomorphine:
Liquid samples are acidified with 2 N HCl; solid samples are extracted with acidulated water and filteration is effected. The filtrate is extracted (3 times) with ether.

The aqueous layer (after removal of the ether layer) is rendered alkaline with 2 N ammonium hydroxide and re-extracted with chloroform (3 times). The combined chloroform layers are washed with water, then dried by filtration through anhydrous sodium sulfate. The dried chloroform layer can be tested for apomorphine.
As apomorphine is usually excreted as a conjugate, therefore prior to estimation, urine or homogenized tissue must be refluxed with 9 N hydrochloric acid for 20 minutes. The solution is then neutralized extracted with ethyl acetate (8).

A modification of the above method of extraction was reported (19).
Tissue to be assayed (1g) was homogenized with 0.1 N HCl (3-4 ml) and ethyl acetate (1 ml) in a homogenizer equipped with pyrax micro-homogenization flasks. The acid homogenate was quantitatively transferred to a volumetric flask, and sufficient 0.1 N HCl was added to make a total volume of 10 ml. An aliquot of the acid homogenate (5 ml) was transferred to a 50 ml glass-stoppered centrifuge tube and was shaken for 2-3 minutes with ethyl acetate (5 ml). The sample was centrifuged, the ethyl acetate was aspirated and discarded and the previous wash step was repeated with another portion of ethyl acetate (5 ml) which similarly discarded. The ethyl acetate washes significantly reduced the fluorescence of tissue blanks. Under acidic condition, apomorphine remained in the aqueous phase (pKa 8.9). The pH of the resulting homogenate was adjusted to 7.0±0.5 by the addition of solid sodium bicarbonate. The apomorphine present in the sample was extracted by shaking with ethyl acetate (15 ml) for 2-3 minutes.

4. Synthesis of Apomorphine

Apomorphine was first prepared in 1869 by the acid treatment of morphine (1). Later, other acid reagents were recommended for the preparation of apomorphine, such as heating morphine with zinc chloride (20) or treating morphine with concentrated hydrochloric acid (21). The following is the general scheme for the synthesis of apomorphine.
Apomorphine is produced upon dehydration of morphine with a strong acid under pressure where morphine loses one molecule of water and undergoes structural rearrangement to form apomorphine (22). This is shown in scheme I.

The first total synthesis of(±)apomorphine was achieved in 1970 (23):
Reissert anion [I] (prepared from isoquinoline) was condensed with 2-nitro-3,4-dimethoxy-benzyl chloride [II] (obtained from vanillin) to form the adduct [III].

A procedure for the generation of the anion [I] employing sodium hydride in N, N-dimethylformamide at room temperature has been reported (24). The product [III] was hydrolyzed with alcoholic potash to furnish [IV] in 90% yield. Treatment of the base [IV] with methyliodide gave the methiodide [V] in a quantitative yield. This was reduced with potassium borohydride to the tetrahydroisoquinoline [VI] in 80% yield. The nitro group of [VI] was reduced with palladium on-charcoal and hydrogen to furnish 1-(2-amino-3,4-dimethoxybenzyl)-2-methyl-1,2,3,4-tetrahydroisoquinoline [VII] in 80% yield. The amine [VII] was treated with copper powder and sodium nitrite to give (±)-apomorphine dimethylether [VIII]. Since [VIII] was isolated as an oil in 41% yield, it was converted into a crystalline picrate and a crystalline percolate which were converted to the hydroiodide salt with 57% hydroiodic acid. Recrystallization from acetone-water yielded white needles of apomorphine dimethyl ether hydroiodide. The latter [VIII] was demethylated by heating with an equimolar mixture of 57% hydroiodic acid and acetic anhydride at reflux for 1 hr., when the reaction mixture was diluted with ether, pure (±)-apomorphine hydroiodide [IX] was collected as white crystalline powder in 93% yield.
(±)-Apomorphine was resolved into (-) and (+) enantiomers (25).

This total synthesis of (±)-apomorphine is presented in scheme II.

Scheme I: The First Synthesis of Apomorphine

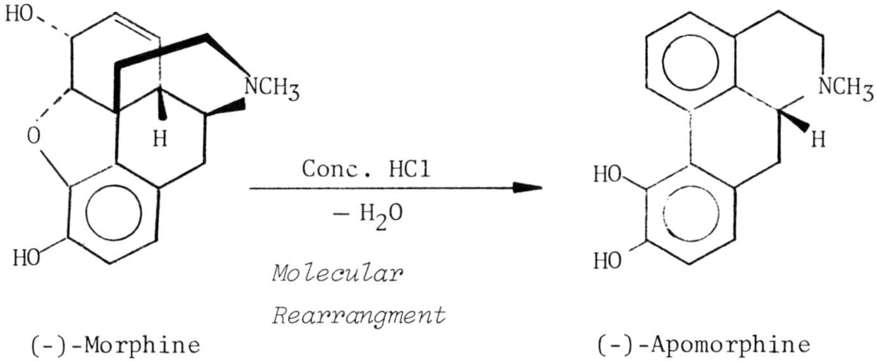

Scheme II: The First Total Synthesis.

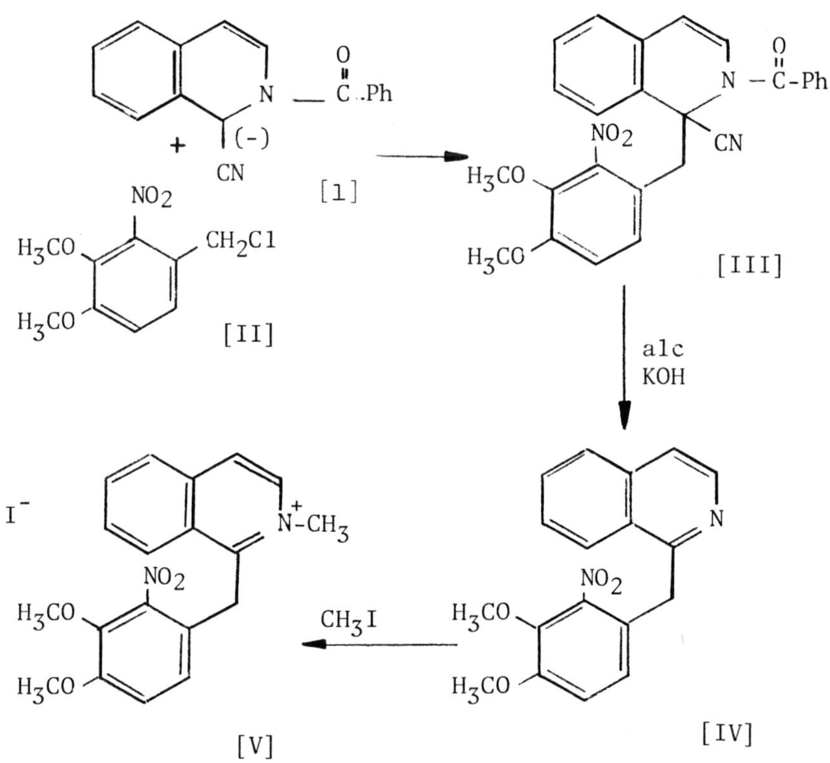

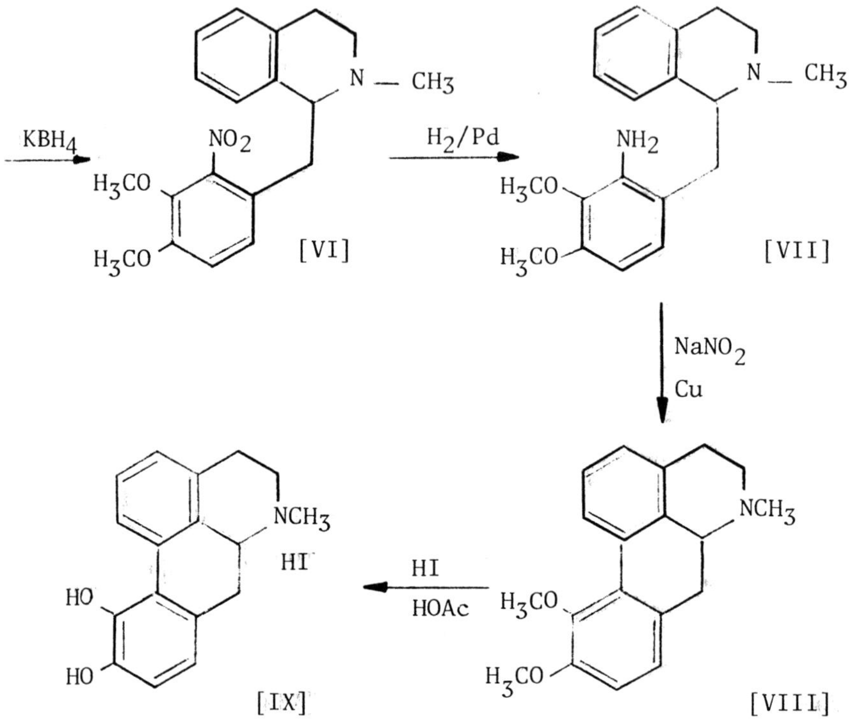
KBH4
NO2
N — CH3
H3CO
H3CO
[VI]
H2/Pd
NH2
N — CH3
H3CO
H3CO
[VII]
NaNO2
Cu
NCH3
H3CO
H3CO
[VIII]
HI
HOAc
NCH3
HI
HO
HO
[IX]

Synthesis of (R)-(-)-apomorphine from thebaine has been reported (26).
The opoid thebaine [I] was N-demethylated to northebaine [II]. Northebaine was treated with concentrated HCl in a pressure bottle to give normorphothebaine [III]. This was N-methylated with methyl iodide in acetonitrile followed by O-demethylation by heating with 48% H Br to give [IV] in a quantitative yield.
Substance [IV] was subjected to treatment with methylene dibromide in alkaline aqueous dimethylsulfoxide to afford [Va]. Removal of the phenolic hydroxyl group at the 2-position of [Va] was achieved in two steps, by formation of the phenyl tetrazolyl ether [Vb], followed by hydrogenolysis over 5% Pd/C in acetic acid at 40°C for 48 hours to give [Vc] in greater than 90% yield.
The removal of the methylenedioxy group in [Vc] was carried out in quantitative yield by treatment with boron trichloride in CH_2Cl_2 to afford (R)-(-)-apomorphine [VI].

This synthesis is presented in scheme III.

Synthesis of (S)-(+)-apomorphine has also been reported (27).

The aporphine alkaloid (S)-bulbocapnine [VII] isolated from the roots of *Corydalis cava*, has been converted to (+)-morphothebaine [VIII] (28,29). By a sequence of reactions described in scheme III, (+)-apomorphine can be produced.

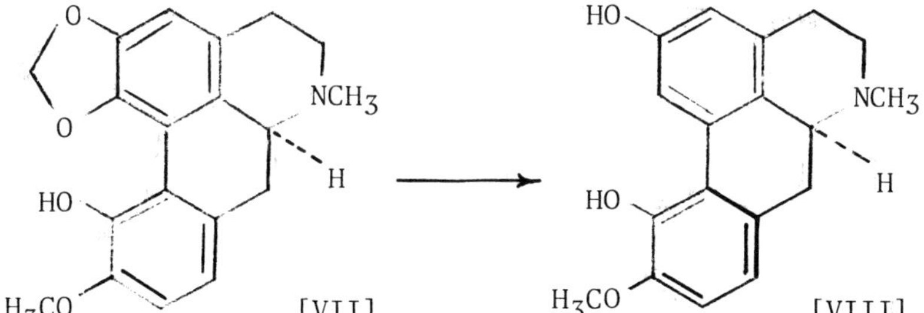

[VII] [VIII]

Preparation of Apomorphine Hydrochloride

Morphine (5 g) and concentrated HCl (50 ml) are heated at 130-140° in a sealed tube (for 3 hours). The content of the tube are dissolved in water and the bases precipitated by the addition of an excess of sodium carbonate and extracted into ether (morphine is quite insoluble in ether). Ether is evaporated to dryness, dilute HCl added the resulting salt is purified by crystallization from hot water (1.8 g) (12).

Scheme III: Synthesis of (R)-Apomorphine from Thebaine

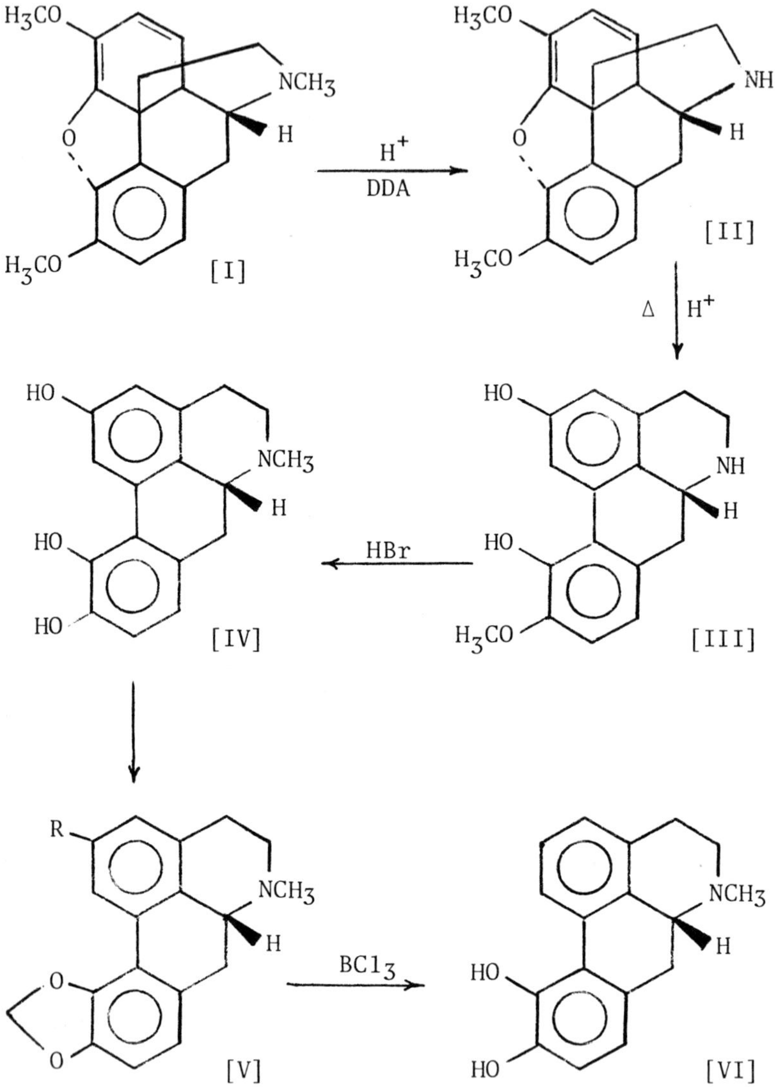

[Va] R = OH

[Vb] R = 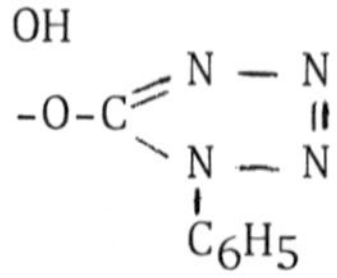

[Vc] R = H

5. Pharmacokinetics

5.1 Drug Absorption

Apomorphine is rapidly and completely absorbed from subcutaneous tissues (30). When apomorphine hydrochloride was administered by subcutaneous injection, by subcutaneous inf usion and by intravenous infusion to 15 patients with parkinsonism, the variation in absorption was high between subjects but low within individuals (30).

5.2 Distribution

Apomorphine administered to monkeys and cats could be detected in the brain tissue within one to two hours (31). The same author reported that about 30% of the dose was found in the brain of monkeys 70 minutes after administration of apomorphine.

Organ-distribution studies of apomorphine after intravenous injection into rats, showed no detectable levels of the alkaloid in the central nervous system, but a high concentration of the bound form was found in the liver, mainly as the glucuronides (32).

5.3 Metabolic Reactions

In rats, rabbits and horses apomorphine is extensively metabolised by conjugation with glucuronic acid.
In rabbits, the glucuronide metabolites have been identified as the O^3- and O^4-β-glucuronides (33,34). In mice apomorphine appears to induce its own metabolism (34).

5.4 Excretion

Following injection into horses, rabbits and rats, apomorphine is excreted in the urine largely as conjugates i.e. glucuronides, which yield the free drug on hydrolysis (32). Normal horses and rabbits excreted little or no free apomorphine.
The excretion of a small percentage in the free form in rats, appears to be dependent on the pH of the urine *(rats with urine of pH 6 ± 0.5, excreted about 4% of the given dose as free apomorphine)* (32).

6. Pharmacology and Therapeutic Uses

This section is written by Dr. Zaki (35):

I. Apomorphine possesses a strong central emetic action by:

a) stimulating the chemoreceptor trigger zone (C.T.Z.) in the area postrima of the medulla oblongata in the periventricular area. This zone sends afferent impulses to the vomiting centre proper which is located in the lateral reticular formation in the medulla near the fasciculus solitarius.

b) stimulating the vestibular apparatus of the internal ear increasing its sensitivity.

Vomiting is preceeded by nausea (stomach flacidity), salivation, increased bronchial secretion with restlessness and tremors. Usually, vomiting is followed by exhaustion and depression.

II. It possesses a potent dopaminergic agonist effect with decreased analgesic potency. Therefore, its emetic effect is counteracted by phenothiazine derivatives that possess potent dopamine - blocking action as well as by drugs that block vestibular stimulation and motion sickness.

III. Subemetic doses possess a sedative expectorant effect modifying the character of the sputum, diaphoretic and sialagogue actions.

IV. Apomorphine possesses a weak respiratory depressant action. Opioid antagonists eg. naloxone ... etc could reverse its depressant actions only e.g. respiratory depression, but do not prevent or slightly affect the excitatory induced effects e.g. vomiting. Therefore , it is not used when the patient is poisoned by CNS depressants or if the patient's respiration is slow or difficult.

Dose:

In poisoning, it is not effective orally and it is only given subcutaneously in a dose of 6 mg. A second dose is never given if the first dose fails. Emesis usually starts within 3-5 minutes and lasts for one hour. It is used as the hydrochloride salt.

Therapeutic Uses:

1) For rapid evacuation of the stomach from oral drug overdosage and toxins provided the CNS is not depressed.

It has an advantage over ipeca due to its rapid onset of action.
N.B. Gastric lavage is much better and safer.

2) To remove excess bronchial secretion and foreign bodies in the upper respiratory tract in children. However, bronchial aspiration is much superior if it is available.
3) In conditioned aversion technics in management of chronic alcoholism and other forms of drug abuse. This is done by giving apomorphine and shortly followed by a small dose of the abused agent, so that nausea and vomiting occur simultaneously. This acts as a conditioned stimulus.

Side-effects:

1) Severe vomiting leads to:
   a) Fluid and electrolyte imbalance and dehydration. Chronic large doses produce renal changes and damage.
   b) When vomiting is severe and prolonged, it leads to coma especially when CNS is depressed.
2) Oesophageal laceration and rupture with intractable vomiting with mediastinitis.
3) Cardiovascular complications due to the Valsalva's mechanism during vomiting eg. arrhythmias, cerebral accidents, ... etc.
4) Intra-ocular complications eg. rupture vessels, ... etc and hernias.

7. Drug Stability

In aqueous solutions, apomorphine is oxidized to various derivatives of quinolindione, which are devoid of emetic properties. Oxidized solutions are emerald green, but the depth of color is not a reliable indication of the extent of oxidation (36,37). The rate of oxidation can be retarded by the addition of dilute hydrochloric acid to adjust the pH of the solution to 3 to 4, and sodium metabisulfite; the solution should be free from dissolved oxygen (36-38).
Apomorphine hydrochloride should not be used if it at once gives an emerald green solution when 1 part is shaken with 100 parts of water. In addition, apomorphine hydrochloride should not be used if a solution so prepared is deeper in color than a 0.005% solution of the sample treated as follows:

Dilute 1 ml with 6 ml of water, add 1 ml of sodium bicarbonate solution and 0.5 ml of 0.1 N iodine, allow to stand for 30 seconds, add 0.5 ml of 0.1 N sodium thiosulfate and dilute to 10 ml with water (34).

8. Preparations and Storage

Apomorphine hydrochloride should be kept in an air-tight containers and protected from light (34,39). Preserve in small tight light-resistant containers from which apomorphine HCl is to be taken for immediate use (13).

Preserve apomorphine hydrochloride tablets in tight light-resistant containers (13).

Apomorphine Injections

A sterile solution of apomorphine hydrochloride with 0.1% sodium metabisulfite in water for injection free from dissolved air. The air in ampoules being replaced by nitrogen or other suitable gas. The ampoules are immediately sealed and sterilized by heating in an autoclave. Maloney (40) recommended that solutions of apomorphine hydrochloride should be sterilized by aseptic filtration rather than by autoclaving as solutions of 1 mg per mL which also containing sodium metabisulfite 0.1%, where green in color after autoclaving, whereas similar solutions prepared by aseptic filtration remained colorless for at least a year when stored in refrigerator and protected from light.

The usual strength is 3 mg of apomorphine hydrochloride per mL (34).

9. Methods of Analysis

9.1 Identification

The following identification tests are mentioned in the British Pharmacopoeia (10) under apomorphine hydrochloride.

- The light absorption in the range 230 to 350 nm of a 0.001% (w/v) solution in 0.1 M hydrochloric acid exhibits a maximum only at 273 nm; A (1%, 1 cm) at 273 nm, about 550.
- To 5 ml of a 1% (w/v) solution add sodium hydrogen carbonate solution until a permanent white precipitate is obtained; the precipitate slowly turns green. Add 0.25 ml of 0.05 M iodine and shake; the precipitate becomes emerald-green. The precipitate is soluble in ether giving a purple solution, in chloroform giving a violet-blue solution and in ethanol (96%) giving a green solution.
- A 1% (w/v) solution yields reaction a characteristic of chlorides. The precipitate on the addition of 0.25 M silver nitrate darkens rapidly on the addition of 6 M ammonia.

The following tests are mentioned in the USP XXI, the NF XVI (13) under apomorphine hydrochloride.

- The infrared absorption spectrum of a potassium bromide dispersion of it, previously dried, exhibits maxima only at the same wavelengths at that of a similar preparation of USP Apomorphine Hydrochloride RS.
- To 5 ml of a solution (1 in 100) add a slight excess of sodium bicarbonate solution (1 in 20); a white or greenish white precipitate is formed. Add 3 drops of iodine TS, and shake vigorously; an emerald green color is produced. Add 5 ml of ether and after vigorous shaking, allow the layers to separate; the ether is colored deep ruby-red, while the water layer retains its green color.
- Dissolve it in nitric acid; a dark purple solution is produced.
- To a solution of it add silver nitrate TS; a white precipitate, which is insoluble in nitric acid is formed. This precipitate soon darkens by reduction to metallic silver, the reduction being accelerated by the addition of 6 N ammonium hydroxide.

Other identification tests are as follows (8).

- Sulfuric acid-formaldehyde test gives with apomorphine a purple color which turns to black (sensitivity : 0.05μg).
- Ammonium bolybdate test gives with apomorphine deep green color which changes to blue green (sensitivity: 0.1μg).

9.2 Microcrystal Tests (8)

- Gold bromide solution gives with apomorphine solution serrated crosses crystals (sensitivity: 1 in 5000).
- With potassium bismuth iodide solution, apomorphine gives rosettes of dark needles (sensitivity: 1 in 5000).

9.3 Titrimetric Determinations

9.3.1 Acid-Base Titrations

Methods for the titration of apomorphine HCl were reviewed (41). The results obtained in H_2O or aqueous acetone medium were compared with those obtained in a medium of ethanol or acetone-methanol.

The USP XXI (13) describes the following acid-base procedure for the determination of apomorphine hydrochloride tablets:

Weigh and finely powder not less than 20 apomorphine HCl tablets. Dissolve an accurately weighed portion of the powder, equivalent to about 50 mg of apomorphine HCl, in 25 ml of H_2O in a separator, add 500 mg of $NaHCO_3$, and completely extract with successive small portions of ether. Combine the ether extracts in a separator, and wash them with three 5-ml portions of H_2O. Shake the combined water washings with 10 ml of ether, and add this ether to the combined ether extracts. Extract the ether solutions with 20.0 ml of 0.02 N H_2SO_4 VS, and wash with three 5-ml portions of H_2O. Combine the acid extract and washings in a beaker, and warm on a steam bath to expel any residual ether. Cool, add methyl red TS, and titrate the excess acid with 0.02 N NaOH VS.

Each ml of 0.02 N H_2SO_4 is equivalent to 6.256 mg of $C_{17}H_{17}NO_2$ HCl, $\frac{1}{2}$ H_2O.

Titration with 0.02 N hydrochloric acid or back titration with 0.02 N sodium hydroxide after using HCl in the presence of methyl red as indicator was recommended for the quantitate analysis of apomorphine hydrochloride (42)
1 ml 0.02 N HCl $\equiv$ 0.006076 g $C_{17}H_{17}O_2N_1$HCl.

9.3.2 Non-aqueous Titrations

Two methods for determination of apomorphine HCl in non aqueous media have been developed (43); i) dissolve 0.3 g of apomorphine HCl in 30 ml of glacial acetic acid by heating. Cool the solution to 30° and add 10 ml of 5% mercuric acetate in glacial acetic acid followed by 20 ml of dioxane. The resulting blue solution is titrated immediately with 0.1 N $HClO_4$ in glacial acetic acid until the color changes to bright red.
ii) Dissolve 0.3 g of the hydrochloride in 30 ml glacial acetic acid. Cool to between 10° and 15° and then add the other two reagents in the same quantities as before. To the resulting colourless solution add 5 drops of crystal violet indicator and titrate at once with 0.1 N $HClO_4$ until the color changes to blue. The same author reported that the results of 12 experiments, expressed as the percentage of the weight of the sample (containing 2% of H_2O, determined by drying over H_2SO_4 to constant weight) gave mean values of 97.43 and 97.65% respectively by methods i and ii and closely similar results were obtained by potentiometric titration. A third method was also mentioned (44); dissolve 0.15 g of apomorphine HCl in 5 ml of anhydrous acetic acid (or 15 ml) + 10 ml of 3% Hg^{II} acetate. Titrate with 0.1 N $HClO_4$ using gentian violet as indicator.

The USP XXI (13) reported a non-aqueous titration method for the assay of apomorphine HCl as follows:
Dissolve about 300 mg of apomorphine HCl accurately weighed, in 100 ml of glacial acetic acid with the aid of heat provided by a steam

bath. Add 0.1 ml of acetic anhydride to the hot solution, and stir for 5 min. Cool to room temperature, add 5 ml of mercuric acetate TS and 0.25 ml of crystal violet TS, and titrate with 0.1 N $HClO_4$ VS to a blue end-point. Perform a blank determination, and make any necessary correction.
Each ml of 0.1 N $HClO_4$ is equivalent to 30.38 mg of $C_{17}H_{17}NO_2$, HCl.

9.3.3 Potentiometric Titration

The BP (10,39) recommends a non-aqueous titration method for the assay of apomorphine hydrochloride and requires the end point to be determined potentiometrically as follows:-
Dissolve 0.25 g in 5 ml of anhydrous formic acid. Add 30 ml of anhydrous glacial acetic acid and 5 ml of mercury (II) acetate solution.

Titrate with 0.1 M perchloric acid to the color change of the indicator that corresponds to the maximum absolute value of dE/dV (where E is the electromotive force and V is the volume of titrant.
Each ml of 0.1 M perchloric acid VS is equivalent to 0.03038 g of $C_{17}H_{17}NO_2$, HCl.

9.4 Electrochemical Determinations

9.4.1 Voltametric Method

Phenanthrene - 9,10-quinone and oxoapomorphine, a product of oxidation of apomorphine in solution of pH 7, were adsorbed on a carbon-paste electrode, and yield well-defined voltametric peaks (45). Owing to the preceding electrochemical reaction, preconcentration results in signal enhancement. Calibration graphs are rectilinear for concentrations from 0.01 to 1 μm.

9.4.2 Oscillographic Polarography

A rapid and sensitive oscillographic polarography for the identification of apomorphine as well as other opium alkaloids was reported (46) papaverine and apomorphine can be used directly without prior oxidation

(C.F. other opium alkaloids which should be oxidized with $K_3Fe(CN)_6$ before examination).

9.4.3 Ion-Selective Electrode

A Crytur valinomycin electrode was used for end point indication in the precipitation titration of many alkaloids including apomorphine (47) with tetraphenylborate; the coefficient of variation (3, 4 or 5 determinations) was < 1.4% for 86 to 270 mg of all alkaloids except physostigmine.

9.5 Spectrophotometric Methods

9.5.1 Colorimetric Determinations

A colorimetric method for quantitative determination of apomorphine was devised (48). The product of oxidation of apomorphine with $HgCl_2$ solution is extracted with amyl alcohol: Mix the sample solution (4 ml containing 60-120 μg of apomorphine) with McIlvaine buffer of pH 6 (1 ml) and saturated aqueous $HgCl_2$ solution (3 drops). Heat in a water bath at 70° for 3-4 minutes, cool to room temperature, shake for 1 minute with isoamyl acetate (10 ml), then centrifuge; measure the extinction of the upper layer at *330 nm*. To determine apomorphine in biological fluids, extract diluted sample with isoamyl acetate at pH 6 to 7 and reextract into 0.05 N HCl before assay.

A review was presented of the use of o- and m- dinitrobenzoic acid; 1,3,5-trinitrobenzene; 3,5-dinitrobenzene sulfonic acid; bis-(2,4-dinitrophenyl) sulfone and picric acid in colorimetric analysis of various compounds that are reducing agents or contain an active methylene group including apomorphine (49).

Apomorphine HCl (300-600 μg) can be determined photometrically in pharmaceutical preparations where morphine HCl does not interfere. The method (50) depends on oxidizing apomorphine with $K_2Cr_2O_7$ and H_2O_2 in acetic acid, extracting the coloured product with $CHCl_3$ and measuring the extinction at specified wavelength.

Another method (51) for micro determination of apomorphine by oxidation with oxygen in alkaline medium was described: heat at 70° on a water bath under reflux a solution of apomorphine HCl (0.06-9.0 mg) in 20 ml of 2.5% NaOH solution in H_2O-dioxane (1:1) and pass oxygen through the solution for 30 minutes, acidify with conc. HCl until a blue color appears, dilute to 50 ml with H_2O-dioxane (1:1) and measure the extinction at *585 nm*. Relative errors ranged from 0 to ± 4%.

9.5.2 Ultroviolet Determinations

Apomorphine injections are assayed by the following method (52).
Dilute injections with sufficient 0.1N hydrochloric acid to produce a solution containing about 0.001% w/v of apomorphine hydrochloride and measure the *extinction* of a 1-cm layer at the maximum at about *273 nm*. Calculate the content of $C_{17}H_{17}NO_2$, HCl, $\frac{1}{2}$ H_2O taking 550 as the value of *E* (1%, 1 cm) at the maximum at about *273 nm*.

A spectrophotometric method for the determination of alkaloidal poisons in tissues including apomorphine HCl has been discussed (53).

The sample of tissue is homogenized with 25 ml of 0.1N HCl, the homogenate is extracted on a water bath with 75 ml of 95% ethanol and 2 ml of 10% Na_2WO_4, the residue being dissolved in 50 ml of McIlvain's buffer at pH 7.0 and extracted with $CHCl_3$ (50 ml). The separated $CHCl_3$ layer is then extracted with 100 ml of 0.1N HCl and the UV absorption is determined. The characteristic UV absorption curve for 30 alkaloids including apomorphine is presented from which its quantity can be calculated.

On the other hand, in two different articles (54,55), the importance of narrow slits in the spectrophotometric determination of extinction coefficients has been demonstrated. The authors stated that care is needed in the B.P. of 1958 and of 1963 spectrophotometric assays of apomorphine HCl to avoid low results.

9.5.3 Spectrofluorimetry

A spectrofluorimetric method for the determination of apomorphine hydrochloride was reported (19).

The method depends on the fact that apomorphine emits a bluish fluorescence under UV light. Florescence spectra of apomorphine can then be determined in ethyl acetate at the wavelengths *270/370 nm.*
The assay involves an immiscble solvent extraction procedure (see 3) which is highly efficient and makes possible the detection of apomorphine in concentrations of 0.1 mcg/g tissue (Brain tissue).
This fluorometric assay is about 40 times more sensitive than the spectrophotometric method currently used (19).
The fluorescence spectra of apomorphine in ethyl acetate (used in the assay) and in water is presented in Fig. 9.

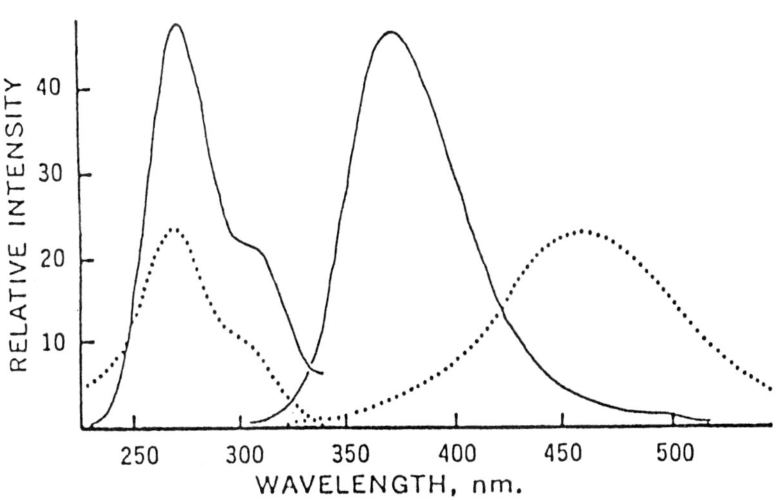

Figure 9: Fluorescence spectra of apomorphine in ethyl acetate (0.33 mcg./ml. maximum, solid lines) and water (1 mcg./ml., dotted lines), after (19).

9.6 Chromatographic Methods

9.6.1 Paper Chromatography

Clarke (8) described the following three systems for the identification of apomorphine:

I) Whatman No. 1 paper sheet 14x6 inches, buffered by dipping in a 5% solution of sodium dihydrogen citrate, blotting and drying at 25° for one hour. A sample 2.5 μl of a 1% solution in ethanol or in 2 N acetic acid is applied to the sheet.
A solvent composed of 4.8 g of citric acid in 130 ml water and 870 ml n-butanol is used (56).
Rf value is 0.33 (location under UV, blue fluorescence; location reagents, iodoplatinate, strong reaction; bromocresol green, strong reaction).

II) Reversed phase paper chromatography, Whatman filter paper No. 1 or No. 3 sheet 17x 19 cm, impregnated by dipping in a 10% solution of tributyrin in acetone and drying in air.
Sample : a 5 μl of 1 to 5% solution in ethanol or chloroform.
Solvent : acetate buffer (pH 4.58).
Time of run 15 to 20 minutes.
Rf value is 0.55 (location as for system I).

III) As II, the solvent is phosphate buffer (pH 7.4) (57).
Rf value 0.11 (location as for system I).

The use of paper chromatography in studying the stability of apomorphine (I) was discussed (58). The alkaloid (I) can be separated from an oxidation product (II) formed in unstabilized aqueous solution (containing 100 to 200 μg of I) with n-butanol-formic acid - H_2O (77:13:10) (Rf values 0.5 and 0.7 for I and II respectively) or n-butanol - acetic acid-water (4:1:5) (R_f values 0.65 and 0.85 for I and II respectively). Dragendorff's reagent gives an orange spot with I and green one with II. For injection solution stabilized with ascorbic acid or mannitol, the best spray reagent is bromophenol blue solution containing borate, which gives yellow spots with ascorbic acid and mannitol, and a green one with I on blue background.

9.6.2 Thin Layer Chromatography (TLC)

The following TLC systems were recommended for the identification of apomorphine:

Chromatogram	Solvent System	Rf value	Ref.
1. Silica gel G *(0.25 mm)*	Methanol - Strong ammonia solution (100:1.5)	0.57	(8)
2. Silica gel G dipped in, or sprayed with 0.1M KOH in methanol and dried.	Methanol - Strong ammonia solution (100:1.5)	0.83	(15)
3. The same as 2	Chloroform-methanol (90:10)	0.21	(15)
4. The same as 2	Cyclohexane-toluene-diethylamine (75:15:10)	0.0	(15)

Location reagent, acidified iodoplatinate spray (8,15).

A method for determination of apomorphine (I) and norapomorphine (II) in urine by TLC fluorescence quenching was discussed (59). A 1.8-mg amount of either of these metabolic products (I and II respectively) was dissolved in a minimum of 0.1 M - citric acid (pH 3) and added to 10 ml of 0.1 M - citric buffer (pH 7) or rat urine (≅ 10 ml). The solution was adjusted to pH 7 with 0.1 M - HCl or - NaOH and extracted with ethyl acetate (3x10 ml). The combined extracts were evaporated to dryness in vacuo and the residues were dissolved in acetone (1 to 2 ml). The solutions (1 to 4 µl containing 0.5 to 6 µg of I or II) were applied to silica gel GF_{254} plates (0.25 mm layers), and chromatograms were developed with acetone - methanol (1:1). The plates were scanned in the fluorescence mode with excitation at 254 nm, and the areas of the recorded peaks were measured by triangulation. The average recoveries of I and II from urine were 88.5 and 41.4% and the coefficient of variations were 3.5 and 2.1% respectively.

Another TLC procedure for studying apomorphine and its analogues was described (60). The standard solution in ethyl acetate (1 to 5 µl) and the sample [0.25-0.5 ml of conc. microsomal extracts in $CHCl_3$ - methanol (1:1)] were applied to a 0.25 mm layers of silica gel GF_{254}. The chromatograms were developed with benzene-methanol (4:1), acetone-methanol (1:1), benzene-ethylacetate-diethylamine (6:3:1) or $CHCl_3$ - acetone (4:1). Spots were located by quenching of the fluorescence in UV radiation (254 nm) and by spraying with diazotised sulfanilic acid or 3,5-dichloro-p-benzoquinone chlorimine.

9.6.3 Gas Liquid Chromatography (GLC)

Many GLC systems have been reported for the identification and quantitative determination of apomorphine in biological fluids and tissues. The following two systems were reported for the identification of apomorphine (8, 15):

1) Column condition : 5% SE-30 on 60-80 mesh Chromosorb W AW (5 ft. x 1/8 inch), stainless steel, maintained at 230°.

   Carrier gas : Nitrogen, 30.7 ml/minute.

   Detector : Flame ionization, hydrogen 22 ml/minute.

   Retention time : 1.28 (relative to codeine).

2) Column condition : 2.5% SE-30 on 80-100 mesh Chromosorb G (acid-washed and dimethyldichlorosilane treated), glass column (2m x 2mm). Column temperature between 100° and 300°.

   Carrier gas : Nitrogen at 45 ml/minute.

   References : n-Alkanes with an even number of carbon atoms.

   Retention Index (RI) : 2530.

The following GLC systems have been recommended for the quantitative determination of apomorphine in biological fluids and tissues.

A gas chromatographic determination of apomorphine in rat urine and faeces was reported (61). The alkaloid was extracted into ethyl acetate and silylated with NO-bis-(trimethyl silyl) acetamide before analysis. The column used was 6 ft x 0.125 in. packed with 3% OV-17 on Gas - Chrom Q (100-120 mesh) and operated at 260° with He as carrier gas (70 ml min^{-1}) and FID. Faeces were freez-dried and hydrolysed before extraction of apomorphine and GLC analysis. In the range 0.05-0.2 mg ml^{-1} recoveries were 80 to 100% (mean 88.5%) with a standard deviation of 3.8%.

Another gas chromatographic procedure using selected-ion monitoring of apomorphine in whole rat brain was discussed (62). The rat brain was homogenised with $HClO_4$ containing N-propyl-norapomorphine as internal standard. After centrifugation of homogenate, the supernatant solution was mixed with ethyl ether - benzene. A portion of the aqueous phase was adjusted to pH 7.0 and extracted with ethyl ether; the organic phase was evaporated to dryness at 30° under N_2 and the drug was derivatised with trifluorobis (trimethyl silyl) acetamide before analysis on a column (1m x 2mm) of 3% OV-17 on Gas-Chrom Q operated at 250°C with He as carrier gas (25 ml min^{-1}) and mass spectrometry detection with selected ion monitoring at m/e 411 and 439. The calibration graph was rectilinear in the range of 0.05 (the detection limit) to 5.0 μg ml^{-1}.
The coefficients of variations were 2.52 (for 4 μg ml^{-1}) to 12.0% (for 0.05 μg ml^{-1}).

Analysis of apomorphine in plasma and tissue by capillary gas chromatography - electron capture detection was stated (63). Plasma or tissue extract was extracted at pH 7.4 with ethyl acetate, the extract was washed with 0.2 M-HCl and the acid layer was washed with hexane, buffered at pH 7.4 and extracted with ethyl acetate. This extract was evaporated under N and the alkaloid was converted into trifluoroacetyl derivative for GLC analysis on a fused-silica column (15m x 0.25mm) coated with SE-54; the column was temperature programmed (170° to 285° at 6° min^{-1}), with N_2 (5 ml min^{-1}) as carrier gas and ^{63}Ni electron capture detector. In pure solutions, sub-pg amounts could be detected, but, in serum, the practical limit of determination was 100 ng ml^{-1}.

Other gas chromatographic methods for determination of apomorphine were also reported (64-66).

9.6.4 High Performance Liquid Chromatography (HPLC)

A high performance liquid chromatographic determination of apomorphine in blood serum was devised (67). Buffered neutral serum (1 ml), containing apomorphine I, N-propylnorapomorphine as internal standard and dithiothreitol (1 ng ml^{-1}) to prevent decomposition of I, was shaken with ethyl ether (rocker device). The residue from evaporation of the ether was dissolved in the mobile phase for HPLC on μ Bondapak - phenyl with detection at *273 nm*. The mobile phase (1 ml min^{-1}) was a mixture (36:9:55) of methanol - acetonitrile - aqueous buffer solution of pH 3.25; the last named being 1 mM in sodium dodecylsulfate, 20 mM in KH_2PO_4 and 30 mM in citric acid. Recoveries of I to 100 μg ml^{-1} of added I exceed 94% and satisfactory accuracy and precision were reported for the range 1.5 - 8.0 μg ml^{-1}.
The limit of determination was 100 ng ml^{-1}.

Another HPLC method for separation of apomorphine and its O-methyl metabolite was discussed (68). A column (30 cm X 4 mm) of μ Bondapak phenyl (preferred) or μ -Bondapak C_{18}, with methanol - acetonitrile buffer solution [0.02 M-Na_2HPO_4 +0.03 M Citric acid (pH 3.25) that is 1 mM in Na dodecyl sulfate] (36:9:55) as mobile phase and detection at *273 nm* were followed.
The separation of apomorphine, its two O-methyl metabolites and the internal standard boldine (69) was accomplished on a column (30cm x 2mm) of μ Bondapak phenyl (10 μm), with citrate/phosphate buffer solution (pH 3.25) - methanol - acetonitrile (55:36:9) as mobile phase which was 0.55 mM in sodium dodecyl sulfate to provide the counter-ion, and detection at *273 nm*. To a 250 μl plasma sample were added 10 μl of 0.15 M ascorbic acid, 40 μl of 15 mM-HCl and

0.75 ml of acetonitrile that contained 1.5 μg of boldine, and the mixture was vortex-mixed and centrifuged. A 0.8-ml portion of the supernatant solution was evaporated to 0.1 ml, the residue was diluted to 0.2 ml with the mobile phase and a 40 ml portion was injected on the column. The peak height ratio of apomorphine and boldine was referred to a calibration graph, which was prepared on the basis of standard solution in plasma of 0.4, 0.8, 1.6, 2.4 and 4.0 μg ml^{-1} of apomorphine.
The absolute recoveries of apomorphine and boldine from plasma were ≅94 and ≅99% respectively.

Comparison of HPLC electrochemical detectors for determination of apomorphine, noradrenaline, adrenaline, dopamine and melatonin revealed that the LC-2A (Bioanalytical systems Inc) showed better sensitivity than the E611 (Brinkmann Instruments) for the determination of low concentrations of these compounds (70).

A method for determination of apomorphine in tablets using HPLC with electrochemical detector was presented (71).
Tablets containing apomorphine were dissolved in the mobile phase [MeoH -acetonitrile -0.05 M KH_2PO_4 (1:3:16) containing methanesulfonate (1 mM) and N-propylnorapomorphine I as internal standard]. The solution was applied on a column (30cm x 4mm) of μBondapak phenyl; elution was effected at 3 ml min^{-1}, and detection with a vitreous carbon electrode operated at 0.5 V vs. a silver -AgCl reference electrode. Ret. times in this system were ≅5.5 min for apomorphine (II) and ≅10 min for I and the calibration graph (based on peak-height ratio) covered the range 10 to 80 μg ml^{-1} of II.
When the method was used to analyse tablets nominally containing 6 mg of II the contents found ranged from 85.5-113.7% of the expected value and the coeff. of variation (n=2 or 3) was never worse than 2.2% (usually better than 1%).

Determination of apomorphine and N-propyl nor apomorphine in plasma using HPLC and fluorescence detection was presented (72). Serum (1 ml) + internal standard (each of the drugs was used as standard for the other) were adjusted to pH 2 and extracted with toluene; the extract was rejected. The pH was adjusted to 7 and the drugs were extracted into ethyl ether; the extract was evaporated under N_2 and the residue was dissolved in 1 ml of mobile phase [acetate buffer solution of pH 3.25 - methanol (3:2)]. Portions of this solution were analyzed by HPLC on a column (15 cm x 4 mm) of μ Bondapak phenyl (10 μm). Elution was effected at 1 ml min^{-1}, and detection was by fluorimetry at *418 nm* (cut-off filter) with excitation at *281 nm*. The method was suitable for determining 0.1 to 1 μg ml^{-1} of each compound. Recovery averaged 86 and 83.6% respectively, and coefficient of variation were 4 to 9% (n=5 to 8).

Two HPLC methods for determination of apomorphine in brain and plasma of mouse (73). and rat (74) were also discussed:

- i) Mouse brain was homogenized with 4 ml of 0.4 M $HClO_4$, 2 μg of N-propylnorapomorphine (I; Int. St.) and 2 mg of $NaHSO_3$; rat plasma (1 ml) was treated with 0.5 μg of I. samples were then passed through Sep. Pak C_{18} cartridges, and I and apomorphine II were eluted with 50% of methanol in solvent A [20 mM -Na_2HPO_4-30 mM -citric acid (pH 3.2)]. The eluate was analysed by HPLC on μ Bondapak C_{18} with an RP-18 guard column, methanol-solvent A (17:33) as mobile phase (0.7 ml min^{-1}) and electrochemical detection. The average recovery was 92±2.8% for 0.3 to 3 μg ml^{-1} of II, and between-day coeff. of variation was 40 ng ml^{-1} in plasma and 0.5 ng g^{-1} in brain tissue. The method was used in pharmacokinetic studies.
- ii) Plasma samples and brain homogenates, plus EDTA, were extracted with ethylacetate, the organic phase was then extracted with 0.1 M - HCl and the aq. phase was subjected to HPLC on a column (30 cm x 3.9 mm) of μ Bondapak CN (10μm) with a mobile phase (1 ml min^{-1}) of acetonitrile - 0.02M KH_2PO_4 (11:89) containing 0.5 mM EDTA (pH 3) and electrochemical detection with a vitreous carbon electrode at 0.77 V vs.

silver-AgCl. The calibration graph was rectilinear for 0.1 to 100 ng of apomorphine; detection limits were 0.3 ng ml^{-1} and 3 ng g^{-1} for plasma and brain tissue, respectively. Recoveries were 80.3 to 82%.

HPLC analysis of 462 basic drugs, including apomorphine, on silica columns using non aqueous ionic eluents was carried out (75). Low wave length UV and fluorescence detection may be used, and fluorescence may be optimized by, for example, post-column pH change or derivatisation of some primary aliphatic amines with O-phthaldialdehyde. The authors stated that electrochemical oxidation can be used for the detection of most analytes. Retention and relative responses (UV), *254 nm* and electrochemical, + 1.2 V) have been generated for the investigated components using a 125 - μm Spherisorb S5W silica column and methanolic ammonium per chlorate (10 mM, pH 6.7) as eluent. This system can be used isocratically in qualitative and quantitative work, when either the w.l. or the applied potential can be adjusted to optimise the response.

The following HPLC system has been described for the identification of apomorphine (15):

<u>Column</u> : Silica (Spherisorb S5W, 5 μm)
Column (12.5 cm x 4.9 mm).

<u>Eluent</u> : A solution containing 1.175g (0.01M) of ammonium perchlorate in 1000 ml methanol, ajusted to pH 6.7 by the addition of 1 ml of 0.1M NaOH solution in methanol.

<u>K´value</u> : 3.7 (tailing peak).

9.7 Other Methods of Determination

9.7.1 Radiometric Assay

A radiometric determination of apomorphine with the use of labelled potassium tetraiodomercurate was described (76). The alkaloid was precipitated from dilute HCl solution by this reagent labelled with ^{131}I. The radioactivity of the precipitate was measured in a well type scintillation counter; a rectlinear relationship between alkaloid concentration and activity was found for 20 to 100 g, up to 1 mg of alkaloid per 0.1 ml of solution.

9.7.2 Enzymatic Radioisotopic Procedure

A method for determination of apomorphine using enzymatic radioisotopic procedure was mentioned (77).

9.7.3 Turbidimetric Determination

The purity of apomorphine HCl was determined by turbidimetry. Sample solution was treated with ethanol and 1% tungstophosphoric acid; after 3 to 5 minutes, the reaction products were treated with H_2O and the absorbance was measured with use of a blue filter. The standard deviation was 0.825% or less for purities of 98.01% or greater (78).

ACKNOWLEDGEMENT

The authors would sincerely like to thank Mr. Uday C. Sharma and Mr. Altaf H. Naqvi for their efforts in typing this manuscript. Mr. Usama Mokhtar for his fine drawings of the spectral figures.

REFERENCES

1. A. Mathiessen and C.R.A. Wright, Proc. Roy. Soc., Ser. B, 17, 455 (1869).

2. R. Pschorr, B. Jaecke and H. Fecht, Chem. Ber., 35, 4377 (1902).

3. R. Pschorr, Ibid., 40, 1984 (1907).

4. Avenarius and R. Pschorr, Ibid., 62, 321 (1929).

5. S. Späth and O. Hromatka, Ibid., 62, 325 (1929).

6. H. Corrodi and E. Hardegger, Helv. Chim. Acta, 38, 2038 (1955).

7. J.G. Grasselli and W.M. Ritchey, "Atlas of Spectral Data and Physical Constants for Organic Compounds", 2nd ed., Vol. II, p. 184, CRC Press Inc., Cleveland, Ohio (1975).

8. E.G.C. Clarke, "Isolation and Identification of Drugs", Vol. 1, p. 199, "The Pharmaceutical Press", London (1978).

9. "The Merck Index", 10th ed., M. Windholz Edit., 769, Merck and Co. Inc., Rahway, N.J. (1983).

10. "The British Pharmacopoeia", p. 38, "Her Majesty's Stationary Office, Cambridge (1980).

11. "Specifications for the Quality Control of Pharmaceutical Preparations", p. 46 and 824, "World Health Organization", Geneva (1967).

12. E.A. Swinyard, Chapter 40 in "Remington's Pharmaceutical Sciences", p. 746, 15th ed., Mack Publishing Co., Easton, PA (1975).

13. USP XXI - NF XVI "The United States Pharmacopeia, The National Formulary", p. 73, United States Pharmacopeial Convention Inc., Rockville, Md (1985).

14. J.M. Kolthoff, Biochem. Z., 162, 289 (1925); through Trans. Faraday Soc., 39, 338 (1945).

15. "Clarke's Isolation and Identification of Drugs", 2nd ed., A.C. Moffat Edit., p. 357, "The Pharmaceutical Press", London (1986).

16. H. Guinaudeau, M. Leboeuf and A. Cave , "The Aporphinoid Alkaloids IV", J. Nat. Prod., 51, 389 (1988).

17. E. Wenkert, B.L. Buckwalter, I.R. Burfitt, M.J. Gasic, H.E. Gottlieb, E.W. Hagaman, F.M. Schell and P.M. Wovkulovich, "Topics in C-13 NMR Spectroscopy", Vol. 2, p. 105, G.C. Levy Edit., Wiley Interscience, New York (1976).

18. G.A. Cordell, "Introduction to Alkaloids, A Biogenetic Approach", p. 399, A Wiley Interscience Publication, John Wiley and Sons, New York (1981).

19. W. Kent Van Tyle and A.M. Burkman, J. Pharm. Sci., 60, 1736 (1971).

20. E.L. Mayer, Chem. Ber., 4, 121 (1871).

21. L.F. Small, B.F. Faris and J.E. Mallonee, J. Org. Chem., 5, 334 (1940).

22. "The United States Dispensatory", 27th ed., A. Osol and R. Pratt Edits., p. 151, J.P. Lippincott Co., Philadelphia (1973).

23. J.L. Neumeyer, B.R. Neustadt and K.K. Weinhardt, J. Pharm. Sci., 59, 1850 (1970).

24. F.D. Popp and J.M. Wefer, Chem. Commun., 207 (1966).

25. W.S. Saari, S.W. King and V.J. Lotti, J. Med. Chem., 16, 171 (1973).

26. V.J. Ram and J.L. Neumeyer, J. Org. Chem., 46, 2830 (1981).

27. J.L. Neumeyer, U.S. pat. 3, 717, 639 (1972).

28. W.A. Ayer and W.I. Tayler, J. Chem. Soc., 472 (1956).

29. K.C. Rice and A. Brossi, Synth. Commun., 8, 391 (1978).

30. T.S. Gancher, W.R. Woodward, B. Boucher and J.G. Nutt, Ann. Neurol., 26, 232 (1989).

31. F. Veit, Arch. exptl. Pathol. Pharmakol., 178, 577 (1935).

32. P.N. Kaul, E. Brochmann-Hanssen and E.L. Way, J. Pharm. Sci., 50, 244; 248 and 840 (1961).

33. P.N. Kaul and M.W. Conway, J. Pharm. Sci., 60, 93 (1971).

34. "The Pharmaceutical Codex", 11th ed., p. 57, The Pharmaceutical Press, London (1979).

35. Mamdouh A. Zaki, Professor of Clinical Pharmacology, King Saud University, Riyadh, Saudi Arabia, Personal Communication (1990).

36. A.M. Burkman, J. Pharm. Pharmac., 15, 461 (1963).

37. A.M. Burkman, J. Pharm. Sci., 54, 325 (1965).

38. P. Lundgren and L. Landersjo, Acta Pharm. Suec., 7, 133 (1970).

39. "The British Pharmacopoeia", p. 45, Vol. I, Her Majesty's Stationary Office, London (1988).

40. T.J. Maloney, Austr. J. Hosp. Pharm., 15, 34 (1985).

41. J.B. Schute, Pharm. Weekbl. Ned., 99, 1253 (1964).

42. D.C. Garrat, "The Quantitative Analysis of Drugs 3rd ed., p. 61, "Chapman and Hall Ltd.", London (1964).

43. A. Paulsen, Medd. Norsk Farm. Selsk., 18 (9), 145 (1956).

44. E. Posagy, Acta Pharm. Hung., 35(2), 75 (1965).

45. H.Y. Cheng, L. Falat and R.L. Li, Anal. Chem., 54(8), 1384 (1982).

46. G. Dusinsky', Českosl. Farmac., 4(8), 400 (1955).

47. K. Vytrask and V. Riha, Cesk. Farm., 26(1), 9 (1977).

48. P.N. Kaul, E. Brochmann-Hanssen and E.L. Way, J. Amer. Pharm. Ass., Sci. Ed., 48 (11), 638 (1959).

49. M.P. Yavorskli, Farm., Zh. (Kiev), 5, 26 (1976).

50. F. Machovicova, Českasl Farmac., 5(1), 24 (1956).

51. K. Rehse and G. Dreke, Z. Anal. Chem., 248 (3-4), 179 (1969).

52. "The British Pharmacopoeia" p. 36, Her Majesty's Stationary Office, London (1973).

53. E. Berman and H.N. Wright, Arch. Ind. Hyg. Occ. Med., 8 (6), 518 (1953).

54. A.R. Rogers, J. Pharm. Pharmac., 11(5), 291 (1959).

55. A.R. Rogers, J. Pharm. Pharmac., 16(6), 433 (1964).

56. A.S. Curry and H. Powell, Nature, 173, 1143 (1954).

57. H.V. Street, J. Foren. Sci., 2, 118 (1962); J. Pharm. Pharmac., 14, 56 (1962).

58. F. Machovicova, O. Mohelska and V. Parrak, Ceskosl. Farmac., 9(5), 243 (1960).

59. R.V. Smith, M.R. Cook and A.W. Stocklinski, J. Chromatog., 87(1), 294 (1973).

60. P.W. Erhardt, R.V. Smith, T.T. Sayther and J.E. Keiser, J. Chromatog., 116(1), 218 (1976).

61. R.V. Smith and A.W. Stocklinski, Anal. Chem., 47(8), 1321 (1975).

62. H. Watanabe, S. Nakano and N. Ogawa, J. Chromatog., 229(1) Biomed. Appl., 18, 95 (1982).

63. H.M. Maksoud, S.H. Kuttab, J.L. Neumeyer and P. Vouras, J. Chromatog., 274; Biomed. Appl., 25, 149 (1983).

64. J.R. Miller, J.W. Blake and T. Tobin, Res. Commun. Chem. in Pharmacol., 15, 447 (1976).

65. D.M. Baaske, J.E. Keiser and R.V. Smith, J. Chromatog., 140, 57 (1977).

66. H. Watanabe, S. Nakano and N. Ogawa, Biomed. Mass Spectram., 7, 160 (1980).

67. R. V. Smith, D.W. Humphrey, S. Szeinbach and J.C. Glade, Anal. Lett., Part B, 12(4), 371 (1979).

68. R.V. Smith, J.C. Glade and D.W. Humphrey, J. Chromatog., 172, 520 (1979).

69. R.V. Smith, A.E. Klein, A.M. Clark and D.W. Humphrey, J. Chromatog., 179(1), 195 (1979).

70. D.W. Humphrey, M.E. Goldman, R.E. Wilcov, C.K. Erickson and R.V. Smith, Microchem. J., 25(2), 186 (1980).

71. R.V. Smith and D.W. Humphrey, Anal. Lett., Part B, 14(8), 601 (1981).

72. R.V. Smith and M.R. De Moreno, J. Chromatog., 274; Biomed. Appl., 25, 376 (1983).

73. R.K. Yang, J.Y.K. Hsieh, K.S. Kendler and K.L. Davis, J. Liq. Chromatog., 7(1), 191 (1984).

74. G. Bianchi and M. Landi, J. Chromatog., 338, Biomed. Appl., 39(1), 230 (1985).

75. I. Jane, H. Mckinnon and R.J. Flanagan, J. Chromatog., 323, 191 (1985).

76. A.M. Nour, A.A. Saleh, A.E.M. Habib, N. Hamad and A.F. Shalaby, Isotopenproxis, 8(7), 274 (1972).

77. J.W. Kebabian, J. Neurochem., 30, 1143 (1978).

78. N.V. Cherepanova, Prod. Khim. Prom. Sti., 6, 44 (1980).

****

ANALYTICAL PROFILE OF ASTEMIZOLE

ABDULRAHMAN M. AL-OBAID

and

MOHAMMAD SALEEM MIAN

Pharmaceutical Chemistry Department,
College of Pharmacy, King Saud University,
P.O. Box 2457, Riyadh-11451,
Saudi Arabia.

Contents

1. Introduction (1, 2)

Astemizole (Hismanal ®) is a long-acting, non-sedating antihistamine which has a major clinical implications for the treatment of allergic diseases such as hay fever, perennial allergic rhinitis, and allergic skin conditions in adults as well as in children. Hismanal ® which was developed recently by Janssen Pharmaceutica in their Belgain Research Laboratories, has a unique properties comparing it to standard antihistamines such as chlorpheniramine. Pharmacologically, hismanal ® is a highly selective and a potent H_1-histamine receptor antagonist exhibiting little blood brain barrier penetration and thus has no central nervous system (CNS) effects. High efficacy, long duration of action permitting once-daily dosage, freedom of drug interactions, freedom of tachyphylaxis and superiority to reference antihistamines are exceptional features of hismanal ®. Chemically, hismanal ® is an aminobenzimidazole derivative and it has no structural resemblance to other known antihistamines such as chlorpheniramine, terfenadine, clemastine, ketotifen, azatadine, etc. The discovery of hismanal ® as an ideal treatment of allergic diseases and the finding of the many advantages of this drug over other antihistamines have long been sought in the field of the antihistamines and thus representing an important advance in pharmaceutical research.

2. Description (3)

2.1 Nomenclature

2.1.1 Chemical Names

2-(p-Fluorobenzyl)-2-[(1-(p-methoxyphenethyl)-4-piperidylamino]benzimidazole.

1-[4-Fluorophenyl)methyl]-N-[1-[2-(4-methoxyphenyl]ethyl]-4-piperidinyl]-1H-benzimidazole-2-amine.

2.1.2 Generic Names

Astemizole.

2.1.3 Trade Names

Hismanal ® (Janssen) and Alermizole (Sintyal, Buenons Airs, Argentina).

2.2 Formulae

2.2.1 Empirical

$C_{28}H_{31}FN_4O$

2.2.2 Structural

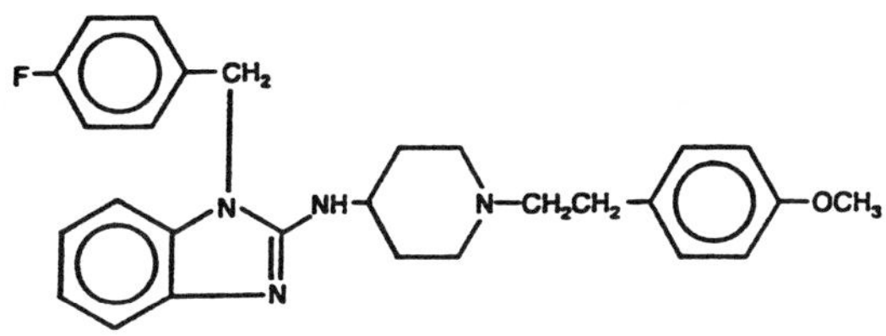

2.2.3 CAS Registry Number

68844-77-9.

2.3 Molecular Weight

458.25

2.4 Elemental Composition

C, 73.32%; H, 6.82%; F, 4.15%; N, 12.22%; O, 3.49%.

2.5 Appearance, Color and Odor

White fluffy odorless powder.

2.6 The Three-Dimensional Structure (4)

The three-dimensional structure of astemizole was determined by X-ray crystalography. The basic aliphatic nitrogen atom of astemizole is incorporated into a piperidine ring, and the aromatic rings are linked by a guanidino function, thus increasing the distance between the aliphatic nitrogen atom and the center of the X group to 5.35 Å. Fig. 1 represents the three-dimensional structure of astemizole.

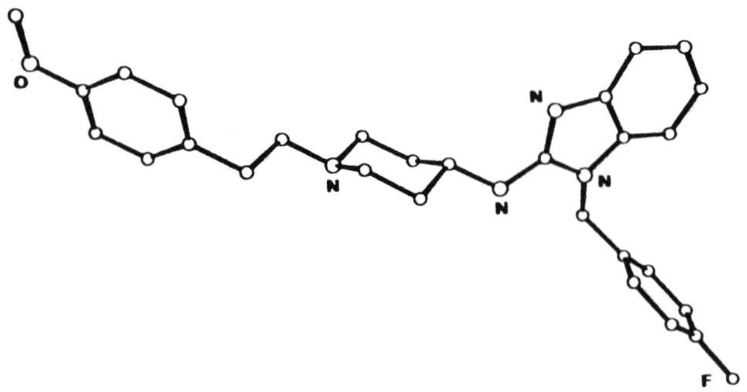

Fig. 1: The Three-Dimensional Structure of Astemizole.

3. Physical Properties

3.1 Melting Range

Astemizole melts at 149.1°. The base melts at 174.9-176.8°C.

3.2 Solubility

Astemizole is freely soluble in organic solvents e.g. chloroform. It is practically insoluble in water.

3.3 Storage and Stability

Astemizole tablets should be stored in a tightly closed container in a cool, dry place at room temperature (15° to 30°C) and away from light.

3.4 Microcrystal Formation

Different types of crystals were formed with different reagents after dissolving astemizole in trifluroacetic acid and nitric acid. The crystals so formed were microscopically examined. Shapes of the crystals are presented in Fig. 2.

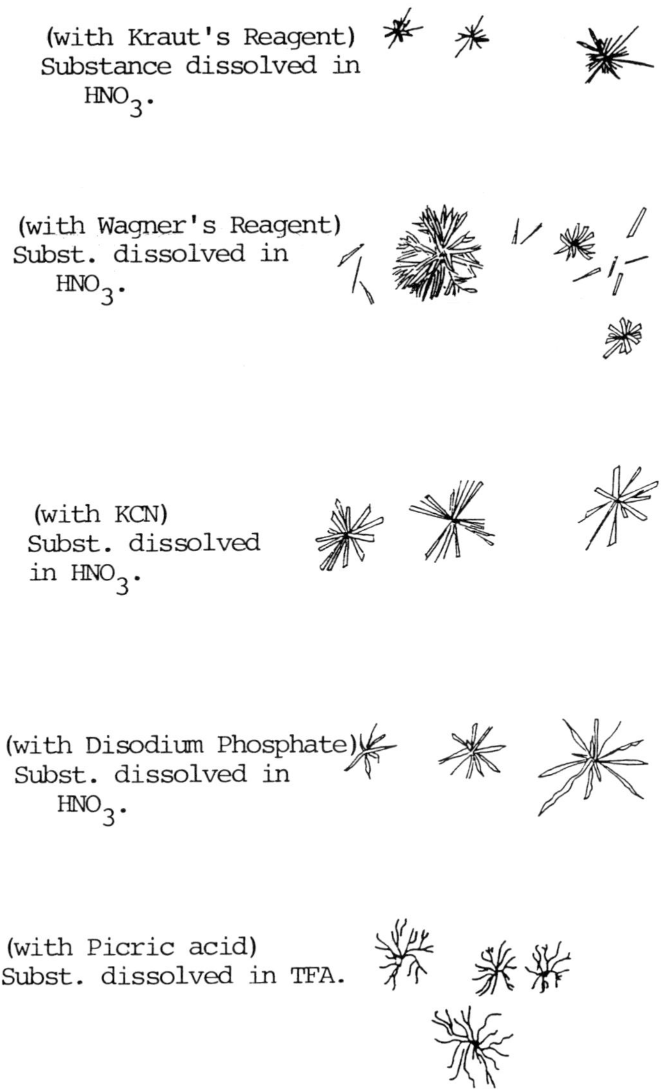

Fig. 2: Microcrystals of Astemizole with different reagents.

3.5 X-Ray Powder Diffraction*

The x-ray diffraction pattern of astemizole was determined (Fig. 3) using Philips full automated x-ray diffraction spectrogoniometer equipped with PW 1730/10 generator. Radiation was provided by copper target (cu anode 2000 W) and high intensity x-ray tube operated at 40 KV and 35 MA. The monochromator was a curved single crystal one (PW 1752/00). Divergance slit and the receiving slit were 1 and 0.1° respectively. The scanning speed of the goniometer used as 0.02-2 θ per second. The instrument is combined with Philips PM 8210 printing recorder with both analogue recorder and digital printer. The goniometer was aligned using silicon sample before use. Values of scattering angle (2θ), interplanner distance (dÅ) and % relative intensity (I/I_o X 100) are listed in Table 1.

Table 1: X-ray diffraction pattern of Astemizole

2θ	dÅ	I/I_o	2θ	dÅ	I/I_o
5.157	17.1373	35.957	24.754	3.5966	16.19
5.449	16.2194	32.629	25.380	3.5092	29.764
7.162	12.3417	12.756	25.970	3.4309	20.027
8.414	10.5081	16.669	26.911	3.3130	25.096
9.825	9.0026	44.338	27.495	3.2440	15.929
12.998	6.8112	13.341	28.043	3.1817	15.111
14.475	6.1192	15.744	28.696	3.1108	8.380
16.750	5.4545	15.082	29.805	2.9975	6.547
16.918	5.2407	11.754	30.212	2.9581	9.505
18.077	4.9072	7.795	30.942	2.8900	8.642
18.865	4.7039	44.245	32.157	2.7835	9.552
19.141	4.5657	54.613	32.765	2.7332	7.980
19.740	4.4973	87.382	33.084	2.7076	9.120
20.642	4.3028	17.747	33.757	2.6552	9.890
21.296	4.1721	100.000	34.685	2.5862	9.628
22.036	4.0336	26.498	34.936	2.5682	10.106
22.859	3.8902	14.419	35.294	2.5429	7.687
24.844	3.5838	10.429	36.162	2.4839	5.992

I/I_o = relative intensity (based on hightest intensity as 100.

dÅ = interplanner distance, 2θ = scattering angle

*Al-Obaid, A.M. and Mian, M.S. Unpublished data. (1990).

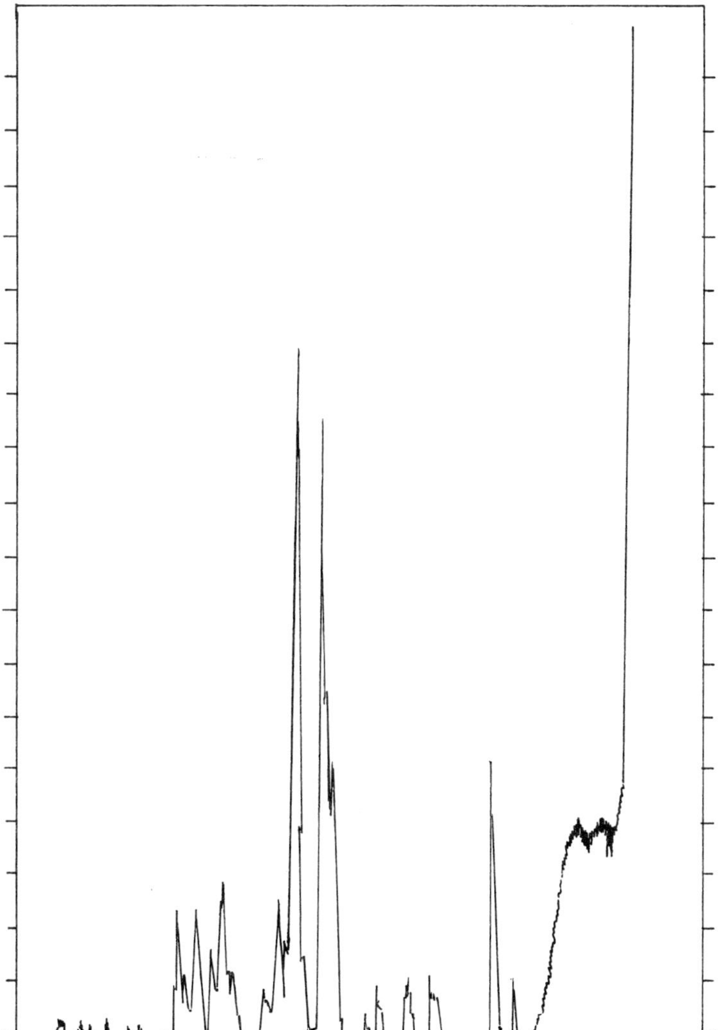

Fig. 3: X-ray diffraction Pattern of Astemizole.

3.6 Differential Scanning Calorimetry (DSC)

The differential scanning calorimetry (DSC) thermal curve for astemizole is given in Fig. 4. The scanning has been run at a rate of 10°C. min^{-1} from 150 to 210°C. Astemizole melts at 175.5°C, the H-value is 50.6 KJ/mole for 100 mole % purity. A DuPont TA-9900 Thermal Analyzer attached to a DuPont Data Unit were used for the DSC-scanning.

3.7 Spectral Poroperties

3.7.1 Ultraviolet Spectrum

The UV spectrum of astemizole in ethanol and 0.1N HCl were scanned from 200 to 400 nm (Figs. 5a and 5b) using LKB 4054 UV/Vis spectrophotometer. Astemizole exhibited the following UV data (Table 2).

Table 2. UV data of astemizole.

λ_{max} nm	Absorbance	Molar absorptivity (ϵ)
In ethanol		
219	2.973	27250.229
249	0.707	6480.293
286	0.942	8634.280
In 0.1N HCl		
209	0.631	57889.908
277	0.197	18073.394

3.7.2 Infrared Spectrum

The IR spectrum of astemizole as KBr disk was recorded on a Perkin Elmer 1310 Infrared Spectrophotometer. Fig. 6 shows the infrared absorption spectrum of astemizole. The structural assignment of astemizole have been correlated with the following frequencies (Table 3) (5).

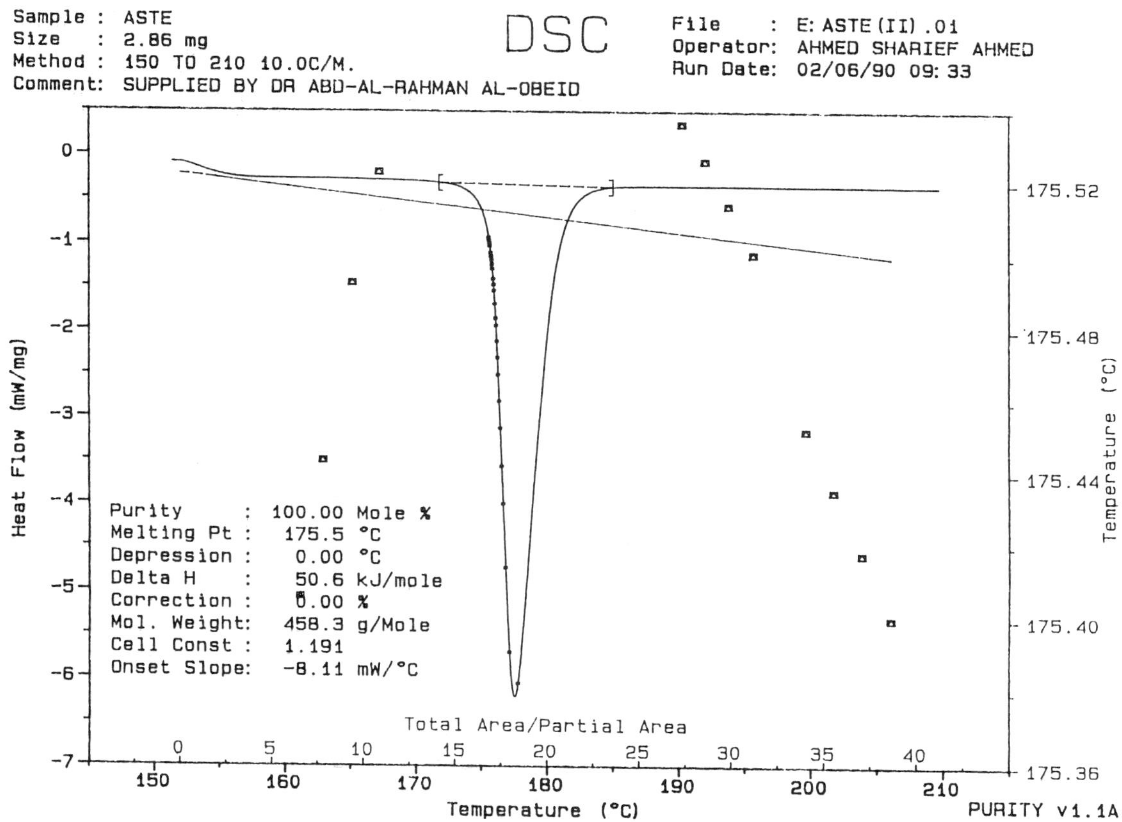

Fig. 4: DSC curve of Astemizole.

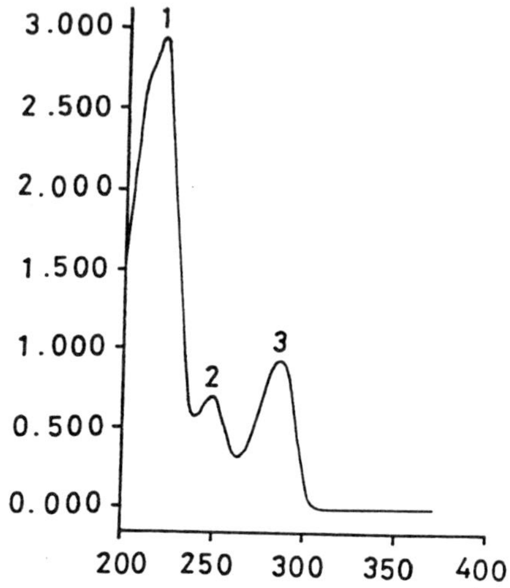

Fig. 5a: UV Spectrum of Astemizole in Ethanol. (50ugm/ml).

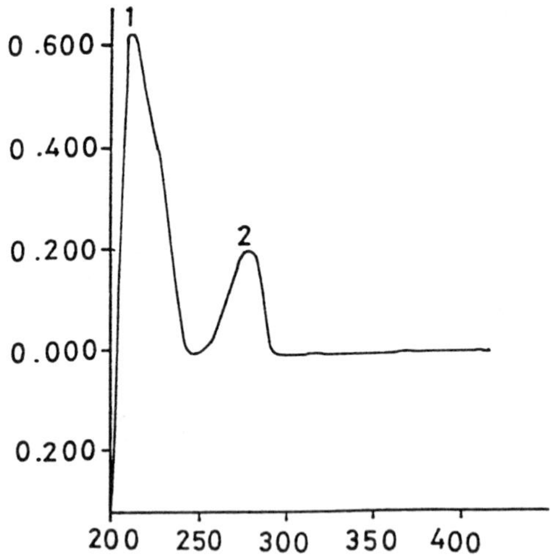

Fig. 5b: UV Spectrum of Astemizole in 0.1N HCl. (5ugm/ml).

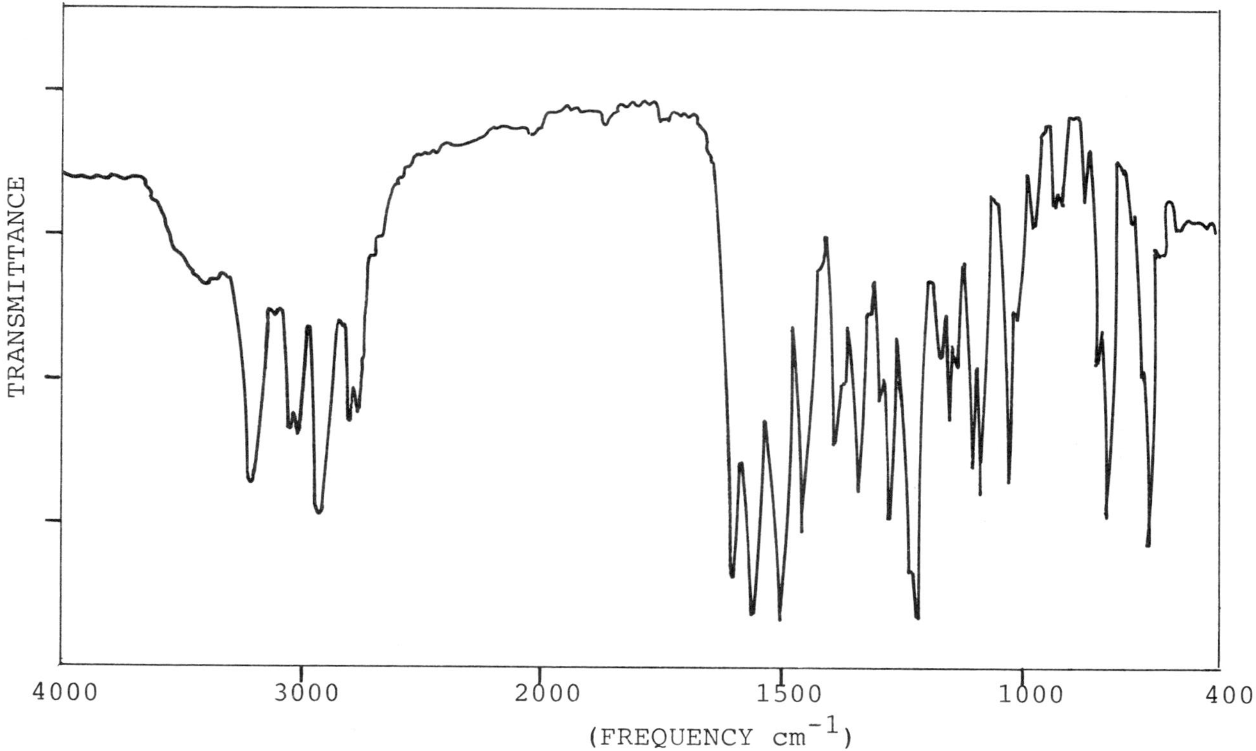

Fig. 6: IR Spectrum of Astemizole (KBr disc).

Table 3. IR characteristics of Astemizole

Frequencies and relative intensities of absorption bands	Approximate description of vibrational modes
3219 m, 15269 m	NH stretch & planar bend
3116 w, 3062 m, 3029 m	methoxyphenyl CH stretch
3062 m, 2029 m	benzimidazole CH stretch
3062 m, 3029 m	fluorophenyl CH stretch
3010 sh, 2830 w, 1466 m, 1445 sh, 1435 sh	methyl stretch & bends
2946 m, 2925 m, 2880 sh, 2849 w, 2830 w, 2801 m	Piperidine methylene stretch
2946 m	Ethyl methylene stretch
1604 m, 1514 s, 1466 m, 1222 m	benzimidazole ring stretch
1604 m, 1569 m, 1514 s, 1435 sh, 1299 m	methoxyphenyl ring stretch
1604 m, 1514 s, 1419 w, 1299 m	fluorophenyl ring stretch
1466 m, 1445 sh, 1435 sh	piperidine methylene bend
1395 m, 1377 m, 1344 m, 1323 w	piperidine methylene wag
1360 w, 1281 m, 1243 m, 1168 sh	piperidine methylene twist
1323 w, 1175 m, 1108 m, 1013 m	methoxypheneyl CH planar bend

Continued /...

Continued (Table 3) ...

1323 w, 1154 m, 110 m, 1013 m	fluorophenyl CH planar bend
1281 m, 1141 m, 1089 m, 1013 m	benzimidazole CH planar bend
1281 m, 1243 m	ethyl methylene twist
1243 m, 413 m	methoxyphenyl CO stretch & bend
1222 m, 384 w, 291 sh	methoxyphenyl CC stretch & bends
1222 m, 434 m	fluorophenyl CF stretch & bend
1222 m, 340 w, 314 w	fluorophenyl CC stretch & bends
1188 w	ethyl metylene twist
1175 m, 847 m	methyl rock
1141 m, 1120 w, 1033 m, 921 m, 873 m, 825 m	piperidine ring stretch
1089 m, 984 m	ethyl skeletal stretch
1050 sh, 1013 m, 976 m, 792 w	piperidine methylene rock
1033 m, 340 w	methoxy CO stretch & bend
1033 m	benzimidazole CN stretch
952 w, 933 m, 847 m, 825 m	methylphenyl CH spatial bend
921 m, 894 w, 847 m, 743 m	benzimidazole CH spatial bend
847 m, 717 w, 515 m, 448 w	fluorophenyl x-sensitive

Continued ...

Continued (Table 3) ...

792 w, 756 m, 579 m, 515 m	methoxyphenyl x-sensitive
774 m, 685 w, 579 m	benzimidazole x-sensitive
774 m	ethyl methylene rock
756 m, 619 w, 486 m	benzimidazole ring planar bend
717 w, 413 m	methoxyphenyl ring spatial bend
717 w, 593 m, 434 m	benzimidazole ring spatial bend
685 w, 413 m	fluorophenyl ring spatial bend
629 w	methoxy- & fluorophenyl ring planar bend.
549 m, 413 m, 384 w	piperidine ring planar bend
523 w	ethyl skeletal bend
448 w, 314 w	piperidine ring spatial bend

(m = medium, w = weak, sh = shoulder and s = strong)

3.7.3 Nuclear Magnetic Resonance Spectra

3.7.3.1 ^{1}H-NMR Spectrum

The ^{1}H-NMR spectra of astemizole in $CDCl_3$ (Fig. 7) and in DMSO-d_6 (Fig. 8) were recorded on a Varian XL200 MHz NMR spectrometer using TMS as an internal reference in both. The following structural assignments have been made (Table 4) (5).

Table 4: PMR characteristics of astemizole

Structure.

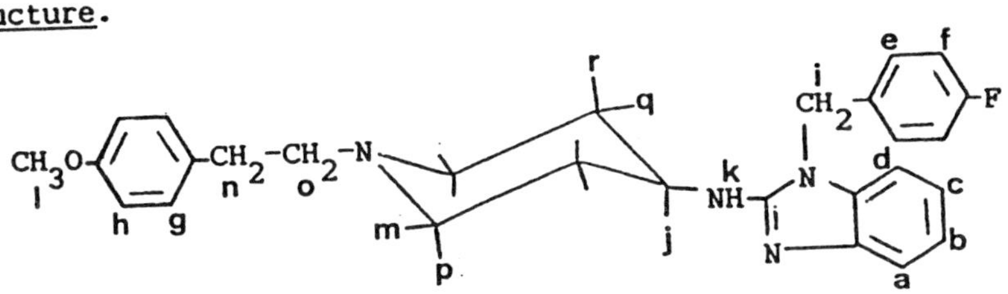

$C_{28}H_{31}FN_4O$

Protons	δ (ppm)	Relative surface	Multiplicity
a	7.52	1	multiplet
b, c, d) e, f)	6.95-7.20	9	multiplet
h	6.82	2	multiplet
i	5.03	2	singlet
j	3.95	1	multiplet
k	3.82	1	doublet
l	3.77	3	singlet
m) n, o)	2.45-2.90	6	multiplet
p, q	2.0-2.30	4	multiplet
r	1.44	2	multiplet

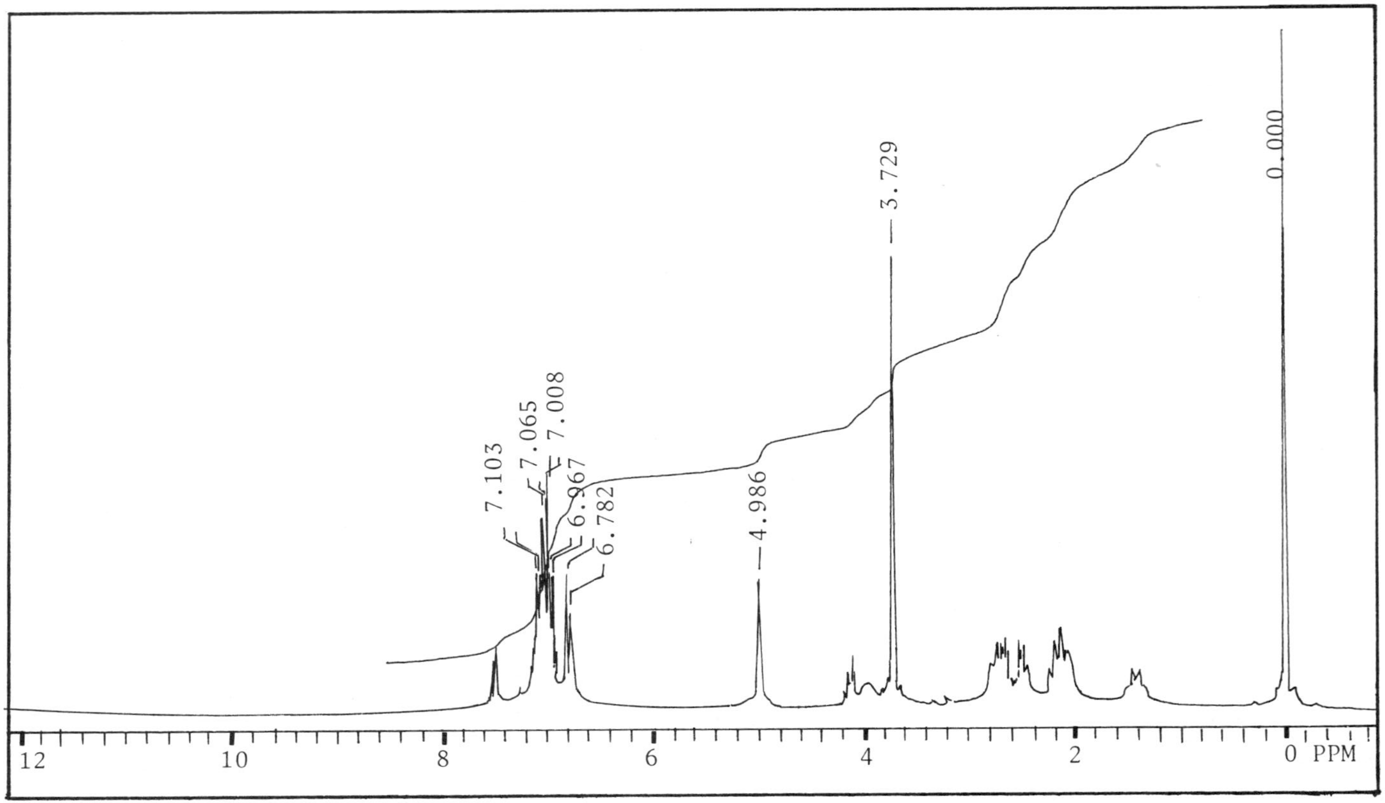

Fig. 7: ^{1}H-NMR spectrum of astemizole in $CDCl_3$.

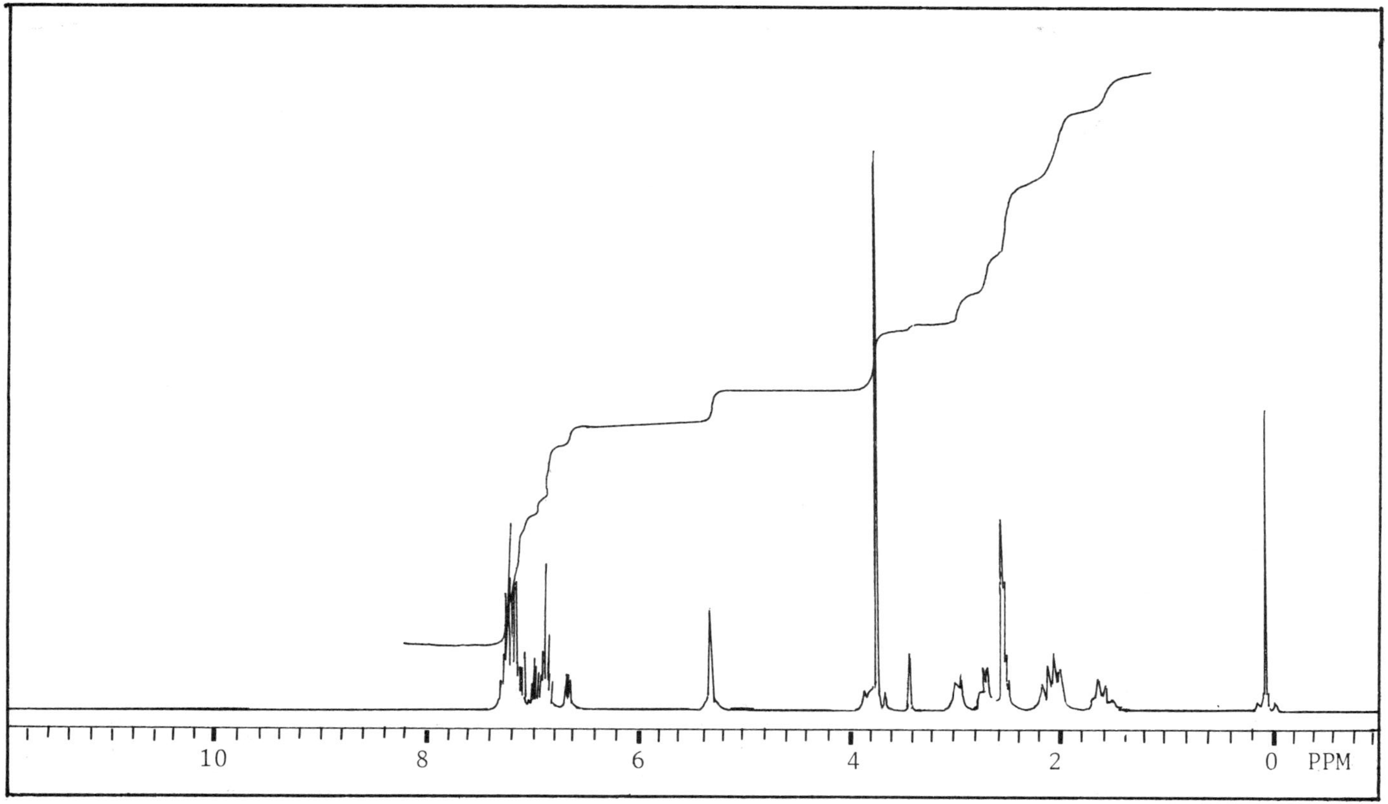

Fig. 8: ^{1}H-NMR spectrum of astemizole in DMSO-d_6,

3.7.3.2 ^{13}C-NMR Spectrum

The ^{13}C-NMR spectrum of astemizole in $CDCl_3$ using TMS as an internal reference was recorded on a Varian XL 200 MHz NMR spectrometer and is presented in Fig. 9.

3.7.4 Mass Spectrum

The mass spectrum of astemizole obtained by electron-impact ionization (Fig. 10) was recorded on a Finnigan MAT 90 mass spectrometer.

The spectrum was scanned from 50 to 500 a.m.a. Electron energy was 70 eV. Emission current 1 mA and ion source pressure 10^{-6} Torr. The spectrum (Fig. 10) shows a molecular ion M^+ at m/z 96 with a relative intensity 100%. The most prominent fragments and their relative intensities are listed in Table 5. (5).

Table 5: The mass fragments of astemizole

m/z	Relative intensity %	m/z	Relative intensity %
458	3.3	121	4.99
337	46.6	109	30%
294	7.49	96	100
241	7.47	94	8.33
216	3	78	2.3
135	7	78	2.3

Mass fragmentation behavior of astemizole is presented in Fig. 11 (5).

4. Synthesis (4,6)

Astemizole and astemizole-like compounds were synthesized in the laboratories of Janssen

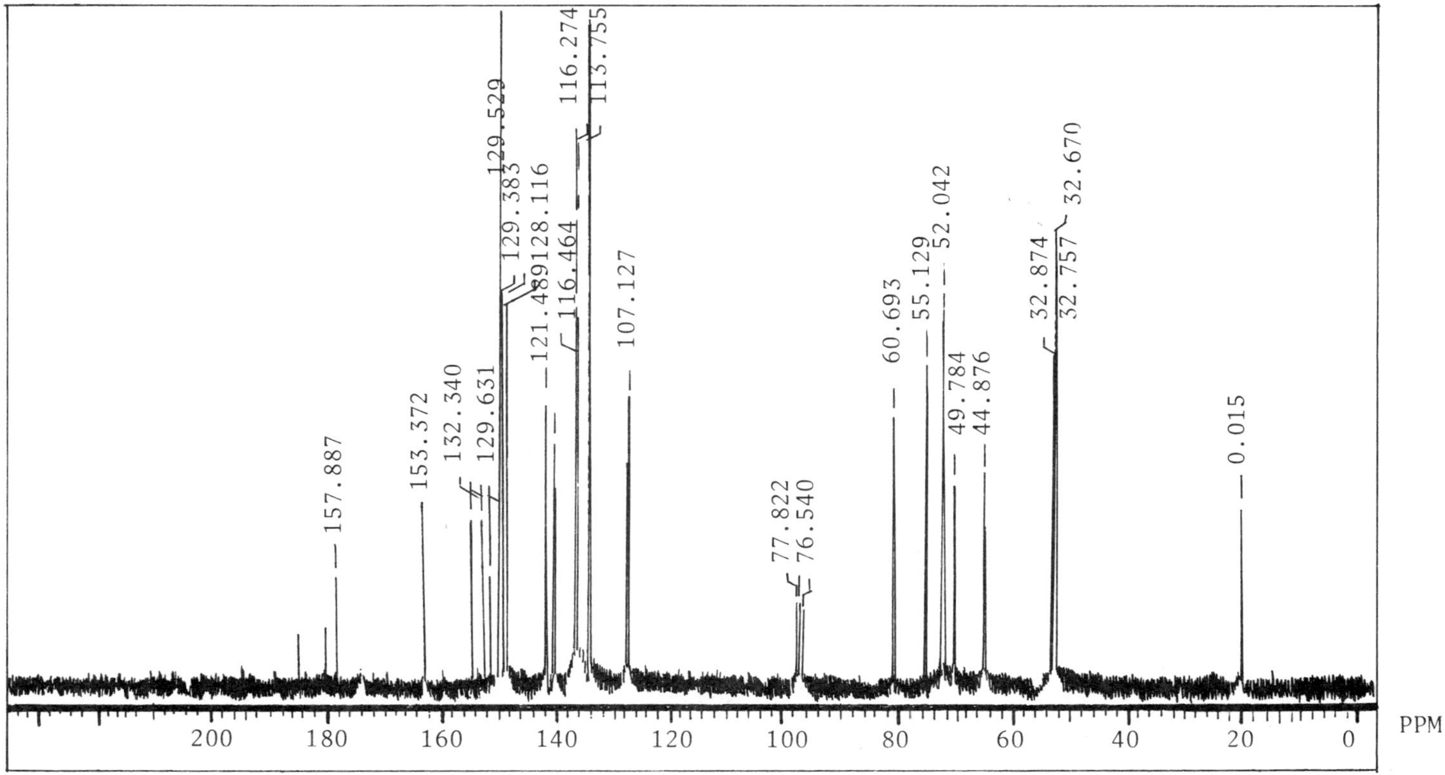

Fig. 9: ^{13}C-NMR spectrum of astemizole in $CDCl_3$.

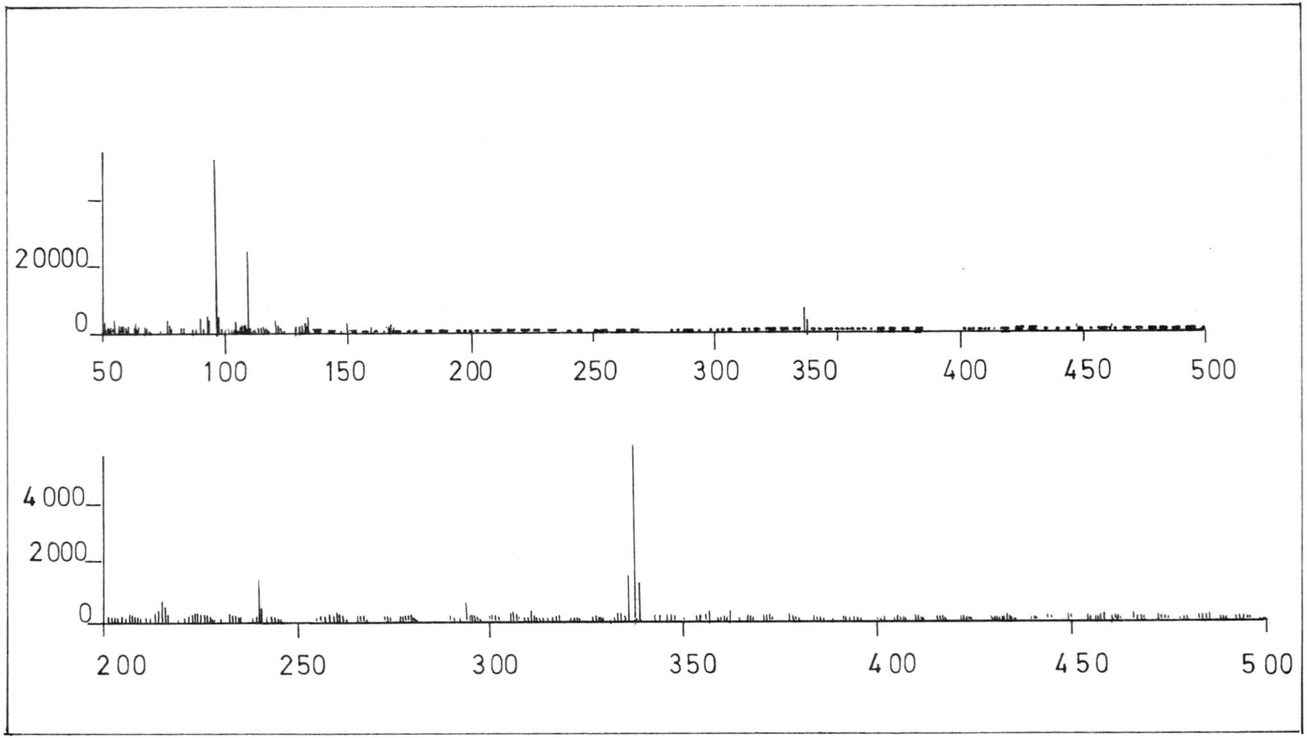

Fig. 10: Mass spectrum of Astemizole.

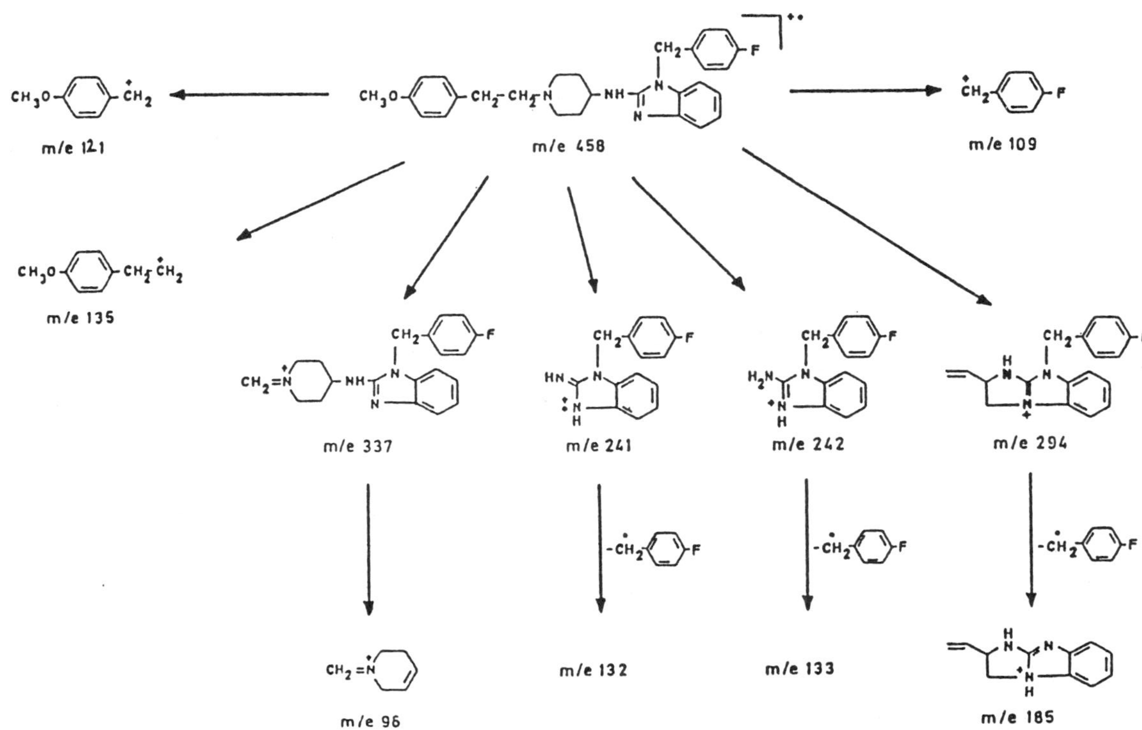

Fig. 11: Mass fragmentation behaviour of Astemizole.

Pharmaceutica, Beerse, Belgium. Two major synthetic pathways were used. Scheme 1 outlines the synthesis of astemizole and related compounds. According to pathway I the addition of protected 4-piperidinl-isothiocyanate (prepared from the condensation of ethyl 4-amino-1-piperidinecarboxylate with carbon disulfide in the presence of sodium hydroxide) to o-phenylenediamine (reaction A) resulted in a thiourea derivative which was cyclodesulfurized (reaction B) to a 2-aminobenzimidazole derivative using HgO in the presence of powdered sulfur as one of the best cyclodesulfurizing agents. Reaction C indicates that the substituent on the endo-nitrogen of the benzimidazole ring was introduced by alkylating 4-(1H-benzimidazol-2-y/amino)-1-piperidinecarboxylate with 1-(chloromethyl)-4-fluorobenzene in the presence of sodium carbonate and N,N-dimethylformamide. Deprotection of the piperidine nitrogen was achieved via reaction D by using a mixture of hydrobromic acid solution 48% in glacial acetic acid to yield 1-[(4-fluorophenyl)methyl]-N-(4-piperidinyl)-1H-benzimidazole-2-amine. Alkylation of the latter compound (reaction E) with p-methoxy-2-bromoethylbenzene in $NaCO_3$/KI using DMF as a solvent gave rise to astemizole.

For the synthesis of many astemizole-like compounds, pathway II was followed. Amines with the desired R group were coupled (reaction F) to give nitroaminoaryl derivatives. Catalytic hydrogenation (reaction G) of these derivatives gave rise to the corresponding diamino-derivatives. The subsequent steps were similar to the reactions described for pathway I.

5. Methods of Analysis

5.1 Elemental Analysis

The elemental analysis of astemizole was done using Perkin Elmer 240 B Elemental analyser and is presented in Table 6.

Synthesis Scheme:
(Astemizole)

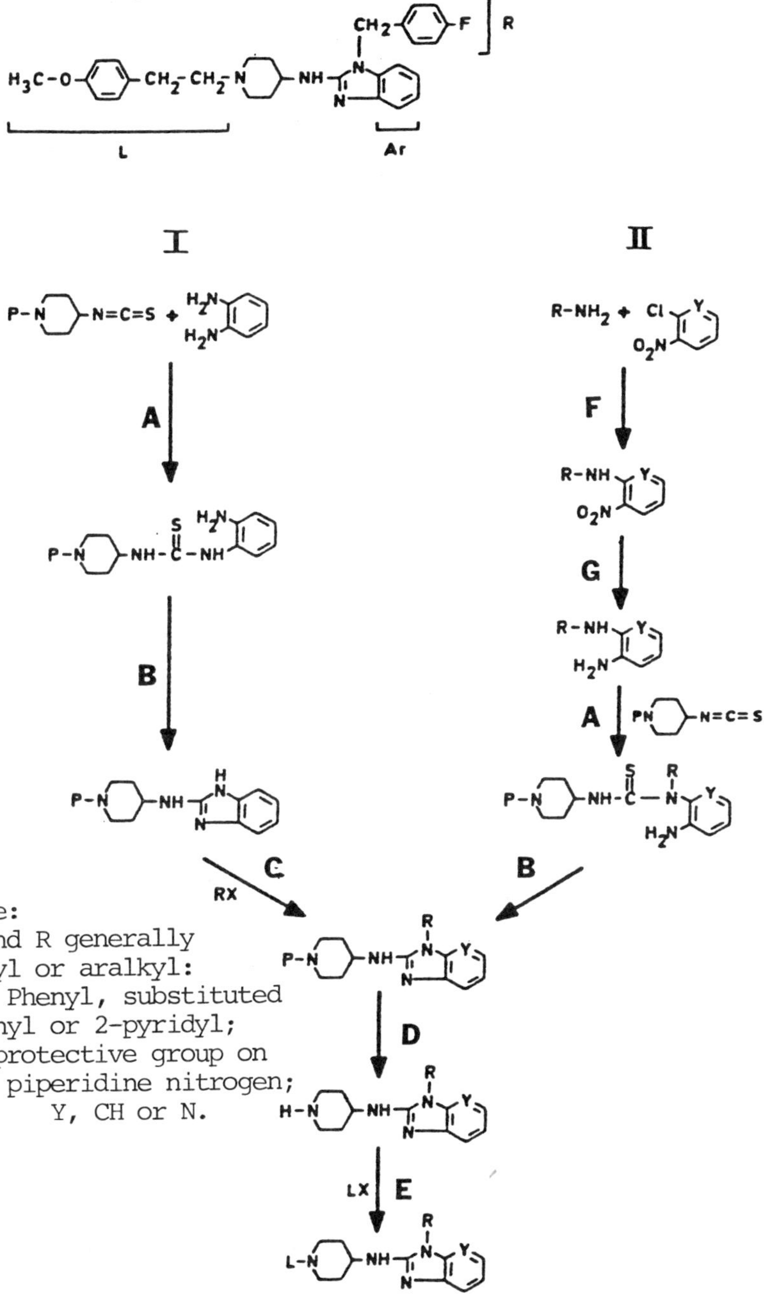

Note:
L and R generally alkyl or aralkyl:
Ar, Phenyl, substituted Phenyl or 2-pyridyl;
P, protective group on the piperidine nitrogen;
Y, CH or N.

Table 6. Elemental analysis of astemizole.

Element	% Expected	% Found
C	73.32%	73.88
H	6.82%	6.88
F	4.15%	ND
N	12.22%	12.76
O	3.49%	ND

ND = not determined.

5.2 Radioimmunoassays (7)

A sensitive and selective radioimmunoassay for astemizole and its metabolites in plasma was described. In this procedure, the N-(1-(4-(2-amino-ethoxy)phenylethyl)piperidine-4-yl) analog of astemizole, coupled to bovine serum albumin was used to raise an antiserum in rabbits. Plasma was treated with methanol and 0.05M phosphate buffer (pH 7.8) before extraction with heptane-isoamyl alcohol (98.5:1.5) for unchanged astemizole and 95:5 for astemizole and its hydroxy-metabolites. The extract (after washing with 0.1 M NaOH) was evaporated, a solution of the residue in methanol was extracted with heptane, and the methane phase was evaporated for use in the Radioimmunoassay (RIA). The diluted (1:600) antiserum, ^{3}H-astemizole and the extraction residue were incubated for 2 h at room temperature, bound and free astemizole were separated by absorption of the free ligand on dextran-coated charcoal, and the activity in the supernatant solution was measured by scintillation counting. Cross-reactivity with demethyl-astemizole was appreciable, allowing determination of this important metabolite at concentration less than 0.05 ng mL^{-1} in plasma. For unchanged astemizole, the detection limit was 0.05 ng ml^{-1}, within- and between- run reproducibility was 7.0 and 6.4%, respectively and correlation with a HPLC method was satisfactory (r = 0.85).

5.3 High-Performance Liquid Chromatography (HPLC, 8)

A high-performance liquid chromatography procedure has been developed for the determination of astemizole and its primary metabolite in plasma and animal tissues. Thus, astemizole, its demethylated metabolite, and its p-ethoxy-analog (internal standard) were extracted from plasma or aqueous tissue homogenate into heptane-isoamyl alcohol (19:1) and re-extracted in presence of conc. aqueous NH_3 into the organic mixture.

The liquid chromatograph used consisted of a Spectra-Physics SP 8750 organizer module fitted with a 50-µl loop Valco valve injector, and an SP 8300 selectable-wavelength UV-visible detector operating at 254 nm. The separations were achieved using a reversed-phase column (15 cm X 2.1 mm I.D.) packed with 5-µm particle-sized RSiL C18HL (Alltech Europe) by the balanced density procedure by means of an air-driven liquid pump (Haskel). The samples were eluted with acetonitrile-water (50:50) at a constant flow rate of 0.6 ml/min. To suppress the ionization of the basic functions of the investigated compounds, 0.05% diethylamine was added to the solvent system. Area integrations, peak height measurements, calculations and plotting of the chromatograms were carried out by a Spectra-Physics Model SP 4100 computing integrator. The detection limits for astemizole and its demethylated metabolite were 1 ng/ml of plasma and 5 ng/g of tissue and extraction recoveries were sufficiently high (71-84%). The method was applied to plasma and tissue samples from dogs after repeated oral administration, and to plasma samples from a volunteer taking a 300 mg oral dose of the drug.

5.4 Determination in Pharmaceutical Dosage Forms (9)

Astemizole was determined in pharmaceuticals by HPLC on an Inert sil ODS-2 column with methanol - acetonitrile (50:50) as the mobile phase and variable wavelength detection (195-700 nm). Trimethoprim was used as the internal standard. The recovery and relative standard deviation were 100.44-102.09 and 0.4471-1.4331%, respectively.

6. Pharmacokinetics

The pharmacokinetics of astemizole have been extensively studied in animals, human volunteers, and patients. Radioimmunoassay methods were used for detection of the drug in such studies. The pharmacokinetic profile of astemizole features two components, the unchanged drug and its active metabolite, desmethylastemizole.

6.1 Onset and Duration (10-12)

The precise onset of action of astemizole in either loading or single doses have not been established. It does appear, however, that in the case of moderate allergic stimulus, results with astemizole are excellent. On the other hand, in the case of greater allergic stimulus, results from several studies indicated that astimizole has no different from placebo or other H_1 antagonists. In one controlled study of hay fever patients the onset of action of astemizole was reported slow with relief to allergic symptoms not occurring for two days (terfenadine produced allevatution of symptoms within 3 hours). Because of its delayed onset of action, astemizole is not as effective for treatment of actue allergic symptoms. The slower action of desmethylastemizole provides sustained antihistaminic protection.

6.2 Absorption (13,14)

Astemizole is subject to rapid and extensive absorption with plasma peak concentrations over a range of doses occurring between 1 and 4 hours and appeared to be linearly related to dose. Astemizole undergoes an extensive first-pass metabolism and a considerable tissue distribution. This is attributed to the finding that peak concentrations of astemizole plus metabolites were several fold greater than those of unchanged astemizole.

Regarding bioavailability of astemizole, it was found that peak plasma concentration, time to peak and area under the plasma concentration-time curve are not significantly different following administration of astemizole tablet and suspension. However, concomitant administration of oral astemizole and food significantly decreased the bioavailability of astemizole and its metabolites.

Steady-state plasma concentrations of unchanged astemizole were seen within 1 to 2 weeks following multi-dose administration of oral astemizole. However, following initiation of therapy, steady-state plasma concentrations of astemizole and its metabolites were seen 4 to 8 weeks. No further change in levels of astemizole nor drug accumulation were seen once the steady-state concentrations was reached. (15)

6.3 Distribution (13,14)

Astemizole has rapid and extensive distribution. The highest tissue concentrations of astemizole and its active metabolites were found in the pancreas, adrenal glands, liver, lung, salivary gland, kidney tissue and testes. The lowest tissue concentrations, on the other hand, were found in subcutaneous fat, skin, muscle, thymus, heart and brain. The plasma protein binding of astemizole was reported to be 96.7%. Following a single-oral administration of astemizole (300 mg, po), the appearent volume of distribution (V_d) was 250 l/kg.

6.4 Metabolism (16)

Astemizole undergoes extensive first-pass metabolism via three major metabolic pathways: oxidative O-demethylation, aromatic hydroxylation at the benzimidazole moiety, and oxidative N-dealkylation at the piperidine nitrogen. Hydroxylated metabolites resulting from the two former pathways were excreted as their glucuronides. The metabolic products of astemizole are desmethylastemizole, which has similar antihistaminic potency to the parent compound, norastemizole, and 6-hydroxydesmethylastemizole, which are less potent and shorter acting. The prolonged duration of therapeutic effect of astemizole may be attributed to the enterohepatic recirculation of desmethylastemizole and conjugated metabolites. Fig. 12 shows a schematic representation of major and minor metabolites of astemizole.

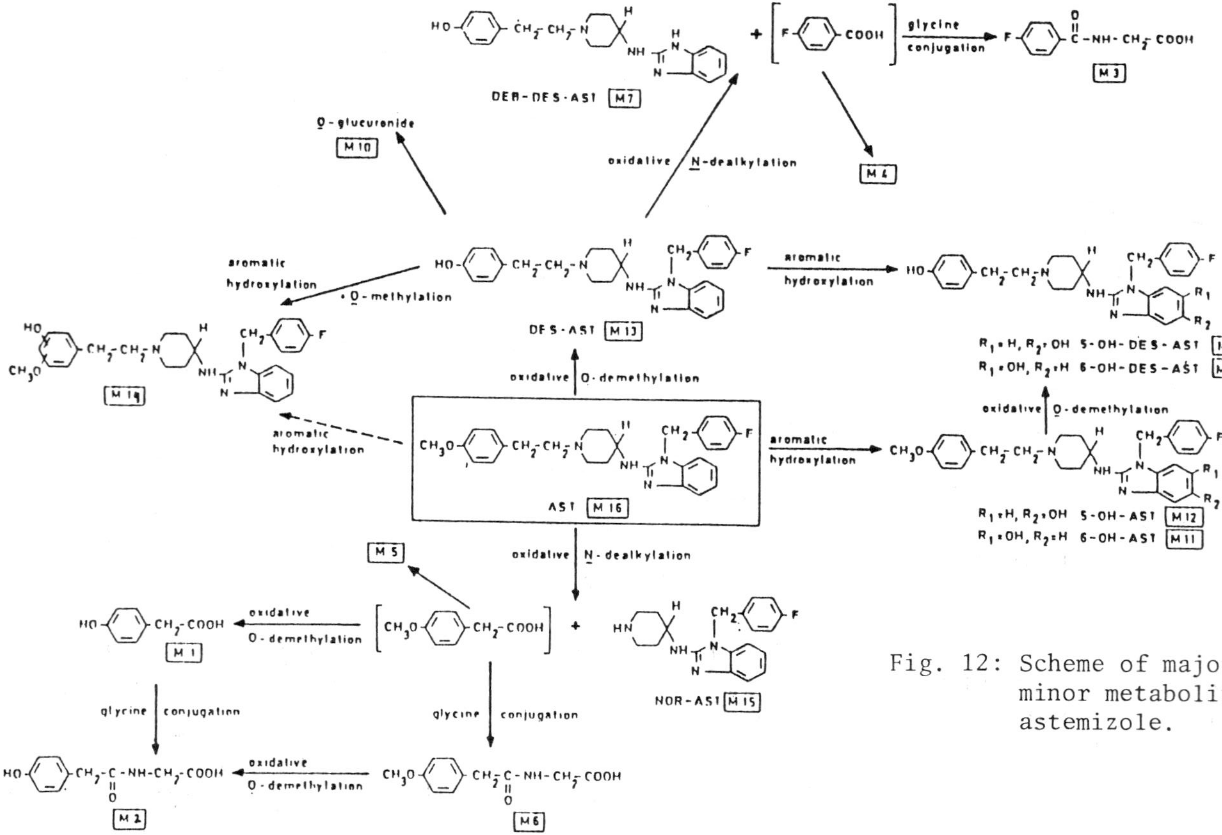

Fig. 12: Scheme of major and minor metabolites of astemizole.

6.5 Excretion (13)

Following oral administration, unchanged astemizole was not recovered in the urine or faeces. All astemizole metabolites are slowly excreted. About five to six percent of an administered dose is recovered in the urine within 14 days as glucuronides of desmethylastemizole and dorastemizole. Fifty-four to 73% of a dose of astemizole is excreted in faeces within 14 days. Each of the three metabolites accounted for 17 to 30% of the excretion.

6.6 Half-life (15)

Estimation of plasma concentrations half life of astemizole was difficult because could not be followed for long enough periods of time. Following a single 300 mg dose of astemizole, the half-life of astemizole is approximately 20 hours. Following a single doses of astemizole 10 mg, 20 mg, or 30 mg orally, the terminal half-life of unchanged astemizole is about 24 hours. on the other hand, the elimination half-lives of desmethylastemizole are 12 and 18 to 20 days following administration of a single 300 mg dose of astemizole and following long-term administration of 10 mg daily, respectively.

7. Cautions

7.1 Contraindication

Patients who are hypersensitive to astemizole or any of the inactive ingredients should not receive the drug.

7.2 Precautions (17)

Astemizole should be used cautiously in pregnant women since there are inadequate data to determine teratogenic effects (potential for harm to the developing fetus) of astemizole. In women of child-bearing potential, it is highly recommended that astemizole should not be used unless adequate contraceptive precautions are taken.

7.3 Side Effects (18)

The most frequent side effects of astemizole include central nervous system (CNS) sedation/depression weight gain, dry mouth and increased appetite. In contrast to reference antihistamines, astemizole does not show any anticholinergic effects in healthy volunteers. It should be pointed out that the occurence of these side effects, particularly sedation is quite similar to those seen in placebo-treated patients and lower than those seen in conventional antihistamines. In addition, it was reported that repeated doses of astemizole are free of psychoperformance effects.

7.4 Drug Interactions (19)

Concomitant administration of astemizole and other drugs has been reported to cause no drug-drug interactions. For example, astemizole does not alter the psychomotor effects of alcohol following acute alcohol intake. Astemizole also does not potentiate or counteract the CNS effects of diazepam when they were given concomitantly. It should be noted, however, that concomitant administration of astemizole and food significantly decreased the bioavailability of astemizole and its metabolites.

8. Clinical Application

8.1 Place in Therapy (20)

Astemizole, a chemically unique H_1-histamine receptor antagonist, is effective in treating both seasonal and perennial allergic rhinitis and appears to be as effective as other H_1-histamine receptor antagonists. The apparent lack of CNS depression and anticholinergic effects with astemizole makes it advantageous to use in patients who demonstrate these adverse effects with other more conventional antihistamines. Astemizole's long duration of action and its once daily dosage also make it advantageous. The only drawback to astemizole therapy at this time appears to be the possibility of increased appetite and weight following prolonged usage.

It was suggested that the major role of astemizole will be as prophylaxis in outpatients with chronic or recurrent allergic conditions (allergic rhinitis, chronic urticaria, allergic conjunctivitis), and recommend addition of the drug to the hospital formulary for institutions treating these outpatients. Due to its delayed onset of action, the drug is not a preferred agent for the treatment of acute allergic symptoms. Although cost information is not yet available, astemizole is more expensive than conventional antihistamine preparations. The drug may be limited primarily to patients with chronic allergic conditions who are intolerant of the side effects of the more conventional antihistamines.

8.2 Pharmacology (21,22)

In animals, pharmacological studies have shown that astemizole to be a histamine H_1-antagonist exhibiting little blood-brain barrier penetration. Because it has a higher affinity for lung histamine H_1-receptors than for cerebral histamine H_1-receptors, astemizole is devoid of central nervous system (CNS) effects. In addition, astemizole has no cholinergic nor β-adrenergic activity. Serotonin antagonism and affinity for α-adrenergic activity of astemizole were reported when higher doses are used.

Astemizole has a considerable safety margin. Large oral doses (2.5 mg/kg) of astemizole in conscious dogs did not affect EEG patterns, respiratory function or behaviour. A minor decrease in blood pressure was observed. There have been only two reports of overdose with astemizole in human. Tachyphylaxis has not been reported to date with astemizole.

8.3 Therapeutic Indications

Comparison studies have shown astemizole to be equal or superior to currently available antihistamines in relizing allergic symptoms of a variety of diseases. Some examples include perennial allergic rhinitis (14), seasonal allergic rhinitics (23), allergic conjunctivitis (14), chronic articaria (24), chronic idiopathic urticaria (25).

9. Dosage

The recommended oral maintenance dosage for adults and children 12 years of age and older is 10 milligrams once daily. To reduce the time to reach steady-state concentrations, a single dose of 30 milligrams may be administered on day 1, 20 milligrams on day 2 followed by the recommended 10 milligrams daily dose. All doses should be taken as on empty stomach, 1 hour before or 2 hours after meals.

When symptoms are severe, 30 milligrams daily can be given for up to 7 days followed by 10 milligrams daily.

In children, age 6 to 12 years, daily oral dose of astemizole 5 milligrams or one half the adult dose per day have been used. Children younger than 6 years of age should receive astemizle in dosage of 0.2 milligram/kilogram daily.

10. Manufacturer Information

Janssen
40 Kingsbridge Road
Piscatway, NJ 08854
USA.

11. Conclusion

Astemizole is a potent long acting H_1-receptor antagonist structurally dissimilar from other conventional antihistamines. A once-daily dosing with 10 mg of astemizole was found clinically free of side-effects and more effective than conventional antihistamine treatment. Astemizole is a long acting, with steady state levels of the drug and its metabolites not seen until 4 to 8 weeks following chronic therapy. Astemizole is devoid of any central activity and thus it should be utilized in place of other more conventional antihistamines when CNS effects of conventional antihistamines need to be avoided. Toxicity studies have shown astemizole to be very safe drug despite of its long duration of action. Astemizole is indicated for the treatment of perennial and seasonal allergic rhinitis, allergic conjunctivits and urticaria.

12. Acknowledgements

The authors are highly indebited and thankful to Janssen Pharmaceutica, Scientific Office Riyadh, Saudi Arabia, for providing the authentic sample of astemizole and extending a very warm cooperation and valuable help. The authors are also very much thankful to Mr. Tanvir A. Butt, Pharmaceutical Chemistry Department, College of Pharmacy, King Saud University for typing the manuscript.

13. References

1. Awouters F.H.L., Niemegeers CJE, Janssen PAJ. Arzneimittel-Forschung, 33, 381-388 (1983).

2. Van Waume J., *et al.* Archives Internationales de Pharmacodynamie et de Therapie, 251, 39-51 (1981).

3. Annual Drug Data Report 1981, vol. 3, p. 22, Ed. J.R. Prous J.R. Press International Publishers, Provena, 385-387, Barcelona, 25-Spain.

4. Janssens, *et al.*, Drug Development Research, 8, 27 (1986).

5. Private Communication, Janssen Pharmaceutica, Scientific Office, Riyadh, Saudi Arabia.

6. Janssens, *et al.*, Eur. Pat. Appli. 5318 corresp. to US Pat. 4,219,559 (1979, 1980) both to Janssen.

7. Woestenborghs, *et al.*, Drug Dev. Res., 8, 63-69 (1986).

8. Woestenborghs, *et al.*, J. Chromatogr. Biomed. Appli., 278, 359-366 (1983).

9. Sadana, G.S. and Potdar, A.J., Indian Drugs, 27(2), 140-2 (1989).

10. Weintraub, M. & Evans, P., Hosp. Formula, 22, 922 (1987).

11. Girard, *et al.*, J. Int. Med. Res., 13, 102-108 (1985).

12. Chapman, P.M. and Rawlins, M.D.; J. Clinical Pharmacology, 13, 593 (1982).

13. Richards, *et al.*, Drugs, 28, 38-61 (1984).

14. Product Information, Astemizole (Mismenal R. Janssen Pharmaceutica, Piscataway, N.J. (1989).

15. HeyKants, *et al.*, Drug Dev. Res., 8, 71-78 (1986).

16. Meuldermans, *et al.*, Drug Dev. Res., 8, 37-51 (1986).

17. Wihl, *et al.*, J. Allergy. Clin. Immunol., 75, 720-27 (1985).

18. Seppälä, T. and Savolaninen, K.; Current Therapeutic Research, 31, 638-644 (1982).

19. Moser, L., Phum, H. and Brückmann, M.; Medizinische Welt, 35, 296-299 (1984).

20. Callier, *et al.*, Curr. Ther. Res., 29, 24 (1981).

21. Krstenansky, P.M. and Chixton, R.J.; Drug Intell. Clin. Pharm., 21, 947-953 (1987).

22. Howarth, *et al.*, Br. J. Clin. Pharmacol., 18, 1-8 (1984).

23. Malmberg, *et al.*, Allergy, 38, 227-231 (1983).

24. Bateman, *et al.*, Eur. J. Clin. Pharmacol., 25, 547-551 (1983).

25. Kailasam, V. and Mathews, K.P., J. Am. Acad. Dermatol., 16, 797-804 (1987).

ANALYTICAL PROFILE OF CEFUROXIME SODIUM

Timothy J. Wozniak and John R. Hicks

Lilly Research Laboratories
Eli Lilly and Company
Indianapolis, Indiana 46285

TABLE OF CONTENTS

1. DESCRIPTION

1.1 Name, Formula and Molecular Weight

Cefuroxime sodium is marketed by Eli Lilly and Company under the trade name of Kefurox® and by Glaxo Limited under the trade name of Zinacef®.

Chemically it is known as either (6R, 7R)-7-[2-(2-furyl) glyoxylamido]-3-(hydroxymethyl)-8-oxo-5-thia-1-azabicyclo-[4.2.0]oct-2-ene-2-carboxylate 72-(Z)-(O-methyloxime) carbamate, sodium salt or (6R, 7R)-3-carbamoyl oxymethyl-7-[(Z)-2-(2-furyl)-2-(methoxyimino) acetamido]-ceph-3-em-4-carboxylic acid, sodium salt.

The CAS Registry Number is 56238-63-2. Cefuroxime sodium has an empirical formula of $C_{16}H_{15}N_4O_8S$ Na and a molecular weight of 446.37

Structure:

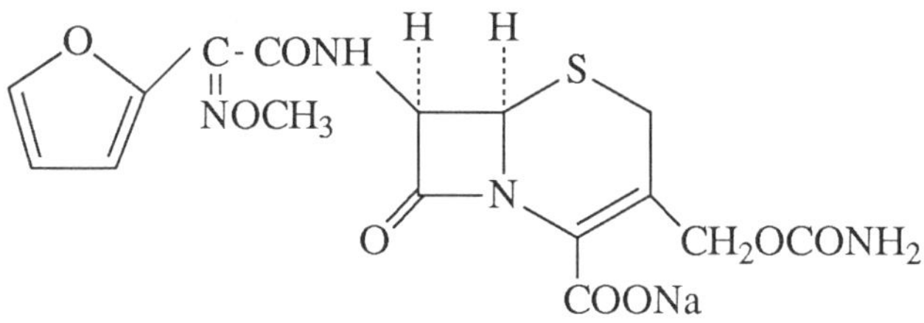

1.2 Appearance, Color and Odor

Cefuroxime sodium is an off-white to white amorphous powder. In a one percent solution in water, cefuroxime sodium is a clear to slightly yellow color. Solutions of a higher concentration exhibit a stronger yellow color.

1.3 History

Cefuroxime sodium is an injectable second generation, semisynthetic cephalosporin antibiotic with excellent activity *in vitro*, enhanced stability to many enterobacterial β-lactamases and favorable pharmacokinetic properties (1-3). The major structural difference between cefuroxime and other cephalosporins is at the methoxyimino group at position 7 on the β-lactam ring and a carbamate group at position 3 on the ring. The methoxyimino group results in stability against hydrolysis by many β-lactamases and the carbamate group results in metabolic stability. It is effective for treating meningitis, lower respiratory tract infections, urinary tract infections, gonorrhea, bone and joint infections, and is approved for surgical prophylaxis. Like other first and second generation cephalosporin antibiotics, cefuroxime sodium is absorbed poorly by the oral route. Cefuroxime sodium is administered as a 750 or 1500 mg parenteral dose.

2. SYNTHESIS

The reaction of diphenylmethyl (6R, 7R)-3-hydroxymethyl-7-(2-thienyl-acetamido)-ceph-3-em-4-carboylate (I) with trichloroacetyl isocyanate in anhydrous acetone gives the corresponding N-trichloroacetylcarbamate (II). Reaction of (I) with phosphorous pentachloride and pyridine in methylene chloride, followed by p-toluenesulfonic acid affords the diphenylmethyl 7-amino-3-carbamoyloxy-methylceph-3-em-4-carboxylate p-toluenesulfonic acid salt (III). The hydrolysis of (III) with trifluoroacetic acid in anisole yields the corresponding acid (IV), which is finally condensed with 2-(2-furyl)-2-methoxyiminoacetic acid (V) by means of phosphorous pentachloride and N, N-dimethylacetamide in acetonitrile containing triethylamine to yield cefuroxime acid (VI). Conversion of the free acid (VI) to cefuroxime sodium is accomplished by reaction of a sterile solution of VI with a sterile-filtered sodium lactate (VII) solution at 50°C (Figure 1). Addition of a seed slurry results in a controlled crystallization process. The resulting solution is centrifuged and the sterile wet cake is dried at 50°C for 24 hours (4,5).

3. PHYSICAL PROPERTIES

3.1 Infrared Spectroscopy

The infrared spectrum for cefuroxime sodium as a potassium bromide pellet is illustrated in Figure 2. The spectrum was recorded on a Nicolet Model 5SXC Fourier Transform infrared spectrophotometer. The major adsorption bands for the infrared frequencies and the corresponding assignments are listed in Table I.

3.2 Nuclear Magnetic Resonance Spectroscopy

The 300 MHz ^1H spectrum of cefuroxime sodium (5 mg/mL) in D_2O is shown in Figure 3. The spectrum was obtained on a Varian Unity spectrometer using the following instrumental parameters: 5 mm $^1H/^{13}C$ dual probe; spectral width, 4811 Hz; 50° pulse width; 64K time-domain data points; acquisition time, 6.8 seconds; 100 scans and probe temperature, 27°C.

The proton-decoupled ^{13}C spectrum of cefuroxime sodium (5 mg/mL) in D_2O is shown in Figure 4. These data were obtained using a 5 mm $^1H/^{13}C$ dual probe; spectral width, 20 KHz; 90° pulse width; 64K time-domain data points; acquisition time, 1.6 seconds; relaxation decay, 2.4 seconds; WALTZ-16 proton decoupling; 4000 scans and probe temperature, 27°C. The spectrum was processed with 1.0 Hz Lorentzian line broadening followed by the addition of 64K zero-fill data points. The chemical shift assignments of cefuroxime sodium for both the ^1H and ^{13}C spectra are shown in Table II.

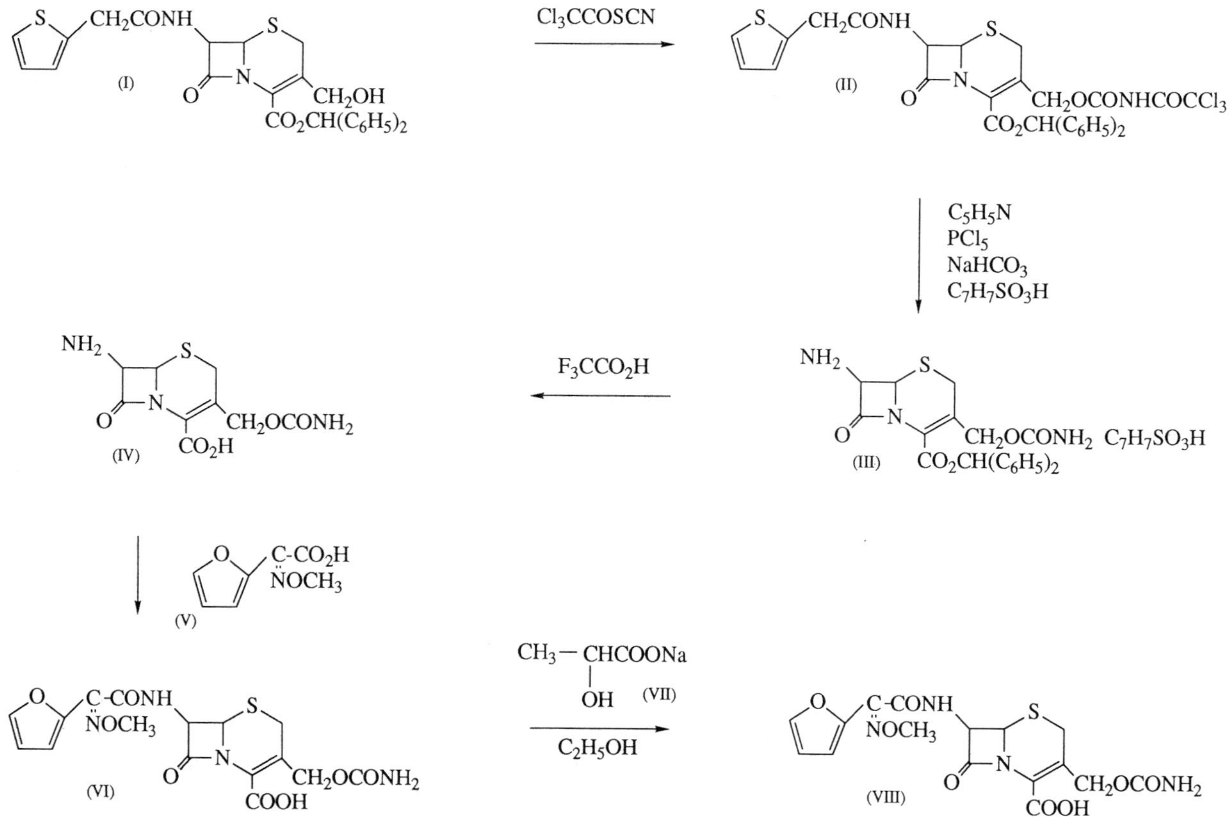

Figure 1. Chemical synthesis of cefuroxime sodium.

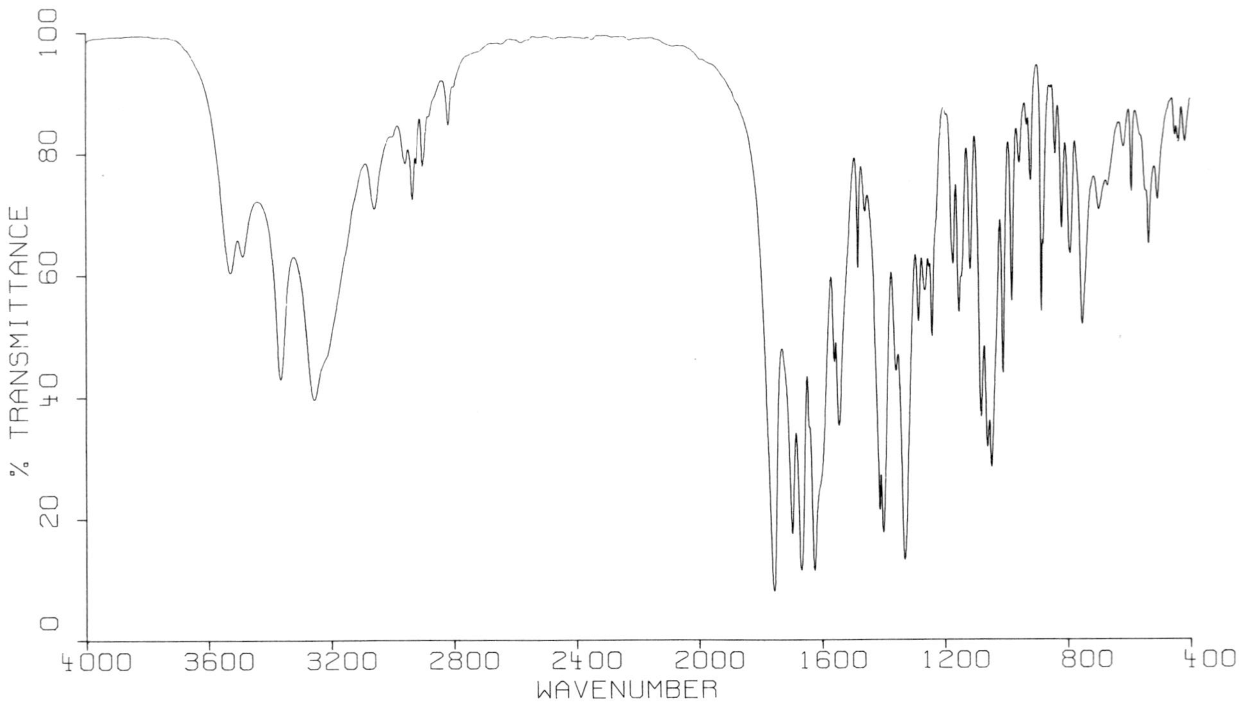

Figure 2. Fourier transform infrared spectrum of cefuroxime sodium.

Table I. Infrared Band Assignments for Cefuroxime Sodium

Wavenumber, cm^{-1}	Infrared Assignment
3500	NH stretch of amide, H-bonded
3368, 3254	NH stretches, symmetric and antisymmetric of carbamate NH_2
3063, 3000	6-H, 7-H stretches in b-lactam ring
2961	CH asymmetric stretch in CH_3
2937, 2926	CH asymmetric stretch in CH_3
2906	CH asymmetric stretch in OCH_3
2879	CH symmetric stretch in CH_3
2856	CH symmetric stretch in CH_2
2820	CH deformation, overtone, CH_3
1758	C=O stretch, β-lactam
1699	C=O stretch, carbamate
1667	C=O stretch, amide I
1642	C=N, oxime
1627, 1412	C=O stretches, asymmetric and symmetric, in CO_2
1560, 1546	NH deformations; also amide II in syn-CH_3O oximes
1603, 1483, 1402	Ring bands in 2-substituted furans
1461	CH_3 deformation in CH_3O
1335	NH_2 bend, in carbamate
1083	C-O stretch, in CH_3O
1063, 1048	C-O and N-O stretches, in CH_2O of carbamate and oxime

Table II. NMR Chemical Shift Assignments of Cefuroxime Sodium

Site	^{1}H	^{13}C
2	3.49, 3.24	25.25
3		113.89
4		134.08
6	5.04	57.22
7	5.60	58.19
8		161.93
9	4.94, 4.83	63.72
11		156.93
12	6.50	
14		164.40
18	9.67	
19		161.61
20		145.02
21		145.50
22	7.83	145.19
23	6.63	111.88
24	6.70	112.75
28	3.89	62.22

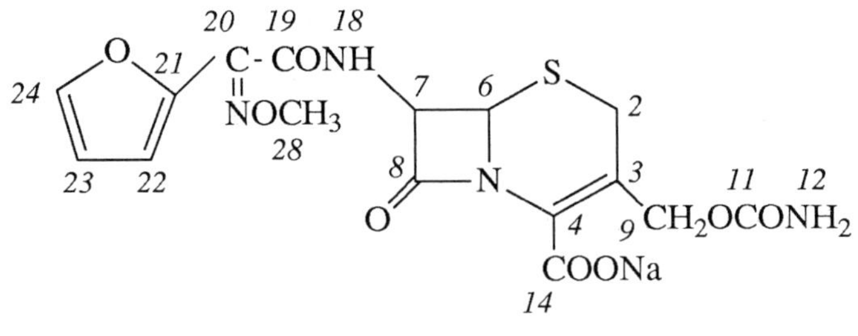

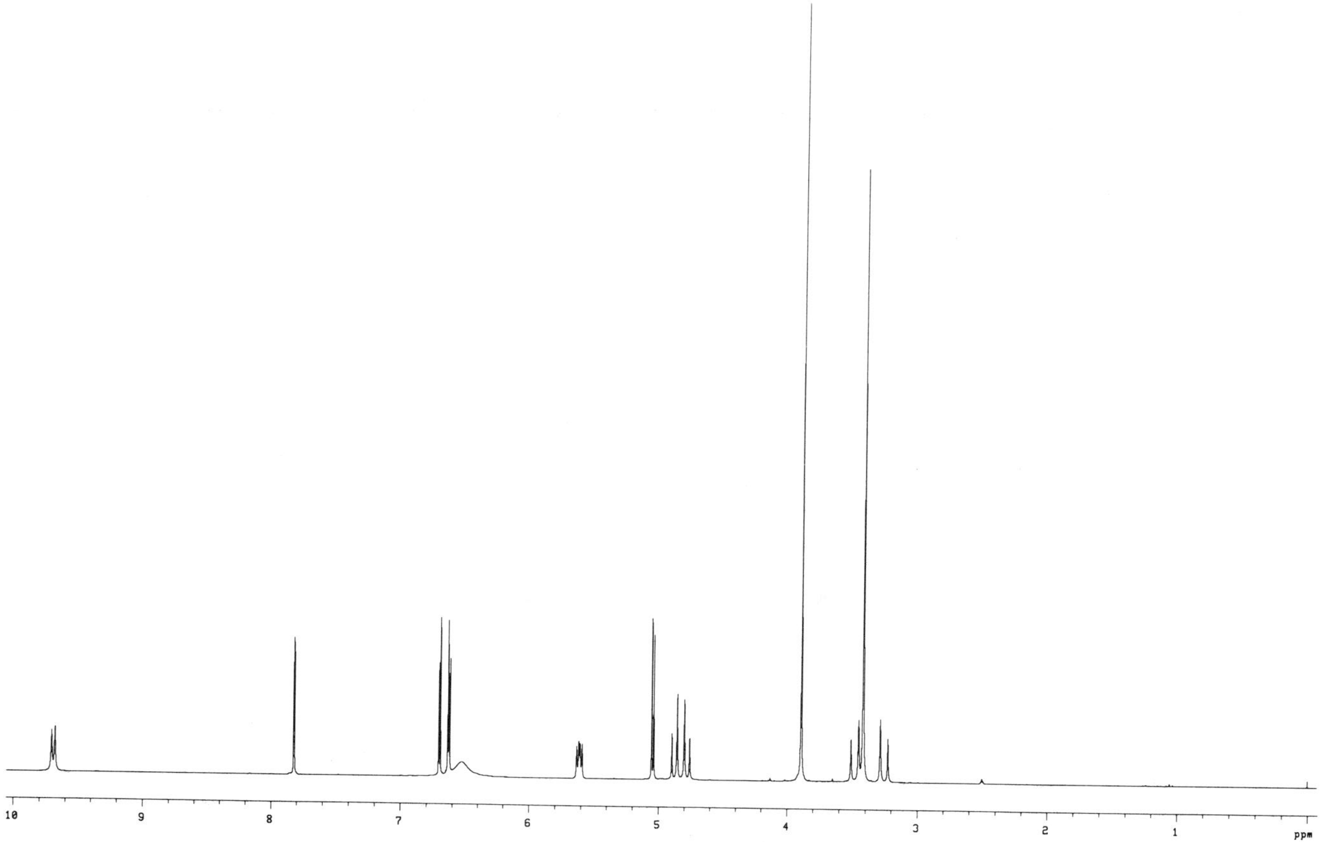

Figure 3. ^{1}H *NMR spectrum of cefuroxime sodium in* D_2O.

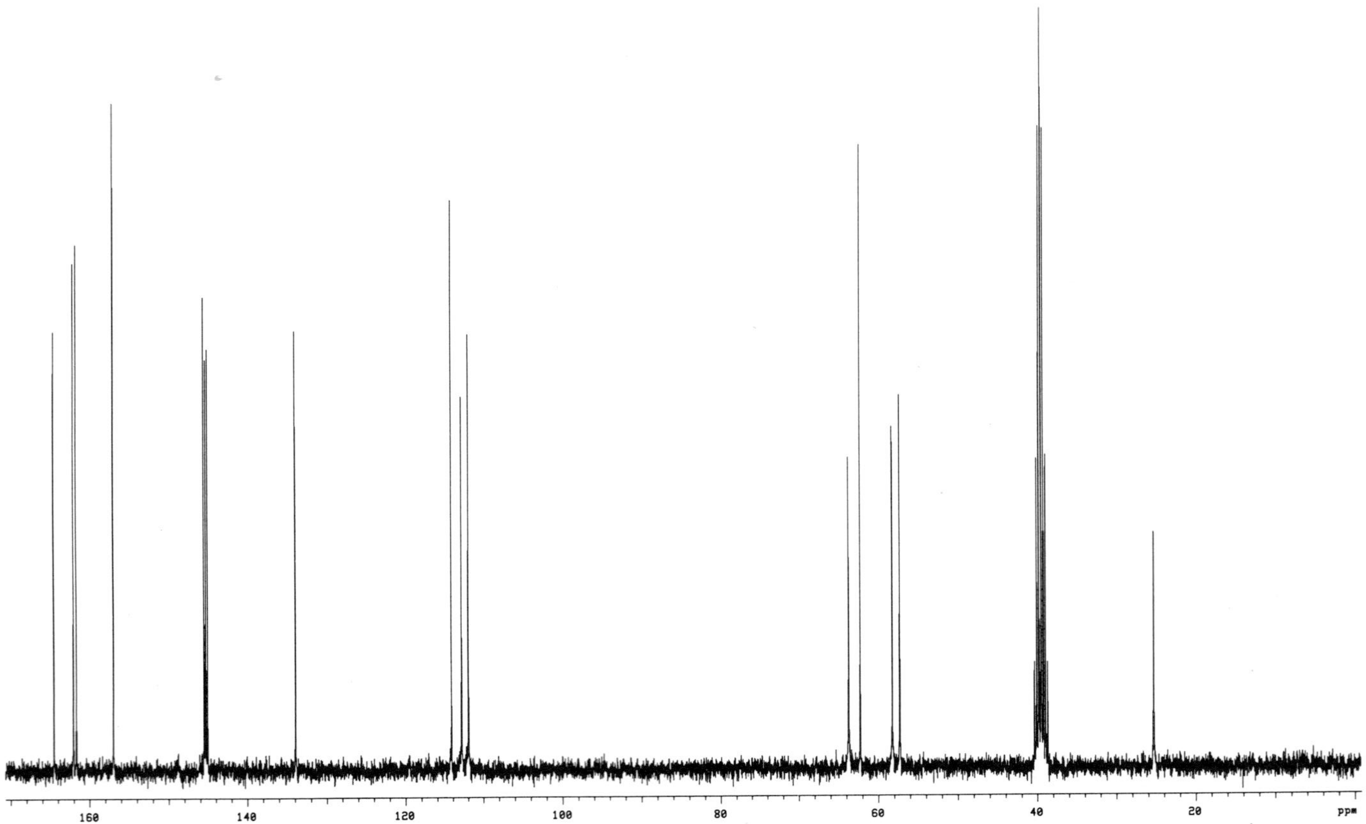

Figure 4. ^{13}C *NMR spectrum of cefuroxime sodium in* D_2O.

3.3 Mass Spectrometry

The fast atom bombardment mass spectrum of cefuroxime sodium is shown in Figure 5. The spectrum was obtained using a VG Model ZAB-3F sector instrument. The spectrum exhibits an intense protonated molecular ion at m/z 447 (M+H). A catonized molecular ion is exhibited at m/z 469 (M+Na). The fragment ion at m/z 408 is analogous to the ion at m/z 386, which is the result of the loss of CO_2NH_3 from the catonized molecular ion. The ion at m/z 342 is due to the loss of CO_2 from the ion at m/z 386. All of the other prominent ions in the mass spectrum are either dispersant ions or catonized dispersant ions resulting from the FAB matrix.

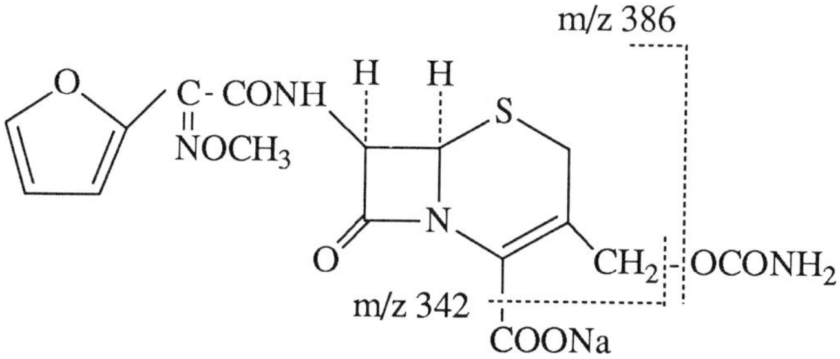

3.4 Ultraviolet Spectroscopy

The ultraviolet spectra of cefuroxime sodium in methanol is shown in Figure 6. Spectral acquisition was performed on a Perkin Elmer Model Lambda 6 spectrophotometer using 1-cm quartz cells. The primary chromophore contributing to the ultraviolet absorbance spectrum is the 3-cephem nucleus. This structural entity exhibits an absorbance maximum at 274 nm with a molecular absorptivity of ε = 17,400 for both methanol and water solutions. The peak absorbances for methanol and water solutions are listed in Table III.

Table III. UV Absorbances and Molecular Absorptivities

	Methanol			Water	
λ	E 1%/1cm	ε	λ	E 1%/1cm	ε
274	400	17,400	274	404	18,000

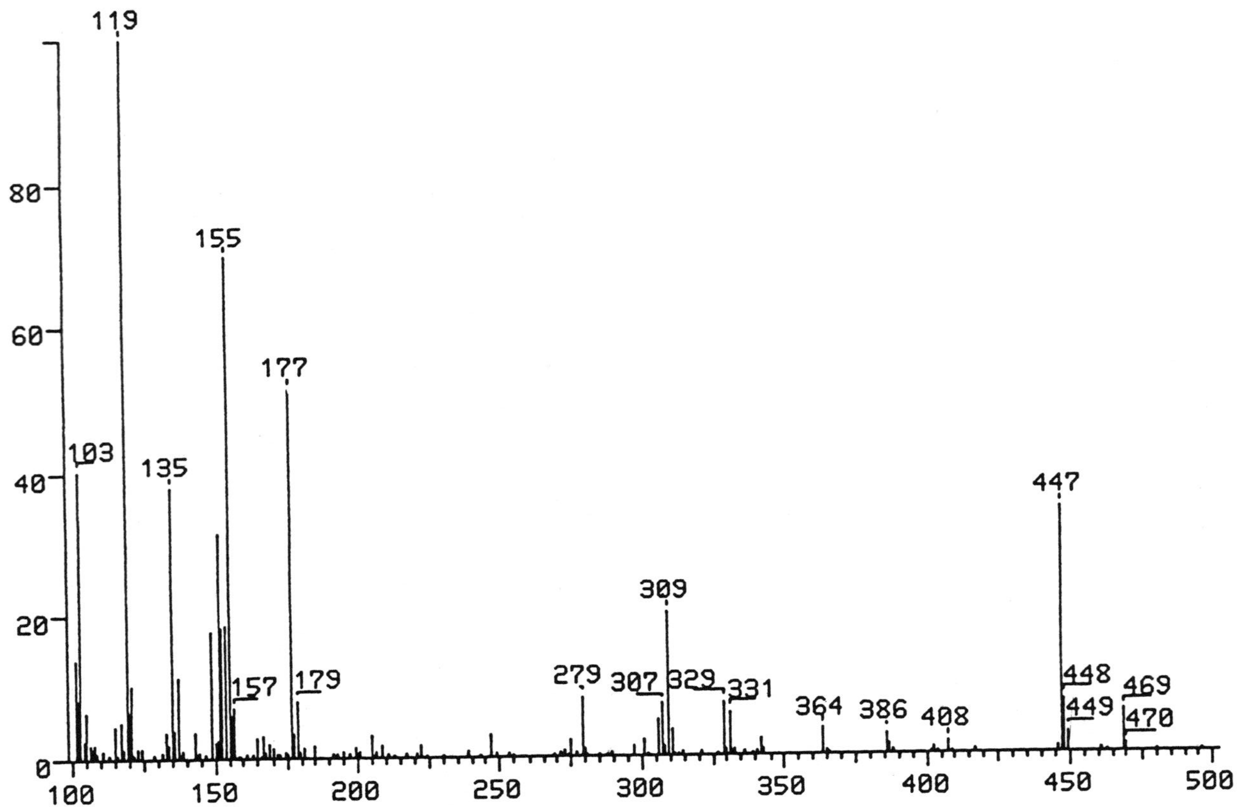

Figure 5. Fast atom bombardment mass spectrum of cefuroxime sodium

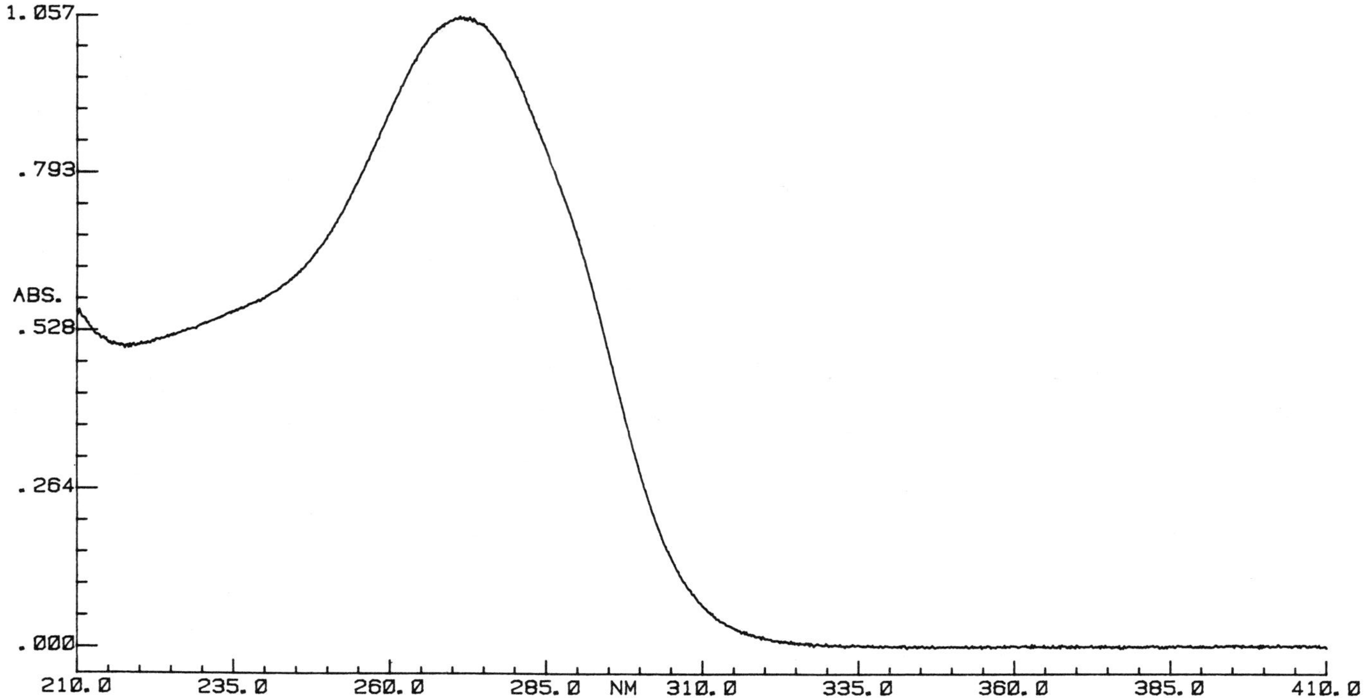

Figure 6. Ultraviolet absorption spectra of cefuroxime sodium.

3.5 Optical Rotation

The specific rotation for a 10 percent solution (w/v) of cefuroxime sodium in water was determined to be +60.0°. The determination was made using the sodium D-line on a Perkin-Elmer 241 MC Polarimeter.

3.6 Differential Thermal Analysis

The DTA thermogram for cefuroxime sodium, at a heating rate of 5°C per minute, shows a broad endotherm at 158°C. This is followed by exotherms at 183°C and 226°C, where the compound begins to decompose (Figure 7).

3.7 Thermogravimetric Analysis

The TGA thermogram for cefuroxime sodium, at a heating rate of 5°C per minute, shows a 0.1% weight loss from 25°C to 89°C, indicating the loss of residual water. A major loss in mass occurs near 185°C which corresponds to the exotherm in the DTA curve for cefuroxime sodium (Figure 8).

3.8 Solubility

Cefuroxime sodium is freely soluble in water and buffered solutions; soluble in methanol; very slightly soluble in ethyl acetate, diethyl ether, octanol, benzene and chloroform. The solubility data for cefuroxime sodium are shown in Table IV.

Table IV. Solubility of Cefuroxime Sodium

Solvent	Solubility (mg/ml)	USP Solubility
Water	≥100.0	Freely Soluble
Buffer - pH 7.0	≥100.0	Freely Soluble
Buffer - pH 4.5	≥100.0	Freely Soluble
Buffer - pH 1.2	≥100.0	Freely Soluble
Methanol	≥50.0 - <100.0	Soluble
Ethyl acetate	<0.5	Very Slightly Soluble
Diethyl ether	<0.5	Very Slightly Soluble
Octanol	<0.5	Very Slightly Soluble
Benzene	<0.5	Very Slightly Soluble
Chloroform	<0.5	Very Slightly Soluble

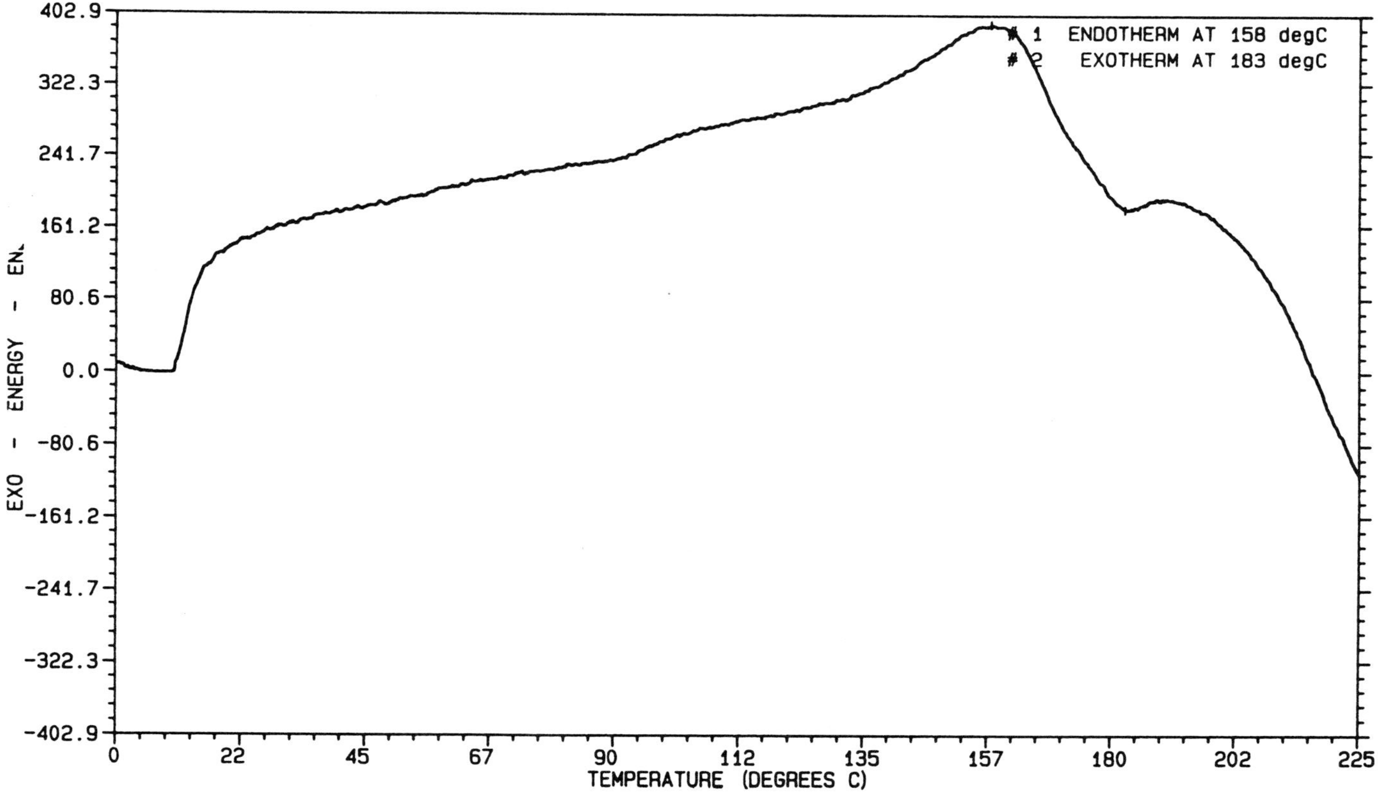

Figure 7. DTA thermogram of cefuroxime sodium.

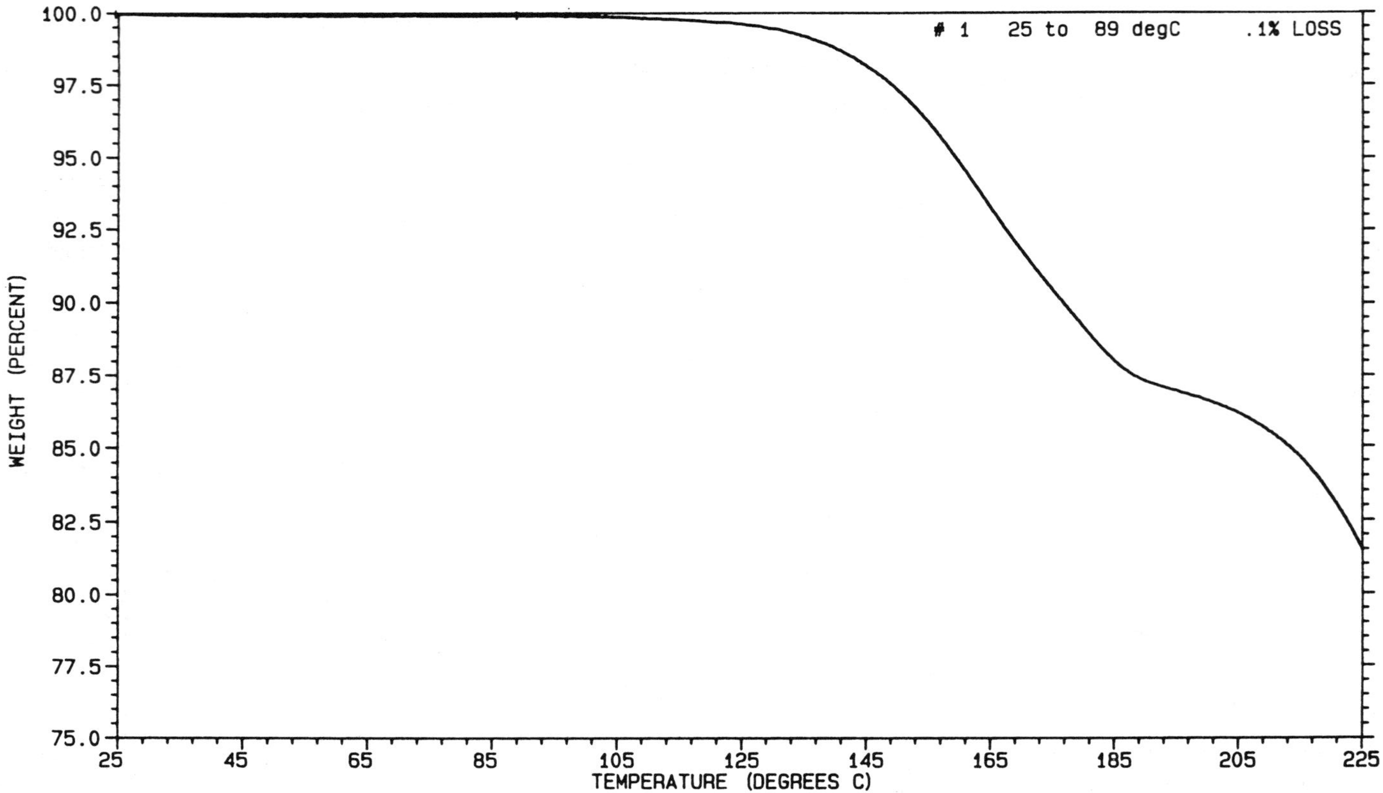

Figure 8. TGA thermogram of cefuroxime sodium.

3.9 Crystal Properties, Polymorphism

The X-ray powder diffraction pattern of cefuroxime sodium is shown in Figure 9. The spectrum was obtained using a Nicolet powder diffractometer using copper Kα irradiation (1.5418 A) with a graphite monochrometer. A total of 4 peaks were detected at scattering angles between 5 and 35 degrees 2-theta. As observed, the sodium salt has a poorly defined diffraction pattern indicating a predominantly amorphous material.

3.10 Ionization Constant, pK_a

The ionization constants for cefuroxime sodium in aqueous media and dimethyl formamide are $pK_a = 2.5$.and $pK_a = 5.1$, respectively.

4. METHODS OF ANALYSIS

4.1 Identity

The identity of the cefuroxime sodium is determined using the specificity of infrared spectroscopy which differentiates it from synthetic intermediates, process related substances or degradation products. Cefuroxime sodium is triturated with potassium bromide and pressed into a transparent pellet for spectroscopic analysis. The identity is confirmed by comparison to a reference standard spectrum obtained under similar conditions.

4.2 Elemental

The following elemental composition was obtained using a Perkin Elmer Model 2400 Analyzer.

Element	% Calculated	% Found
C	43.05	42.56
H	3.39	3.46
N	12.55	12.16
O	28.67	28.25
S	7.18	7.22
Na	5.15	5.08

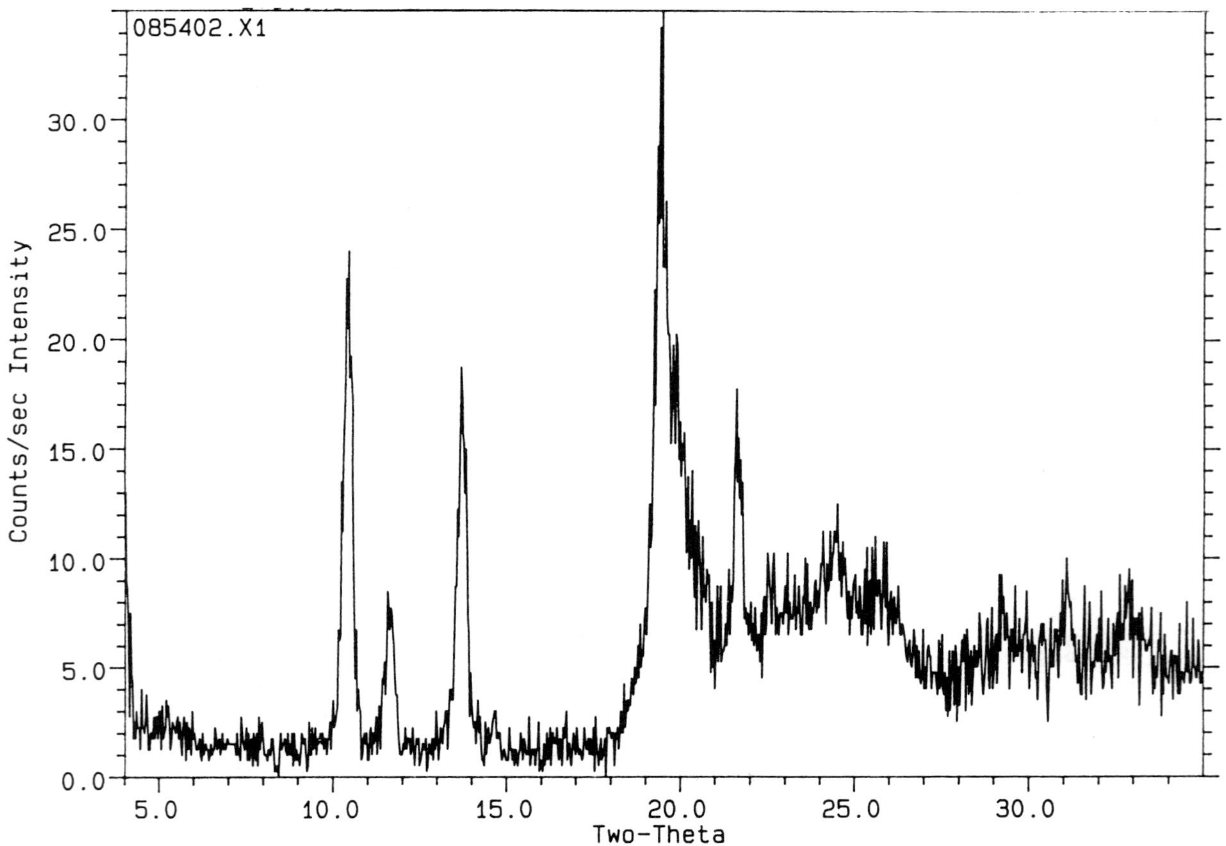

Figure 9. Powder x-ray diffraction pattern of cefuroxime sodium.

4.3 Chromatography

Conditions for quantifying cefuroxime sodium bulk drug substance and parenteral formulations have been optimized using reversed- phase HPLC. The mobile phase consists of a sodium acetate buffer (0.1M, pH 3.4) and acetonitrile (90:10). A Regis HiChrom Reversible® hexyl column (250 x 4.6 mm; 5 micron particle size) is used in conjunction with a flow rate of 2 mL/min. Detection is obtained with an UV detector set at 254 nm. The method is stability-indicating as indicated by its ability to separate cefuroxime and known degradation products. Similar conditions, as detailed in the cefuroxime sodium USP monograph and the CFR, provide appropriate selectivity for the determination of cefuroxime potency (6, 7).

Quantification is performed using a standard curve over the range of 0.08 to 0.12 mg/mL of USP reference standard. Samples are prepared by accurately weighing sufficient sample to produce a 0.1 mg/mL analysis solution. These solutions should be used within one hour of preparation or stored under refrigerated conditions for up to twenty-four hours to prevent degradation. Figure 10 shows the separation of cefuroxime sodium and the internal standard, 5-methylresorcinol monohydrate.

Gradient reversed-phase HPLC methodology is used to quantify cefuroxime sodium and its potential process related substances. A Beckman Ultrasphere® column (250 x 4.6 mm; 5 micron particle size) is used in conjunction with a flow rate of 1 mL/min and detection at 254 nm. The mobile phase components are a sodium acetate buffer (0.1M, pH 3.4, mobile phase A), and HPLC-grade acetonitrile (mobile phase B). Following a 5 minute isocratic period at 90%A:10%B, a linear gradient to 10%A:90%B over 20 minutes is initiated. This is followed by an isocratic period of 5 minutes before recycle. The sample is analyzed using the high-low approach at a high sample concentration of approximately 10 mg/mL (8). Figure 11 shows the separation of cefuroxime sodium from its major process related substances and degradation products (3'-hydroxy cefuroxime, 3'-acetoxy cefuroxime, cefuroxime anti-isomer and cefuroxime lactone).

4.4 Iodometric

The iodometric assay for cefuroxime sodium involves hydrolysis in aqueous base which results in the cleavage of the β-lactam ring. This is followed by oxidation of the hydrolysis products with iodine, and the titrimetric determination of the iodine consumed. Since the unhydrolyzed drug substance does not react with iodine, an unreacted sample is used as a blank to compensate for iodine consuming impurities. Since the specificity is limited to any intact β-lactam ring, it cannot be considered as a stability-indicating assay. This is a problem for cefuroxime sodium since several of the degradation products retain the β-lactam ring, but also have considerable antibiotic activity (9-11).

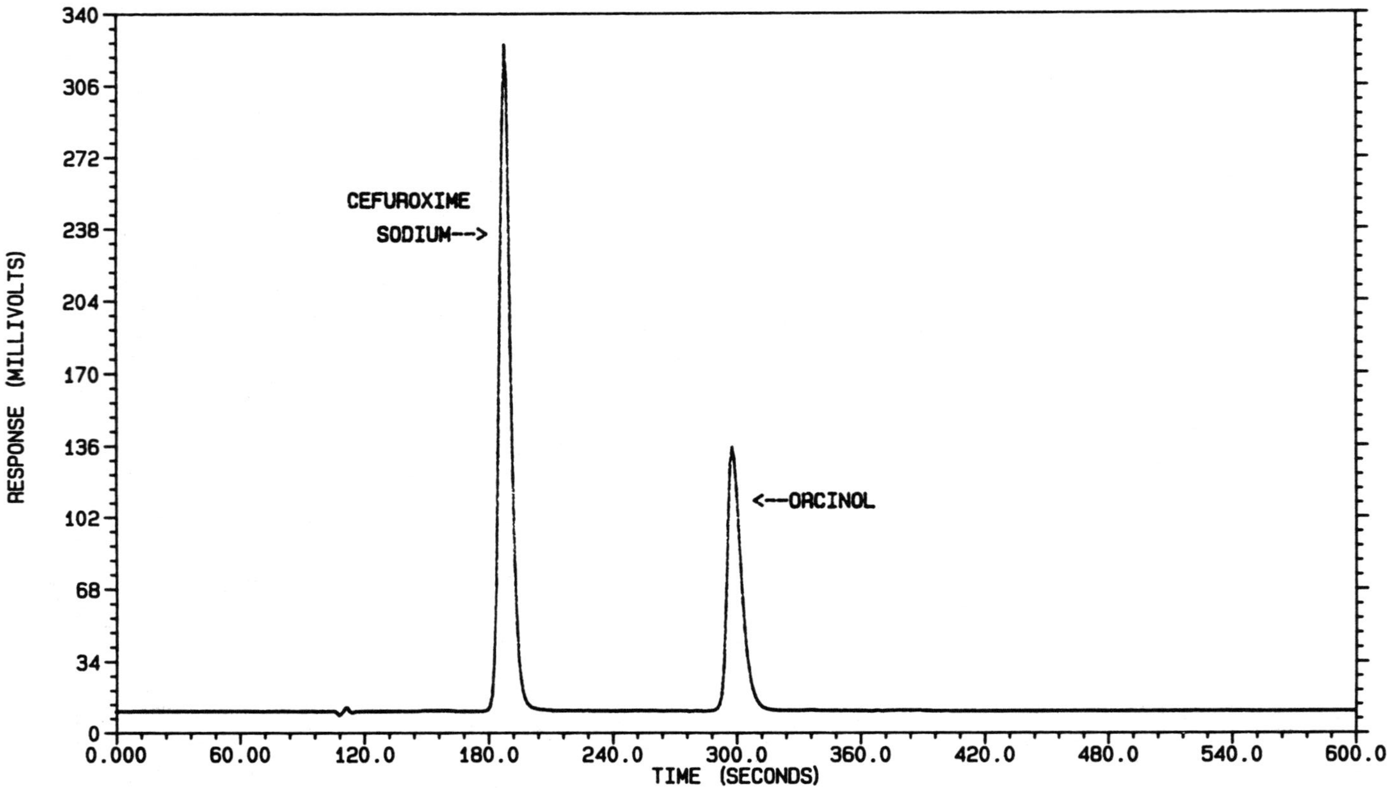

Figure 10. HPLC chromatogram of stability-indicating assay for cefuroxime sodium.

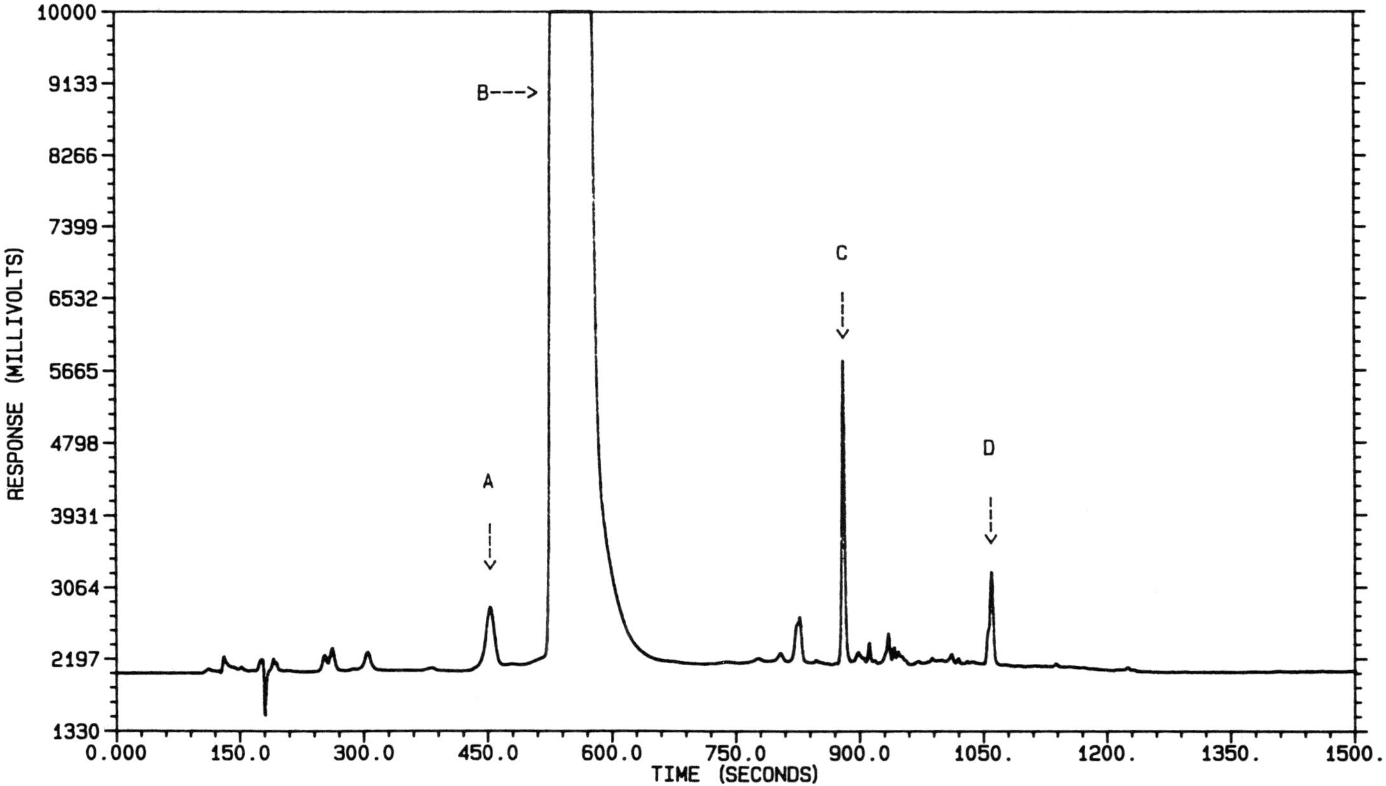

Figure 11. *HPLC chromatogram of related substance assay for cefuroxime sodium (peak A: 3'-hydroxy cefuroxime, peak B: cefuroxime sodium, peak C: 3'-acetoxy cefuroxime and peak 4: cefuroxime lactone).*

4.5 Hydroxylamine

Cephalosporins containing the β-lactam functionality can be assayed by hydroxylamine. Reaction with hydroxylamine results in cleavage of the β-lactam ring to form the corresponding hydroxamic acid. The hydroxamic acid reacts with ferric ion under acidic conditions to form a colored complex which is measured at 480 nm. While not a stability-indicating assay, this test will determine β-lactam ring content (12, 13).

4.6 Microbiological

4.61 Turbidimetric

A turbidimetric assay with *E. coli* ATCC 10536 may be used for the bulk drug substance and the dosage forms. On the day prior to assay, antibiotic medium 3 (14) is inoculated with the organism and the broth culture is incubated for approximately 16 hours on a rotary shaker at room temperature. Immediately prior to assay, a sufficient volume of medium 3 is inoculated with 0.05 mL of the broth culture for each 100 mL of medium. Samples are diluted with potassium phosphate buffer (0.01M, pH 6.0) and 0.1 mL of each sample is added to assay tubes containing 10 mL of inoculated medium. Dose response concentrations are in the range of 0.02 to 0.04 μg/mL of assay medium. The test is incubated in a 37°C water bath until tubes containing no antibiotic have a light transmittance of 20 percent at 530 nm. The test is terminated by heating for 2-3 minutes in an 80°C water bath, and the percent transmittance is determined for each tube at 530 nm.

4.62 Agar Plate Diffusion

Agar diffusion assays of the bulk drug substance and dosage forms are performed with *E. coli* ATCC 10536. The auger plate system consists of 10 ml of agar medium 2 (15) as the base layer and 5 mL of medium (16) as the seed layer. Dose response concentrations are in the range of 1.0 to 20.0 μg/mL of assay medium. The agar for the seed layer is inoculated with 0.5 mL of an overnight culture of *E. coli* per each 100 mL of medium. Tests are incubated at 30°C for 16 to 18 hours.

Microbiological assays for cefuroxime sodium in serum, urine and tissues is performed by agar plate diffusion assays with *E. coli* MB3804. The agar plate system is a 10 mL single layer plate consisting of trypticase soy agar which has been modified by the addition of 1 g of dextrose and 10 g of sodium acetate per liter of medium. The agar medium is inoculated at 0.1 percent (v/v) with a broth culture which has been adjusted to 25 percent light transmittance at 530 nm. Standard curve concentrations are prepared in the sample diluent at concentrations of 0.1 to 1.0 μg/mL. Tests are incubated at 37°C for 16 to 18 hours.

5. STABILITY - DEGRADATION

5.1 Potential Routes of Degradation

Cefuroxime sodium was exposed to stress conditions of heat, acid, base and photodegradation. In all of these studies, the compound was evaluated by specific HPLC procedures. The conclusions of these studies demonstrate that appropriate storage of the bulk drug substance and solutions is necessary to insure product integrity.

Cefuroxime sodium bulk drug substance was found to be stable when exposed to thermal stress of 60°C for 7 days. After 21 days, both the 3'-hydroxy cefuroxime and 3'-acetoxy cefuroxime degradation products were observed at levels of approximately 7 percent (Figure 12).

When exposed to three different pH solutions at two separate temperatures (i.e., pH 5, pH 7 and pH 8 at temperatures of 4°C and 25°C), cefuroxime was found to be less stable. Three primary degradation products are observed at varying levels depending on the conditions. These are 3'-hydroxy cefuroxime, 3'-acetoxy cefuroxime, cefuroxime anti-isomer and cefuroxime lactone. At 4°C, all three solutions showed an average degradation of 10 percent over a 21 day period. When exposed to the harsher temperature condition of 25°C, the solutions also exhibited a non-pH dependent degradation of approximately 50 percent after 7 days.

Irradiation of the bulk drug substance in both clear and amber glass vials resulted in degradation to 3'-acetoxy cefuroxime. After 30 days, the sample stored in clear glass had degraded approximately 15 percent while the sample stored in amber glass only degraded 5 percent.

5.2 Solid-State Stability

Cefuroxime sodium bulk drug substance and formulations in sterile vials are stable for at least 24 months in the dry state, when protected from light.

5.3 Solution Stability

Following reconstitution with sterile water for injection, cefuroxime sodium solutions containing 90-100 mg of cefuroxime per mL or suspensions containing 200-220 mg per mL are stable for 24 hours at room temperature or 48 hours at 5°C. Parenteral infusion solutions, which have been reconstituted with sterile water for injection, 5% dextrose injection or 0.9% sodium chloride injection, at concentrations of 7.5-15 mg/mL are stable for 24 hours at room temperature or 7 days at 5°C.

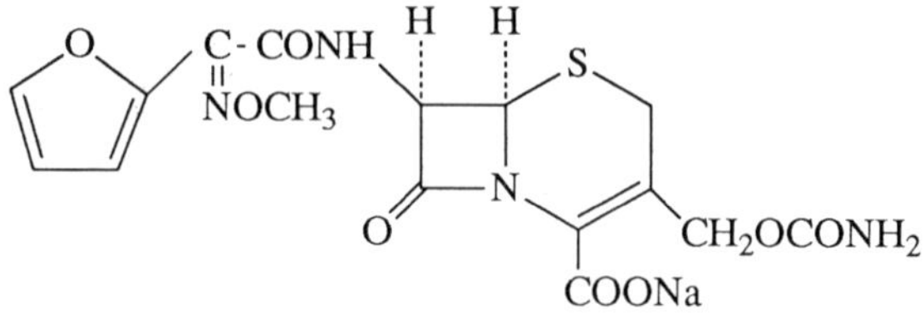

cefuroxime sodium

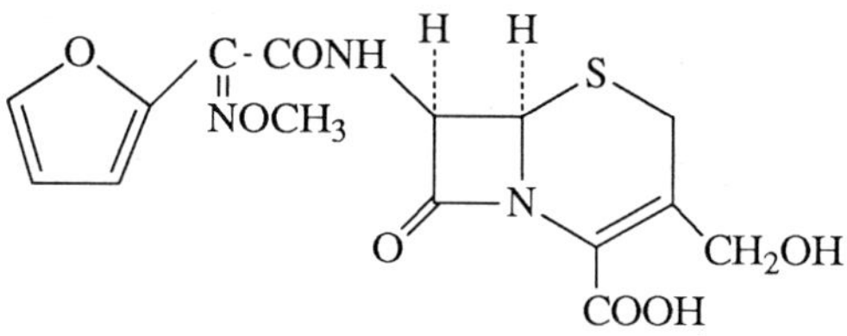

3′ - hydroxy cefuroxime

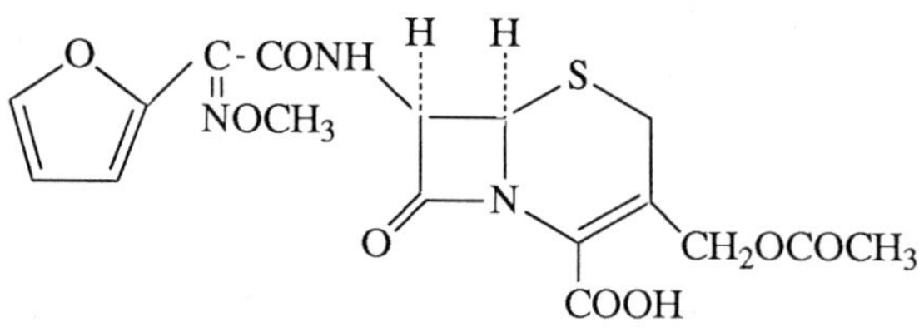

3′ - acetoxy cefuroxime

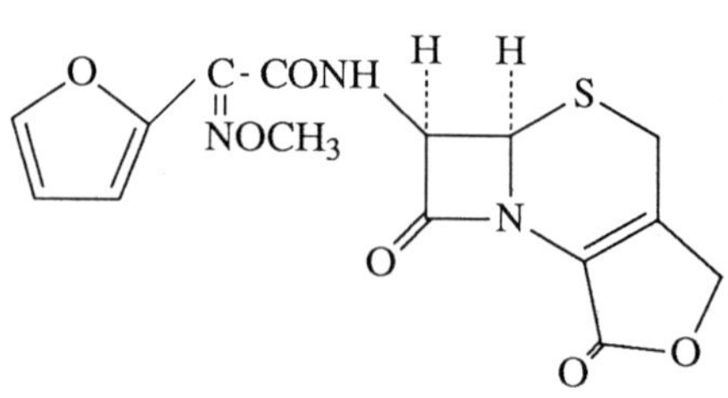

cefuroxime lactone

Figure 12. Potential degradation products of cefuroxime sodium.

Cefuroxime sodium is also chemically and physically compatible with parenteral solutions of 0.9% sodium chloride, lactated Ringer's, 5% dextrose, 10% dextrose, 10% invert sugar or 0.1M sodium lactate. These solutions, at concentrations of 1-30 mg/mL, are stable for 24 hours at room temperature or 7 days at 5°C.

Cefuroxime sodium is potentially incompatible with some drugs, including aminoglycosides, but the compatibility depends on concentrations, specific diluents, pH and temperature.

6. DRUG METABOLISM AND PHARMACOKINETICS

6.1 Absorption

Cefuroxime sodium is poorly absorbed from the gastrointestinal tract because it is highly ionized at physiological pH (3). An oral dose of 0.5 g yielded plasma levels of only 8 µg/mL, while an equivalent intravenous dose resulted in plasma levels of 800 µg/mL. Therefore, it must be given by intravenous or intramuscular routes of administration. Peak serum concentrations were attained within 15-60 minutes of dosage. Mean peak serum concentrations of cefuroxime and the areas under the concentration-time curve (AUC) were not significantly different for intravenous or intramuscular administration. The AUC was found to be proportional to the dose administered (3, 17).

6.2 Distribution

Cefuroxime sodium is widely distributed into body tissues and fluids, including the kidneys, heart, gallbladder, liver, prostatic tissue, uterine and ovarian tissue, bone, bile, adipose tissue, ascitic fluid, synovial fluid, pericardial fluid and pleural fluid. The apparent volume of distribution of cefuroxime in healthy adults ranges from 5-9 L/m^2. Approximately 35 percent of the cefuroxime is bound to serum proteins (3).

Cerebral spinal fluid (CSF) concentrations are generally low for patients with uninflamed meninges, but therapeutic levels can be attained for patients with inflamed meninges. Following a dose of 1-2 g every eight hours, the cefuroxime concentrations for patients with uninflamed or inflamed meninges were 0.1 or 10 µg/mL, respectively (17).

Intravenous administration of cefuroxime as prophylactic treatment during heart valve surgery has been shown to prevent prosthetic valve endocarditis (18). Eight patients were given 1.5 g intravenous doses of cefuroxime as bolus injections prior to surgery. The low degree of protein binding in plasma and the large volume of distribution resulted in serum concentrations well above the minimum inhibitory concentration over the course of the surgical procedure.

Prophylactic antimicrobial therapy is critical in patients undergoing joint replacement. Cefuroxime was shown to diffuse into bone at concentrations acceptable for chemoprophylaxis (19). Twenty-one patients were administered 750 mg of cefuroxime sodium intravenously at induction of anaesthesia, prior to knee or hip replacement. The mean bone concentrations of cefuroxime were in excess of the MIC for *Staphylococcus aureus*, the most common infecting pathogen for this type of surgery.

6.3 Elimination

In adults with normal renal function, the serum half-life is between 1-1.5 hours. Patients with renal impairment have prolonged serum half-lives ranging from between 2-16 hours. Cefuroxime is not metabolized and is excreted unchanged primarily in urine by both glomerular filtration and tubular secretion. In adults with normal renal function, 90-100 percent of the dose is eliminated in the urine within 24 hours (3).

7. REFERENCES

1. O'Callaghan, C.H., Sykes, R.B., Griffiths, A. and Thornton, J.E. (1976). Antimicrob. Agents Chemother. 9, 511.
2. Greenwood, D., Pearson, N.J. and O'Grady, F. (1976). J. Antimicrob. Chemother. 2, 337.
3. Foord, R.D. (1976). Antimicrob. Agents Chemother. 9, 741.
4. Cook, M.C. (1975.) German Patent 2,439,880; U.S. Patent 3,974,153
5. Wilson, E.M. (1984). Chem. and Ind., 19 March, 217.
6. USP XXII, *Official Monograph: Sterile Cefuroxime Sodium* (1990).
7. Coomber, P.A., Jeffries, J.P. and Woodford, J.D. (1982). Analyst 107, 1451.
8. Inman, E.L.and Tenbarge, H.J. (1988). J. Chromatogr. Sci. 26, 89.
9. Marrelli, L.P. (1972). *Cephalosporins and Penicillins: Chemistry and Biology*, Academic Press, Inc., New York.
10. Alcino, J.F. (1946). Ind. Eng. Chem. Anal. Ed. 18, 619.
11. Alcino, J.F. (1961). Anal. Chem. 33, 648.
12. Ericsson, H.M.and Sherris, J.C. (1971). Acta. Pathol. Microbiol. Scand. Supplement B, 217.
13. Ford, J.H. (1947). Anal. Chem. 19, 1004.
14. § CFR 436.102 (b) (3).
15. § CFR 436.102 (b) (2).
16. § CFR 436.102 (b) (1).
17. Ryan, D.M., O'Callaghan, C.H. and Muggleton, P.W. (1976). Antimicrob. Agents Chemother. 9, 520.
18. Lippert, S., Josephsen, S.D., Jendresen, M., Sorenson, T.S. and Gutschik, E. (1989). J. Antimicrob. Chemother. 24, 775.
19. Leigh, D.A. (1989). J. Antimicrob. Chemother. 23, 877.

8. ACKNOWLEDGMENTS

The authors wish to express their sincere thanks to the following people who have provided information for portions of this chapter: S. Maple and K. McCune for the NMR spectra and interpretations; J. Gilliam for the mass spectra and assignments; C. Underbrink for the infrared spectra and assignments; G.A. Stephenson for the X-ray diffraction spectra and assignments, D.R. Long for the thermal analysis data, and A.D. Kossoy for the pK/partition coefficient data.

CELIPROLOL HYDROCHLORIDE

David J. Mazzo
Cynthia L. Obetz
Judith E. Shuster

Department of Analytical and Physical Chemistry
Rhône-Poulenc Rorer Central Research
500 Virginia Drive
Fort Washington, PA, USA 19034

1. INTRODUCTION

1.1 Therapeutic Category

Celiprolol hydrochloride is a highly cardioselective (β_1), beta-adrenergic blocker with partial β_2 agonist activity (vasodilating) but no quinidine-like membrane stabilizing activity (1-8). Its high degree of selectivity does not change throughout the full dose range. As is common with other β-blockers, the exact contributions of these biochemical mechanisms to the overall pharmacological effect cannot be precisely assigned.

Once-daily administration of celiprolol hydrochloride produces reductions in blood pressure. The usual dose range is 200 or 400 mg QD, but doses of 600 mg have been investigated and been found safe and effective. Celiprolol hydrochloride is equal to other β-blockers in controlling blood pressure and heart rate increases during exercise (9-13). It is comparable or superior to other β-blockers (4,9,10,12,14-17) and calcium channel blockers (11) in angina. Celiprolol hydrochloride is not intended to be used in patients with asthma. Therapeutic doses in extended studies have been associated with frequently significant favorable shifts in serum lipid factors, including increase in HDL-C, decreases in LDL-C and triglycerides, and decrease in total cholesterol (18-20). The compound is currently marketed, however, as being lipid neutral (i.e., inducing no significant alteration in total serum cholesterol or triglycerides). Common antihypertensive therapy side effects such as headache, impotence, cold extremities, fatigue, swollen ankles, and dizziness are generally at placebo levels and less frequent than with other β-blocking drugs. Impotence is so rarely seen that it has been reported as "very few cases" in studies involving from 1,200 to 5,400 patients (21-22). Comparative instances of the seven most common side effects in 3,200 patient-exposures to drug (23) are shown below.

The Incidence (percent) of the Most Frequently Occurring Drug-Related Adverse Experiences

Body System	Celiprolol (n=2036)	Placebo (n=658)	Atenolol (n=198)	Propranolol (n=247)	Chlorthalidone (n=121)
Bradycardia	0	0	0	2	0
Dizziness	2	2	4	3	4
Dry mouth	1	0	1	0	0
Dyspepsia	0	0	0	2	1
Fatigue/tiredness	2	1	3	3	1
Gastritis	0	0	2	1	0
Headache	3	3	4	3	5

1.2 History

Celiprolol hydrochloride was first synthesized in the laboratories of Chemie Linz, AG (now known as Hafslund Nycomed Pharma), Linz, Austria. The first patent and first USA patent citations may be found in References 24 and 25. The first non-patent literature reference to celiprolol hydrochloride synthesis appeared in 1983 (26). Celiprolol hydrochloride has gained a wide acceptance as a safe and efficacious β-blocker since its introduction in pharmaceutical formulations. A search of Chemical Abstracts (1966-1990) and Pharmaceutical Abstracts (1974-1990) yielded over 95 unique bibliographic citations for works dealing with the chemical characterization and analysis of celiprolol hydrochloride. Over 250 citations dealing with the pharmacologic effects of celiprolol hydrochloride have been found to date.

2. DESCRIPTION

2.1 Chemical Name, Formula, Molecular Weight

Celiprolol hydrochloride has been known as RHC-5320A REV-5320A, RG-5320A and ST-1236 HCl. The accepted chemical name for celiprolol hydrochloride is: N'-[3-acetyl-4-[3-(1,1-dimethylethyl)amino]-2-hydroxyproxy]phenyl]-N,N-diethylurea, monohydrochloride. The CAS registry number is 57470-78-7. Other names which have been used for celiprolol hydrochloride include the monohydrochloride salts of 3-[3-acetyl-4-[3-(tert-butylamino)-2-hydroxypropoxy]phediethylurea and N-[3-acetyl-4-(3'-tert-butylamino-2'-hydroxypropoxy)]phenyl-N'-diethylurea. The chemical structure of celiprolol hydrochloride is shown in Figure 1. Its molecular formula is $C_{20}H_{33}N_3O_4$ HCl and its molecular weight is 415.96 g/mole.

2.2 Definition

Celiprolol hydrochloride possesses an asymmetric carbon atom in chemical structure (the 2 position of the propoxy group) and is provided as the racemate of its R and S enantiomers. It has trade names of Selecor®, Selectol® and Celectol®. Celiprolol, when referenced in this manner, indicates the free base (CAS registry number 56980-93-9).

2.3 Appearance, Color, Odor

Celiprolol hydrochloride is a white, odorless, crystalline powder.

3. SYNTHESIS

Celiprolol has been prepared through a series of synthetic steps beginning with p-phenetidine (1-amino-4-ethoxybenzene) and diethylcarbamoyl chloride (26, 27). The synthetic scheme is shown in Figure 2. The starting materials are reacted in the presence of base ($KHCO_3$ or NaOH) and the organic layer containing I is reacted with acetyl chloride and aluminum chloride in dichloromethane. After the addition of water, the resulting suspension is filtered to yield II. Dried compound II is then reacted with epihalohydrine (epichlorohydrine or epibromohydrine) in the presence of base and a catalyst to produce III which is then brominated (IV) and further reacted with tert-butylamine

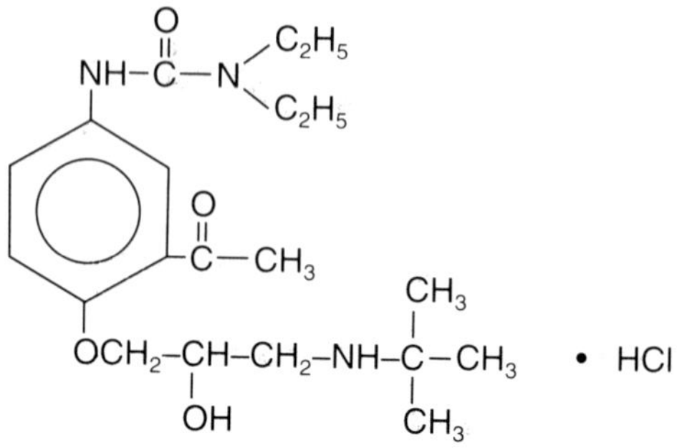

Figure 1. Chemical structure of celiprolol hydrochloride.

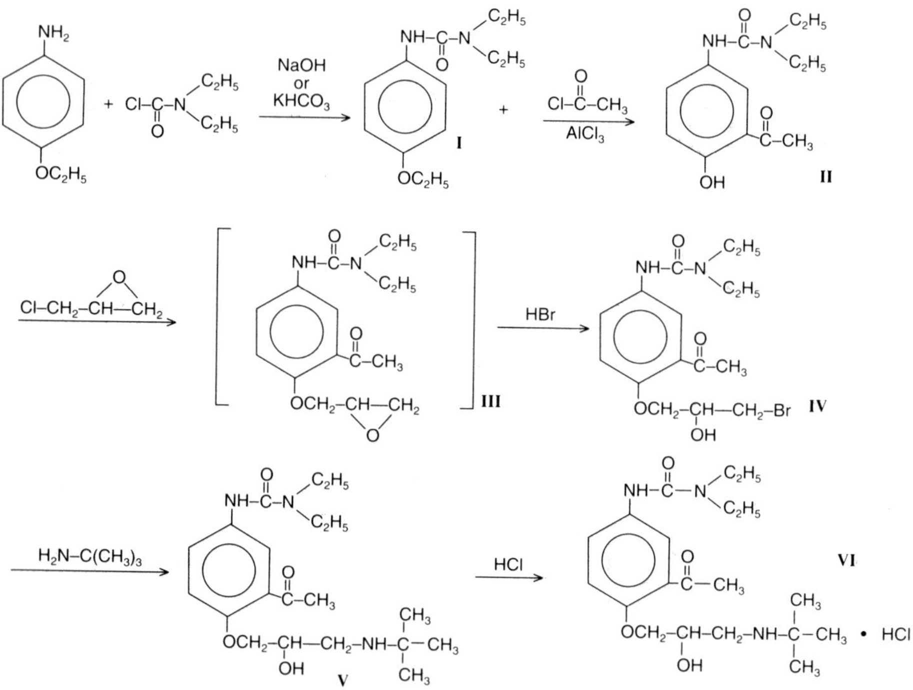

Figure 2. General scheme for the synthesis of celiprolol hydrochloride.

to yield V (celiprolol base). Compound V isreacted with hydrochloric acid to produce celiprolol hydrochloride (VI).

Alternately (d), compound V may be produced directly from compound II by reacting II with epichlorohydrine, adding water and distilling to remove residual epichlorohydrine. Tert-butylamine is reacted with the residue and the product (V) is extracted with dichlorethane. Celiprolol hydrochloride is then created by mixing compound V with HCl.

4. PHYSICAL PROPERTIES

4.1 Infrared Spectrum

The infrared spectrum of celiprolol hydrochloride taken in a KBr pellet is shown in Figure 3 (28). A Nicolet Model 740 Fourier transform infrared spectrophotometer was used to acquire the spectrum. Frequency assignments for some of the characteristic bands are listed in Table I.

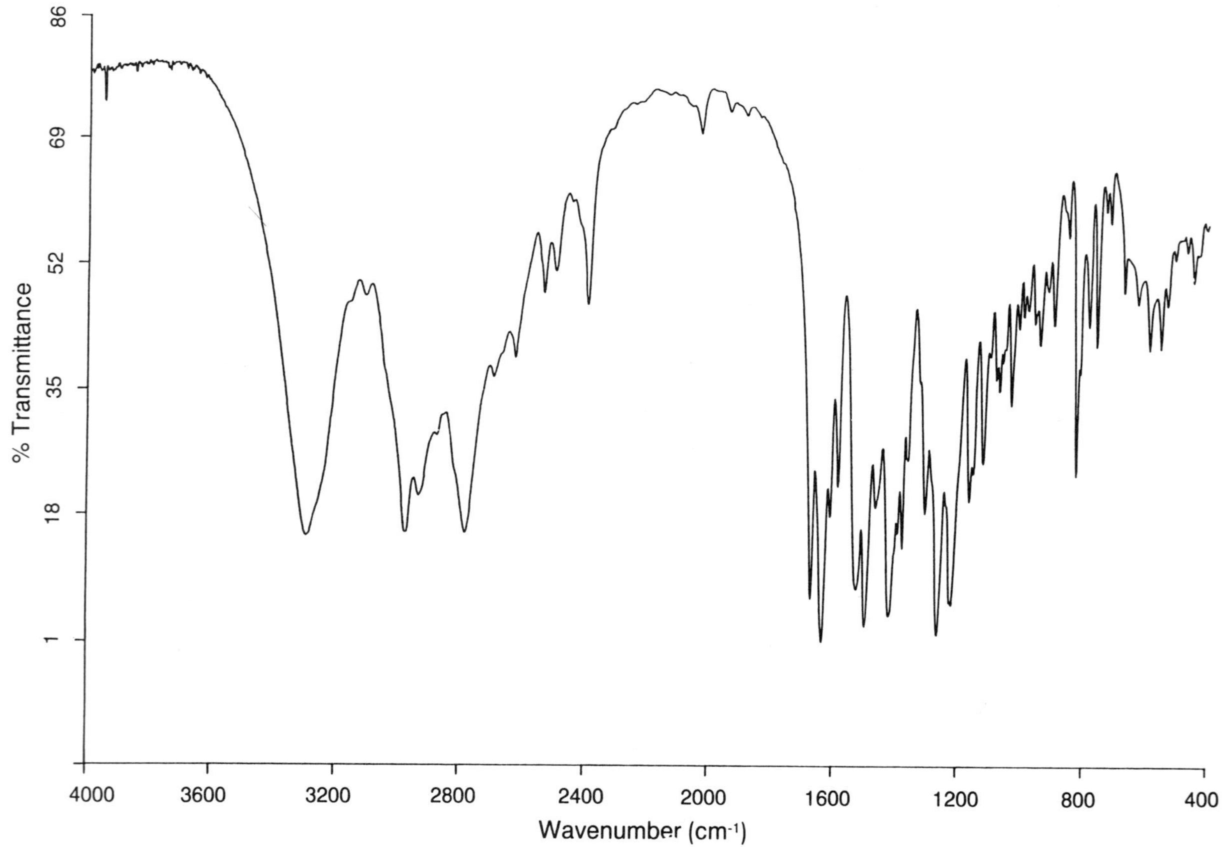

Figure 3. Infrared spectrum of celiprolol hydrochloride taken in a KBr disk.

TABLE I

Vibrational Assignments for Selected Celiprolol Hydrochloride Infrared Absorptions

Frequency (cm^{1})	Relative Intensity*	Assignment
3295	s	O-H and urea N-H stretch
2974	s	C-H stretch
2935	s	C-H stretch
2779	s	C-H stretch
2626	w	amine HCl N-H stretch
2536	w	amine HCl N-H stretch
2497	w	amine HCl N-H stretch
2391	m	amine HCl N-H stretch
1674	vs	acetyl C=O stretch
1636	vs	urea C=O stretch
823	s	aromatic C-H out of plane deformation
591	w	aryl methyl ketone C-CO-C in plane deformation

*m=medium; s=strong; v=very; w=weak.

The infrared spectrum of celiprolol hydrochloride taken in a mineral oil mull is shown in Figure 4 (28). The spectra shown in Figures 3 and 4 are both consistent with the chemical structure of celiprolol hydrochloride.

4.2 Nuclear Magnetic Resonance Spectrum

All spectra were obtained using a Varian VXR 200 Fourier transformer NMR Spectrometer (29) Figure 5 shows the structural numbering applied to celiprolol hydrochloride to facilitate NMR spectral assignment.

All NMR spectra presented are consistent with the chemical structure of celiprolol hydrochloride.

4.2.1 Proton NMR Spectrum

The proton magnetic resonance spectrum of celiprolol hydrochloride was obtained from an approximately 50 mg/mL solution using deuterated methanol as a solvent. The spectrum is shown in Figure 6 and the spectral assignments are listed in Table II. A listing of instrument parameters used in obtaining this spectrum is provided in Table III.

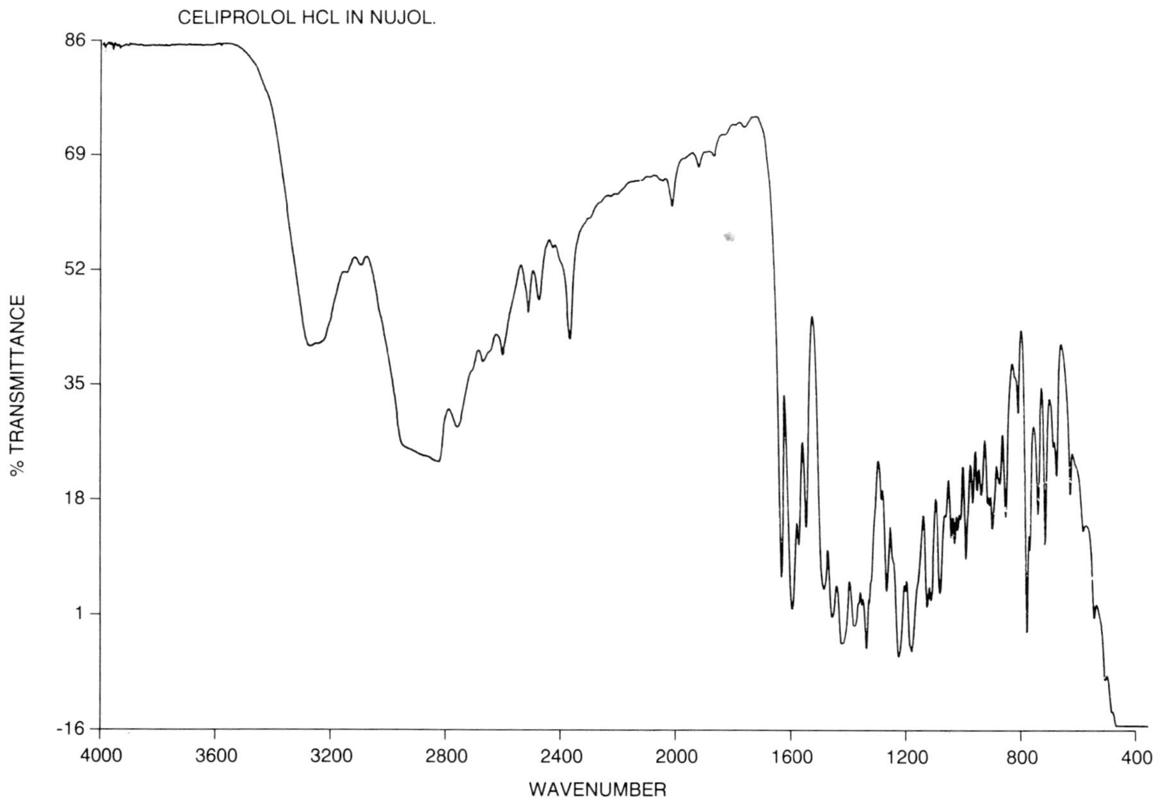

Figure 4. Infrared spectrum of celiprolol hydrochloride taken in a mineral oil mull.

TABLE II

Proton NMR Spectral Assignments of Celiprolol Hydrochloride

Proton	Chemical Shift (ppm)
H-2	7.79
H-5	7.08
H-6	7.57
H-10	3.41
H-11	1.19
H-12, H-13	4.08 - 4.34
H-14	3.13 - 3.38
H-17	1.44
H-19	2.63

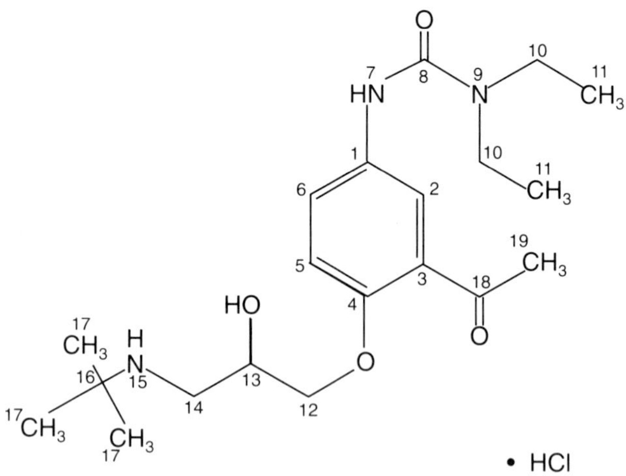

Figure 5. Chemical structure of celiprolol hydrochloride showing element numbering used to facilitate NMR spectral assignment.

TABLE III

Parameter Table for the ^1H NMR Spectrum of Celiprolol·HCl

Observe:

Nucleus	1.250	
Spec. Width	4000.0	Hz
Acq. Time	2.000	sec
Pulse Width	7.0	μsec
Freq.	200	MHz
Offset	800	Hz
Delay	1.000	sec
Transients	64	

Decouple:

Nucleus	1.250	
Mode	NNN	
Modulation:Mode	C	
Offset	-43.9	Hz
Power	20	db
Freq.	200	Hz

Plot/Processing:

FN	16	K		
Width	2040.8	Hz/ppm	Start	-40.0 Hz/ppm

Experiment:

Pulse Sequence	STD1H
Solvent	CD_3OD

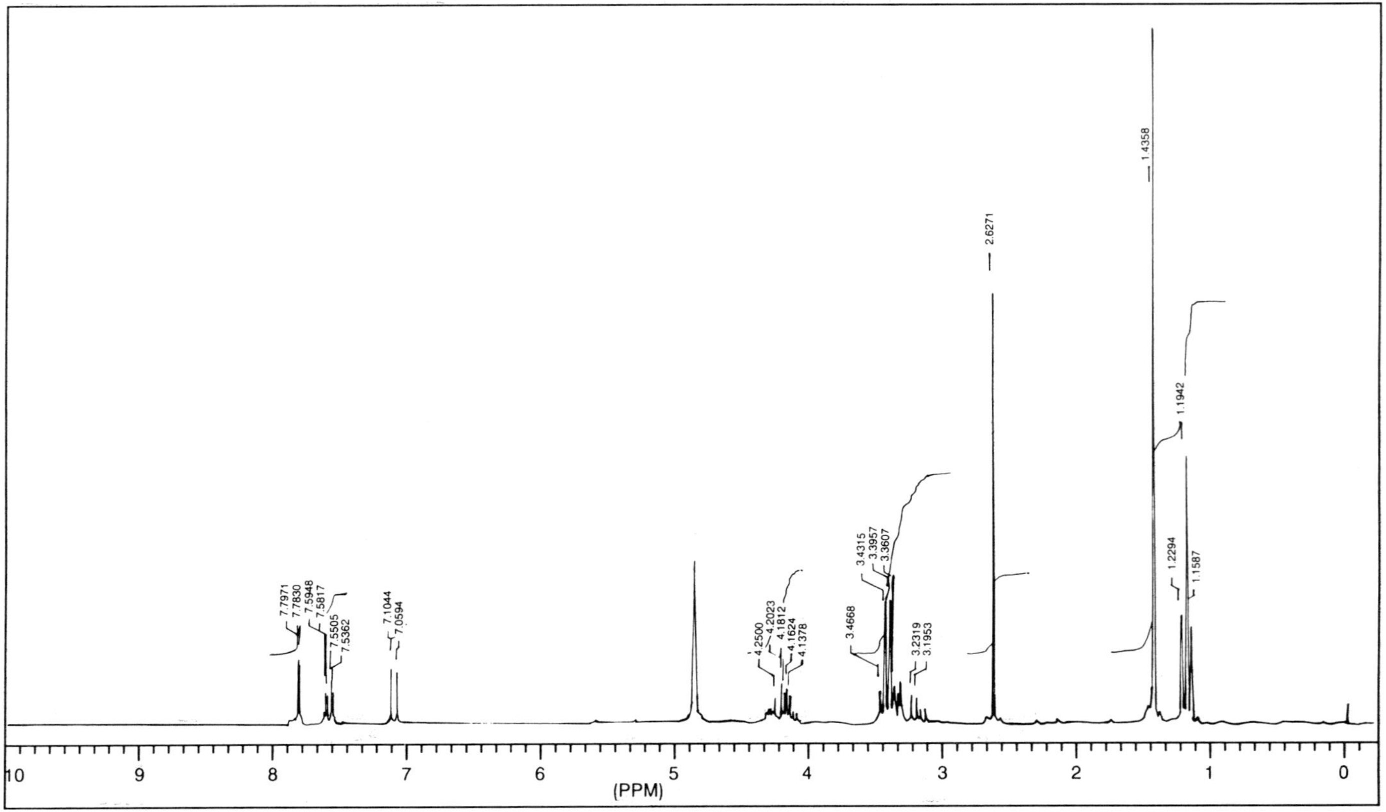

Figure 6. Proton NMR spectrum of celiprolol hydrochloride.

4.2.2 Carbon Magnetic Resonance Spectrum

The ^{13}C NMR spectrum shown in Figure 7 was obtained using a 200 mg/mL solution of celiprolol hydrochloride in deuterated methanol. The ^{13}C NMR spectrometer parameters and spectral assignments are listed in Tables IV and V, respectively. Note that in Figure 6 spectrally equivalent atoms are given the same numerical designation.

TABLE IV

Parameter Table for ^{13}C Spectrum of Celiprolol·HCl

Observe:

Nucleus	13.250	
Spec. Width	11001.1	Hz
Acq. Time	0.399	sec
Pulse Width	6.0	μsec
Freq.	50	MHz
Offset	0	Hz
Delay	2.000	sec
Transients	1024	

Decouple:

Nucleus	1.250	
Mode	YYY	
Modulation:Mode	S	
Pulse Width	17.5	μsec
Offset	-43.9	Hz
Power	5	db
Freq.	9900	Hz
Power Mode	64.0	

Plot/Processing:

FN	16	K			
LB	2.000	Hz			
Width	10407.5	Hz/ppm	Start	0	Hz/ppm

Experiment:

Pulse Sequence	STD13C
Solvent	CD_3OD

TABLE V

^{13}C NMR Spectral Assignments of Celiprolol Hydrochloride

Carbon No.	Chemical Shift (ppm)
1	134.57
2	125.36
3	128.43
4	157.57
5	114.65
6	129.18
8	155.01
10	42.42
11	14.12
12	72.52
13	66.76
14	46.04
16	58.49
17	25.79
18	201.38
19	31.41

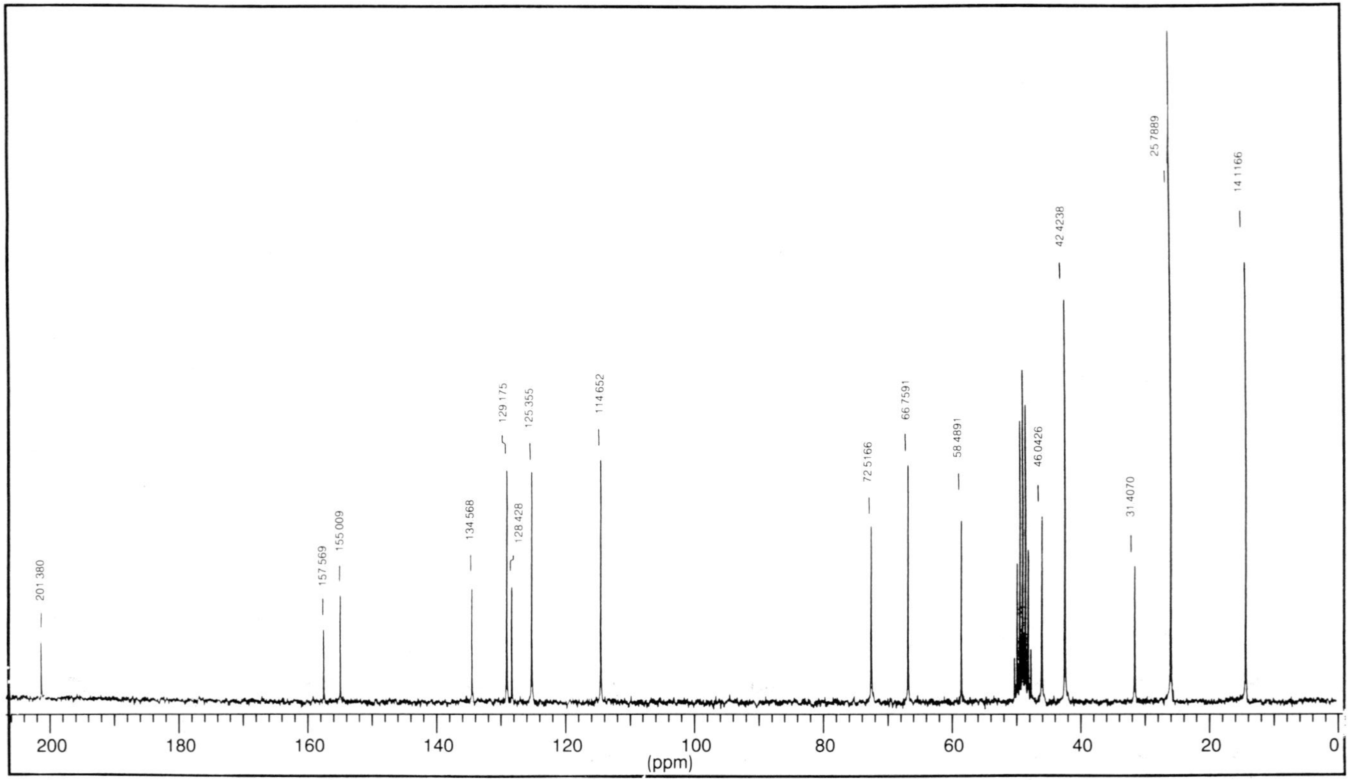

Figure 7. Carbon 13 NMR spectrum of celiprolol hydrochloride.

4.2.3 Proton Homonuclear Magnetic Resonance Spectrum

The proton-proton homonuclear correlated (COSY) NMR spectrum of celiprolol hydrochloride is shown in Figure 8. The spectrum was acquired using an approximately 200 mg/mL solution of celiprolol hydrochloride with deuterated methanol as a solvent. The spectrometer parameters used in obtaining this spectrum are listed in Table VI.

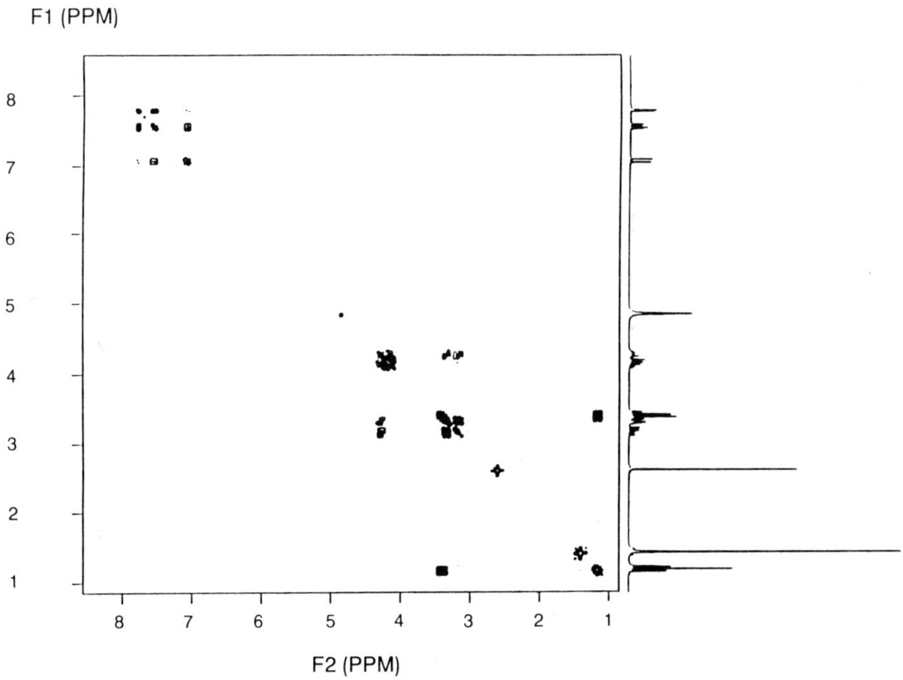

Figure 8. Proton-proton homonuclear correlated (COSY) NMR spectrum of celiprolol hydrochloride.

TABLE VI

Parameter Table for the Homonuclear Correlated (COSY) Spectrum of Celiprolol·HCl

Sample Celiprolol·HCl

Observe:

Nucleus	1.250	
Spec. Width	1542.0	Hz
Acq. Time	0.332	sec
Pulse Width	14.0	μsec
Freq.	200	MHz
Offset	-100	Hz
Delay	1.000	sec
Transients	32	

Decouple:

Nucleus	1.250	
Mode	NNN	
Modulation:Mode	C	
Offset	-43.9	Hz
Power	20	db
Freq.	200	Hz

Plot/Processing:

FN	1	K	RE	0.021	
LB	2.000	Hz	AF	0.083	
Width	1542.0	Hz/ppm	Start	172.0	Hz/ppm

Experiment:

Pulse Sequence	COSY
Solvent	CD_3OD

4.2.4 Proton-Carbon Heteronuclear Magnetic Resonance Spectrum

The ^{1}H-^{13}C heteronuclear correlated NMR spectrum of celiprolol hydrochloride is shown in Figure 9. The spectrum was obtained using an approximately 200 mg/mL solution of celiprolol hydrochloride with deuterated methanol as a solvent. The spectrometer parameters used to obtain this spectrum are listed in Table VII.

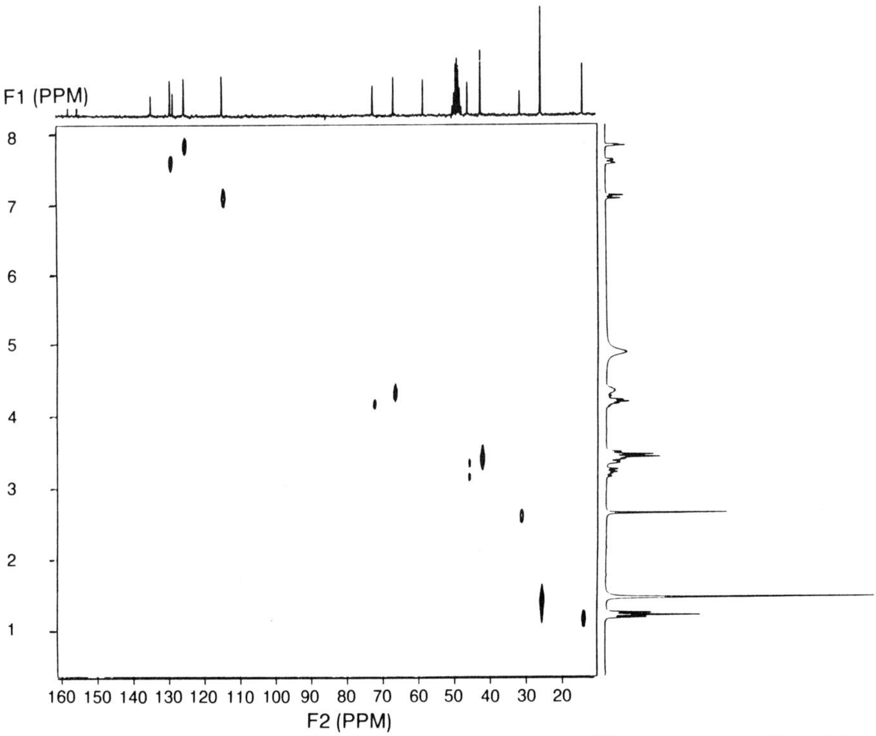

Figure 9. Proton-carbon 13 heteronuclear correlated NMR spectrum of celiprolol hydrochloride.

TABLE VII

Parameter Table for the ^{1}H-^{13}C Heteronuclear Correlated Spectrum of Celiprolol·HCl

Observe:

Nucleus	13.250	
Spec. Width	7587.3	Hz
Acq. Time	0.067	sec
Pulse Width	11.1	μsec
Freq.	50	MHz
Offset	-600	Hz
Delay	1.000	sec
Transients	192	

Decouple:

Nucleus	1.250	
Mode	NNY	
Modulation:Mode	CCE	
Pulse Width	13.0	μsec
Offset	-200.0	Hz
Power	5	db
Freq.	9900	Hz
Power Mode	64.0	

Plot/Processing:

FN	1	K	RE	0.004	sec
AF	0.017	sec			
Width	7587.3	Hz/ppm	Start	519.7	Hz/ppm

Experiment:

Pulse Sequence	HETCOR
Solvent	CD_3OD

4.3 Ultraviolet-Visible Spectrum

The ultraviolet-visible absorption spectrum for celiprolol hydrochloride was obtained using a solution with a concentration of 10.4 mg/mL with 0.1 N HCl (aq) as a solvent (30). The spectrum (Figure 10) exhibits characteristic absorption bands with maxima at 231 nm ($E_{1\%}$ = 660) and 324 nm ($E_{1\%}$ = 60). Absorptivity values for alkaline and neutral solvent systems are listed in Table VIII. All spectral characteristics were determined using a Beckman DU-6 spectrophotometer and a 1 cm spectral pathlength cell.

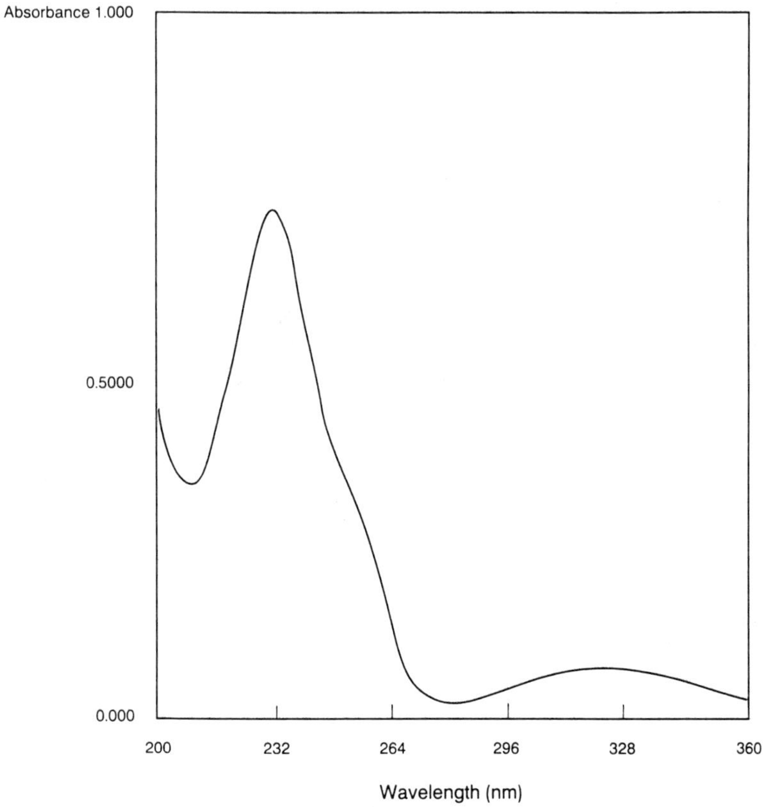

Figure 10. Ultraviolet absorption spectrum of celiprolol hydrochloride in 0.1 N HCl (aq.).

TABLE VIII

Ultraviolet Spectrum Absorption Maxima of Celiprolol Hydrochloride in Various Solvents

Solvent	λ_{max} (nm)	$E_{1\%}$
H_2O	231	652
	324	57
0.01 N HCl (aq)	231	660
	324	60
0.01 N NaOH (aq)	231	640
	324	60
Methanol	232	775
	329	58

4.4 Mass Spectrum

A mass spectrum of celiprolol hydrochloride taken in either the chemical ionization (CI) mode or the electron impact (EI) mode is shown in Figure 11 or 12, respectively (31). Both spectra were obtained using a Finnigan 4500 mass spectrometer. Isobutane was the reagent gas for the CI spectrum. The EI spectrum was obtained at 70 eV. The molecular ion and fragmentation pattern are consistent with the chemical structure of celiprolol. While no molecular ion corresponding to the free base (m/e = 379) was observed, a strong $[M+H]^+$ ion (m/e = 380) was seen. Principal mass spectral fragmentations of celiprolol hydrochloride are shown in Figure 13.

4.5 Celiprolol hydrochloride, having one chiral carbon is optically active. Since the drug substance is provided as the racemic mixture, however, a solution of celiprolol hydrochloride does not rotate the plane of polarized light. In other words, the optical rotation of a 5% (w/v) solution of celiprolol hydrochloride in 0.1 N HCl (aq) taken at 231 nm and 25°C $[A^{25}_{231}]$ is 0°.

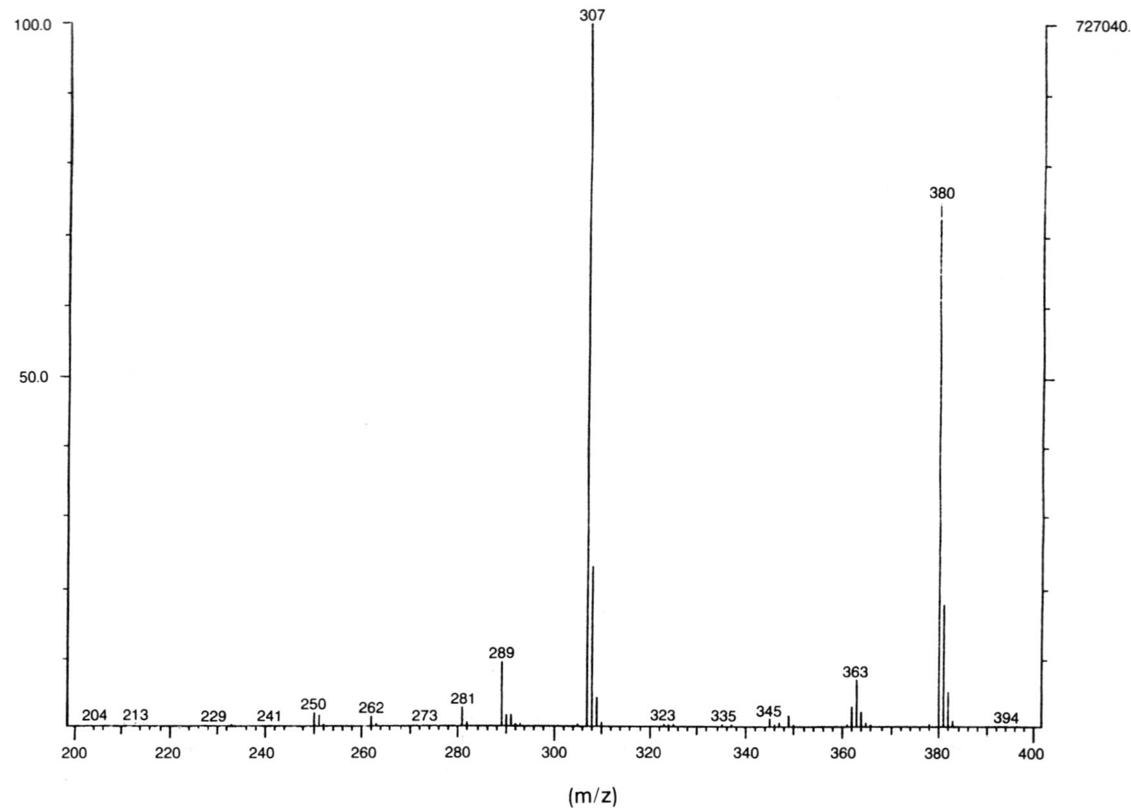

Figure 11. Mass spectrum of celiprolol hydrochloride taken in the chemical ionization mode.

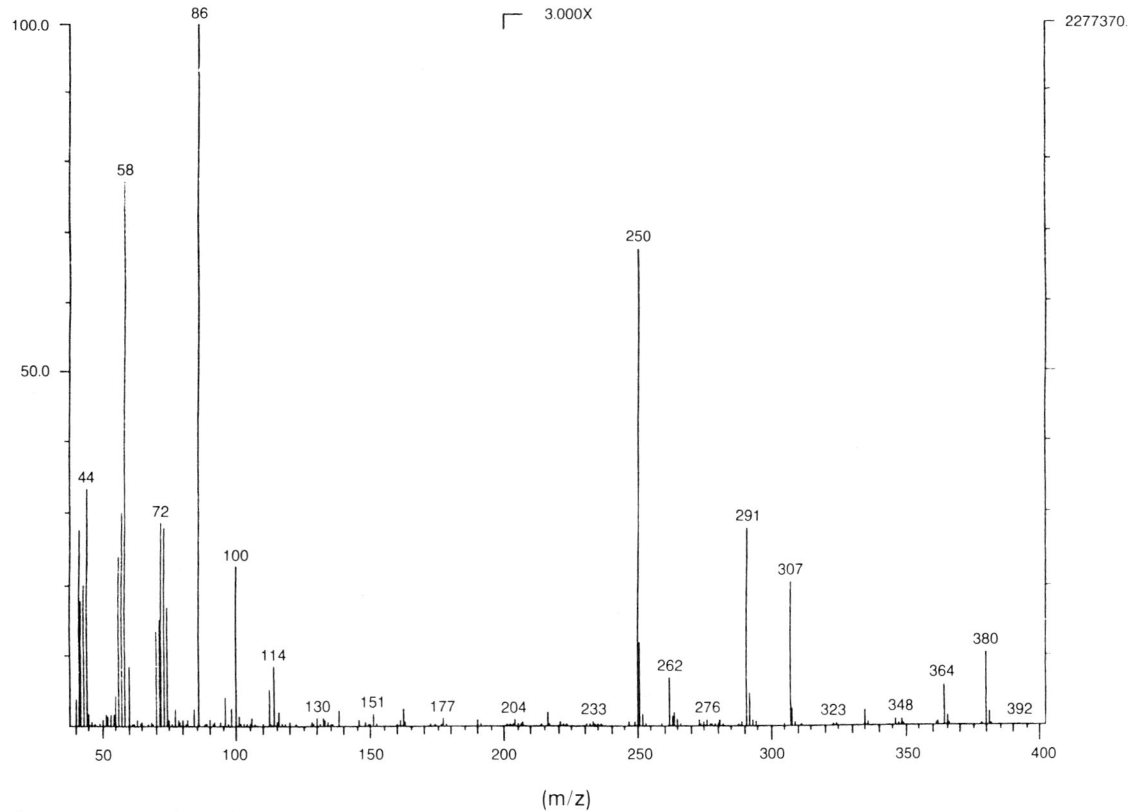

Figure 12. Mass spectrum of celiprolol hydrochloride taken in the electron impact mode.

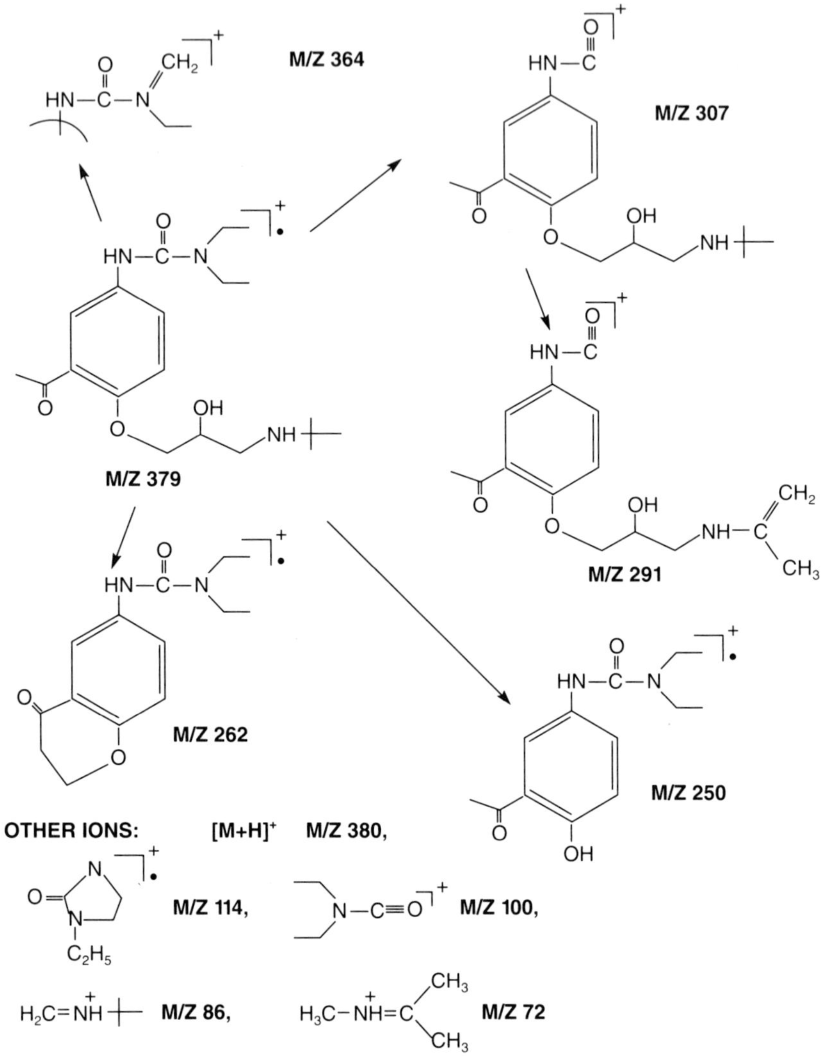

Figure 13. Principal mass spectral fragmentations of celiprolol hydrochloride.

4.6 Electroanalytical Behavior

Celiprolol hydrochloride is electrochemically active and it undergoes an irreversible 2 electron oxidation. A cyclic voltammogram of an approximately 24 ppm solution of celiprolol hydrochloride in aqueous 0.1 M KCl is shown in Figure 14. The $E_{\frac{1}{2}}$ of this solution is approximately +1.06 volts (vs SCE). The voltammogram was obtained using an EG&G Princeton Applied Research Model 273 potentiostat with a model 5210 Lock-In Amplifier. The working electrode was glassy carbon, the reference electrode was saturated calomel and a platinum auxiliary electrode was used. The mechanism of electro-oxidation is illustrated in Figure 15 (32).

4.7 Thermoanalytical Behavior

4.7.1 Melting Point

The melting point of celiprolol hydrochloride varies between 197°C and 200°C (with decomposition) as determined by the USP class Ia capillary method.

4.7.2 Loss on Drying

Weight loss is typically <0.5% for an accurately weighed sample of celiprolol hydrochloride heated to constant weight at 105°C under vacuum.

4.7.3 Differential Thermal Analysis Behavior

The differential thermal analysis (DSC) behavior of celiprolol hydrochloride is shown in Figure 16 (33). The thermogram was obtained using a Perkin Elmer Series 7 DSC scanning from 40°C to 250°C at 5°C/minute. A primary endotherm corresponding to melting is observed at a peak onset temperature of ~200°C.

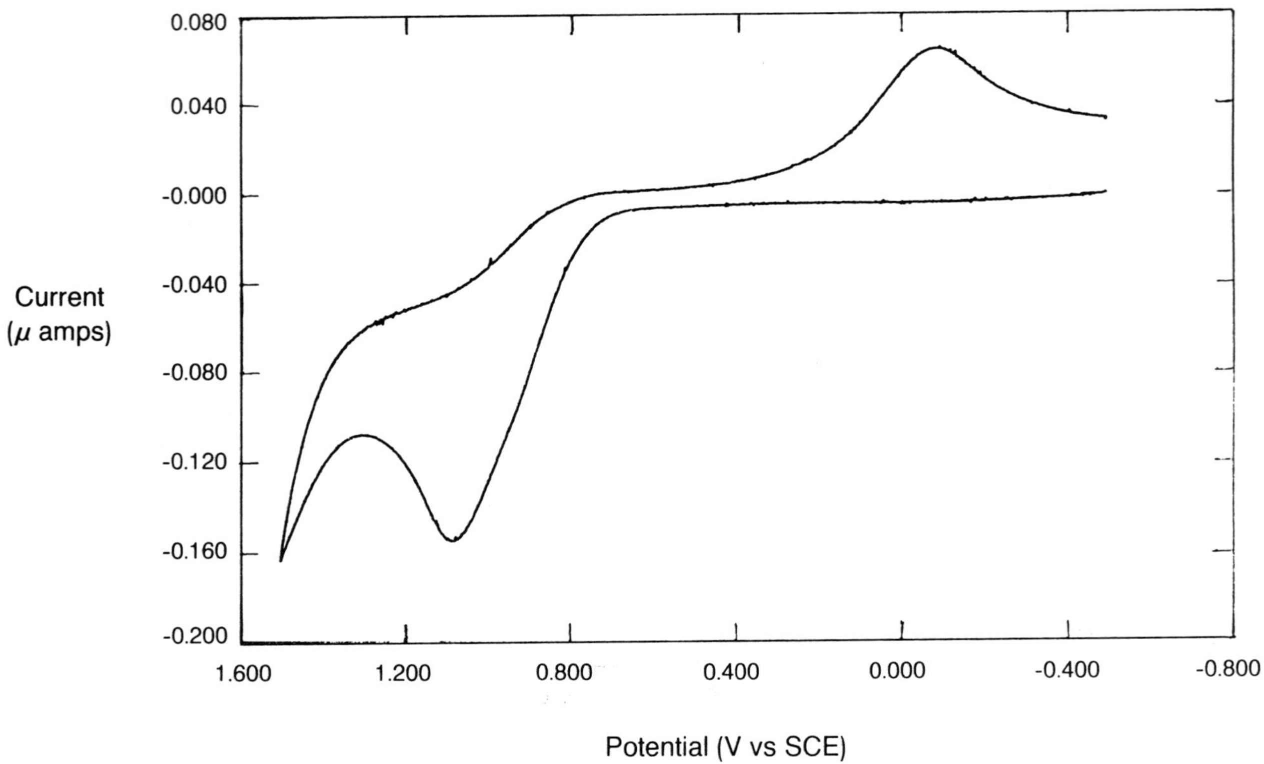

Figure 14. Cyclic voltammogram of celiprolol hydrochloride in 0.1 N KCl (aq.) obtained using a glassy carbon working electrode, saturated calomel reference electrode and a platinum auxiliary electrode.

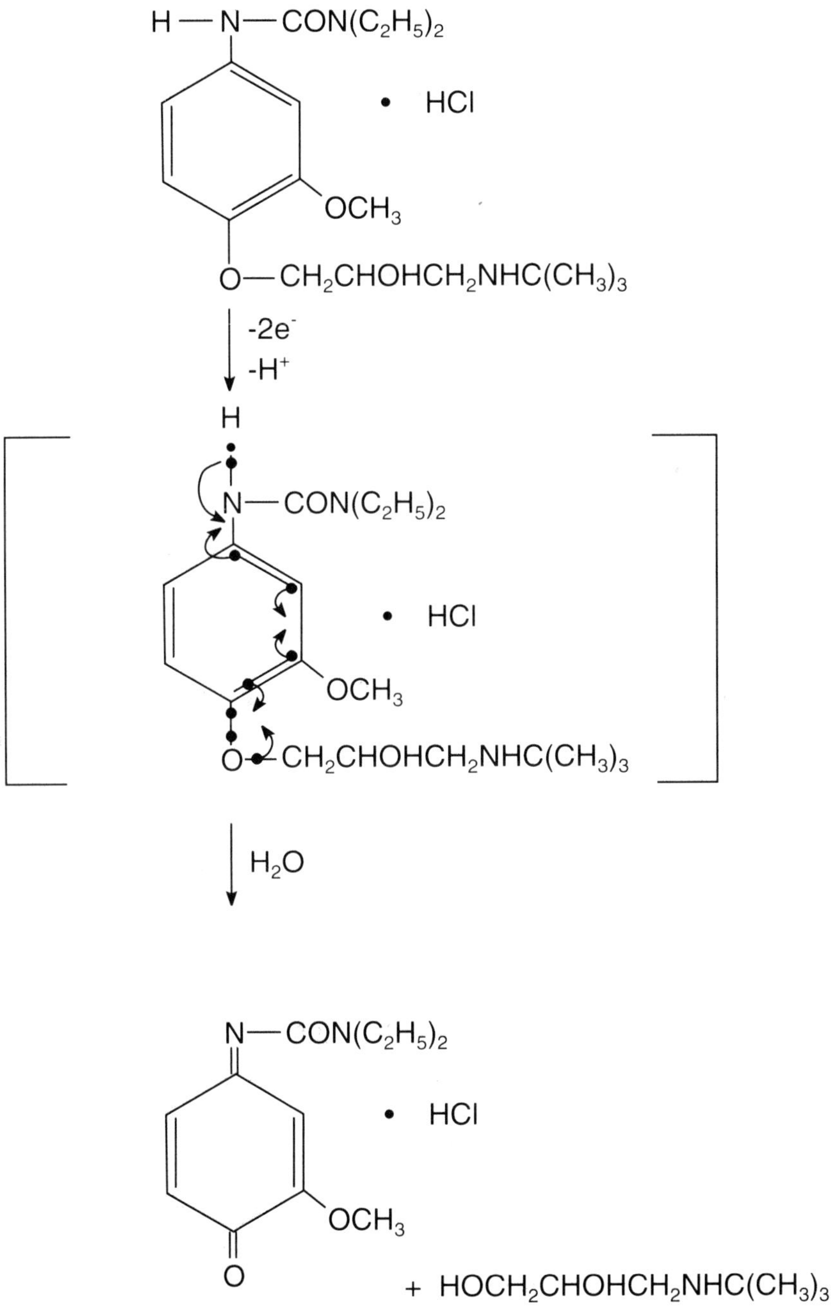

Figure 15. Mechanism of electro-oxidation of celiprolol hydrochloride.

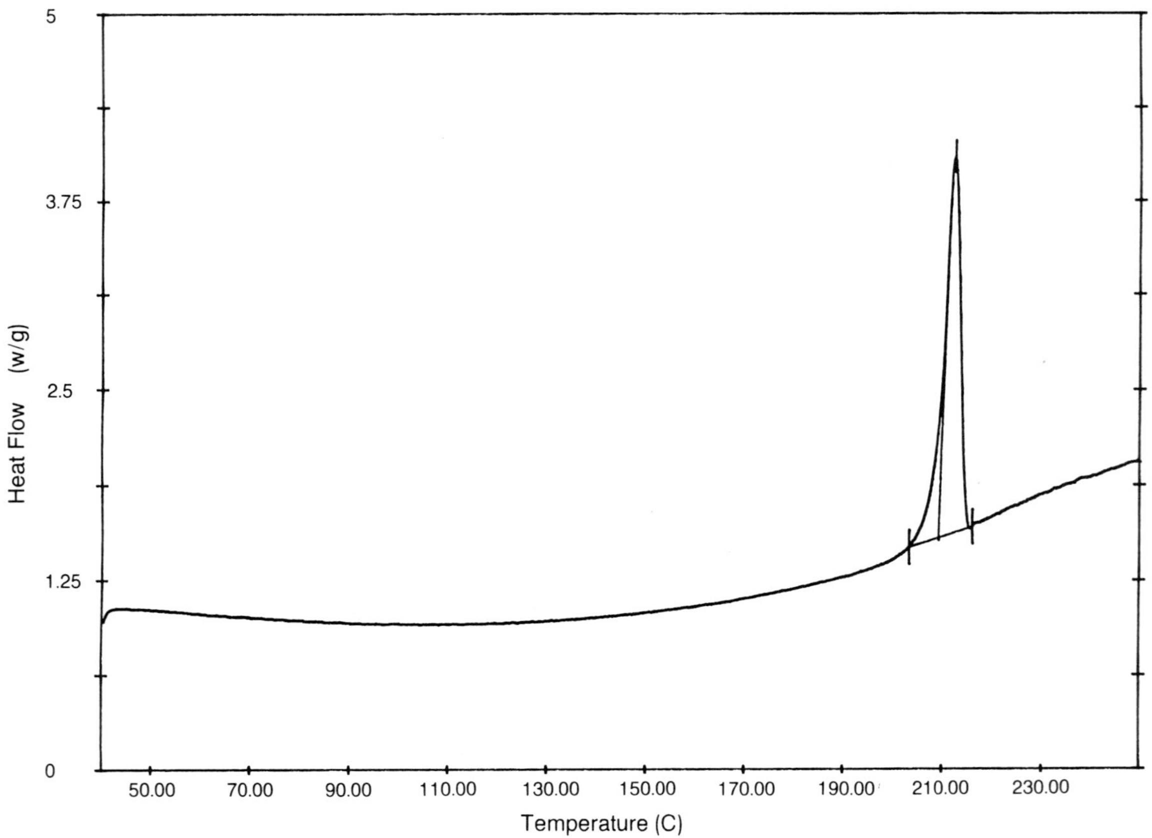

Figure 16. Differential thermoanalytical behavior or celiprolol hydrochloride.

4.7.4 Thermogravimetric Analysis Behavior

Thermogravimetric analysis of celiprolol hydrochloride indicates no weight loss until a temperature of 195°C is reached. Significant weight loss is observed above 200°C due to decomposition/vaporization. A typical TGA curve is shown in Figure 17. The curve was obtained using a Perkin Elmer Series 7 TGA scanning from 40°C to 250°C at 10°C/minute. A 5.232 mg sample of celiprolol·HCl was used (33).

4.7.5 Residue on Ignition

Typically <0.1% residue is obtained when a sample (1 gram) of celiprolol hydrochloride is ignited in a platinum crucible in the presence of concentrated H_2SO_4 in a muffle furnace at 800°C until no visible residue remains.

4.8 Solubilities

The solubility of celiprolol hydrochloride in several solvents at room temperature (~25°C) is shown in Table IX (26).

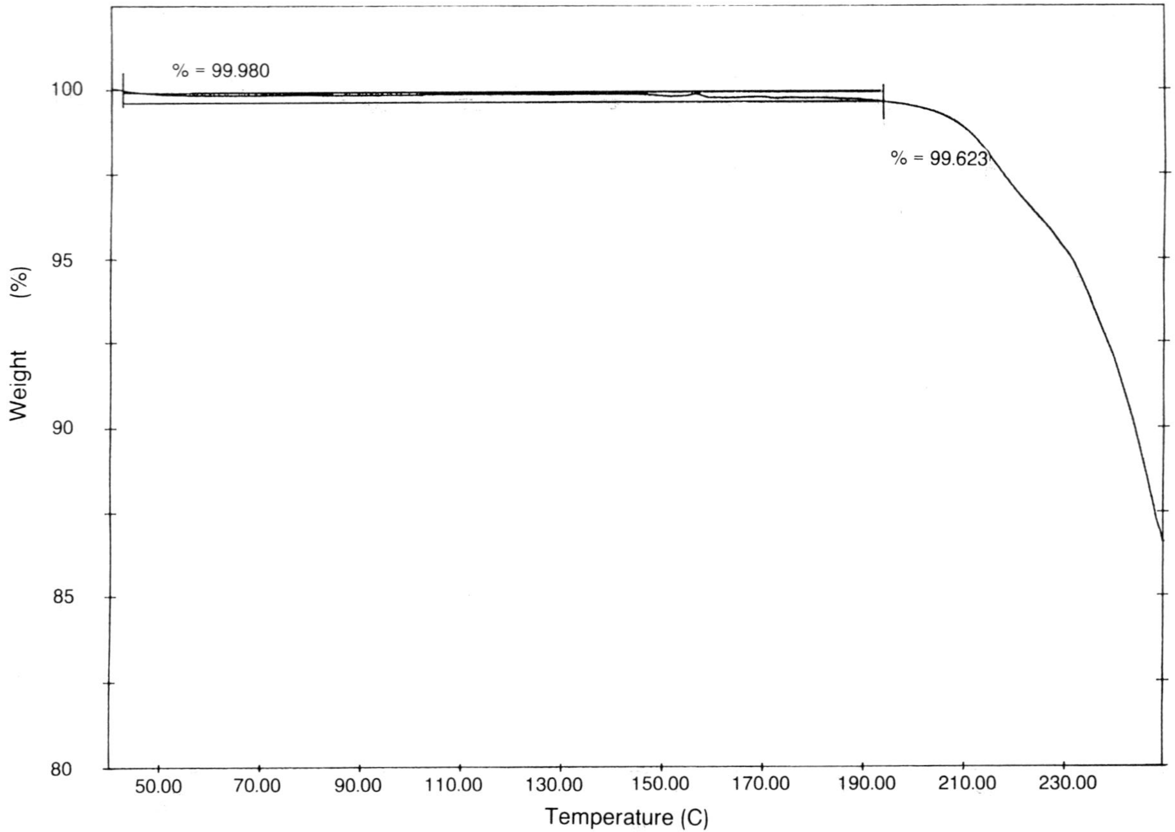

Figure 17. Thermogravimetric behavior of celiprolol hydrochloride.

TABLE IX

Solubility of Celiprolol Hydrochloride

Solvent	Solubility (g/100mL)
H_2O	15.1
CH_3OH	18.2
CH_3CH_2OH	1.61
$CHCl_3$	0.42

4.9 Crystal Properties

4.9.1 Polymorphism

Celiprolol hydrochloride exists as a crystalline powder with several polymorphic forms having been identified (34). Form I is the typical polymorph for the hydrochloride salt. It is obtained from organic solvents under a wide variety of conditions. Form I has the appearance of long, thin needles. An amorphous substance will be formed with Form I, however, if crystallization proceeds too rapidly. Form III is obtained when celiprolol hydrochloride is crystallized from aqueous solutions. Form III typically has a granular appearance often mixed with coarse lath-shaped crystals. Form III is readily distinguished from Form I by its lower melting point (~150°C vs ~200°C for Form I and III, respectively) and distinguishing infrared absorption spectrum.

4.9.2 X-ray Crystallography

Attempts to produce single crystals of celiprolol·HCl for x-ray crystallographic have been unsuccessful. Suitable crystals of the free base, celiprolol, have been obtained and studied, however (35). Crystallographic information was obtained using a Enraf-Nonius CAD-4 diffractometer using CuKα radiation. Tables X and XI list select crystallographic parameters from this study.

TABLE X

Fractional Coordinates and Estimated Standard Deviations (in parentheses) for Celiprolol

Atom	x	y	z
O1	0.6319(3)	0.3361(1)	-0.2063(1)
O2	0.3707(3)	0.0425(1)	0.0746(1)
O3	0.2756(3)	0.3285(1)	0.0986(1)
O4	0.1463(2)	0.5327(1)	0.2022(1)
N1	0.6922(3)	0.1933(2)	-0.2586(1)
N2	0.5511(3)	0.1975(2)	-0.1545(1)
N3	0.2583(3)	0.4050(2)	0.3261(2)
C1	0.3374(3)	0.3010(2)	0.0341(1)
C2	0.3650(3)	0.2014(2)	0.0253(1)
C3	0.4384(3)	0.1722(2)	-0.0375(2)
C4	0.4812(3)	0.2360(2)	-0.0921(2)
C5	0.4484(4)	0.3338(2)	-0.0834(2)
C6	0.3792(4)	0.3648(2)	-0.0209(2)
C7	0.6262(3)	0.2462(2)	-0.2060(2)
C8	0.7710(4)	0.2455(3)	-0.3152(2)
C9	0.6729(5)	0.2909(3)	-0.3788(2)
C10	0.7100(5)	0.0872(3)	-0.2572(2)
C11	0.6257(6)	0.0382(3)	-0.3235(3)
C12	0.3183(4)	0.1234(2)	0.0780(2)
C13	0.2039(5)	0.1412(3)	0.1326(2)
C14	0.2477(4)	0.4292(2)	0.1107(2)
C15	0.1944(4)	0.4357(2)	0.1901(2)
C16	0.3170(4)	0.4081(3)	0.2515(2)
C17	0.3586(4)	0.3699(3)	0.3907(2)
C18	0.2791(7)	0.3933(6)	0.4626(2)
C19	0.3782(8)	0.2611(4)	0.3818(4)
C20	0.5079(5)	0.4203(4)	0.3974(3)

TABLE XI

Bond Distances (Å), Angles (°), Selected Torsion Angles (°) and Estimated Standard Deviations (in parentheses) for the C, N and O Atoms in Celiprolol

Bond distances

Atoms		Distance	Atoms		Distance
O1	C7	1.242(3)	C2	C3	1.393(4)
O2	C12	1.217(3)	C2	C12	1.501(4)
O3	C1	1.356(3)	C3	C4	1.379(4)
O3	C14	1.432(3)	C4	C5	1.393(4)
O4	C15	1.429(3)	C5	C6	1.376(4)
N1	C7	1.354(4)	C8	C9	1.498(5)
N1	C8	1.461(4)	C10	C11	1.494(6)
N1	C10	1.472(4)	C12	C13	1.492(5)
N2	C7	1.355(4)	C14	C15	1.513(4)
N2	C3	1.413(3)	C15	C16	1.524(4)
N3	C16	1.452(4)	C17	C20	1.520(6)
N3	C17	1.468(4)	C17	C19	1.522(6)
C1	C6	1.381(4)	C17	C18	1.527(6)
C1	C2	1.407(4)	C17	C18	1.537(6)

Bond angles

Atoms			Angle	Atoms			Angle
C1	O3	C14	119.1(2)	O1	C7	N2	121.5(3)
C7	N1	C8	117.8(3)	N1	C7	N2	117.5(3)
C7	N1	C10	125.4(3)	N1	C8	C9	114.4(3)
C8	N1	C10	116.2(3)	N1	C10	C11	112.9(4)
C7	N2	C4	127.8(3)	O2	C12	C13	118.6(3)
C16	N3	C17	116.6(3)	O2	C12	C2	119.7(3)
O3	C1	C6	123.9(3)	C13	C12	C2	121.6(3)
O3	C1	C2	117.0(2)	O3	C14	C15	105.5(2)
C6	C1	C2	119.0(3)	O4	C15	C14	108.5(2)
C3	C2	C1	118.1(3)	O4	C15	C16	110.2(2)
C3	C2	C12	117.1(2)	C14	C15	C16	111.0(3)
C1	C2	C12	124.8(3)	N3	C16	C15	109.6(3)
C4	C3	C2	122.9(3)	N3	C17	C20	113.4(3)
C3	C4	C5	117.9(3)	N3	C17	C19	108.6(3)
C3	C4	N2	117.8(3)	N3	C17	C18	104.7(3)
C5	C4	N2	124.3(3)	C20	C17	C19	110.4(4)
C6	C5	C4	120.3(3)	C20	C17	C18	108.7(4)
C5	C6	C1	121.8(3)	C19	C17	C18	110.9(5)
O1	C7	N1	121.0(3)				

TABLE XI (Continued)

Bond Distances (Å), Angles (°), Selected Torsion Angles (°) and Estimated Standard Deviations (in parentheses) for the C, N and O Atoms in Celiprolol

Selected torsion angles

Atoms				Angle	Atoms				Angle
O1	C7	N1	C8	.3(6)	C1	C2	C12	C13	16.5(5)
O1	C7	N1	C10	171.1(3)	C2	C1	O3	C14	179.0(3)
O1	C7	N2	C4	-5.5(5)	C3	C4	C5	C6	1.8(5)
O2	C12	C2	C3	15.3(4)	C3	C4	N2	C7	-166.4(3)
O2	C12	C2	C1	-166.0(3)	C3	C2	C12	C13	-162.2(3)
O3	C1	C6	C5	176.6(3)	C5	C4	N2	C7	15.8(5)
O3	C14	C15	O4	-172.6(2)	C6	C1	O3	C14	1.6(5)
O3	C14	C15	C16	66.2(3)	C7	N1	C8	C9	-76.7(4)
O4	C15	C16	N3	65.8(4)	C7	N1	C10	C11	115.0(4)
N1	C7	N2	C4	176.0(3)	C8	N1	C10	C11	-74.1(5)
N2	C7	N1	C8	178.8(3)	C9	C8	N1	C10	111.7(4)
N2	C7	N1	C10	-10.4(5)	C15	C16	N3	C17	174.0(3)
N2	C4	C3	C2	-178.1(3)	C16	N3	C17	C20	51.0(5)
N2	C4	C5	C6	179.6(4)	C16	N3	C17	C19	-72.1(5)
N3	C16	C15	C14	-174.0(3)	C16	N3	C17	C18	169.3(4)
C1	O3	C14	C15	-174.3(3)					
C1	C6	C5	C4	-1.4(6)					

4.10 Hygroscopicity

The hydrochloride salt of celiprolol is a stable anhydrous compound. No hydrates of celiprolol hydrochloride have been reported. Anhydrous celiprolol hydrochloride is non-hygroscopic when stored at approximately ~25°C in the presence of relative humidity ranging from ~30% to ~80%.

4.11 Dissociation Constant

The pKa of celiprolol base obtained by potentiometric titration in water at 25°C is approximately 9.7 (36). The native pH of a saturated aqueous solution of celiprolol hydrochloride is ~6.8 (30).

4.12 Distribution Constant

The distribution constant for racemic celiprolol hydrochloride was determined in an octanol/aqueous phosphate buffer (pH=7.4) system at 37°C. The value of the partition coefficient (organic/aqueous) is 0.8. This value was obtained using a spectrophotometric method measuring equilibrium concentrations in both phases. Distribution constants were similarly determined for chloroform/water and chloroform/pH=7.4 aqueous phosphate buffer and were found to be 0.14 and 2.42, respectively (36).

5. METHODS OF ANALYSIS

5.1 Identification Tests

5.1.1 Ultraviolet Spectrophotometry

Identification testing of celiprolol hydrochloride drug substance and the tablet granulation, cores or coated tablets can be accomplished by UV absorbance spectrophotometric analysis. Celiprolol hydrochloride is extracted from the tablet matrix using water and 2.0 mL of concentrated hydrochloric acid. The samples are then centrifuged and the supernatant is diluted appropriately in 0.01 N hydrochloric acid to obtain the ultraviolet spectrum. The solution is scanned from 200 nm to 400 nm and the resulting spectrum exhibits the same absorption characteristics at the same wavelengths as a similarly prepared and concomitantly measured celiprolol hydrochloride reference standard. Quantitatively, the equimolar sample and standard solution will exhibit UV absorbance maxima at ~324 nm and ~231 nm which differ by no more than 3%.

5.1.2 Infrared Spectrophotometry

Infrared spectroscopy can be used to identify celiprolol hydrochloride. The infrared absorption spectrum of celiprolol is obtained as a potassium bromide disc. A disc is made from 1 mg of sample in about 200 mg of potassium bromide. The infrared absorption spectrum of a potassium bromide dispersion of celiprolol hydrochloride exhibits maxima only at the same wavelength as that of a similar preparation of celiprolol hydrochloride reference standard.

5.1.3 Elemental Analysis

The identity of celiprolol hydrochloride may also be determined by an elemental weight percent determination of carbon, hydrogen, nitrogen and chlorine. The elemental analysis of a celiprolol sample is performed and the results are compared to the following theoretical values (36).

Element	Theory (%)
C	57.75
H	8.24
N	10.10
Cl	8.52

5.1.4 High Performance Liquid Chromatography

A stability-indicating HPLC procedure is used for the identity testing of celiprolol hydrochloride. The retention time of the major peak in the sample chromatogram should correspond to that of the celiprolol hydrochloride reference standard (approximately 7 minutes). The HPLC procedure is described under Section 5.4.2. This method is suggested as a secondary identification procedure.

5.2 Titrimetric Analysis

Celiprolol hydrochloride may be determined by a quantitative non-aqueous titration in glacial acetic acid. A celiprolol hydrochloride solution is prepared in glacial acetic acid with a small amount of mercuric acetate. The solution is titrated potentiometrically with standardized 0.1 N perchloric acid titrant using a combination glass/calomel electrode (36).

5.3 Spectrophotometric Analysis

5.3.1 Direct Ultraviolet Spectrophotometry

Celiprolol hydrochloride is quantitatively analyzed by acquiring the ultraviolet spectrum in 0.04 N hydrochloric acid. The absorption of celiprolol at 232 nm is used for the quantitative determination. The amount of celiprolol·HCl is determined by comparison of the absorbance reading with that of a standard solution of known concentration taken at 232 nm using a suitable ultraviolet spectrophotometer (37).

5.3.2 Colorimetry

Quantitation of celiprolol in aqueous solutions may be accomplished by colorimetry. The urea group of celiprolol is hydrolyzed and the resulting aromatic amine is converted via diazotisation and coupling to a blue azo dye. The celiprolol·HCl solution can be monitored at 565 nm and its concentration determined by comparison to the absorbance at 565 nm of a celiprolol·HCl standard of known concentration similarly prepared (38).

5.3.3 Specific Rotation

Celiprolol hydrochloride is synthesized as a racemate. The dextrorotatory enantiomer has a specific rotation of $(\alpha)_D^{24°} = +7.7°$ (methanol), and the levorotatory enantiomer has a specific rotation of $(\alpha)_D^{24°} = -7.7°$ (methanol). Racemic celiprolol·HCl solutions will not rotate the plane of polarized light.

5.4 Chromatographic Analysis

5.4.1 Thin-layer Chromatography

Three celiprolol·HCl specific normal-phase thin-layer chromatographic methods are available which separate celiprolol from its impurities. The methods use a silica gel $60F_{254}$ plate (250 μm particle size, E. Merck) with visualization using shortwave UV light at 254 nm and iodine staining. The TLC systems are listed in Table XII. Standard and sample solutions are prepared in methanol to produce a 1% (w/v) solution of celiprolol·HCl. The chamber is allowed to equilibrate with mobile phase for one hour. Ten microliters of standard and sample solution are applied to the plates and then developed for 10 cm (36). Detection is accomplished by shortwave UV light and an iodine chamber.

Another thin-layer chromatographic method utilizes a silica gel GF plate and a solvent system comprised of ether:methanol:ammonium hydroxide (80:20:5). Shortwave UV light is used to visualize the analyte.

A third TLC method uses a silica gel GF plate and a solvent system comprised of acetone:chloroform: acetic acid:water (30:35:20:15). Shortwave UV light is used to visualize celiprolol·HCl.

Celiprolol has also been separated from 19 other β-blockers on a silica gel $60F_{254}$ plate. Ascending development of the separation was achieved with the following systems:

Solvent System	Rf
chloroform/methanol (90:10)	3
ethyl acetate-methanol-concentrated ammonia (85:10:5)	33
methanol	15
methanol-n-butanol (60:40): 0.1 M NaBr	71
methanol/concentrated ammonia (100:1.5)	40*
chloroform/methanol (90:10)	2*
acetone	3*

*silica impregnated with 0.1 M KOH and dried

Detection for most of the β-blockers (including celiprolol) was by UV-visualization at 254 nm (39).

TABLE XII

Thin Layer Chromatography of Celiprolol·HCl

SYSTEM*		Rf VALUE
1	Ethyl Acetate, Ethanol, Ammonia 2M 50:25:20	0.55
2	Diethylether, Methanol, Ammonia Conc. 80:20:5	0.60
3	Acetone, Chloroform, Acetic Acid, Water 35:30:20:15	0.53

*Separating Layer - Silica Gel $60F_{254}$ (E. Merck) 250 micron particle size.

5.4.2 High Performance Liquid Chromatography

A stability-indicating high performance liquid chromatographic method (40) is available for celiprolol·HCl. The drug substance is separated from its related substances (Figure 18) using a Radial-pak cyano column (8 mm ID x 100 mm, 10 μm particle size) and a mobile phase consisting of a 0.05 M monosodium phosphate buffer (adjusted to a pH of 4.0 ± 0.05 with phosphoric acid) and acetonitrile (60:40, v/v) at a flow rate of 2.2 mL/minute. Detection of the analytes is achieved at ambient temperature using UV absorbance at 254 nm. The standard and sample solutions are prepared in a diluent of HPLC grade water and acetonitrile (50:50) and 20 μL are injected onto the column. The retention time of celiprolol is approximately 7 minutes. All related substances are eluted within 30 minutes and may also be quantitated. A typical HPLC chromatogram is shown in Figure 19. The celiprolol related substances are eluted with the following times:

A-1362	3 minutes
A-1354	4 minutes
St-1429	5 minutes
St-1414	8 minutes
St-1453	19 minutes
St-1444	26 minutes

Another HPLC method (41) which uses anthracene as an internal standard is available for celiprolol hydrochloride. The chromatographic conditions are very similar to the previous method. Separation is achieved at ambient temperature with a Waters Radial-pak cyano column (8 mm ID x 100 mm, 10 μm particle size) and a mobile phase composed of 0.05 M monosodium phosphate (adjusted to a pH of 4.0) and acetonitrile (60:40 v/v) at a flow rate of 2.5 ml/minute. Detection is achieved by UV absorbance at 254 nm. The standard and sample solutions are prepared in a diluent of HPLC grade water and acetonitrile (2:8) and 20 μL is injected onto the

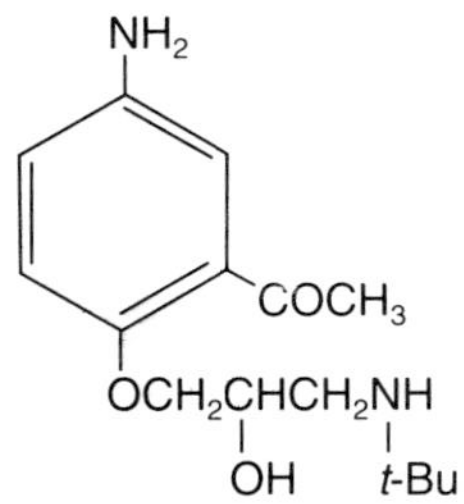

ST 1429

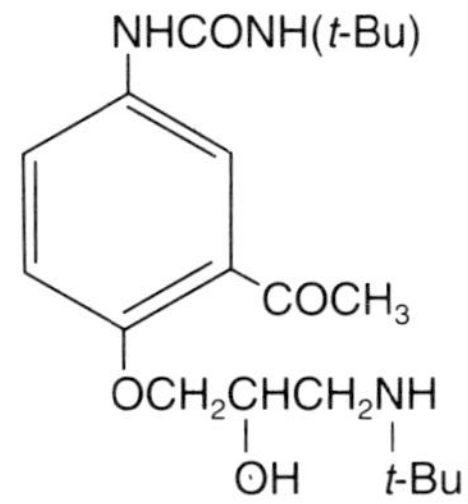

ST 1414

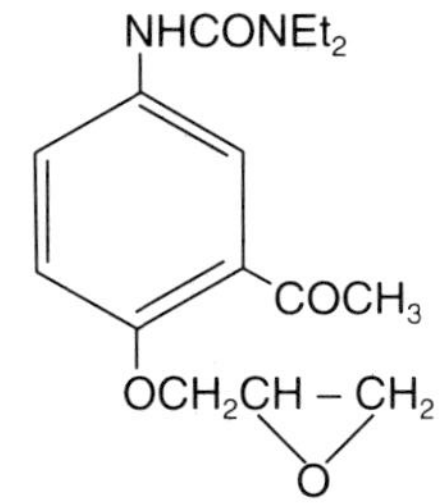

A 1362

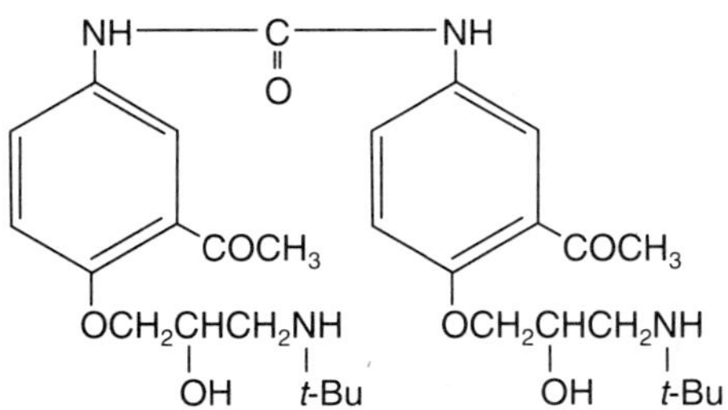

ST 1444

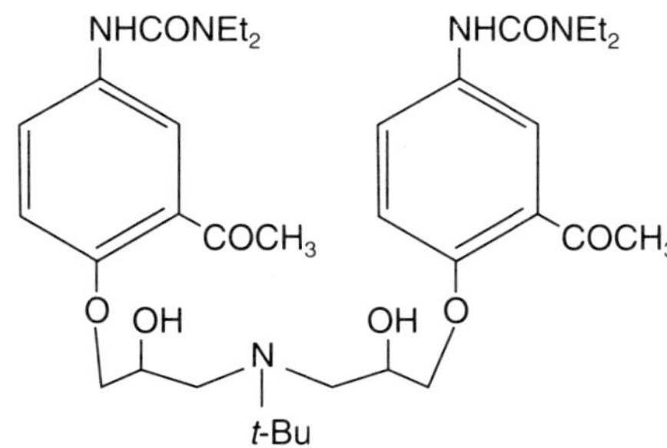

ST 1453

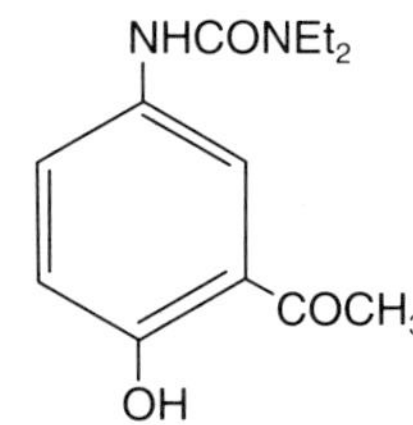

A 1354

Figure 18. Chemical structure of the related substances of celiprolol hydrochloride.

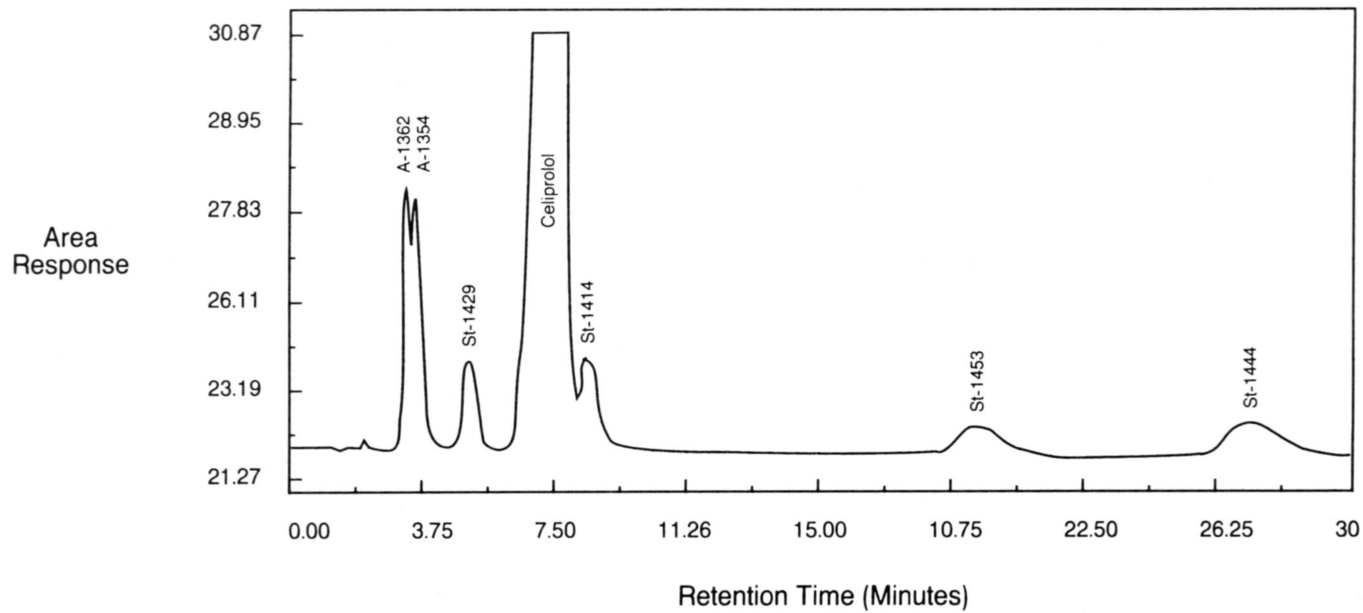

Figure 19. Typical HPLC chromatogram of celiprolol hydrochloride. See section 5.4.2 for chromatographic system parameters.

column. Celiprolol has a retention time of 7.0 minutes and the anthracene internal standard of 11.0 minutes. Under these conditions, celiprolol is separated from its known potential impurities.

Celiprolol·HCl can also be separated from its synthetic precursors, organic impurities and potential decomposition products using a gradient HPLC system employing a Whatman Partisil C8 RAC (100 mm x 4.6 mm I.D., 5 μm particle size) analytical column. Detection is achieved by UV absorbance at 254 nm. Celiprolol and its related substances are eluted with a mobile phase under the following gradient conditions: 15 minutes at 72% A (28% B) and then a 20 minute linear gradient from 72% A (28% B) to 20% A (80% B) [A is 36 mM triethylamine, 50 mM acetic acid (pH 5.0); B is methanol]. The eluent is pumped at ambient temperature at 1.5 mL/minute. This method is capable of detecting and quantitating possible impurities at a level of 0.2% relative to celiprolol·HCl on a weight basis (42).

Another reversed-phase HPLC method for celiprolol uses a C-18 chromatographic column with fluorescence detection (335nm/472nm). The limit of detection is 5 ng/mL (43). The separation of celiprolol into its enantiomers can be achieved using the following two high performance liquid chromatographic methods. The first method allows celiprolol to be separated into its respective enantiomers by HPLC with the use of diasteomeric tartaric acid monoesters. The derivatization reactions are performed by adding several nanograms to several milligrams of free celiprolol to 1 mL or less of dry, aprotic solvent, an excess of trichloroacetic acid and about a three-fold excess of the chiral reagent, the tartaric acid anhydride. The reaction vessel is sealed and placed in an oven at 50°C for several hours. Two different chromatographic procedures may then be used. Separation of celiprolol enantiomers is accomplished with O,O-Diacetyl (R,R)-Tartaric acid anhydride using a Spherisorb RP-18 HPLC column (250mm x 4.6 mm I.D., 5 μm particle size) and a mobile phase of 2% acetic acid in water (adjusted with ammonia to pH 3.7):methanol (50:50) at a flow rate of 1.5 mL/min. Detection is by UV absorbance at 254 nm. The capacity factors of the R and S enantiomers are 0.65 and 1.50, respectively. A second tartaric acid monoester, O,O-Dibenzoyl tartaric acid anhydride has also been used to separate the celiprolol enantiomers. The chromatographic system consists of a Polygosil RP-18 analytical column (250 mm x 4.6 mm I.D, 7 μm particle size), and a mobile phase of 2% acetic acid in water (adjusted with ammonia to a pH of 3.7:methanol (35:65) at a flow rate of 1.5 mL/minute. Detection is by UV absorbance at 254 nm. The capacity factors of the R and S enantiomers are 1.13 and 3.0, respectively (44).

Another chromatographic method available for the analysis of the celiprolol enantiomers uses no derivatization to separate the enantiomers by high performance liquid chromatography using a cellulose tris-3,5-dimethylphenyl carbamate coated on a silica gel analytical column (Chiralcel-OD®) (250 x 4.6 mm I.D., 10 μm particle size). A mobile phase of hexane, 2-propanol and diethylamine (80:20:0.7) at a flow rate of 1 mL/minute is used. A resolution between the celiprolol enantiomers of 2.94 was achieved while maintaining the system at 23°C. Detection was achieved by UV absorbance at 232 nm at a sensitivity of 0.01 AUFS. The following concentration of samples was injected: 10 nmole for racemic celiprolol and 5 nmole for (+) - R - celiprolol and (-) - S - celiprolol enantiomers. Enantiomeric separation is dependent upon hydrogen bonding between the hydroxy group of the β-blocker and the carbonyl group of the triphenylcarbamate derivative. The retention time of the (-) - S - celiprolol was 10.02 minutes while that of the (+) - R - celiprolol was 13.43 minutes. This method can be applied to the determination of enantiomeric purity of celiprolol in bulk drug and pharmaceutical formulations (45).

6. STABILITY - DEGRADATION

6.1 Solid State Stability

Celiprolol hydrochloride is a stable compound in the solid state, i.e. no change in celiprolol content and no significant degradation have been observed in samples stored in amber glass containers for five years under room temperature or 45 C conditions. The stability of celiprolol hydrochloride has been studied for only limited periods of time in clear glass vials. These limited duration studies indicate that, when stored in clear glass vials, celiprolol hydrochloride is minimally stable for a) 1 year under room temperature and 35 C conditions, b) 2 months when stored under 35°C/85% relative humidity conditions, and c) stable 3 months under 45 C storage conditions. It is stable in high density polyethylene containers for a minimum of 3 months when stored at 35°C/85% relative humidity and for a minimum of 6.5 years when stored at room temperature. Exposure to light (130 ft. candles) for 1 month caused a slight yellow discoloration but no degradation (46).

6.2 Solution Stability

Celiprolol hydrochloride is stable in distilled water for 2 weeks at 37 C but it degrades substantially in dilute HCl and dilute NaOH. The degradation process follows pseudo first order kinetics. The $t_{\frac{1}{2}}$ = 50 days for 0.1 N HCl solutions and 25 days for solutions in 0.1 N NaOH. The degradation product in acid and base was identified as the anilino hydrolysis product, ST-1429. In borate and citrate buffer solutions at neutral pH, the drug was stable for 1 month at room temperature. At 45°C significant degradation was observed after 1 month. The major degradation products were identified as ST-1429 and ST-1444 (Figure 20) (36).

7. BIOPHARMACEUTICS AND METABOLISM

7.1 Absorption and Bioavailability

Celiprolol·HCl is rapidly absorbed from the gastrointestinal tract (47). The oral bioavailability is dose-dependent, reflected in non-linear increases in area under the curve (AUC) values over the therapeutic dose range, 100-600 mg (48-52). In

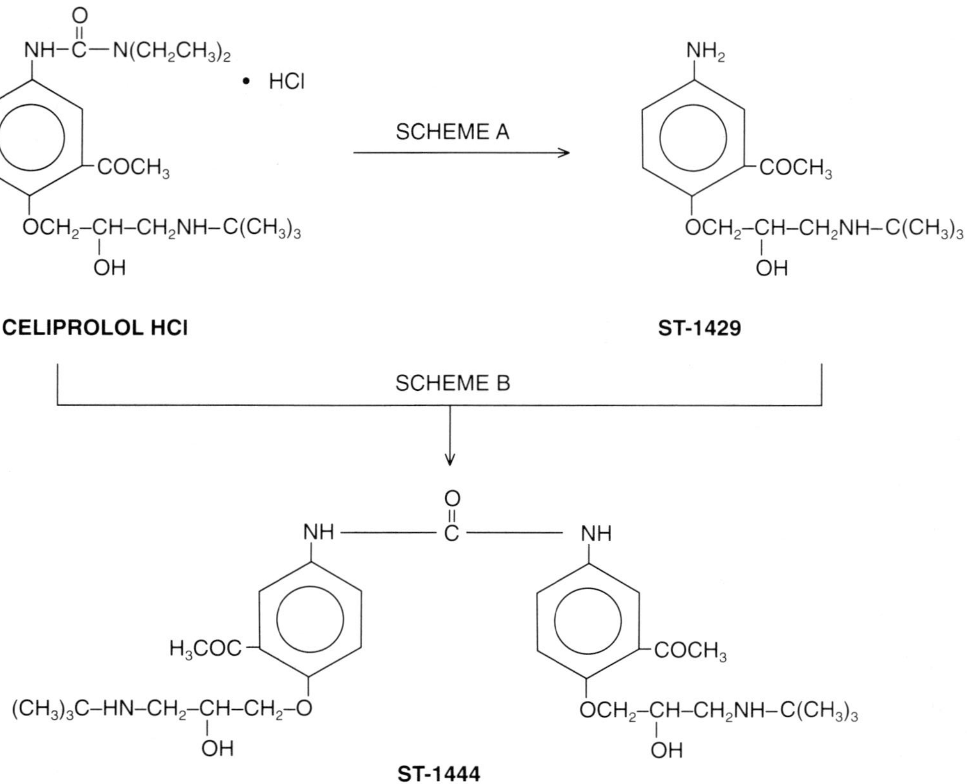

Figure 20. Principal potential solution-state degradation products of celiprolol hydrochloride.

contrast to acute oral doses, results after repeated administration suggest a decrease in non-linearity (49). Its oral bioavailability is 30-70% and is dose-dependent (49-51). Twenty five percent of an administered dose is protein bound (51). The presence of food decreases the absorption of celiprolol·HCl (51). It was shown in healthy males that the bioavailability of oral celiprolol·HCl is significantly reduced when taken in the fed rather than the fasted state (53). The presence of food in the gastrointestinal tract was also found to reduce the absorption of celiprolol·HCl in rats, the effect having been more pronounced at a higher dose. Adequate bioavailability is obtained if celiprolol·HCl is taken 1 hour before or 2 hours after meals (50). In contrast to oral celiprolol·HCl, bioavailability is dose-proportional following parenteral administration of 5 to 20 mg (49-51).

7.2 Metabolism

Celiprolol·HCl is minimally metabolized with only a very low percentage of the administered dose being excreted as three metabolites (49,51,52,53). None of the metabolites are known to be pharmacologically active (52). It is not affected by significant first pass metabolism (51). After an oral dose the majority is excreted in the feces (84%) with 11% excreted in the urine (51,52). The predominant route of elimination is renal excretion for celiprolol·HCl administered parenterally (49,52). After an IV dose, 50% is excreted in the urine and 31% in the feces.

Rats with carbon tetrachloride induced hepatic damage exhibited compensatory renal elimination sufficient to maintain normal blood levels. In addition, rats with uranyl nitrate induced renal impairment exhibited compensatory shift to biliary elimination almost sufficient to normalize blood levels (55).

7.3 Pharmacokinetics

Pharmacokinetic studies in over 500 patients have shown that celiprolol·HCl is well absorbed, is largely unmetabolized, and is excreted in urine and bile (54). Peak plasma levels occurred between 2-3 hours post dose (47,52). The pharmacokinetic half-life ($T_{½}$) was approximately 5 hours when administered orally (47,49,51). Steady state is achieved within 2-3 doses when administered daily (49,52). Peak pharmacodynamic activity occurs at approximately 2-4 hours and is still seen at 24 hours postdose (47). Renal clearance, 120-180 mL/minute, is independent of dose or route of administration. A study was conducted in healthy males which showed that celiprolol·HCl and diltiazem·HCl could be co-administered in the treatment of hypertension without fear of a pharmacokinetic drug interaction (56).

A study of the pharmacokinetics in the elderly showed that there was no apparent change in bioavailability or disposition of the drug after 7 days of dosing. There is a nonlinear increase in C_{max} with increasing dose which parallels that of the area under the curve. This identical phenomenon was seen in young healthy volunteers. There was no tendency to excessive accumulation of celiprolol on repeated dosing in this population which had normal renal and hepatic function (47).

8. DETERMINATION IN BIOLOGICAL MATRICES

Celiprolol hydrochloride can be analyzed in blood, plasma and urine by radiolabelled drug assays or by HPLC and colorimetric analytical methods.

A specific and sensitive (10 ng/mL in plasma) assay for celiprolol in blood, plasma and urine using HPLC was developed. The method uses a 250mm x 4.6mm i.d. Spherisorb ODS (5 μm particle size) column and an aqueous mobile phase that is composed of 55% acetonitrile, 6% 0.1M phosphate buffer and 39% water. This method uses an internal standard of acebutolol. The method is also able to detect two unidentified presumed metabolites. The method can achieve a sensitivity of 10 ng/mL using UV detection at 237 nm or 5 ng/mL using florescence detection with 335 nm excitation and 472 nm emission (57).

A sensitive HPLC method was developed by Takacs (58) for the analysis of celiprolol in plasma and urine. Extraction of celiprolol from a highly alkaline solution is accomplished using ethyl acetate. It is then reextracted in 0.01 N sulfuric acid and analyzed on a Micro-Pac MCH-10 column using a mobile phase consisting of 55% acetonitrile, 6% buffer (pH 3.3) and 39% water. Using fluorescence detection, a sensitivity of 10 ng/mL was achieved. Celiprolol may also be analyzed by a colorimetric assay. The drug is extracted from a strongly alkalinized biological medium and hydrolyzed at the urea group. The aromatic amine is diazotized and coupled to a blue azo dye. The absorbance is measured at 565 nm (59).

9. DETERMINATION IN PHARMACEUTICALS

9.1 Dissolution Testing

Dissolution testing of celiprolol·HCl is accomplished using the rotating paddle method (USP Method II) with 900 mL of deionized water heated to 37°C$\pm$ 0.5°C as the dissolution medium and a paddle rotation speed of 50$\pm$ 2 rpm. Samples are introduced into the dissolution vessel and then withdrawn at appropriate intervals through a 0.45 μm Millex® filter. The dissolution medium is assayed for celiprolol hydrochloride against appropriate standards using an UV absorbance assay. Not less than 80% of the labeled amount of celiprolol hydrochloride is dissolved in 30 minutes (60).

9.2 Potency

9.2.1 Assay

9.2.1.1 High Performance Liquid Chromatography

The assay of finished product tablets is accomplished with an HPLC method. The chromatographic conditions are identical to those employed for the bulk drug determination and are listed in Section 5.4.2. This analytical method will distinguish celiprolol from its impurities and two known degradation products.

Twenty tablets are ground to a fine powder and an accurately weighed portion of the powder is dispersed in the solvent of water/acetonitrile (50/50, v/v). The sample is sonicated for 10 minutes and cooled to room temperature. After dilution to volume with the water/acetonitrile diluent, the sample is mixed and then centrifuged or filtered through a 0.45 μm membrane filter to remove any dispersed solids. The supernatant liquid is injected into the HPLC system. Tablets must contain not less than 90.0 percent and not more than 110.0 percent of the labeled amount of celiprolol hydrochloride.

9.2.2.2 Nuclear Magnetic Resonance (61)

^{1}H NMR spectroscopy has been used to quantitate celiprolol in tablets.

Celiprolol·HCl tablets are ground and the powder is added to an appropriate amount of internal standard, sodium acetate trihydrate. Distilled water is used to suspend the powder and the suspension is centrifuged for 20 minutes. The clear supernatant is used for the spectrum. The methyl groups of celiprolol resonate in the upfield region of a 1H NMR spectra. A suitable NMR spectometer is used to collect the spectra. The spectrum of a mixture of the pure drug and internal standard are compared to that of the sample from the tablet to evaluate the interference from the extracted excipients on the signals of interest. The amount of active ingredient was calculated by comparing the signals for the drug and internal standard.

9.2.3 Dosage Uniformity

Tablets must meet the U.S.P. (62) requirement for dosage uniformity testing. The requirement for uniformity of the dosage form is met if the individual assay of each of 10 dosage units lies between 85.0% and 115.0% of the label claim and the relative standard deviation of all 10 assays is less than or equal to 6.0%. Twenty additional dosage units must be tested if one unit is outside of this range and no unit is outside of the range of 75.0% to 125.0% of the label claim or if the relative standard deviation is greater than 6.0 % or both. The 30 dosage units tested meet the requirements if no more than one unit is outside the 85.0% to 115.0% range, none are outside of 75.0 to 125.0% range and the relative standard deviation does not exceed 7.8%.

The content uniformity assay is performed by assaying ten tablets individually. The samples are prepared by dissolving each tablet in a diluent of acetonitrile/water (50:50, v/v). The samples are centrifuged or alternatively filtered through a 0.45 μm membrane filter and diluted to an appropriate concentration. The chromatographic procedure for potency assay is then utilized.

9.2.4 Stability Testing

Stability testing of celiprolol hydrochloride tablets is accomplished by an HPLC assay and dissolution testing. The conditions for the HPLC assay are described in the section for potency assay (9.2.1.1). The dissolution test methodology is described in Section 9.1. Based upon data generated in accelerated and ambient temperature stability studies, celiprolol hydrochloride tablets have been shown to be stable for a minimum of 3 years.

9.3 Disintegration Test

A disintegration test is performed for coated and uncoated celiprolol hydrochloride tablets. The apparatus is described in USP XXII under section [701] Disintegration. The test is accomplished by placing a tablet in each of the six tubes of the basket. For uncoated tablets, the amount of time needed for the tablets to disintegrate completely in water maintained at 37 ± 2°C is recorded. The disintegration test needs to be repeated for 12 additional tablets if one or two tablets fail to disintegrate completely. Of the 18 tablets, not less than 16 of the tablets must disintegrate completely.

For coated tablets having a soluble external coating, immerse the basket in water at room temperature for 5 minutes. Then, immerse the tablets in simulated gastric fluid test solution at 37 ± 2°C for 30 minutes. The time is recorded if disintegration is complete. If disintegration is not complete, replace the simulated gastric fluid with simulated intestinal fluid test solution at 37 ± 2°C as the immersion fluid. Immerse the tablets for 30 minutes longer than the time specified in the monograph. The same requirements for acceptance needed for the uncoated tablets must be met.

ACKNOWLEDGEMENT

The authors express their grateful appreciation to Ms. Barbara Lynch for her assistance and patience as she typed this manuscript.

D.J.M./C.L.O./J.E.S.

References

1. H. Pittner, Arzneim-Forsch. 33(1A), 13-29 (1983).
2. J.A. Barrett, R.D.Smith, P.S. Wolf and T.P. Pruss, Drug Dev. Res. 9(2), 159-169 (1986).
3. H. Pittner, Wein. Klin. Wochenschr 97(Suppl. 162), 3-21 (1985).
4. G. Carboni, E.A. Rodrigues, A. Hains, P. Dasgupta, A. Lahiri and E. Raftery, Clin. Sci. 72 (Suppl. 16), 9P-10P (1987).
5. C. Diehm, J. Int. Med. Res. 16(Suppl. 1), 34A-38A (1988).
6. G. Mancia, American Heart Journal 116(5:2), 1405-1411 (1988).
7. O.-E. Brodde, R. Schemuth, M. Brinkmann, X. L. Wang, A. Daul and U. Borchard, Arch. Pharm. 333(2), 130-138 (1986).
8. A. Daul, X.L. Wang, U. Borchard, K.D. Bock and O.-E. Brodde, J. Cardiovasc. Pharm. 8 (Suppl. 4), S93-S96 (1986).
9. M. Niederberger, W. Ebm and S. Panzer, Arzneim-Forsch. 33(1A), 63-66 (1983).
10. N. Jackson, P. Lee and S.H. Taylor, J. Cardiovasc. Pharm. 8 (Suppl. 4), S145-S147 (1986).
11. H. Douard, B. Mora and J.P. Broustet, J. Cardiovasc. Pharm. 12 (Suppl. 6), S210 (1988).
12. B. Silke, S.P. Verma, M.A. Frais, G. Reynold and S.H. Taylor, Clin. Pharm. Ther. 39(1), 5-14 (1986).
13. P. Dorow, J. Int. Med. Res. 16(Suppl. 1), 23A-26A (1988).
14. B. Silke, S.P. Verma, M.A. Frais, G. Reynolds and S.H. Taylor, Br. J. Clin. Pharm. 20(3), 245P (1985).
15. B. Silke, S.P. Verma, M.A. Frais, G. Reynolds and S.H. Taylor, . Cardiovasc. Pharm. 8(Suppl 4), S138-S144 (1986).
16. J.M. McLenachan, I.N. Findlay, E. Henderson, J. Wilson and H.J. Dargie, Am. J. Cardiology 61(5), 52C (1988).
17. J.E. Soberman and W.H. Frishman, American Heart Journal 116 (5:2), 1422-1425 (1988).
18. J.M. Herrmann and E.O. Mayer, American Heart Journal 116(5:2), 1416-1421 (1988).
19. J.M. Herrmann, F. Von Heymann and G. Freischeutz, J. Int. Med. Res. 16(Suppl. 1), 39A-46A (1988).
20. C. W. Menzies, D.C. Quail and J. Horobin, Eur. Heart J. 9 (Suppl. 1), 115 (1988).
21. J.M. Herrmann and E. Zieseniss, J. Hypertens. 6 (Suppl. 4), 5756 (1988).

22. J.M. Herrmann, E. Zieseniss and W. Haase, Eur. Heart J. 9 (Suppl. 1), 114 (1988).
23. K.D. Lamon, Am. Heart J. 116 (5, Suppl. 2), 1438-1440 (1988).
24. V. G. Zölss, et al, Belgium Patent No. 823,411 (to Chemie Linz AG, Linz, Austria), June 16, 1975.
25. V. G. Zölss, et al, United States of America patent No. 4,034, 009 (to Chemie Linz Ag, Linz, Austria), July 5, 1977.
26. V. G. Zölss, Arzneim. - Forsch. Drug/Res. 33 (I), 2-4 (1983).
27. W. Studt and D. Taube, Department of Process Research and Development, Rhône-Poulenc Rorer Central Research, King of Prussia, PA, U.S.A., personal communication.
28. E. Orton, Department of Analytical and Physical Chemistry, Rhône-Poulenc Rorer Central Research, King of Prussia, PA, U.S.A., personal communication.
29. C. Strohbeck, Department of Analytical and Physical Chemistry, Rhône-Poulenc Rorer Central Research, King of Prussia, PA, U.S.A., personal communication.
30. J. Fernandez and A. Daruwala, Department of Analytical and Physical Chemistry, Rhône-Poulenc Rorer Central Research, Ft. Washington, PA, U.S.A., personal communication.
31. S.-Y. Tang and B.-H. Hsu, Department of Analytical and Physical Chemistry, Rhône-Poulenc Rorer Central Research, King of Prussia, PA, U.S.A., personal communication.
32. R. Engerer, Department of Analytical and Physical Chemistry, Rhône-Poulenc Rorer Central Research, Ft. Washington, PA, U.S.A., personal communication.
33. J. Omana and A. Daruwala, Department of Analytical and Physical Chemistry, Rhône-Poulenc Rorer Central Research, Ft. Washington, PA, U.S.A., personal communication.
34. A. Burger, A.W. Ratz and G. Zölss, Acta Pharm. Technol. 34(3), 147-151 (1988).
35. N. Kumar, C. Strohbeck, Department of Analytical and Physical Chemistry, Rhône-Poulenc Rorer Central Research, King of Prussia, PA, U.S.A. and H.L. Ammon, Department of Chemistry and Biochemistry, University of Maryland, College Parke, MD, U.S.A., "The Structure of Celiprolol," preprint of an article submitted to Pharmaceutical Research (1990) and personal communication.
36. M.J. O'Hare and J.E. Zarembo, Revlon Health Care, Tuckahoe, NY, U.S.A., personal communication.

37. John C. Egoville, Department of Analytical and Physical Chemistry, Drug Development, Rhône-Poulenc Rorer Central Research, Ft. Washington, PA, U.S.A., personal communication.
38. Von F. Takacs and D. Hippmann, Arzneim-Forsch. 33 (No. 1a), 5-12, (1983).
39. H. Schutz and T. Meister, Arzneim-Forsch./Drug Res., 40(6), 651-653 (1990).
40. Aroonsri Anavil, Revlon Health Care, Tuckahoe, NY, U.S.A., personal communication.
41. Aroonsri Anavil and Karl Blumberg, Revlon Health Care, Tuckahoe NY, U.S.A., personal communication.
42. C.M. Won, Department of Analytical and Physical Chemistry, Drug Development, Rhône-Poulenc Rorer Central Research, Ft. Washington, PA, U.S.A., personal communication.
43. J. N. Buskin, R. A. Upton, F. Sorgel, R. L. Williams, E. Lang and L.Z. Benet, J. Chromatogr., 230, 454 (1982).
44. W. Lindner and Ch. Leitner, J. Chromatogr. 316, 605-616 (1984).
45. Hassan Y Aboul-Enein and M. Rafiqul Islam, Analytical Letters 23(1), 83-91 (1990).
46. R. Witt, Department of Analytical and Physical Chemistry, and L. Schneider, Department of Pharmaceutics, Rhône-Poulenc Rorer Central Research, Ft. Washington, PA, U.S.A.
47. R.J. Norris, E.H. Lee, D. Muirhead, and S.W. Sanders, J. Cardiovasc. Pharma. 8 (Suppl. 4), S91-S92 (1986).
48. S. Mangat, L. Klunk, and P. Grebow, Pharmacologist, 27(3), 181 (1985).
49. H.D. Doshan, B.M. Berger, R. Costello, W. Applin, F.S. Caruso, and E.S. Neiss, Clinical Pharmacology and Therapeutics, 37(2), 192 (1985).
50. F.S. Caruso, H.D. Doshan, P.H. Hernandez, R. Costello, W. Applin and E.S. Neiss. British Journal of Clinical Practice Symp., Supp 39(6) (Suppl. 40), 12-16 (1985).
51. J.G. Riddell, D.W.G. Harron and R.G. Shanks, Clinical Pharmacokinetics 12(5), 305-320 (1987).
52. J.G. Riddell, R.G. Shanks and R.N. Brogden, Drugs 34(4), 438-458 (1987).
53. J. Eff, M. Connolly, D. Magner, S. Pollack, and K.C. asseter, J. Clin. Pharmacol. 30, 831-862 (1990).
54. T.P. Pruss, K.D. Lamon and N.S. Hagen, The Journal of International Medical Research 16 (Suppl. 1), 17-22A (1988).
55. L. Klunk, J. Knipe, S. Capello and P. Grebow, Pharmacologist 27(3), 181 (1985).

56. S. Pollack, M. Connolly, J. Miksic, J. Eff, and J. Ziemniak, J. Clin. Pharm. 30(9), 845 (1990).
57. J.N. Buskin, R.A. Upton, F. Sorgel and R.L. Williams, J. Chromatogr. 230(2), 454-460 (1982).
58. D. Hippman and F. Takacs, Arzneim - Forsch./Drug Research 33(I), 8-12 (1983).
59. F. Takacs, Scientia Pharmaceutica 50(4), 272-273 (1982).
60. S.M. Mamrosh, Department of Analytical and Physical Chemistry, Drug Development, Rhône-Poulenc Rorer Central Research, Ft. Washington, PA, U.S.A., personal communication.
61. M. A. Iorio, A. Mazzeo Farina and A. Doldo, Journal of Pharmaceutical & Biomedical Analysis, 5(1), 1-10 (1987).
62. The United Stated Pharmacopeia, Twenty-Second Revision, The United States Pharmacopeial Convention, Inc., Rockville, MD, 1989, pp. 5-6, 1617-1619.

Analytical Profile of Iodoxamic Acid

Davide Pitrè, Alexandra Davies and Maurizio Grandi

Davide Pitrè, Faculty of Pharmacy, University of Milan, Viale Abruzzi n° 42, Milan, Italy

Alexandra Davies and Maurizio Grandi, Research & Development Division, Bracco Industria Chimica Spa, Via Folli n° 50, Milan, Italy

1. DESCRIPTION

1.1 Nomenclature

Benzoic acid, 3,3'-[(1,16-dioxo-4,7,10,13-tetraoxahexadecane-1,16-diyl)diimino]bis-[2,4,6-triiodo-

3,3'-[Ethylene-bis-(oxyethyleneoxyethylenecarbonylimino)]-bis-[2,4,6-triiodobenzoic acid]
CAS:31127-82-9

1.1.2 Generic name

- Iodoxamic Acid (USAN, INN, BAN)

- Iodoxamate Meglumine (USAN) compound with 1-deoxy-1-(methylamino)-D-glucitol (1:2)
  CAS : 51764-33-1

1.1.3 Trade names

- Endobil (Bracco, Italy)
- Endomirabil (ByK-Gulden, FRG)
- Cholovue (Squibb, USA)
- Cholegrafin (Takeda, Japan)

1.2 Formula, Molecular Weight and Iodine Content

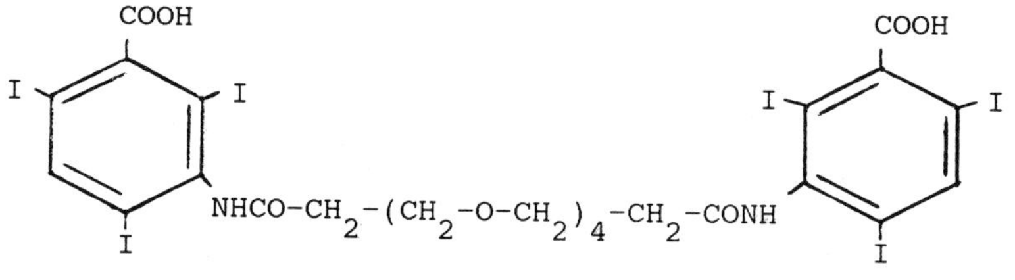

$C_{26}H_{26}I_6N_2O_{10}$ Mol. wt = 1287.9

Organically bound iodine : 59,12%

1.3 Appearance, Color, Odor and Taste

White powder, odorless, white a slightly bitter taste (1).

2. PHYSICAL PROPERTIES

2.1 Spectra

2.1.1 Ultraviolet Spectrum

The ultraviolet spectrum of iodoxamic acid (Fig 1) was determined in methanolic solution with a Cary mod. 219 spectrophotometer. Spectral data are presented in Table 1.

Table 1
UV spectral data

Solvent	λ max (nm)	ε max	$E^{1\%}_{1cm}$	Ref.
Methanol	238.5	69200	537	(2,3)
0.1 N HCl	239	61200	472	(3)
1.0 N HCl	239	-	472	(2)
0.1 N NaOH	234	71100	552	(2,3)
1.0 N NaOH	234	-	552	(2)
Water pH 4÷11	238	-	520	(2)

2.1.2 Infrared Spectrum

The infrared spectrum of iodoxamic acid in KBr discs (Fig.2) was recorded using a Perkin Elmer mod. 882 infrared spetrometer. The spectral assignments reported in Table 2 are in accordance with previously reported spectral data (2,3).

Table 2
IR spectral data for iodoxamic acid

Wavenumber (cm^{-1})	Assignments
3250,3215	ν NH, Amide groups
2905-2870	ν CH, Alkane groups

(cont'd)

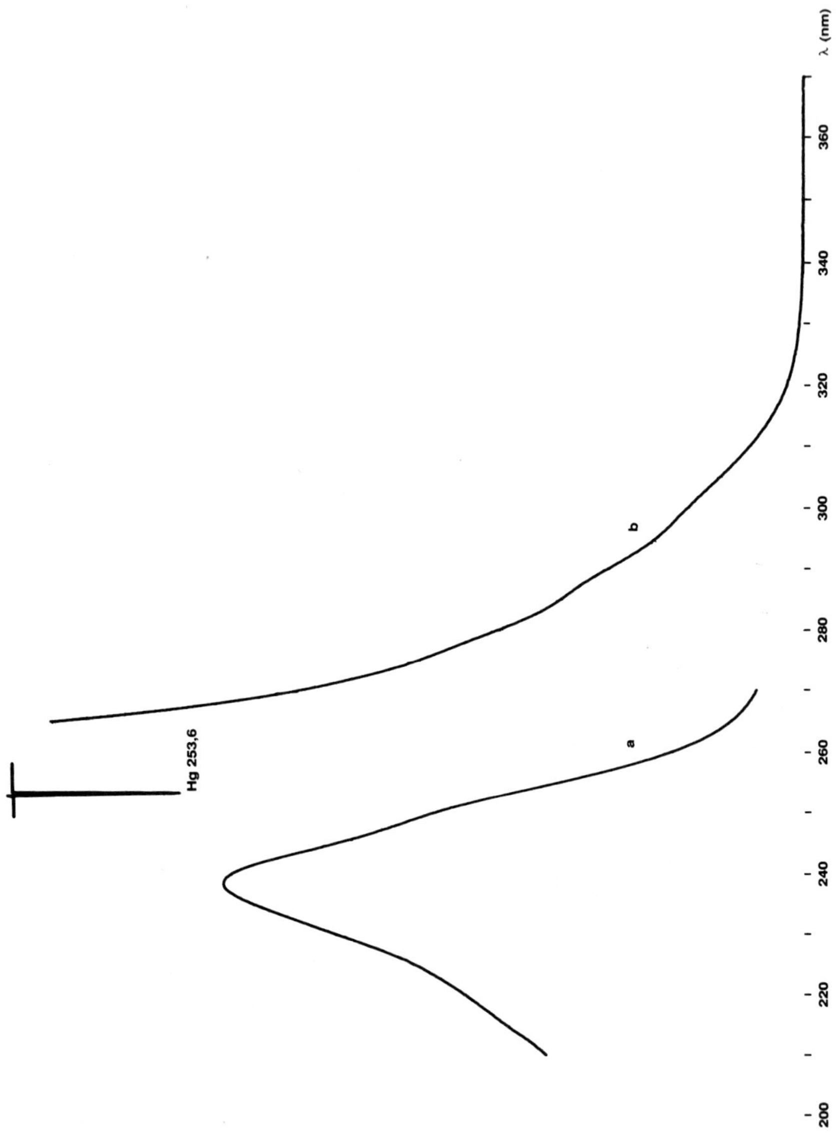

Figure 1 : Ultraviolet spectrum of iodoxamic acid

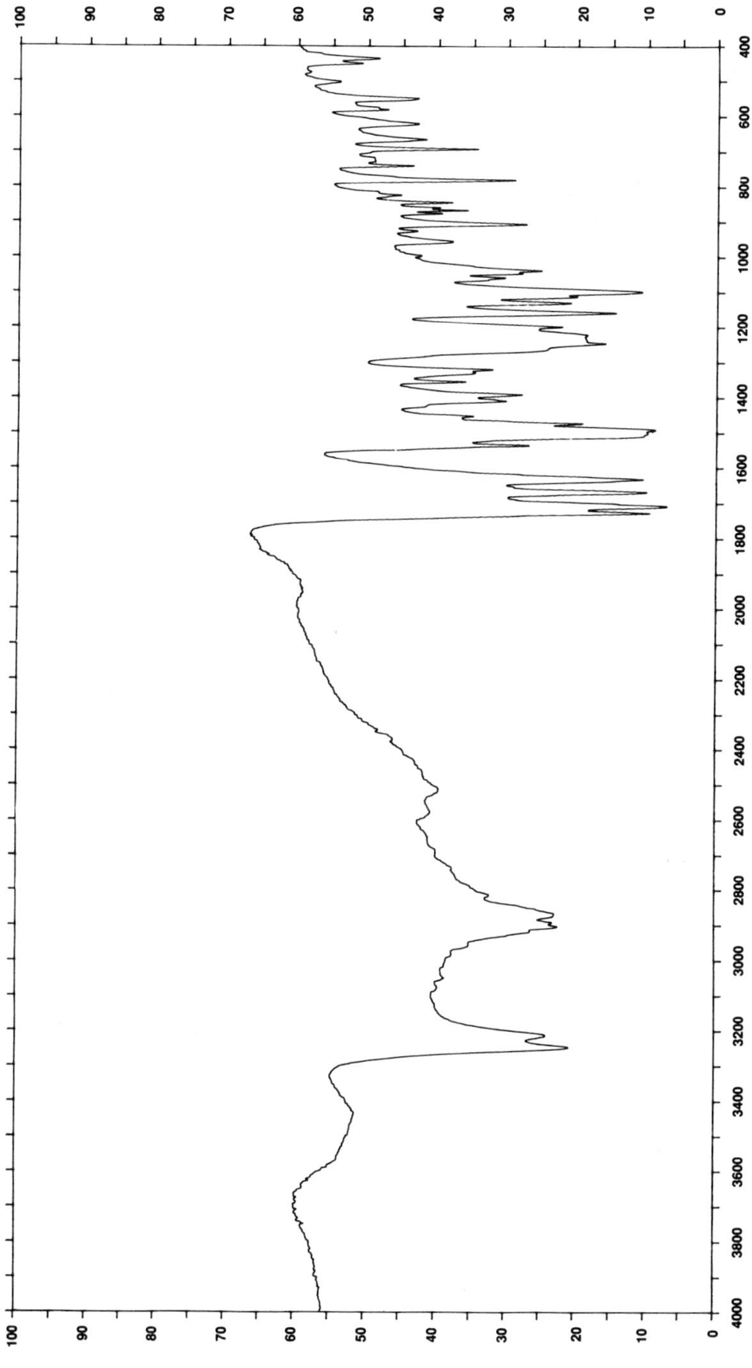

Figure 2 : Infrared spectrum of iodoxamic acid

Table 2 (cont'd)

Wavenumber (cm^{-1})	Assignments
ca. 2500	νOH, Carboxylic acid
1730-1710	νC=O, Carboxylic acid
1670-1635	νC=O, Amide (1st band)
ca. 1500	δNH+νCN, Amide (2nd band)
1320	νCN+δNH, Amide (3rd band)
1250	δOH, Carboxylic acid
1160-1100	νC-O, δCH_2-O-CH_2 group
960	δOH out of plane, carboxylic acid
905	δCH out of plane, penta substituted aromatic
780	δCH_2(rock), $-CH_2(CH_2-O-CH_2)_4-CH_2$ group

2.1.3 Nuclear Magnetic Resonance Spectra

2.1.3.1 ^{1}H-NMR

The ^{1}H-NMR spectrum of iodoxamic acid (Fig. 3) was recorded in DMSO solution with a Bruker AC-200 spectrometer, operating at 200 MHz. Chemical shifts and assignments are reported in Table 3. In order to confirm the assignments of the 4 mobile (2 amidic and 2 carboxylic) protons, a spectrum of the product's dimethylester was recorded operating at 90 MHz; chemical shifts and assignments are reported in Table 4.

Table 3
^{1}H-NMR data for iodoxamic acid in DMSO

Chemical shift δH(ppm, TMS)	Multiplicity	nr. of protons	Assignments
2.55	t	4	$2CO-\underline{CH_2}-CH_2-O-$
3.52	d	12	$6\ CH_2O$
3.74	t	4	$2CO-CH_2-\underline{CH_2}-O-$
8.35	s	2	2 Ph-H
9.95	s	2 exch.	2 Ph-<u>NH</u>CO

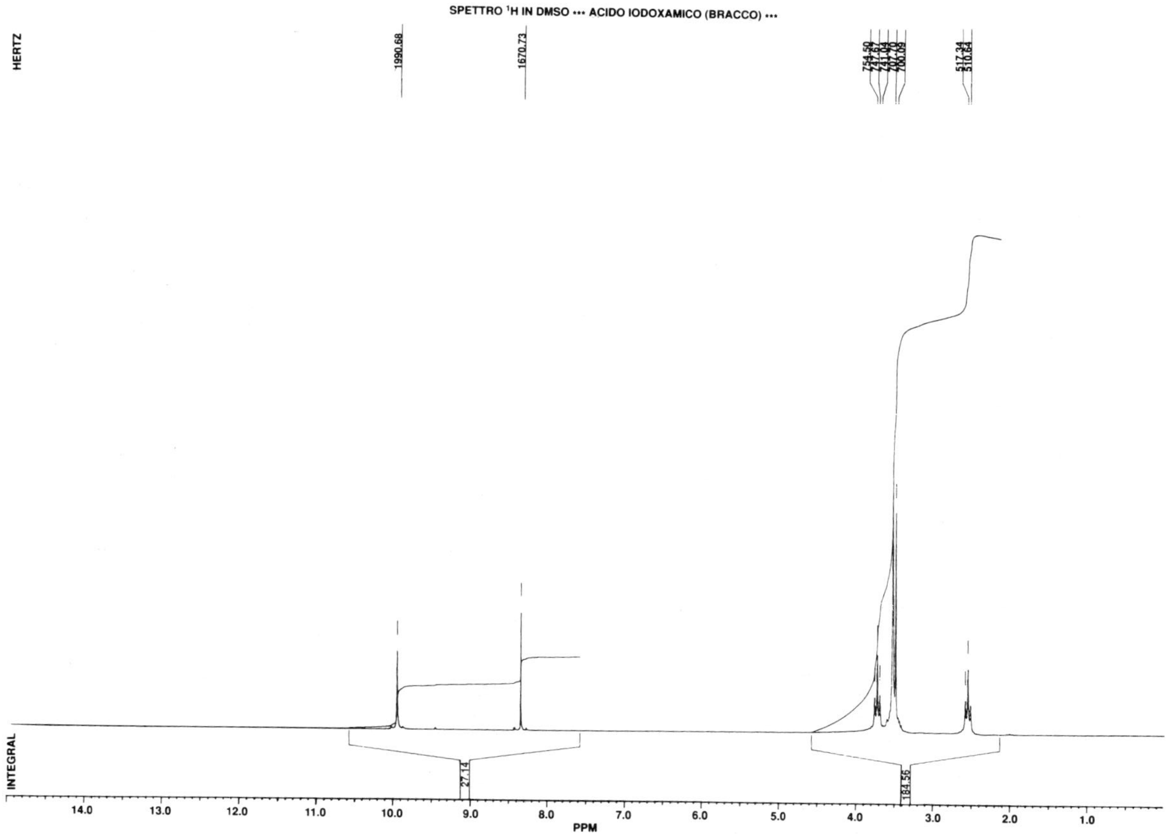

Figure 3 : ¹H-NMR spectrum of iodoxamic acid in DMSO

Table 4
^{1}H-NMR Spectral data for iodoxamic acid dimethylester (DMSO, 90MHz)

Chemical shift δ_H (ppm, TMS)	Multiplicity	nr. of protons	Assignments
2.56	t	4	$2CO-\underline{CH}_2-CH_2-O-$
3.52	d	12	$6\ \underline{CH}_2O$
3.77	t	4	$2CO-CH_2-\underline{CH}_2-O-$
3.92	s	6	$2COOCH_3$
8.43	s	2	2 Ph-H
9.98	s	2 exch.	2 Ph-NHCO

2.1.3.2 ^{13}C-NMR

The ^{13}C-NMR spectrum of iodoxamic acid was recorded in DMSO solution with a Bruker AC 200 spectrometer operating at 50 MHz. The all decoupled spectrum is reported in Fig. 4. Chemical shifts and assignments are reported in Table 5.

Table 5
^{13}C-NMR data for iodoxamic acid in DMSO

Line	Intensity	δ_C (ppm, TMS)	Assignments
1	7.94	169.4	C=O
2	4.81	168.7	C=O
3	5.58	148.4	Aromatic C3
4	3.64	147.1	Aromatic C5
5	5.94	142.7	Aromatic C1
6	3.99	101.6	Aromatic C6
7	6.17	98.7	Aromatic C4
8	4.09	91.0	Aromatic C2
9	10.58	69.9	$PhNHCOCH_2CH_2OCH_2CH_2O\underline{CH}_2$
10	10.46	69.8	$PhNHCOCH_2CH_2OCH_2\underline{CH}_2$
11	9.06	69.7	$PhNHCOCH_2CH_2O\underline{CH}_2$
12	5.94	66.4	$PhNHCOCH_2\underline{CH}_2O$
13-17	-	----	DMSO
18	4.16	36.5	$PhNHCO\underline{CH}_2CH_2O$

#	CURSOR	FREQUENCY	PPM	INTENSITY
1	2013	8522.674	169.3575	7.942
2	2035	8490.260	168.7134	4.805
3	2722	7467.221	148.3841	5.576
4	2764	7404.511	147.1380	3.637
5	2915	7179.886	142.6744	5.936
6	4303	5114.113	101.6246	3.987
7	4400	4969.249	98.7459	6.171
8	4662	4579.606	91.0032	4.092
9	5376	3515.879	69.8654	10.584
10	5378	3513.387	69.8159	10.461
11	5383	3505.878	69.6667	9.062
12	5493	3341.919	66.4086	5.937
13	6376	2027.659	40.2924	1.020
14	6390	2006.902	39.8800	2.151
15	6404	1985.952	39.4636	2.496
16	6418	1964.929	39.0459	2.021
17	6432	1943.947	38.6290	1.002
18	6505	1835.136	36.4667	4.164

SPETTRO ^{13}C IN DMSO ••• ACIDO IODOXAMICO (BRACCO) •••

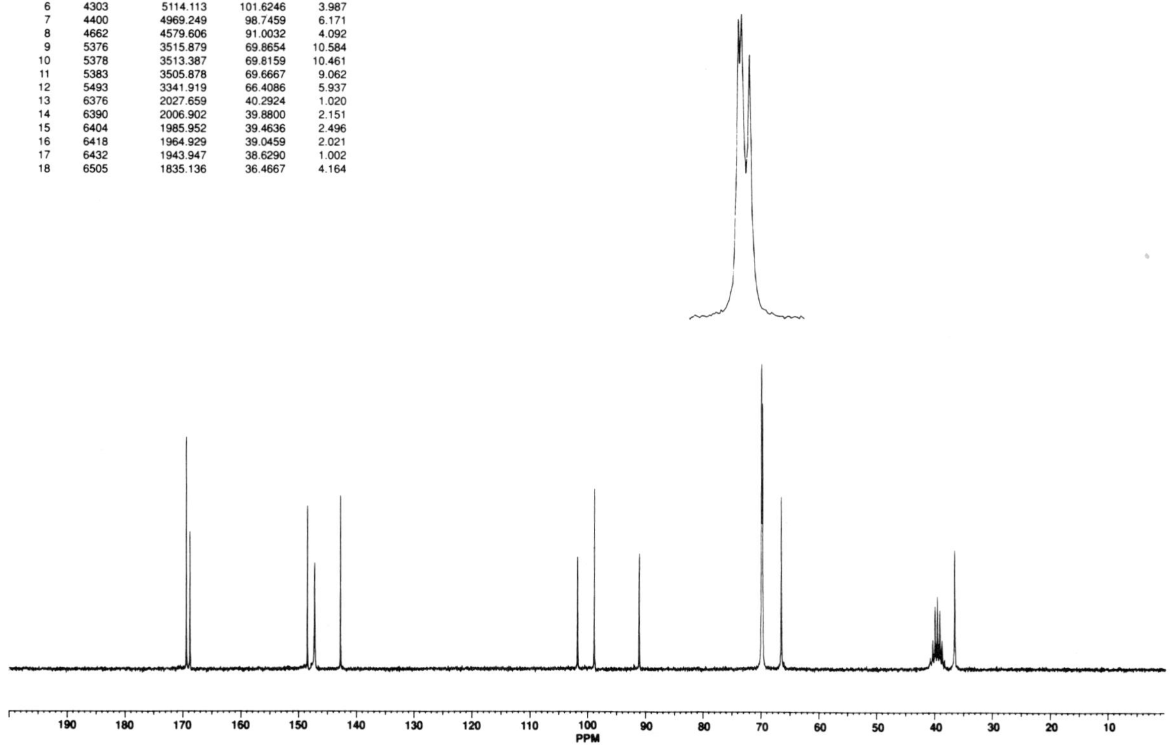

Figure 4 : ^{13}C-NMR spectrum of iodoxamic acid in DMSO

2.1.4 Mass Spectrum

Iodoxamic acid, like all hexaiodinated products, is characterized by a low volatility, a high molecular weight and thermal lability. Its spectrum was therefore recorded using the FAB/MS technique. Analysis was performed on samples mixed with glycerol (4) using a VG Micromass 70-70 E spectrometer. The spectrum (Fig.5) shows a quasi-molecular ion peak $[M+H]^+$ and a base peak of $[C_6H_{11}O_2]$ at m/z 115. The most important fragmentation sequence is that shown in Figure 6, in which the $[M+H]^+$ peak at m/z 1289 decomposes, according to the process observed in similar polyiodinated compounds (5,6), by sequential losses of iodine atoms with H substitution. The completely deiodinated species $[C_{26}H_{32}N_2O_{10} + H]^+$ at m/z 533 is responsible for most of the fragmentation pattern above m/z 300.

2.2 Solid State Properties

2.2.1 X-Ray Powder Diffraction

The X-ray powder diffraction pattern of iodoxamic acid Reference Standard was determined using a Philips Powder Diffractometer with nickel-filtered copper radiation under the following conditions:

Tube: Cu, 50 KV, 30mA;
Filter: Ni;
Slits : 1°-0.1-1°;
Detector: proportional + discriminator;
Scale : 1×10^3 cps;
Scanning speed: 1/4° 2θ x min;
Paper speed: 300 mm/h;
Time constant: 2 sec;
Specimen holder: Niskanen + internal standard.

Data are reported in Table 6.

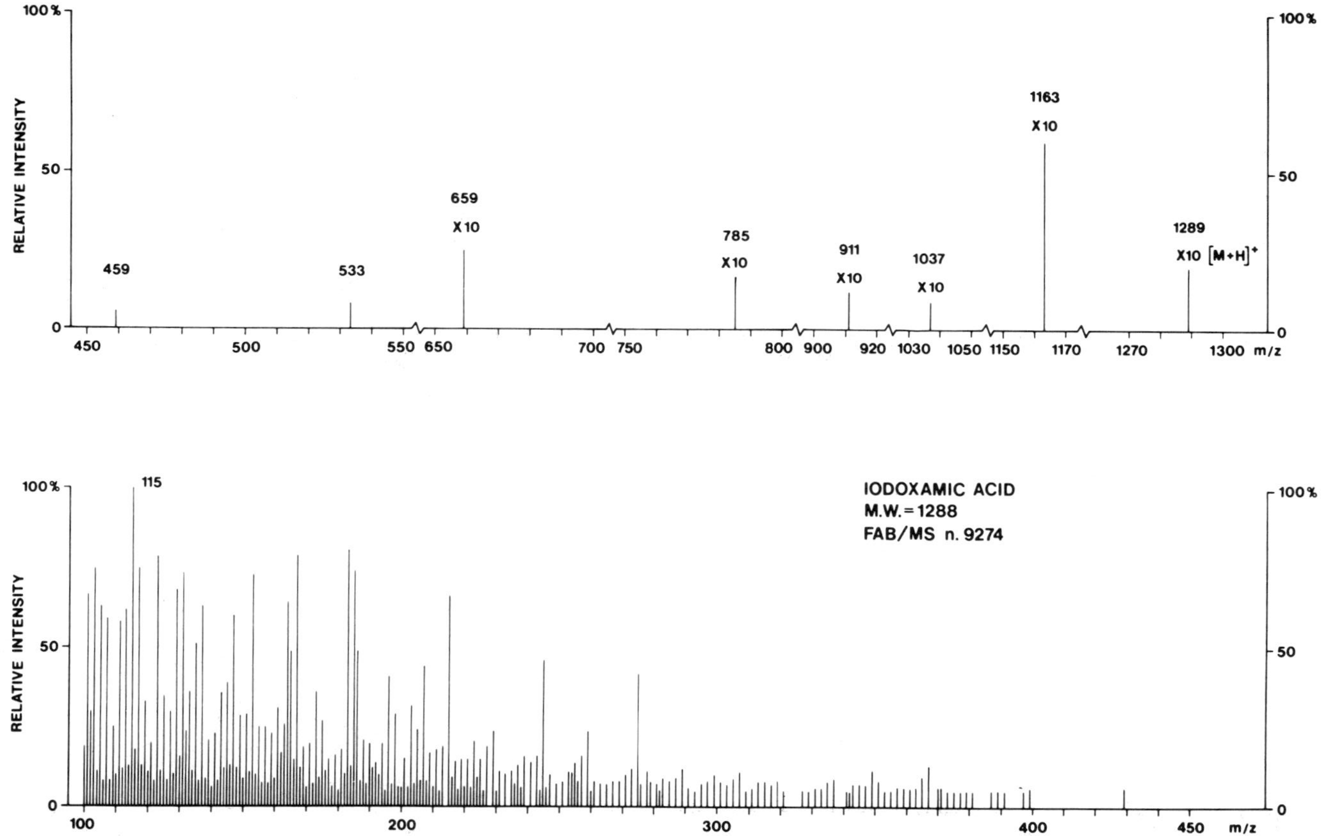

Figure 5 : FAB/MS spectrum of iodoxamic acid in glycerol

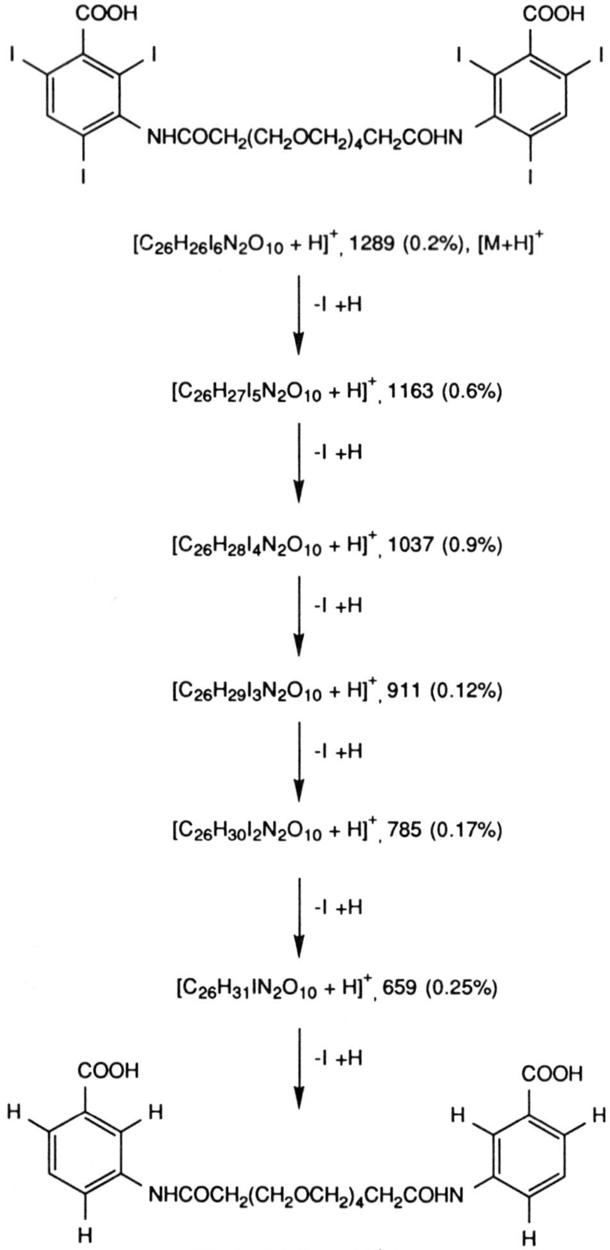

Figure 6 : Fragmentation scheme of iodoxamic acid

Table 6
X-Ray Powder Diffraction Pattern

d(Å)*	I% rel.	d(Å)*	I% rel.
7.36	20	3.19	8
7.03	4	3.18	8
6.87	3	3.15	20
6.51	5	3.10	15
6.37	30	3.08	23
6.21	11	3.02	15
6.15	9	2.95	27
6.07	4	2.93	18
5.90	11	2.83	3
5.80	15	2.76	25
5.48	7	2.74	15
5.35	27	2.70	19
5.28	13	2.65	12
5.06	18	2.61	6
4.95	4	2.55	9
4.89	5	2.49	13
4.76	10	2.46	13
4.65	13	2.45	14
4.57	95	2.40	4
4.51	7	2.37	11
4.41	100	2.33	4
4.30	8	2.30	7
4.16	48	2.28	12
3.98	72	2.26	12
3.91	14	2.24	3
3.83	22	2.21	5
3.77	65	2.17	6
3.64	12	2.13	5
3.63	15	2.12	5
3.61	15	2.10	5
3.54	11	2.09	4
3.47	24	2.02	6
3.36	30	1.96	2
3.29	5	1.94	3
3.26	7	1.90	4
		Plus other lines < 2	

* interplanar distances = $n\lambda/2\sin\theta$ $\lambda = 1.54051$ Å

2.2.2 Differential Thermal Analysis

Differential thermal analysis was carried out using a Mettler Ta 2000 calorimeter with a heating rate of 5°C/min. The thermogram reported in Figure 7 shows an endothermic transition due to melting at 206°C and an exothermic decomposition at about 240°C with loss of iodine.

2.2.3 Melting range

The product crystallized from ethanol has a melting range of 223°-225°C (2,7).

2.3 Solution Properties

2.3.1 Solubility

The solubility values of iodoxamic acid in water and in various organic solvents were determined spectrophotometrically at 21°C (2) (Table 7).

Table 7
Solubility of iodoxamic acid at 21°C

Solvent	Solubility (g/100 ml)
Water (pH = 3.2)	0.04
0.01 N HCl	0.04
0.1 N HCl	$5 \cdot 10^{-3}$
1 N HCl	$2.5 \cdot 10^{-3}$
Ethanol	1.4
Chloroform	0.02
Ethyl ether	$8 \cdot 10^{-4}$

The solubility in water of the sodium and N-methylglucamine salts resulted greater than 100 g /100 ml (2).

The water solubility values at 20°C as a function of pH have also been reported (3) (Table 8).

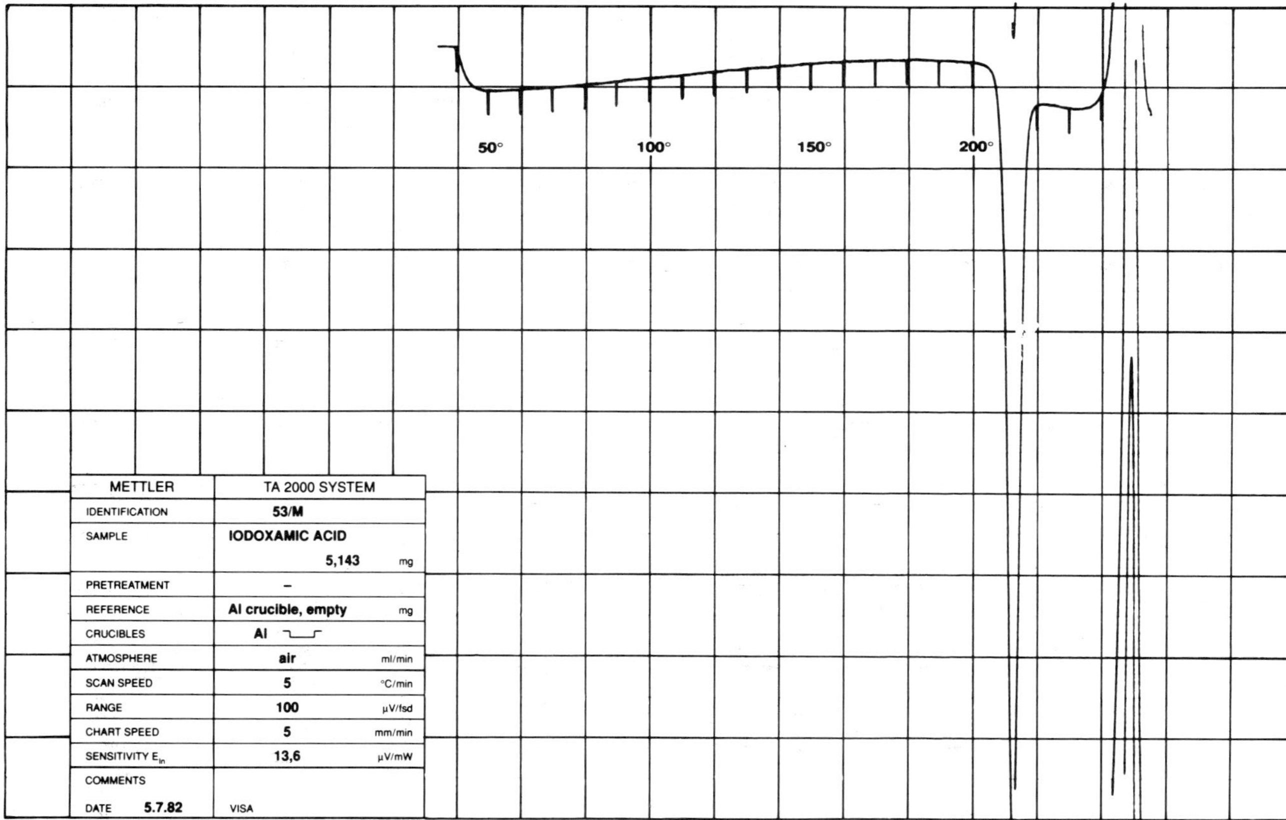

Figure 7 : DTA curve of iodoxamic acid

Table 8
Solubility of iodoxamic acid in water at 20°C as a function of pH

pH	Solubility mg/100ml
2.1	0.5
3.2	75
4.0	$2.1 \cdot 10^3$
4.7	$1.4 \cdot 10^4$

2.3.2 Ionization Constants

The ionization constants of iodoxamic acid were determined by potentiometric titration in the methylcellosolve (MCS)-water (3) and ethanol-water systems (2). In both cases 2 equivalents of alkali were required per mole of acid and the resulting curves showed only one inflection point. Data are reported in Table 9.

Table 9
Apparent ionization constants of iodoxamic acid

Solvent	T (°C)	pK_1^*	pK_2^*	pK_{av}^*
80% MCS	25	4.06	5.07	---
30% EtOH	21	----	----	2.96
50% EtOH	21	----	----	3.22
70% EtOH	21	----	----	3.67

For the 80% MCS system, extrapolation to water solvent yields values of $pK_1 = 1.8$ and $pK_2 = 2.8$ (8,9).

2.3.3 pH Value

The pH value of a saturated solution at 20°C is of 3.85.

2.3.4 Osmotic Properties

The osmotic properties of aqueous solutions of the sodium and N-methylglucamine salts of iodoxamic acid were determined at 37°C using a Knauer vapor pressure osmometer calibrated with standard aqueous solutions of D-mannitol (10). Values are reported in Table 10 and 11 while in Fig.8 the osmolality values are plotted against the concentration values of the two salts.

Table 10
Osmometric data of aqueous solutions of iodoxamic acid sodium salt.

Concentrations		Values found at 37°C		
mol/l	mgI/ml	$\varphi \nu m^{*}$ (osmol/kg)	φ^{**}	osmotic pressure (atm)
0.005	3.8	0.015	0.97	0.37
0.010	7.6	0.029	0.965	0.75
0.025	19.1	0.069	0.90	1.74
0.050	38.1	0.132	0.85	3.35
0.100	76.1	0.242	0.76	6.16
0.250	190.4	0.584	0.67	14.85
0.350	266.4	0.847	0.66	21.6
0.500	380.7	1.306	0.64	33.2

* osmolality
** osmotic coefficient

Table 11
Osmometric data of aqueous solutions of iodoxamic acid N-methylglucamine salt.

Concentrations		Values found at 37°C		
mol/l	mgI/ml	$\varphi \nu m^{*}$ (osmol/kg)	φ^{**}	osmotic pressure (atm)
0.005	3.8	0.015	0.97	0.37
0.010	7.6	0.028	0.93	0.72

(cont'd)

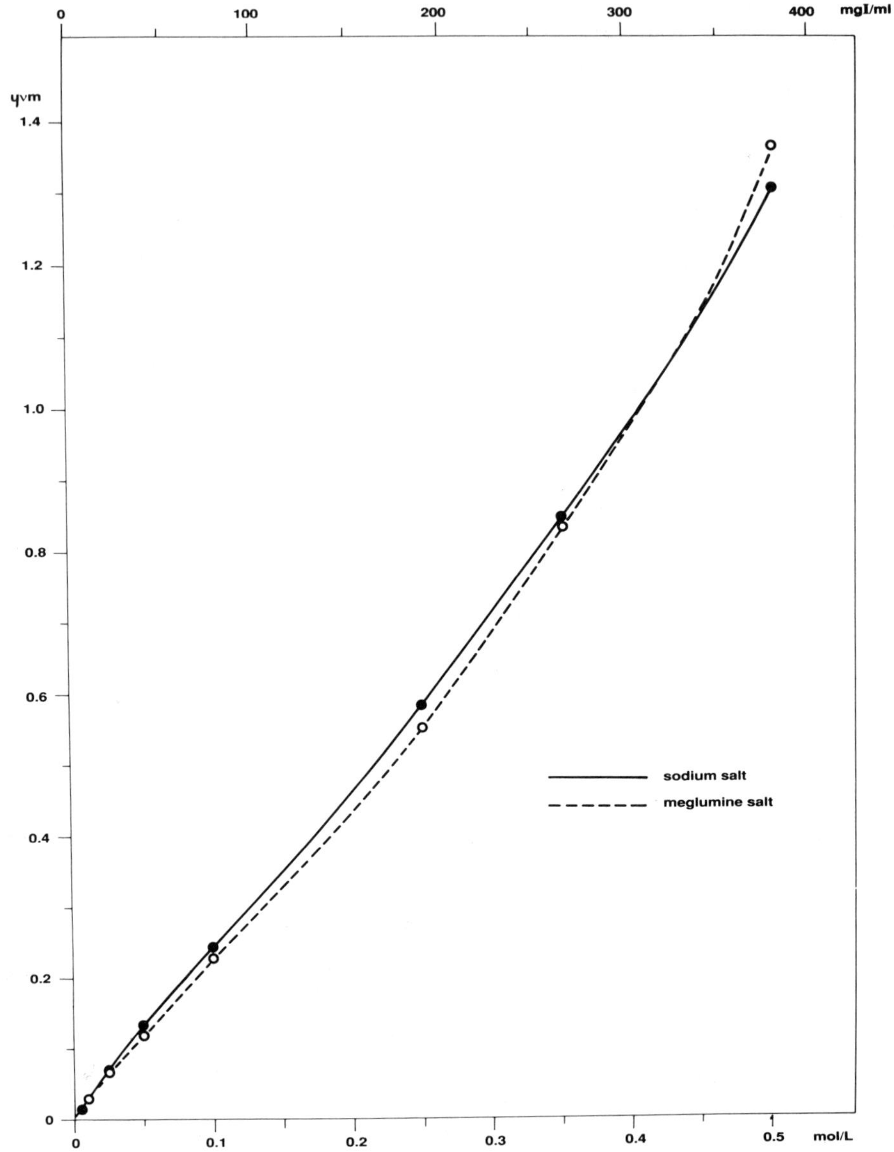

Figure 8 : Osmolality curves of iodoxamic acid sodium salt and methylglucamine salt

Table 11 (cont'd)

Concentrations		Values found at 37°C		
mol/l	mgI/ml	$\varphi \cdot m$ * (osmol/kg)	φ **	osmotic pressure (atm)
0.025	19.1	0.065	0.85	1.66
0.050	38.1	0.117	0.74	2.98
0.100	76.1	0.227	0.69	5.78
0.250	190.4	0.550	0.585	14.0
0.350	266.4	0.832	0.57	21.2
0.500	380.7	1.365	0.545	34.7

* osmolality
** osmotic pressure

2.3.5 Refraction Index

The refraction index of a series of aqueous solutions of the sodium and N-methylglucamine salts of iodoxamic acid were determined at 20°C with an Abbe-ATAGO refractometer in the 0.005÷0.5 mol/l concentration range (C_M). The following relationships were found:

a) sodium salt

$$n_D^{20} = 2.1437 \cdot 10^{-1} \cdot C_M + 1.3328 \pm 4.596 \cdot 10^{-4}$$

$$r = 0.999943$$
$$s_b = \pm 9.377 \cdot 10^{-4}$$
$$s_a = \pm 2.220 \cdot 10^{-4}$$

b) N-methylglucamine salt

$$n_D^{20} = 2.6899 \cdot 10^{-1} \cdot C_M + 1.3328 \pm 2.378 \cdot 10^{-4}$$

$$r = 0,999990$$
$$s_b = \pm 4.851 \cdot 10^{-4}$$
$$s_a = \pm 1.148 \cdot 10^{-4}$$

2.3.6 Density

The density values of aqueous solutions of the sodium and N-methylglucamine salts of iodoxamic acid were determined at 20°C in the concentration range 0.005 ÷ 0.5 mol/l (C_M) according to F.Pregl (11).

The following correlations were found:

a) sodium salt

$$d_{20}^{20} = 8.1585 \bullet 10^{-1} \bullet C_M + 9.9806 \bullet 10^{-1} \pm 1.007 \bullet 10^{-3}$$

$r = 0.999980$
$s_b = \pm 1.964 \bullet 10^{-3}$
$s_a = \pm 4.382 \bullet 10^{-4}$

b) N-methylglucamine salt

$$d_{20}^{20} = 8.8810 \bullet 10^{-1} \bullet C_M + 9.9791 \bullet 10^{-1} \pm 1.066 \bullet 10^{-3}$$

$r = 0.999981$
$s_b = \pm 2.077 \bullet 10^{-3}$
$s_a = \pm 4.635 \bullet 10^{-4}$

2.3.7 Viscosity

Iodoxamic acid is currently available on the market as N-methylglucamine salt solutions at two concentrations: 45 mgI/ml (bottles for i.v. infusion) and 183 mgI/ml (ampoules). Their viscosity data are reported in Table 12 (12).

Table 12
Viscosity data

Concentration mgI/ml	Viscosity (mPa•s)	
	20°C	37°C
45	1.34	0.91
183	4.75	2.84

2.3.8 Surface Tension

The surface tension values of commercially available iodoxamic acid N-methylglucamine salt solutions are reported in Table 13 (12).

Table 13
Surface Tension

Concentration mgI/ml	Surface Tension (dyne/cm)	
	20°C	37°C
45	69.3	66.8
183	66.3	63.2

2.3.9 Partition Coefficient

Partition coefficient values of iodoxamic acid in various solvent systems are reported in Table 14.

Table 14
Iodoxamic acid partition coefficient values

System	Temp.	P	Ref.
n-octanol/0.1 N HCl	20°C	26.1	3
" /water (pH 3)	20°C	2.9	3
" /phosphate buffer (pH 7)	20°C	0.01	3
Chloroform/0.1 N HCl	--	2.5	2
" /water (pH 4)	--	0.1	2
Ethyl acetate/0.1 N HCl	--	12.5	2
" /water (pH 4)	--	0.02	2

3. SYNTHESIS

The iodoxamic acid synthesis is outlined in Scheme 1 (13).

SCHEME 1
Synthesis of iodoxamic acid

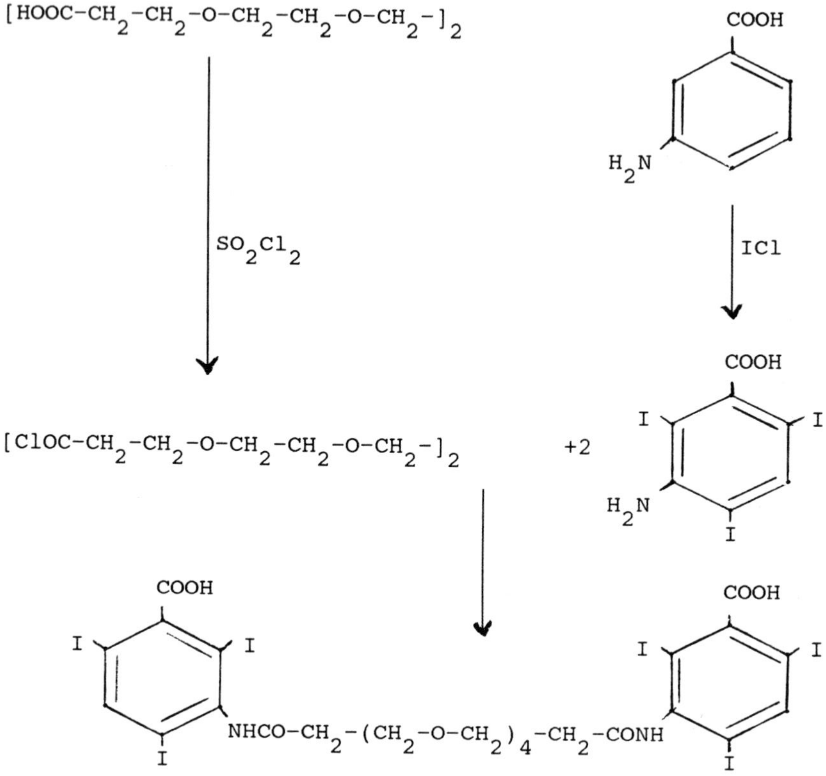

Preparation of ^{125}I- and ^{131}I- labelled iodoxamic acid can be carried out according to a fast and efficient melt-labelling procedure in acetamide (14).

4. STABILITY

Iodoxamic acid is stable both in the solid state and in solution if stored under normal conditions and protected from light (2,7).

5. METHODS OF ANALYSIS

5.1 Elemental analysis

The elemental analysis of iodoxamic acid working standard is shown in Table 15.

Table 15
Elemental analysis

Element	% calc.	% found
C	24.25	24.26
H	2.04	2.04
I	59.12	58.90
N	2.18	2.26

5.2 Identification tests

Violet vapours are evolved by haeating iodoxamic acid in a tube over an open flame (1).
Identification of the product is carried also by IR, UV and TLC (1).

5.3 Organically bound iodine

Organically bound iodine is determined by potentiometric titration with 0.1 N $AgNO_3$ after reductive dehalogenation with Zn and NaOH or with $NaBH_4$ (1).

5.4 Acidimetric titration

Iodoxamic acid can be titrated potentiometrically in $MeOH/H_2O$ with 0.1 N NaOH (1).

5.5 Chromatography

5.5.1 Paper Chromatography

Paper chromatography of iodoxamic acid was carried out using Toyo Roshi n.2 or n.562 filter paper with the following solvent systems (15):

a) butanol-acetic acid-water 3:2:1 (v/v)

b) butanol-acetic acid-water 4:1:5 (v/v)

5.5.2 Thin layer chromatography

The identification and separation of iodoxamic acid form its by-products was carried out using Merck F 254 silica gel precoated plates; UV detection (254 nm) was performed directly or after treatment with 1% aqueous starch solution. The data are summarized in Table 16.

Table 16
TLC data for iodoxamic acid

Solvent * System	Rf	Reference
I	0.41	16
II	0.15	16
III	0.35	17
IV	0.50	18,2
V	0.28	18
VI	0.27	15
VII	0.67	15
VIII	0.16	2
IX	0.48	2
X	0.50	2
XI	0.01	2

* Solvent systes :

I	dioxane-water-25% NH_3	80:20:4
II	ethylacetate-methanol-acetic acid	70:20:10
III	propanol-28% NH_3	4:1
IV	butanol-acetic acid-water	3:1:2
V	methyl ethyl ketone-acetic acid	20:5
VI	methyl ethyl ketone-acetic acid	3:1
VII	butanol-acetic acid-water	3:2:1
VIII	butanol-acetic acid-water	6:0.5:2
IX	isopropanol-butanol-NH_3	35:35:30
X	isobutanol-acetic acid-water	3:1:2
XI	ethyl acetate-acetic acid-water	6:0.5:2

5.5.3 High Pressure Liquid Chromatography

A High Pressure Liquid Chromatography (HPLC) method was developed (19) for the determination of iodoxamic acid using a Hibar RP18-5μm 4x250 mm column (E. Merck, Darmstadt) and an isocratic mobile phase consisting in a 60:40 (v/v) mixture of 0.05M sodium acetate in water (pH = 6) and methanol. Elution was carried out at 35°C with a flow rate of 1.5 ml/min; the retention time of iodoxamic acid is of 4.1 min (UV detection at 238 nm). The method resulted very precise (SD = ± 0.62%, n = 5) and it allowed the separation of two potential impurities: 3-amino-2,4,6-triiodo-benzoic acid (retention time 2.03 min) and 4,7,10,13-tetraoxadecandioic acid-mono-(3-carboxy-2,4,6-triiodo-N-phenylamide) (retention time 3.75 min). The detector responses was linear up to 400 μg/ml.

5.6 Polarography

In pH 1 ÷ 13 buffer solutions iodoxamic acid shows a reduction wave with one peak. Upon addition of 0.01% gelatin the polarogram changes into a wave with 2-3 steps. In the pH 6-7 range, the wave height is proportional to the concentration and the sum of the diffusion current constants Kd is of 14 $\mu AmM^{-1} \bullet mg^{-2/3} \bullet sec^{\frac{1}{2}}$ (2).

5.7 Paper electrophoresis

Paper electrophoresis of iodoxamic was carried out (15) using Toyo Roshi N.50 or N.526 filter paper with a voltage gradient of 10 v/cm for 60 min. or 6 v/cm for 90 min. using the following buffers:

a) 0.1 M sodium acetate, pH 7.7
b) pyridine-acetic acid-water, 1:10:89 (v/v), pH 3.4

5.8 Analysis of impurities

5.8.1 Free aromatic amine

The determination of free aromatic amine is carried out according to a method based on the

Bratton and Marshall reaction (20). It consists in diaxotization of the aromatic amine, elimination of the excess of sodium nitrite with ammonium sulphamate, coupling of the diazo compound with N-(1-naphtyl)ethylenediamine and colorimetric quantitative determination of the compound thus obtained (1). Analysis can be carried out also by an automated procedure developed for other radiological contrast media (21).

5.8.2 Free iodine and free halides

Free iodine is detected by extraction with chloroform of an aqueous solution of iodoxamic acid. Chloroform must show no violet colouring (1). The free halides test is carried out on the mother liquors obtained after precipitation with diluted HNO_3 of a solution of the ammonium salt of iodoxamic acid. Free halides are determined by turbidimetry using 0.1 N $AgNO_3$. Iodide ions can be titrated potentiometrically with 0.001 N $AgNO_3$ (1).

6. METABOLISM AND PHARMACOKINETICS

6.1 Metabolism

Studies in dogs (18), rats (15), monkeys (17,22) and humans (16,18) showed that iodoxamic acid is almost completely excreted as the unchanged species following intravenous administration. No metabolites were found in plasma and bile specimens while small amounts were detected by TLC in urine from dogs (18), monkeys (17,22) and humans (16,18). Only one metabolite could be isolated from human urine by ion-exchange column chromatography and was in sufficient amount to be identified as 3-[3-(α-hydroxyethoxy) propionamido]-2,4,6-triiodo-benzoic acid (18). TLC of human feces showed the presence of only one not identified metabolite (18).

No inorganic iodide was detected in human urine specimens (18). Trace amounts of iodide ions were detected by TLC, paper partition chromatography and paper electrophoresis in bile and urine specimens collected between 4 and 24 h after injection of iodoxamic acid in rats (15).

6.2 Pharmacokinetics

The pharmacokinetic behaviour of iodoxamic acid has been described following intravenous injection in animals (12,15,17,18,22-26) and humans (27-31,37).

In all the animals species examined the compound is eliminated mainly by the biliary route as shown in Table 17.

Table 17
Elimination of Iodoxamic acid in different animal species

Animal	dose (mg/kg)	Time post-injection (h)	Excretion as % administered dose		Ref.
			Bile	Urine	
Rabbit	100	3	77.2	16.6	12,23
Cat	100	3	75.5	17.8	12,23
Rat	200	24	89	8.4	15
Dog	100	6	74	21	18

In dogs, iodoxamic acid shows a maximum biliary transport value (Tm) of 0.914 ± 0.291μmol•Kg^{-1}•min^{-1} (23). In humans, Strata et al. (27) reported that 72 h following intravenous administration, between 59 and 72% of the administered dose, which corresponded to 3.66 g of iodine, was excreted by the fecal route and 13% by the urinary route, most of which was excreted during the first 4 h. Different compartmental models have been proposed to describe the kinetics of the compound (27-29). The pharmacokinetic parameters reported depend on the administered dose, as does the percentage of protein binding and the urinary excretion of the compound (29,31). Following intravenous administration of 200 mg/kg to rats, blood levels were found to decline biphasically with apparent half-lives of 6.4 min for the first 30 min and 91 min for the following 150 min; when the dose was raised to 4500 mg/kg, the half-life of the initial phase increased to 21 min (15). A plasma half-life of 1.1 ± 0.1 h has been reported following administration of a dose corresponding to 70 mgI/kg to humans; the urinary excretion was of over 10% in 24 h and the maximum biliary transport value (Tm) resulted of 0.35 $mgI{\bullet}kg^{-1}{\bullet}min^{-1}$ (33).

6.3 Protein binding

Biliary contrast agents frequently show affinity for plasma proteins. The extent of binding of iodoxamic acids to animal (15,17,22-24) or human (30,32,33) plasma proteins was estimated mainly using the equilibrium dialysis method. Also the ultrafiltration method (31) and circular dichroism measurements (32) were used to investigate this parameter. In all cases a high affinity for plasma proteins was found, though lower than that of some other cholecystographic contrast media (23,31).

The compound was reported to bind solely to plasma albumin and binds to at least two binding sites (17,22,23,33), the values depending on its plasma concentration. Lin et al. (17), in an in vitro experiment carried out using monkey blood samples, found unbound fractions of iodoxamic acid varying between 6.1 and 41.2 percent for plasma concentrations increasing from 42 to 912 μM. Shibota et al. (15) reported that about 70 percent of the compound was bound to rat plasma proteins when the plasma drug concentration was about 0.5μmol/ml, corresponding to plasma levels 5 minutes after injection of a 200 mg/kg dose; at lower concentrations, binding increased to about 90 percent.

6.4 Acute Toxicity

The intravenous acute toxicity values (LD_{50}) of iodoxamic acid were determined in various animal species. The data are reported in Table 18 (12,23).

Table 18
Intravenous acute toxicity data (LD_{50})

Animal	LD_{50} in mg free acid/kg b.w. (confidence limits p = 0.05)
Mouse	4850 (4550-5165)
Rat	4800 (4383-5256)

(cont'd)

Table 18 (cont'd)

Animal	LD_{50} in mg free acid/kg b.w. (confidence limits p = 0.05)
Rabbit	5830 (5300-6413)
Dog	4850 (4292-5480)

Data have also been reported as regards the LD_{50} of the compound following intracarotid injection; the values resulted of 2928 mgI/kg in the rat and of 5398 mgI/kg in the rabbit (34).

7 DETERMINATION OF IODOXAMIC ACID IN BODY FLUIDS AND TISSUES.

The methods used for the assay of iodoxamic acid in body fluids and tissues are based on the determination of the iodine concentration in the specimens. The following techniques have been reported :

a) an automated colorimetric method (18,23,26-30)
b) fluorescent excitation analysis (17,22,24)
c) oxygen flask combustion of the sample followed by colorimetric determination or potentiometric titration of iodide ions (35).
d) a modification of the thiosulfate titration method described by Zac and Boyle (25,36);
e) measurement of ^{125}I or ^{131}I radioactivity of the labelled compound using a scintillation spectrometer (15,22,24,31,33).

8. REFERENCES

1. E. Felder, U. Tiepolo and M. Zingales, Il Farmaco Sci. Ed., 28, 607, (1973)

2. Y. Miyake and Y. Asahi, J.Takeda Res. Lab., 33, 73, (1974).

3. E. Felder, D. Pitrè and M. Grandi, Il Farmaco Sci. Ed., 28, 925, (1973).

4. D. Pitrè, M. Grandi, P. Traldi and R. Self, Arch. Pharm., 317, 1037, (1984).

5. A. Clerici, P. Traldi, M. Grandi and D. Pitrè, Biomed. Mass. Spectrom., 9, 257, (1982).

6. M. Grandi, D. Pitrè, A. Clerici, P. Traldi and I.A.S. Lewis, Biomed. Mass. Spectrom, 10, 17, (1983).

7. U. Tiepolo, Bracco SpA, Personal Communication

8. W. Simon, Angew. Chem., 76, 722, 1964).

9. P. Sommer, Dissertation, Eidgenoessische Technische Hochschule, Zürich, Juris- Verlag 1961.

10. D.F.Burger, J.Phys. Chem., 67, 2590, (1963).

11. Pregl-Roth, Quantitative Organische Mikroanalyse, 7th Edition, Springer Verlag, 310, (1953).

12. E. Felder, D. Pitrè, G. Rosati and P. Tirone, Radiol. Med., 60, 209, (1974).

13. E. Felder, D. Pitrè, L. Fumagalli and E. Lorenzotti, Il Farmaco Sci. Ed., 28, 912, (1973).

14. H. Sinn, and W. Maier- Borst, Int. J. Appl. Radiat. Isotop., 30, 511, (1979).

15. M. Shibota, K. Hamajo and T.J. Fujita, J. Takeda Res. Lab., 33, 96, (1974).

16. W. Mützel, V. Taenzer and R. Wolf. Invest. Radiol., 11, 598, (1976).

17. S.K. Lin, A.A. Moss, R. Motson and S. Riegelman, J. Pharm. Sci., 67, 930, (1978).

18. D. Pitrè and E. Felder, Il Farmaco Sci. Ed., 29, 793, (1974).

19. M. Cabrini, Bracco SpA, Personal Communication.

20. A. Bratton and J. Marshall, J. Biol. Chem., 128, 537, (1939).

21. E. Felder, D. Pitrè and M. Grandi, J. Pharmacol. Sci, 64, 684, (1975).

22. A.A. Moss, S.K. Lin and S. Riegelman, Invest. Radiol., 15, S132, (1980).

23. G. Rosati, P. Schiantarelli and P. Tirone, Il Farmaco Sci. Ed., 28, 996, (1973).

24. S.K. Lin, A.A. Moss and S. Riegelman, J. Pharm. Sci., 68, 1430, (1979).

25. R.N. Berk, P.M. Loeb, A. Cobo-Frenkel and J.L. Barnhart, Radiology, 119, 529, (1976).

26. W. Zaunbauer, E.M. Grandjean, G. Paumgartner and W.A. Fuchs, Schweiz. Med. Wochenschr., 111, 157, (1981).

27. A. Strata, G. Magnani, F. Tirelli and G. Rosati, J. de Radiologie, 6 (Suppl. 1), 40, (1975).

28. G. Segre, G. Rosati, J. de Radiologie, 56, 44, (1975).

29. M. Arditi, G. Molino, A. Biondi, A. Fornara, C. Fava and F. Bonati, Rays, 10, 81, (1985).

30. J. Kutzner, S. Heymann-Schramm and W. Grimm, Rö-Bl, 34, 310, (1981).

31. V. Taenzer, U. Speck and R. Wolf, Fortschr. Rontgenstr., 126, 262, (1977).

32. K.J. Fehske and W.E. Müller, J. Pharm. Sci., 70, 549, (1981).

33. W.E. Müller and A.E. Stillbauer, Pharmacology, 22, 269, (1981).

34. P. Tirone and G. Rosati, Il Farmaco Sci. Ed., 28, 1011, (1973).

35. M. Kan, H. Kashiwagi, T. Terao and K. Terao, J. Takeda Res. Lab., 33, 87, (1974).

36. B. Zak and A.J. Boyle, J. Amer Pharmacol. Assoc. Sci. Ed., 41, 260, (1952).

37. B. Mahner, R. Kruger, V. Siehr and U. Wilhelmi, Fortschr. Roentgenstr., 125, 556, (1976).

IPRONIAZID PHOSPHATE

F. Belal and H. Abdel-Aliem

Faculty of Pharmacy
University of Mansoura
Mansoura 35516
EGYPT

Contents

I- Historical :

The discovery of Iproniazid constitutes a notabl case of serendipity of great importance in the chemotherapy of mental diseases. In 1951 the anti-tuberculous isoniazid was discovered by two independent groups of workers. In 1952 the research was continued for other synthetic tuberculostats. In 1953 in tests of iproniazid and isoniazid against tuberculosis at Sea View Hospital at Staten Island , it was noted that " central nervous stimulation is apparently also to be listed among the side effects of iproniazid ". Inasmuch as this undesirable side effect was more marked with iproniazid in doses used for tuberculosis , isoniazid was preferred and used widely against tuberculosis , while iproniazid was almost abandoned. However , iproniazid was continued to be used because it contributed to the healing process in addition to its tuberculostatic activity. In 1957 , at a regional meeting of the American

Psychiatric Association held in New York , it had been reported that iproniazid had helped some chronically depressed hospitalized psychotics who had not been **helped** by any other therapy. Similar results were also reported in the same meeting , after which iproniazid found widespread use in treating depression (1).

II- Description

II-1- Names

II-1-1- Chemical Name :

1- Isonicotinyl-2-isopropylhydrazine , phosphate.

II-1-2- Chemical Name as cited in the Chemical Abstracts :

Up to 1971 : Isonicotinic acid, 2- iso-propylhydrazide [54-92-2]

the phosphate salt [305-33-9]

Starting from 1972 :

4- Pyridinecarboxylic acid, 2-(1-methylethyl) hydrazide

II-1-3- Generic Name : Iproniazid Phosphate.

II-1-4- Proprietary Name : Marsilid, Euphazid.

II-2- Formulas

II-2-1- Structural Formula:

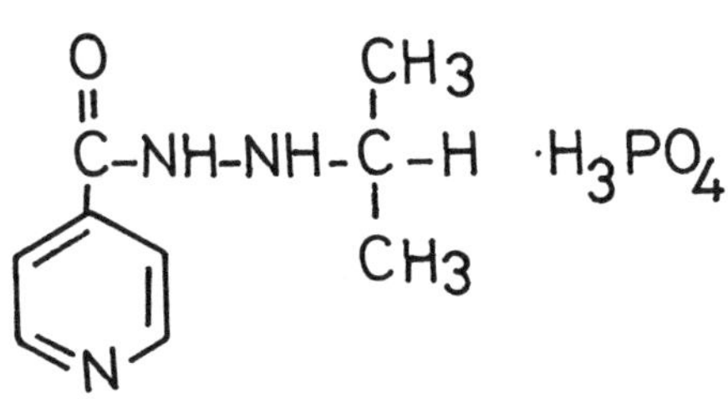

II-2-2- Molecular Formula : $C_9H_{13}N_3O \cdot H_3PO_4$

II-3- Molecular weight : 277.22

II-4- Appearance , color, odor : white to slightly

yellowish powder.

III- Phyisco- Chemical Properties

III-1- Melting range : 175-184°C.

III-2- pH of a 5% solution in water : 2.7-3.6

III-3- Optical Rotation : the substance exhibits no optical activity.

III-4- Powder Fineness (through Sieve No.60 USP): minimum 95%

III-5- Polymorphism : Based on IR spectra and melting points , the substance is of one crystalline form.

IV- Physical Properties

IV-1- Infrared Spectrum (IR) :

The IR spectrum (Fig.1) of bulk reference standard iproniazid phosphate was obtained from a pellet made by dispensing 1.1 mg of iproniazid phosphate in 300 mg of KBr (2). A list of the assignments made for some of the characteristic bands is shown in the following table.

Wavelength (Cm^{-1})	Assignment
3120	N-H Bond
2980 and 2820	C-H Bond
2448	N^{+}—H bond
1660	Amide C=O
1570	Heteroaromatic
1370	$-CH_3$ Symmetrical
1325	C-H deformation
1130	C-N
1070	phosphate

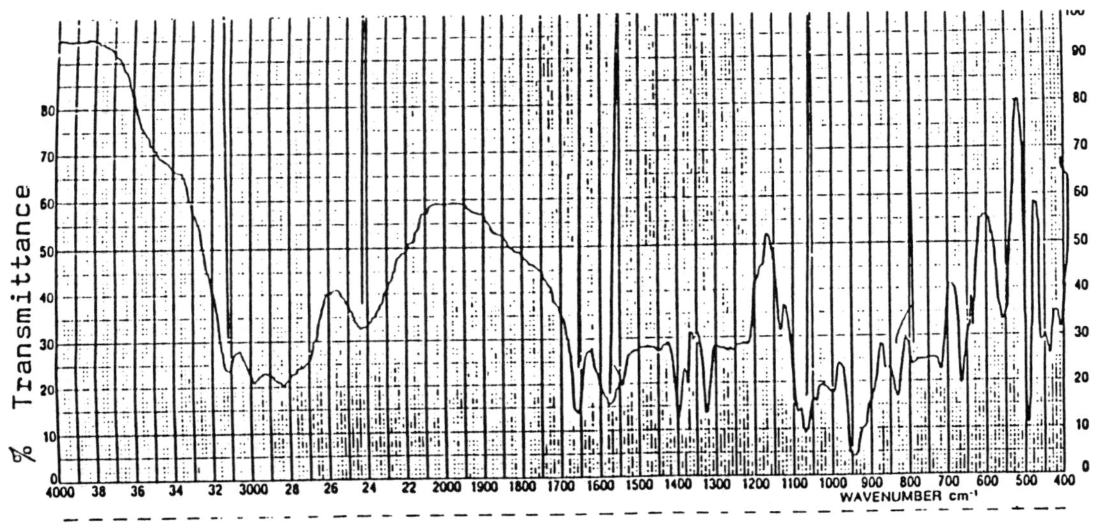

Fig. 1 - Infrared Spectrum of Iproniazid Phosphate.

IV-2- Nuclear Magnetic Resonance Spectrum (NMR) :

The integrated 400 MHz ^1H-NMR spectrum of iproniazid phosphate (Solvent D_2O) is shown in Fig.2. The spectral assignments are presented in the following table :

Chemical Shift	Multiplicity	Number	Assignment
1.23	sharp singlet	6	$-CH_3$
3.44	quartet	1	-CH
8.2	doublet	2	aromatic
8.22	doublet	2	protons

The 2 N-H protons , being weakly acidic , are replaced by deuterium ions. The sharp singlet (un-integrated) at ppm 4.82 is due to the formation of H_2O as a result of this replacement process . Therefore these 2 protons did not appear in the spectrum.

IV-3- Mass spectrum (MS) :

The mass spectrum of iproniazid phosphate is shown in Fig.4. A molecular ion was observed at m/e 179. The base peak in spectrum at m/e 123 is formed by double rearrangement , i.e., one hydrogen migrates from a nitrogen atom and another hydrogen is transferred from the isopropylic CH (3). Structural assignments are listed in the following table :

m/e	% Relative Intensity	Assignment
31	30	$NH=NH_2^+$
43	40	$CH_3-\overset{+}{C}H-CH_3$
51	39	$[C_4H_3]^+$
58	42	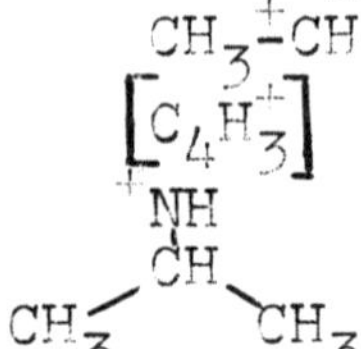

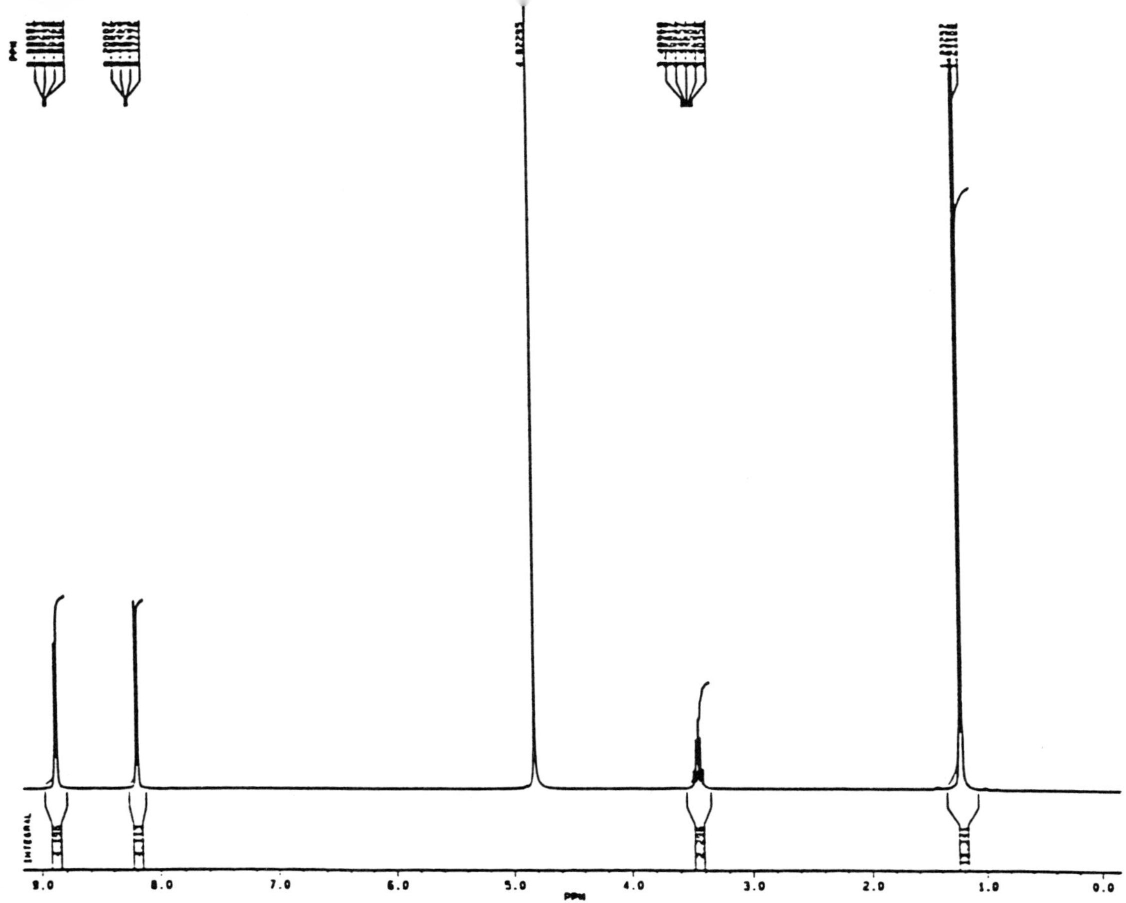

Fig. 2- NMR Spectrum of Iproniazid Phosphate

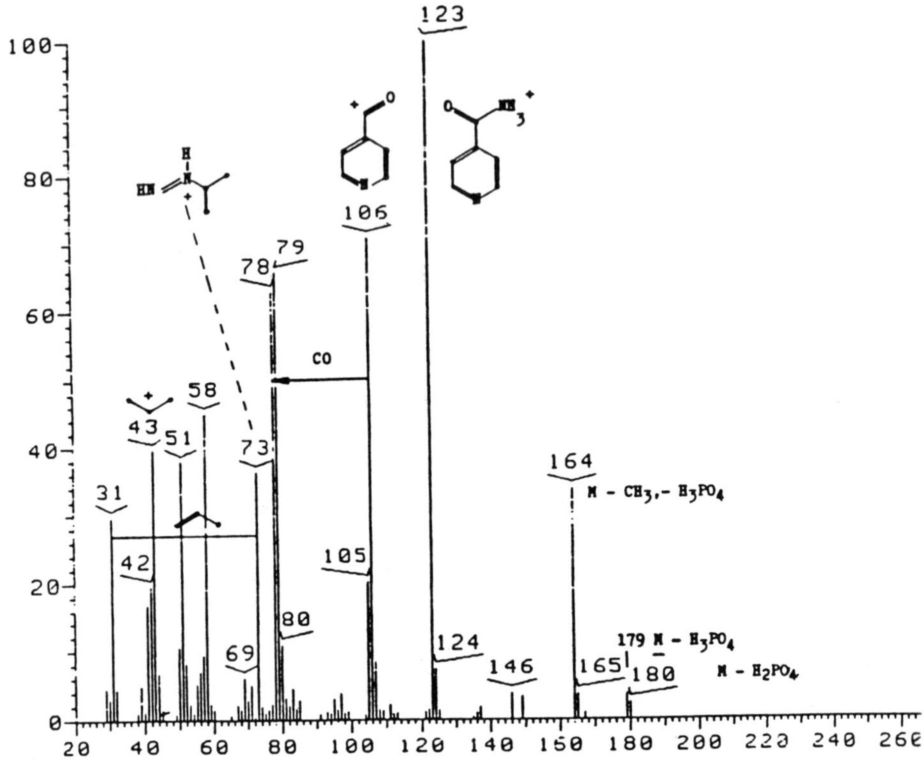

Fig. 3 - Mass Spectrum of Iproniazid Phosphate.

m/e	% Relative Intensity	Assignment
73	38	$NH{=}\overset{+}{N}H{-}CH(CH_3)_2$
78	72	
79	73	
106	66	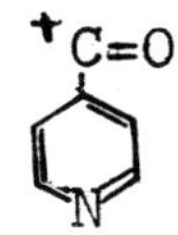
123	100	$C_5H_4N{-}C(=\overset{+}{O}H){-}NH_2$
146	5	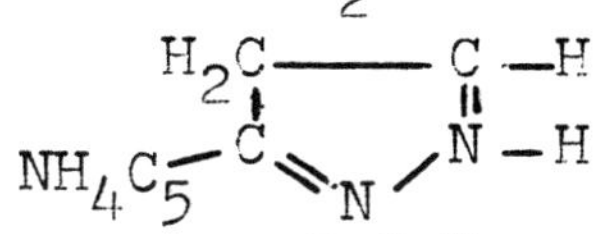
146	38	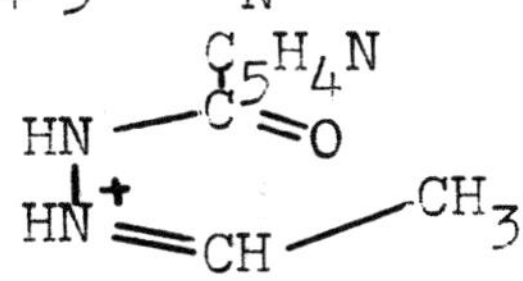
179	5	$M{-}H_3PO_4$

IV-4- Ultraviolet Spectrum (UV) :

The UV spectrum of iproniazid phosphate (2.5mg in 100 ml ethanol and 0.1 N HCl) is shown in Fig.4 . Spectral data in these solvents and additional solvents are abridged in the following table (4&5).

Solvent	λ max.	$A^{1\%}_{1\ cm}$	Molar absorptivity
Methanol	265 nm	166	4600
Water	264 nm	176	4880
HCl,0.1 N	268 nm	184	5090
KOH,0.1 N	244 nm	113	3133
	272 nm	121	3350
	308 nm	135	3740

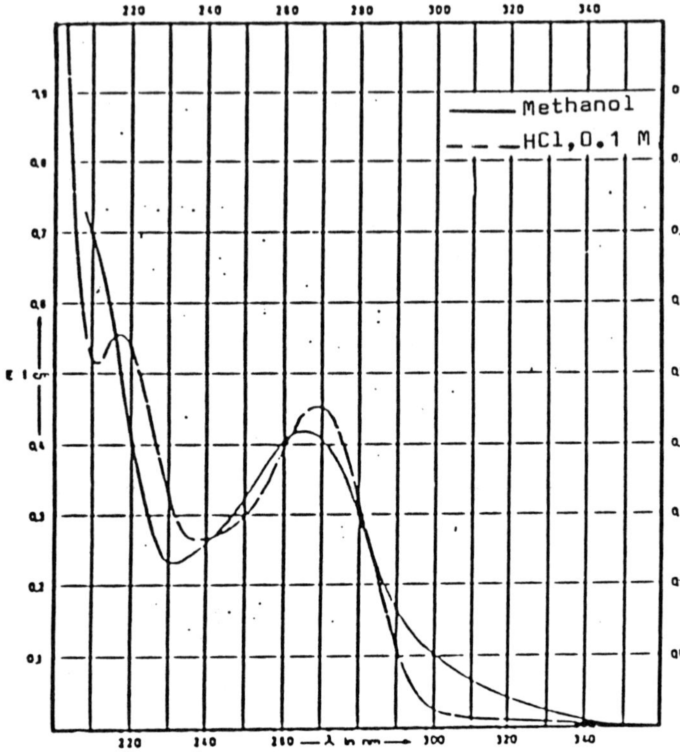

Fig. 4 - Ultraviolet Spectrum of Iproniazid Phosphate (25 ug / ml)

IV-5- Luminescence Characteristics of Iproniazid phosphate :

The excitation, fluorescence and phosphorescence characteristics of iproniazid phosphate have been studied in ethanol at 77K (6). Using 300 nm as λ_{max} excitation, no fluorescence was produced, rather, phosphorescence at 400 nm was observed. This phosophorescence, however, was very feeble and its life time was less than 0.2 sec. which rendered it of no analytical value.

IV-6- Solubility :

The solubilities of iproniazid phosphate at room temperature in different solvents are listed in the following table (2)

Solvent	Solubility (mg/ml)
Water	approx.188
Methanol	approx.21
Ethanol (96%)	approx. 9
Chloroform	approx. 0.6
Diethylether	<0.1
Hexane	<0.1

IV-7- Dissociation Constants :

The first and second dissociation constants of iproniazid phosphate were potentiometrically determined in 1.0 M NaCl solution at 20 ± 0.5 °C and were found to be 4.35×10^{-3} and 1.2×10^{-4} respectively (7). Spectrophotometric determination of the dissociation constants at 370 nm, under the same experimental conditions gave the following values : 6.37×10^{-3} and 1.18×10^{-4} for the first and second dissociation constants respectively.

IV-8- Acid-Base Equilibrium :

The ionisation constants of iproniazid phosphate in aqueous solutions were determined spectrophotometrically , and found to be as follows :

$$K_1 = 10.83 \pm 0.07$$
$$K_2 = 3.52 \pm 0.06$$
$$K_3 = 1.80 \pm 0.08$$

IV- 9- Thermogravimetric Analysis :

Thermogravimetry can be used to determine moisture or residual solvents in iproniazid phosphate . Heating a standard sample for 4 hours at 105 °C , the % loss was max. 1 % (2). When the residue of ignition (sulphated ash) was tested for heavy metals , the latter was max. 20 ppm. (2).

V- Stability :

V-1- Possible Decomposition Products :

a- Isonicotinic acid hydrazide 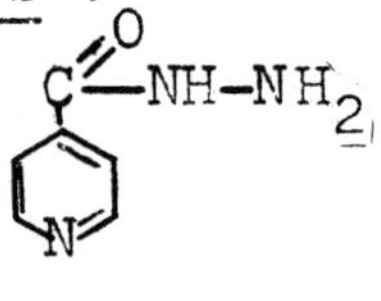

b- Iproden base (1- Isonicotinyl-2-isopropylidenehydrazide) 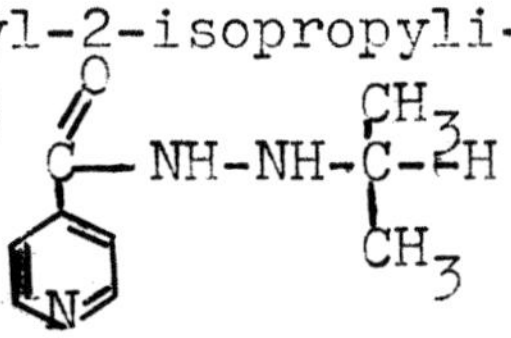

C- 1-Isonipecotyl-2-isopropylhydrazine

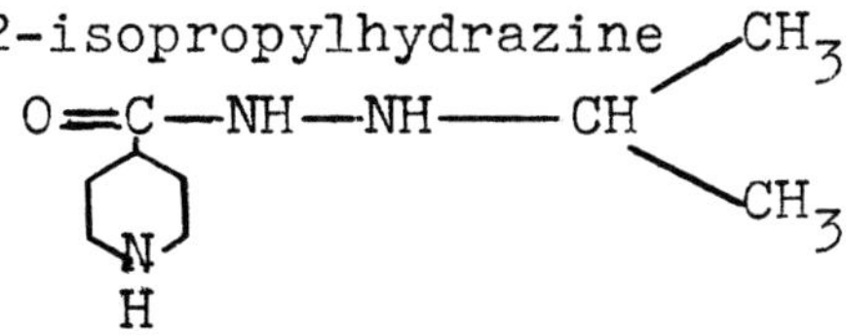

V-2-Stability Against Heat and Humidity :

Stability studies revealed that iproniazid was not decomposed upon keeping at room temperature and 35 °C , dry and at 85% relative humidity for 3 months (2).

V-3-Stability Against Light :

Both iproniazid phosphate powder and a 0.5% solution in ethanol 96% were light-insensitive. No change in appearance or decomposition was observed upon irradiation with a xenon lamp for 6 hours(2).

V-4-Stability Against Oxygen :

Similarly no change in appearance or decomposition was observed when a 0.5% solution in ethanol 96 % was saturated with oxygen for 24 hours in the dark(2).

V-5-Stability Against Acid / Base :

The effect of acid/base was studied by keeping 1 % solutions of iproniazid phosphate in buffer solutions (pH 1 to 13) and in water at room temperature and at 35 °C protected from light. After 24 hours the solutions were tested by TLC. Initial analysis was carried out on a freshly prepared solution. At room temperature no decomposition was found in the range of pH 3 to 12 . The results are shown in the following table.

Temp.	Initial Analysis	pH 1	pH 2	pH3-12	pH 13	Own pH = 3.1
Room Temperature % Decomposition	$\leq$ 1 %	$\leq$ 2.4%	$\leq$ 1%	$\leq$ 1%	$\leq$ 1.8%	$\leq$ 1%
35 °C % Decomposition	Initial Analysis	pH 2-6	pH7-13	Own pH = 3.1		
	$\leq$ 1.2 %	$\leq$ 4.4 %	$\leq$ 2.2%	$\leq$ 1.6 %		

VI- Pharmacology

VI-1- Indications :

The beneficial effects of iproniazid phosphate are : development of euphoria , relief of pain, overcoming anorexia , weight gain in debilitated patients , temperature reduction, wound healing , marked control of tuberculosis in bones and joints in addition to decalcification of osseous structure without weakness (10).

VI-2-Dosage :

The usual safe dose for moderately severe conditions is 2 mg/Kg body weight at 8 hour intervals. The maximum dosage of 4 mg / Kg body weight is indicated in severe conditions and should be restricted to relatively short period , such as 1 to 2 weeks (10).

VI-3- Toxicity

Isopropylhydrazine , a metabolite of iproniazid was reported to be oxidised by Cytochrome P-450 enzymes in humans and rats liver microsomes to a highly reactive alkylating agent. Covalent binding of this metabolite to liver macromoles paralleled hepatic liver necrosis (11).

VI-4- Metabolism

The metabolic fate of C^{14}- iproniazid labeled in the isopropyl moiety in laboratory animals was reported by Koechlin and Iliev (12). It was shown that , in rats , mice and rabbits C^{14}- iproniazid was metabolised predominantly by the cleavage oxidation of the isopropyl substituent to C^{14}-acetone and further to $C^{14}O_2$. While part of the isonicotinyl moiety of iproniazid was excreted by the rat as N_2- acetyl isonicotinic acid hydrazide, the major portion was recovered in the form of free isonicotinic acid, Fig. 5.
Formation of isopropylhydrazine as a metabolic intermediate was surmised from the finding that after high doses , the rate of isonicotinic acid formation exceeded that of acetone plus CO_2 . However , its direct

demonstration in tissues or urine was not possible. Comparison of the fate of iproniazid in man with that of rats revealed identical major pathways of oxidative degradation, but characteristic differences of the degradation rates (13). In humans, C^{14}- iproniazid was metabolised by oxidation of the labeled

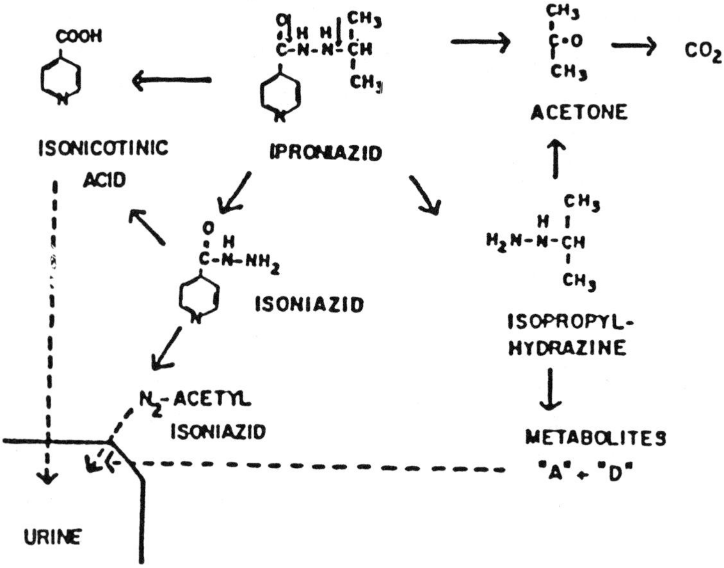

Fig.5- Metabolic Fate of Iproniazid

isopropyl substituent resulting in the elimination of 50 to 60 % of the radioactive dose as pulmonary CO_2 within 24 hours. The oxidation rates differed markedly between individuals as did patterns of the radioactive urinary excretion products.

VII- Synthesis

The following approaches were adopted for the synthesis of iproniazid :

A- Catalytic Reduction of the isopropylidene derivative (14). A more convenient method based on reacting isonicotinylhydrazine with acetone in presence of isopropanol , followed by catalytic reduction was also described (Scheme I , A).

B- Isonicotinylhydrazine is reacted with chlorine followed by reaction with isopropylhydrazine in presence of ammonia (15) as shown in Scheme I , B.

C- Isopropylhydrazine is allowed to react with isopropyl bromide as shown in Scheme I , C(16).

VIII - Crystallographic Data

VIII-1- Crystal Structure

The crystal structure of iproniazid phosphate has been determined by X-ray diffraction techniques. The colorless crystals are triclinic with space group $P\bar{1}$ and cell dimensions of $a = 9.063$, $b = 10.084$, $C = 7.883$ Å, $\alpha = 94.02$ $\beta = 98.84$ and $\gamma = 116.67$ with $Z = 2$.
The structure was determined by Patterson and Fourier methods. The crystal consists of an iproniazid cation and an $H_2PO_4^-$ anion held together by a system of hydrogen bonds (17). The hydrogen bonding together with the boundary thermal ellipsoids are illustrated in Fig. 6.

VIII- 2- Microscopic Crystal Tests :

Sublimation , fusion and crystallization (from 50 % alcohol acidified with 6 N HCl

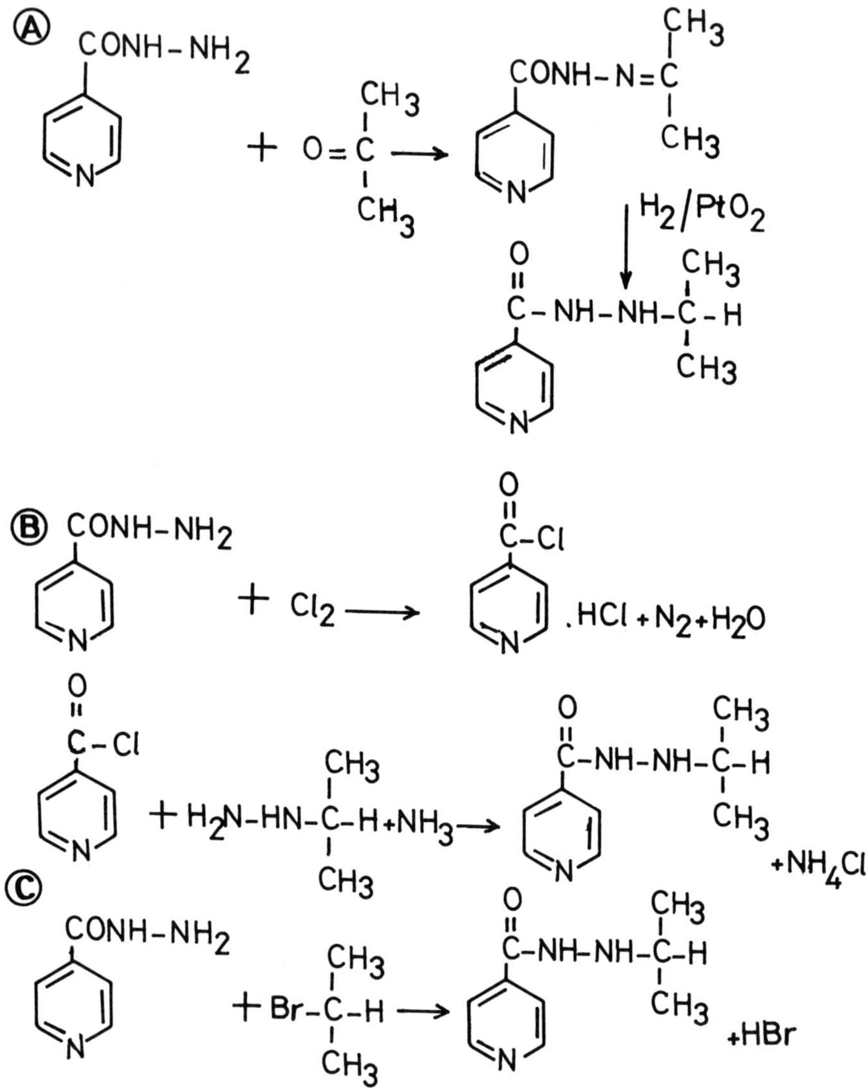

Scheme I - Synthesis of Iproniazid.

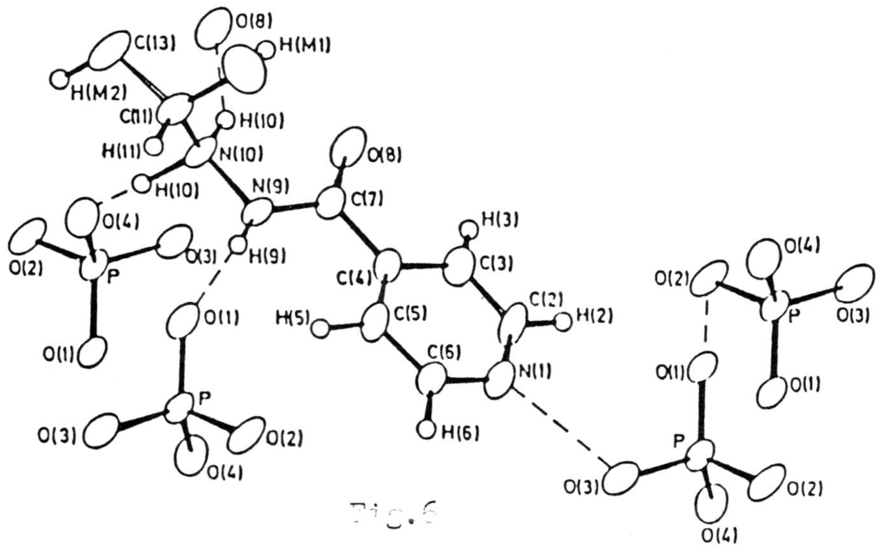

Fig.6

The boundary thermal ellipsoids are shown together with the hydrogen bonding between the phosphate and iproniazid groups

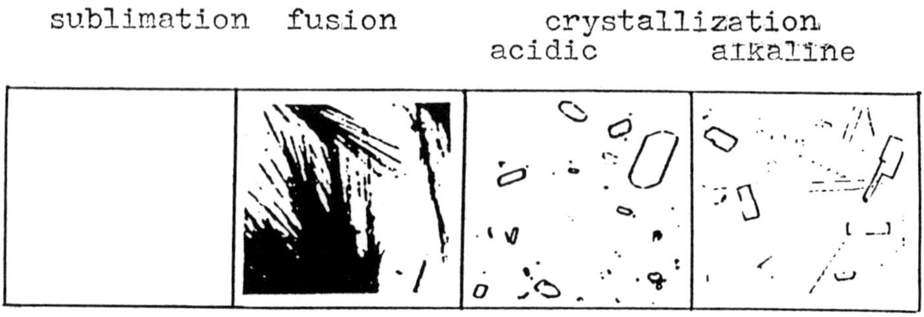

Fig.7- Microscopic Crystal Test of Iproniazid.

and alkalinised with conc. ammonia respectively) tests were performed on iproniazid phosphate crystals (18) . It was found that iproniazid phosphate is not sublimiable . The other preparations were photographed at 75X, 150X and 75X respectively , Fig.7.

IX - Qualitative Analysis

The following color tests were used for the identification of iproniazid phosphate :

1- With sodium nitrosoferricyanide , an orange color is produced and turns bluish-green upon heating (20).

2- Nitration test : the sample is heated with KNO_3 powder and H_2SO_4 on a boiling water bath then cooled , diluted with water and extracted with ethylether . Treating the ether extract with drops of 3 % ethanolic KOH produces a yellow color (21).

3- Micro color tests : these tests are carried out either on a micro-drop (0.02 ml containing 20 µg) or a very small amount (0.5 mg) picked up at the end of a micro-spatula (21).

a- Frochd's Test : with 0.5 % aqueous solution of ammonium molybdate an orange color is produced and changes into black upon warming.

b- Mandlin's Test : with 0.5 % aqueous solution of ammonium vanadate a purple color is produced with the powder.

c- Micke's Test : with 0.5 % aqueous solution of selenous acid , H_2SeO_3, an orange color changing into reddish -brown is produced with the powder , the solution gives a pink color.

X - Quantitative Analysis :

X-1- Titrimetric Methods :

Beltagy et al. described a method based on treating with ICl in strong acid medium and titrating the liberated iodine with KIO_3(22). Another oxidimetric method involving titration with 0.1 N ceric nitrate and potentiometric detection of the end point was reported (23). Gauter et al. (24) described another cerimetric method based on oxidation with 0.1 N $CeSO_4$ and back titration of the residual reagent with As_2O_3, the end point was visually detected using O-phenanthroline/Fe^{++} indicator.

X-2- Spectrophotometric Methods :

Kraemer et al. (25) reported on the use of UV spectrophotometry for the determination of iproniazid. Several colorimetric methods were utilized for the determination of iproniazid phosphate in pharmaceutical preparations or in biological fluids. Cleavage of the pyridine ring with ClSCN (26), ClCN (27), BrCN (28) followed by coupling with barbituric acid or 2-thiobarbituric acid and subsequent measurement of the produced polymethine dye were reported by many workers. The drug in protein-free , oxalate-treated plasma is converted into isonicotinic acid through reaction with $KMnO_4$, then allowed to react with BrCN and NH_3. The color produced is spectrophotometrically measured. (29). Direct colorimetric methods using molybdic acid (30), tetrazolium salt (31), ammonium vanadate (32), epichlorhydrin (33), sodium nitritopentacyanoferrate (III) (34), ferricyanide (35) were frequently described.

X-3- <u>Fluorometric Methods</u> :

Marjorei et al. (36) described a fluorometric method for the determination of iproniazid in serum . The method is based on adsorption of iproniazid from a protein-free filtrate of the serum on an amberlite XE-64, elution with 0.75 N HCl , neutralization, then adjustment to pH 8.7 using buffer solutions, followed by reaction with H_2O_2 . The fluoroscence at 415 nm (λ max. activation 320 nm) is quantitative for 0.05 to 10 µg/ ml serum. An indirect fluorometric method for the determination of iproniazid in human platelets was described by Zaitsu et al.(37).

The method is based on measuring benzaldehyde formed from the substrate benzylamine through reaction with 1,2-diaminonaphthalene producing the fluorophore 2-phenylnaphtho [1,2-d] imidazole. The method assesses the inhibitory potency of iproniazid even with 50 ng of platelet protein.

X-4- Polarographic Methods :

Amer et al. (38) described the polarographic behavior of iproniazid in 0.1 N HCl and buffered solutions up to pH 8. Irreversible, diffusion-controlled cathodic waves were produced. Quantitative measurements should be performed in buffered solutions of pH 6.5-7.5. The anodic wave produced in phosphate buffers pH 11.25 with $E_{1/2}$ value of -0.18 V vs SCE was found to be suitable for analytical purposes (39). An indirect polarographic method based on oxidation with $KMnO_4$ under defined experimental conditions , leading to the transient formation of N-oxides , which are polarographically active , was reported (40).

X-5- Paper Chromatography :

El-Darawy and Mobarak (41) reported the following method for the identification of iproniazid :

Adsorbent	System	R_f(x 100)	Detection
1- Carboxymethyl cellulose ion-exchange paper	- Water	40	UV, iodoplatinate, dragendorff reagent, $FeCl_3/Fe(CN)_6^{-3}$
	- 0.01 M NaCl	50	
	- water-acetone-formamide (10:1:1)	92	

(Cont.)

Adsorbent	System	R_f(x 100)	Detection
			phosphomolybdic acid
2- Ecteola cellulose	water	92	5% ethanolic phosphomolybdic acid.

X-6- Thin Layer Chromatography (TLC) :

Several papers dealing with the TLC identification and determination of iproniazid phosphate were published . A brief description of these methods is presented in the following table :

Adsorbent	Eluting Solvent	R_f(x 100)	Detection	Ref.
1- Silica Gel G.	-$CHCl_3$:NH_3:MeOH(2:1:1)	75	ninhydrine, amm. molybdate + oxalic acid , $FeCl_3$ + fluorescine.	42
	-MeOH:$CHCl_3$ (1:1)	73		
2- Silica Gel G.	- MeOH	63	10 % $CuSO_4$ + 2% NH_4OH (5:1)	43
	- BuOH:HOAc:H_2O (3:1:1)	57		
	- $CHCl_3$: MeOH (1:1)	68		
3- Silica Gel G.	- Acetone : MeOH : NH_3 (50:50:1)	65	Tetrazolium red	44
4- Silica Gel G.	-MeOH: 12M NH_3 (100:1.5)	70	Foline-Ciocalteu reagent, FPN reagent, cinnamaldehyde, Mandelin reagent,p-dimethylaminobenzaldehyde	45
	-Cyclohexane:diethylamine: benzene (75:20:15)	5		
	-$CHCl_3$: MeOH(9:1)	40		
	-Acetone	45		
	-Benzene : EtOH: 12M NH_3 (95: 15 : 5)			
5- Silica Gel	- $CHCl_3$: EtOH: Et_2O (5: 4 : 1)		ammonium vanadate	46

	Adsorbent	Eluting Solvent	R_f(x 100)	Detection	Ref.
6-	a-Silica Gel prepared with 0.1 M. KOH	-Cyclohexane :benzene: diethylamine (75:15:10)	4	1% I_2 in MeOH, Mandlin reagent, 0.5 % Fast red G	47
		- Methanol	64		
		- Acetone	34		
	b-Silica Gel prepared with 0.1 M. $NaHSO_4$	- MeOH	34		
		- EtOH	22		
7-	Cellulose MN 300 + silica gel HF_{254}(1:1)	-$CHCl_3$:12N NH_3:MeOH (90 : 1 10)	61	UV, phosphomolybdic acid , $KMnO_4$ Foline-Ciocalteu reagent, Amm.$AgNO_3$ Iodoplatinate, I_2, Tetrazolium red, amm.molybdate + oxalic acid,$FeCl_3$/$Fe(CN)_6^{-3}$, Dithiocarbamate, Cinnamaldehyde.	48
		- Benzene :MeOH: diethylamine (90 : 10 : 1)	30		
		-$CHCl_3$:1-Hexanol : 13N NH_3 (90 : 10 : 0.2)	16		
		-$CHCl_3$:EtOAc:13N NH_3(50:50:1)	18		
		-$CHCl_3$:Acetone:HOAc (90:10:1)	9		
		-Benzene: Acetone : diethylamine (50:50:1)	40		
		-$CHCl_3$: Acetone: HOAc (50 :50 :1)	29		

	Adsorbent	Eluting Solvent	R_f(x 100)	Detection	Ref.
8-	Silica Gel G. 60 F_{254}	$CHCl_3$: EtOH (5:1)	62	Ninhydrine	49
9-	Silica Gel	-EtOĀc:Cyclohexane:p-Dioxane: MeOH:H_2O:NH_3 (50:50:10:10: 1.5:0.5)	25, 50	Ninhydrine at low temp., 0.5 % H_2SO_4	50
		- The same as above but ratio of H_2O to NH_3 is 0.5 : 1.5	60, 74		
		- EtOĀc:Cyclohexane:NH_3: MeOH:H_2O (70:15:2:8:0.5)	62, 80		
		-EtOĀc:Cyclohexane:MeOH:NH_3 (70:15:10:5)	36 67		
10-	Silica Gel	- $CHCl_3$:MeOH (8:2)	75	Foline-Ciocalteu reagent, phospho-molybdic acid.	51
		- $CHCl_3$:Acetone:Diethylamine (5:4:1)	31		
		- Cyclohexane:$CHCl_3$:Diethyl-amine (4:3:1)	6		

X -7- Gas Chromatography (GC) :

Iproniazid was determined in pharmaceutical preparations using a glass column (2 m X 2 mm) packed with Ankrom AS coated with 3% XE 60. The carrier gas was N_2 and injection temp. 200 °C, (52). The same authors (53) reported on the use of QFL/ Chromosorb and G (silanized) columns.In another attempt , they used 2% GE-SE 30 Aeropacked columns coupled with flame ionization detector (54). The sample injected was 1 µl of 1% trimethylsilyl derivative .
De Sagher and co-workers had studied 4 column systems packed with : versamide-930 , versamid 900, XE-60 and OV-225 for the GC determination of iproniazid, and they were all equally satisfactory . Nitrogen was the carrier gas in a flow rate of 5 $ml.min.^{-1}$. The 2-butyl analogs were used as internal standards. The same group of workers (56) reported on an improved column system for the GC determination of iproniazid as a result of a comparative study performed on 13 liquid phases . Later on , they also utilized a column packed with XE-60 coated on gas Chrom Q for the determination of iproniazid in human urine.

ACKNOWLEDGEMENT

The author is thankful to Hoffmann-La-Roche , Basle , Switzerland for supplying a reference sample of iproniazid phosphate and its analytical data.

REFERENCES

1- J.H.Votiz (1979) , J.Chem. Educ.56,35.
2- F.Hoffmann-La-Roche Ltd, Basle (Switzerland).
3- R.M.DeSagher, A.P. De Leenheer, A-E. Claeys and A.A. Cruyl (1975), Biomed.Mass Spec.2,82
4- " UV and IR Spectra of Some Important Drugs" Editor : H.W.Dibben (1987), Editio Canter Aulendorf, FRG.
5- J. Kracmar and J. Kracmarova (1979), Pharmazie , 34,27.
6- L.A.Gifford , J.N.Miller , J.W.Bridges and D.T. Burns (1977) , Talanta 24,273.
7- B.Salvesen and J.H. Glendrange (1966), Medd.Norsk.Farm.Selskep 28,272, C.A. 1967,66,68866 e.
8- B.Salvesen and S.A. Bikill (1965), Medd.Norsk.Farm Selskep. 27,135, C.A. 1966,64,9507 f.
9- S. Zommer (1969), Rocz.Chem. 43,1023, C.A.1969,71,95485 q.
10- D.A. Bosworth (1959), Ann. NY, Acad.Sci.80,809.
11- S.D.Nelson, J.R. Mitchell,W.R, Snodgrass and J.A.Timbrell (1978), J. Pharmacol. Expl. Ther. 206,574.
12- B. Koechlin and V.Iliev (1959) , Ann.NY. Acad. Sci. 80,864.
13- B.A. Koechlin, M.A. Schwortz and W.E. Oberhaensli (1962),J.Pharmacol.Expl.Ther.138,11
14- H.H.Fox and J.T.Gibs (1953),J.Org.Chem. 18,994
15- Laboratories Bonaplata , S.A. Span.249,575, June 25,1958, C.A.1960, 54,12163 h.

16- F.H.McMillan , F. Leonard, R.I.Miltzer and J.A. King (1953), J.Am.Pharm. Assoc.42,457.

17- P.C. Chich and G.J. Palenik (1971) , J. Chem. Soc. (A) 576.

18- P. Rojeswaran and P.L.Kirk (1961) , Bull. Narcotics 13,21.

19- M.Brandstäter-Kuhnert and A. Kofler (1959), Sci. Pharm. 27, 73.

20- A.Roux and J. Roux-Matignon (1963) , Ann.Pharm. Fr. 21,255.

21- P. Rajeswaran and P.L.Kirk (1962) , Bull. Narcotics 14,15.

22- Y.A.Beltagy, A.S.Issa and M.S.Mahrous (1978), Talanta 25,349.

23- S.Pinzauti , V.Dal Pinz,E.La Porta (1974), Farmaco. Ed. Prat.29,136, C.A.1974,81,29576 a.

24- J.A. Gautier, F.Pellerin and G.Baylocq (1966) Ann. Pharm. Fr. 24,405.

25- J. Kraemer , S.M. Alvarez,J.Kraemarova, B. Moravcova and H.Doslova (1977), Cesk Farm.26, 345, C.A.1978,88,177099 g.

26- A.M.Aliev ,A.S.Beisenbekov (1973),Pharmatisya 22,35. C.A. 1973,79,129120 z.

27- P.Mensard (1966), Arch.Pharm. (Athens)22,105. C.A.1967,66,118877 g.

28- M.B.Jacobs , G.Devaux and J. Fauquet (1965), Sci. Pharm. Proc. 25th, 2,77, C.A. 1968,69,109870 j.

29- S.H.Rubin,L.Drekter ,J.Scheiner and E.De Ritter (1952), Diseases of the Chest 21,439, C.A.1952,48,6686 g.

30- R.J.Colarusso;M.Schmall;E.G.Wollish and E.G.E. Safer (1958) , Anal.Chem.30,451.

31- J.Rieder and M.Roth (1963) , Biochem.Pharmacol.12,445.

32- A.F.Farbushnyi ; Ya.A. Mikhailovskii and E.D. Kuzmenko (1974), Farm.Zh.29 , 60. C.A. 1975, 82, 160302 w.

33- H.J.Roth and H.O. Schrinpf (1960) , Arch.Pharm.293,22.

34- K.B. Bjornesjo and B. Jarnulf (1962) , Scand.J. Clin. Lab. Invest. 14 , 408, C.A. 1963, 59, 11871 h.

35- M.B.Jacobs (1953), J. Am. Pharm. Assoc. 42,340.

36- T.H. Majorie , J.W. Rippon , L.E. Decker and J. Bernsohn (1962) , Anal. Biochem. 4, 85.

37- K. Zaitsu , H. Nagai, K. Ohtsub , S. Eto , K.Kohashi and Y. Ohkura (1978) , Chem. Pharm. Bull. 26,3471.

38- M.Am. Amer , A.K.S. Ahmed and M.M. El-Laithy (1974) , Egypt. J. Chem. 17, 367.

39- J. Brandys (1966) , Diss. Pharm. Pharmacol. 18 , 319. C.A. 1967, 66, 22267 i.

40- F. Pellerin and H. El-Makkawi (1972), Ann. Pharm. Fr.30 , 187, C.A. 1972 , 77, 42391 y.

41- Z.I. El-Darawy and Z.M. Mobarak (1973) , Pharmazie 28, 37.

42- A. Alessandro , F. Maria and S. Settecase (1967), Farmaco. Ed. Prat. 22 ,437, C.A. 1976 , 66, 94053 c.

43- K.C Guven and B. Tekinalp (1968) , Ecsacilik Bul. 18 , 43, C.A. 1968 , 69 , 46112 v.
44- J.J. Thomas and L. Dryon (1964) , J. Pharm. Belg. 19,481.
45- I. Zingales (1967) , J. Chromatogr. 31,404.
46- Ya.A. Mikhailovskii and E.D. Kuzmenko (1974), Farm. Zh. 29, 60, C.A. 1975, 82 ,160302 w.
47- W.W. Fike (1966) , Anal.Chem. 38 , 1697.
48- R.M.De Sagher, A.P. De Leenher and A.E. Claeys (1975) , J. Chromatogr. 106, 357.
49- M.C. Dutt and T.T. Poh (1981) , J. Chromatogr. 206, 267.
50- J.H.J. Kaiha (1972), J.Pharm. Sci. 61, 679.
51- E. Schmid, E. Hoppe , Chr. Meythaler , Jr and L. Zicha (1963) , Arzeni. Forsch. 13, 969.
52- A.Calo , C. Cardini and V. Quercia (1968), Boll.Chim. Farm. 107 , 296, C.A. 1969 , 69 , 80244 v.
53- A. Calo ,C.Cardini and V. Quercia (1968), J. Chromatogr. 37, 194.
54- C. Cardini , V. Quercia and A. Cola (1968) , J. Chromatogr. 37, 190.
55- R.M. De Sagher , A.P. De Leenher and A.P. Claeys (1976) , J. Chromatogr. Sci. 14, 302.
56- Ibid (1976), 14 , 206.
57- R.M. De Sagher , A.P. De Leenher and A.P. Claeys (1976) , J. Pharm. Sci. 65, 879.

LACTOSE, ANHYDROUS

Harry G. Brittain*, Susan J. Bogdanowich, David E. Bugay,

Joseph DeVincentis, Geoffrey Lewen, and Ann W. Newman

Bristol-Myers Squibb Pharmaceutical Research Institute

Bristol-Myers Squibb Company

P.O. Box 191

New Brunswick, NJ 08903

CONTENTS

1. Description

1.1 Name, Formula, and Molecular Weight

Lactose is a sugar obtained from milk, and can be obtained either as a crystalline anhydrous or monohydrate species. The systematic chemical name is 4-O-ß-D-galactopyranosyl-α-D-glucopyranose. It can also be identified as 4-(ß-D-galactosido)-D-glucose.

The Chemical Abstracts identification number is CAS-63-42-3.

Anhydrous lactose can be obtained as either the α-lactose or ß-lactose modification, although the α-form is most often used in pharmaceutical applications.

α-**lactose:**

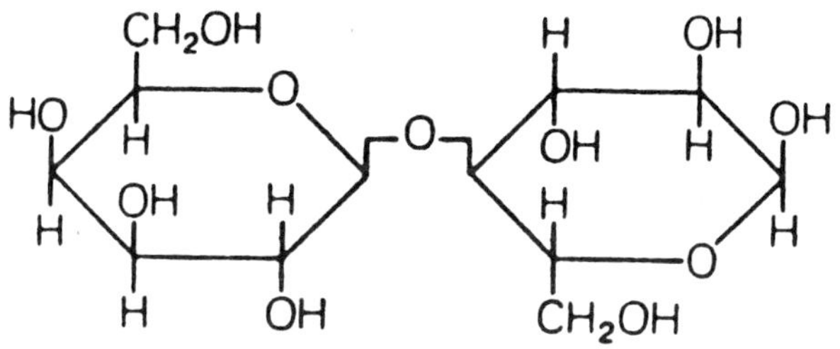

ß-**lactose:**

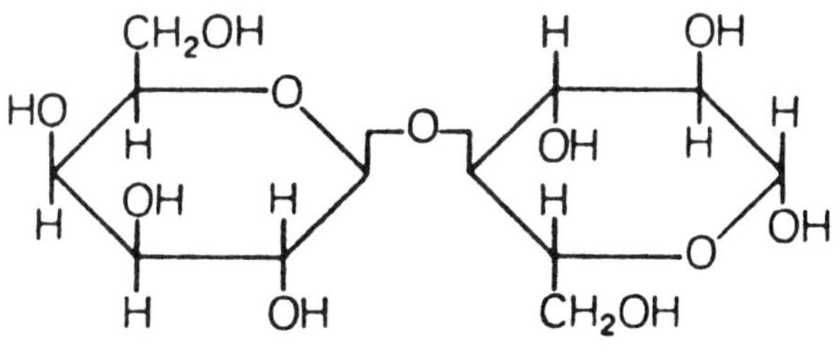

$C_{12}H_{22}O_{11}$ M.W. = 342.30

The elemental composition is C 42.10%, H 6.48%, and O 51.42%.

1.2 Appearance

Lactose is a white, or creamy white, hard crystalline material, which is readily milled into a powder. It is odorless, has a slightly sweet taste, and tends to absorb odors from the atmosphere. The material can be spray-dried, yielding a granular, highly flowable material. A summary of the characteristics of anhydrous lactose is available [1].

1.3 General Chemical Properties

Lactose reduces Fehling's solution, forms an osazone, and exists in α- and ß-forms which undergo mutarotation. Upon acid hydrolysis, (+)-lactose is converted into equal amounts of D-(+)-glucose and D-(+)-galactose. Oxidation of lactose with bromine water yields an acid which may be hydrolyzed to galactose and gluconic acid. It is therefore concluded it is the glucose portion of the sugar which contains the aldehyde associated with the reducing property.

1.4 Uses and Applications

Lactose is one of the most commonly used excipients in direct compression formulations [2]. The compound is normally used as a diluent, whose purpose is to make up the required bulk of a solid dosage form when the amount of drug itself is not adequate. Lactose is probably the most widely used diluent in tablet formation, since it is found to be stable in the presence of many drug substances. Anhydrous lactose is particularly useful since it can be purified to such a degree that it does not undergo the Maillard reaction, which leads to browning and discoloration.

Formulations containing lactose as the major ingredient generally exhibit good product release rates, and tablet disintegration is not strongly affected by the hardness.

The main disadvantage of anhydrous lactose is its hygroscopicity, since it tends to pick up moisture upon exposure to high humidities.

Anhydrous lactose is used primarily in direct compression formulations, and such products ordinarily require protection from adventitious moisture. Lactose will form Schiff base products with compounds containing basic primary amine functionalities, and should not be used in formulations containing such groups.

2. Method of Preparation

Lactose is present in about 5% concentration in all mammalian milk, with the exception of the California sea lion. Whey is first obtained through coagulation of milk, which results in the removal of all milk proteins. Evaporation of the whey at temperatures below 95°C causes the less soluble α-anomer to precipitate. The crude lactose is purified to the desired level by successive recrystallization. This process normally yields the monohydrate form, but the anhydrous phase can be obtained through controlled dehydration. Since the final properties of anhydrous lactose are determined by its method of drying, this last step is of utmost importance in the production of a pharmaceutically acceptable excipient.

α-lactose monohydrate is obtained by crystallization from a supersaturated solution at temperatures below 93.5°C [3], while ß-lactose is obtained by crystallization above this temperature [4]. ß-lactose is only obtained as a crystalline anhydrous modification [5]. The α-lactose monohydrate may be dehydrated either thermally or through the use of anhydrous organic solvents [6]. An essentially non-hygroscopic form of anhydrous lactose can be obtained using thermal dehydration above 130°C [7].

3. Physical Properties

As would be anticipated for any pharmaceutical excipient, the physical properties of anhydrous lactose represent the most important characteristics of measure.

3.1 Particle morphology

Typical photomicrographs of anhydrous lactose are shown in Figure 1. Anhydrous lactose is seen to consist of particles which are aggregates of small, needle-like microcrystals. Significant difference in particle morphology exist between the anhydrous, hydrous, and spray-dried lactose materials [8]. The angular nature of the crystals making up anhydrous lactose is responsible for the poor flow characteristics normally exhibited by this solid.

3.2 Crystallographic Properties

The x-ray powder diffraction pattern of α-lactose anhydrous is shown in Figure 2, while the corresponding table of D-spacings and relative intensities is found in Table I. The most commonly used anhydrous lactose is normally obtained from commercial sources as the α-phase, which is easily distinguished from any of its pseudopolymorphs [8]. Variable temperature x-ray diffraction studies can be carried out on anhydrous lactose, but the observed powder pattern does not change with applied temperature.

3.3 Thermal methods of analysis

As shown in Figure 3, the differential scanning calorimetry thermogram of anhydrous lactose consists mainly of the melting endotherm around 235°C. This melting phenomenon is accompanied by decomposition of the compound. Should any monohydrate contamination be present, then this will be observed as a very minor dehydration endotherm around 135°C. The thermogravimetry weight loss for anhydrous lactose would be 0.0% (by definition), but small losses less than 0.5% can be observed due to the presence of small amounts of monohydrate. Anhydrous lactose is observed to thermally decompose at higher temperatures than does either hydrous or spray-dried lactose, indicating that the anhydrous form is the most stable form of the material [1].

Figure 1. Scanning electron photomicrographs of anhydrous lactose (Sheffield DT grade), obtained at 500x (upper) and 1000x (lower).

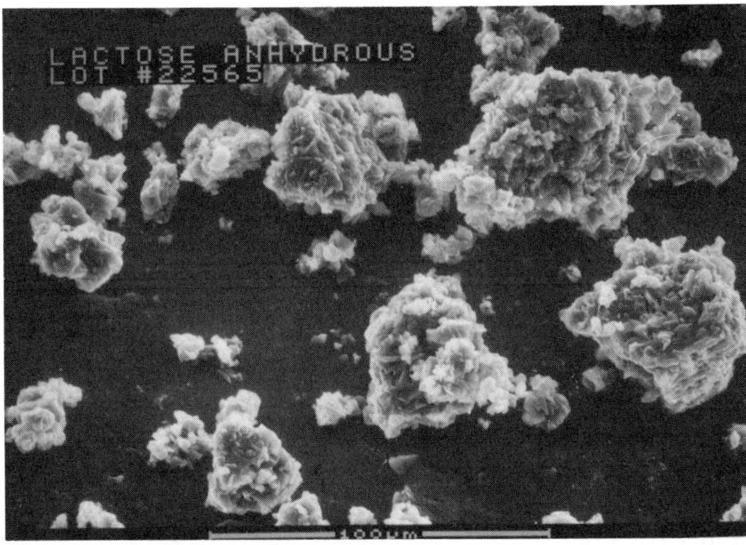

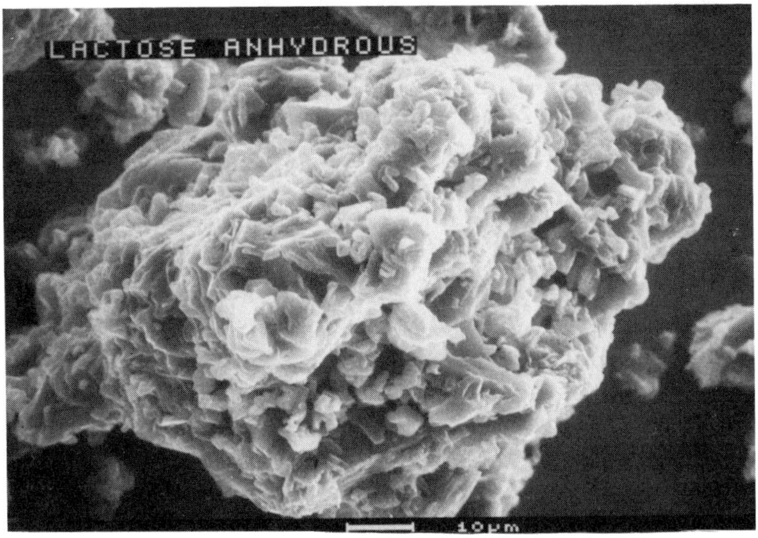

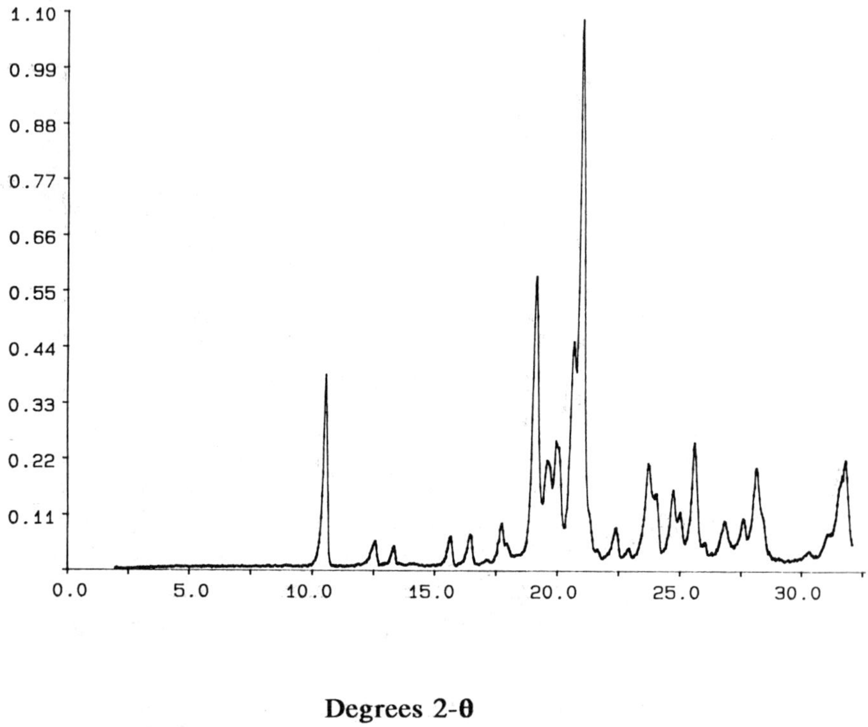

Figure 2. Powder x-ray diffraction pattern of anhydrous lactose (Sheffield DT grade). The intensity scale is presented in arbitrary units.

Table I

Crystallographic Data for Anhydrous α-Lactose

Angle (degrees 2-θ)	D-Spacing (Angstroms)	Relative Intensity (I/I_{max})
6.8100	12.9694	0.13
9.9700	9.8506	0.22
10.4275	8.4768	35.83
12.3975	7.1396	2.18
13.1925	6.7108	3.77
13.9600	6.3387	0.61
15.5050	5.7104	6.14
16.3025	5.4328	2.78
17.5500	5.0493	7.79
18.9785	4.6738	51.68
19.4325	4.5642	14.69
19.8175	4.4764	15.95
20.4650	4.3362	40.19
20.8675	4.2535	100.00
21.4950	4.1307	3.42
22.2300	3.9958	7.73
22.7150	3.9115	2.33
23.5200	3.7795	17.61
23.8875	3.7206	13.40
24.5225	3.6272	14.61
24.8400	3.5815	9.89
25.4275	3.5001	21.92
26.6425	3.3432	9.20
27.3850	3.2542	7.73
27.9375	3.1911	18.55
28.9625	3.0804	1.62
30.1075	2.9658	2.82
30.7550	2.9048	4.81
31.3575	2.8504	16.28
31.6250	2.8269	20.61

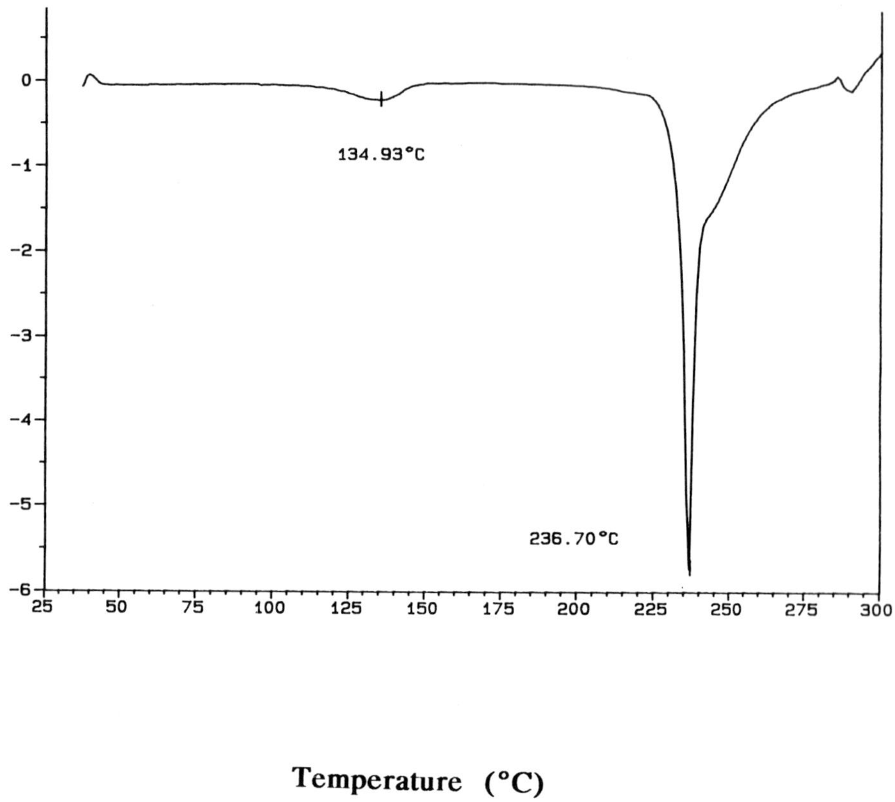

Figure 3. Differential scanning calorimetry thermogram of anhydrous lactose (Sheffield DT grade). The intensity scale is presented in units of heat flow (W/g).

3.4 Particle size distribution

Anhydrous lactose may be obtained in a variety of particle size distributions, depending on the manufacturer and isolation process. For the sake of example, the size distributions (obtained using optical microscopy and image analysis) of several grades of anhydrous lactose produced by Sheffield Products are summarized in Table II. The DT grade is the coarsest of the different grades, with 60M grade being somewhat finer. The 80M and impalpable grades are differentiated only by the percent retained on a 200 mesh screen. These differences are reflected in the particle size distributions of Table II.

3.5 Surface area

The adsorption and desorption isotherms obtained for anhydrous lactose were found to be equivalent, indicating no existence of hystersis. All isotherms were of the Type III described by Brunauer [9], which are typical for powdered, relatively nonporous solids.

As would be anticipated, the surface area of a given lactose sample is a function of its particle size. The near equivalence between the DT and 60M grades is evidenced by their equal surface areas of 0.4 m^2/g. In a similar manner, the near equivalence of the 80M and impalpable grades is evidenced by their equal surface areas of 0.7 m^2/g.

These findings are consistent with the observations made by electron microscopy, which indicate the existence of little surface detail.

3.6 Density

The true density of anhydrous lactose is approximately 1.55 g/mL. The bulk densities of the four anhydrous lactose grades characterized in the present study did not differ greatly, and the following data were obtained:

Table II

Particle Size Distributions Obtained for Various Sheffield Grades of Anhydrous Lactose, Obtained Using Optical Microscopy (aggregate species larger than 50 μm have been excluded)

Band Size (μm)	DT	60M	80M	Impalpable
0 - 5	7.1	13.4	38.2	22.4
5 - 10	32.9	26.5	25.3	24.7
10 - 15	28.7	19.8	12.5	20.1
15 - 20	14.8	12.4	8.2	10.9
20 - 25	9.4	7.7	4.6	5.3
25 - 30	3.5	4.7	3.3	4.3
30 - 35	1.3	3.0	1.6	4.3
35 - 40	1.0	1.3	1.6	2.6
40 - 45	0.8	2.7	1.3	2.0
45 - 50	0.5	1.7	1.0	0.7

Sheffield Grade	Bulk Density (g/mL)	Tapped Density (g/mL)
DT	0.59	0.76
60M	0.55	0.74
80M	0.45	0.64
Impalpable	0.44	0.62

As had been noted in the discussion of particle size and surface area, the bulk and tapped densities of the DT and 60M grades would not permit their distinction. In the same vein, the density data would not permit the differentiation between the 80M and impalpable grades.

3.7 Powder Flow characteristics

Although the bulk flow rates associated with anhydrous lactose are not large, the material does possess some ability to be aerated and therefore fluidized. The relative flow rates of the various lactose grades were obtained through a 11 mm orifice, and these data are summarized in Table III. To obtain these flow rates, gentle stirring of the powder bulk is required. Under these conditions, the mass flow rates of the DT and 60M grades were roughly equivalent, as were the mass flow rates of the 80M and impalpable grade materials.

Further characterization of the powder flow characteristics can be obtained using Carr's method for the evaluation of powder flowability [10]. These results are also summarized in Table III, and are consistent with the existence of moderate powder flow. No single parameter was found to dominate the flow characteristics, and it is concluded that all the variables measured under the Carr system are required to evaluate the powder characteristics. However the overall flowability and floodability indices group the DT and 60M grades as being equivalent to each other, and classify the 80M and impalpable grade materials as being equivalent in their powder characteristics.

Table III

Flowability and Floodability Data Obtained for Various Sheffield Grades of Anhydrous Lactose

Property	DT	60M	80M	Impalpable
Angle of repose	49°	44°	55°	55°
Angle of spatula	58°	69°	74°	75°
Compressibility	21%	26%	30%	28%
Cohesion	31	40	64	40
Angle of fall	28°	30°	30°	36°
Dispersibility	10%	6%	14%	23%
Angle of difference	21°	14°	25°	14°
Flowability index	50	48	34	32
Floodability index	66	56	60	58
Mass flow rate, through a 11 mm orifice (g/sec)	1.5	2.3	0.08	0.10

3.8 Hygroscopicity

Anhydrous α-lactose is readily converted to its monohydrate pseudopolymorph upon exposure to high relative humidities [11]. Anhydrous ß-lactose is essentially non-hygroscopic.

3.9 Solubility

The solubility of anhydrous lactose has been determined according to the USP definitions for solubility [12]:

Solvent	Solubility
water	freely soluble
ammonia	soluble
acetic acid	soluble
methanol	very slightly soluble
chloroform	insoluble
diethyl ether	insoluble

3.10 Spectroscopy

The electronic structure of lactose is such that it has no electronic absorption bands above 180 nm, and therefore does not exhibit useful ultraviolet absorption or circular dichroism spectra. Similarly, non-derivitized lactose does not exhibit native fluorescence.

The mid-IR spectrum of anhydrous lactose (obtained using diffuse reflectance) is shown in Figure 4. The IR spectrum of this modification exhibits significant differences when compared to its pseudopolymorphs [8]. The absorption due to bound crystalline water (3524 cm^{-1}) in hydrous lactose is lacking in the anhydrous phase.

The near-IR spectra obtained for the same sample of anhydrous lactose is shown in Figure 5. Since near-IR spectra normally consist of

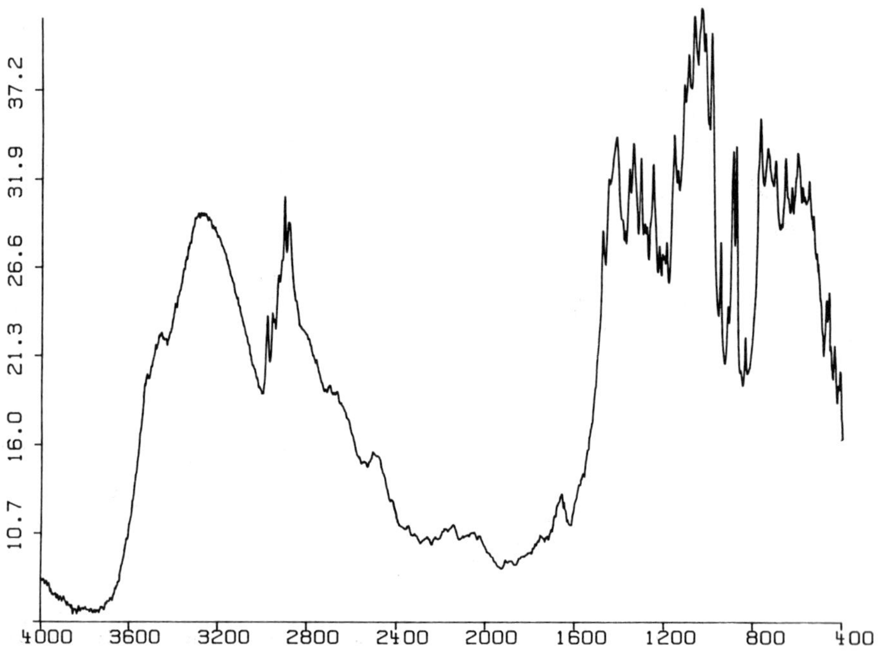

Energy (cm^{-1})

Figure 4. Mid-infrared absorption spectrum of anhydrous lactose (Sheffield DT grade), obtained using diffuse reflectance. The intensity scale is presented in Kubelka Munk units.

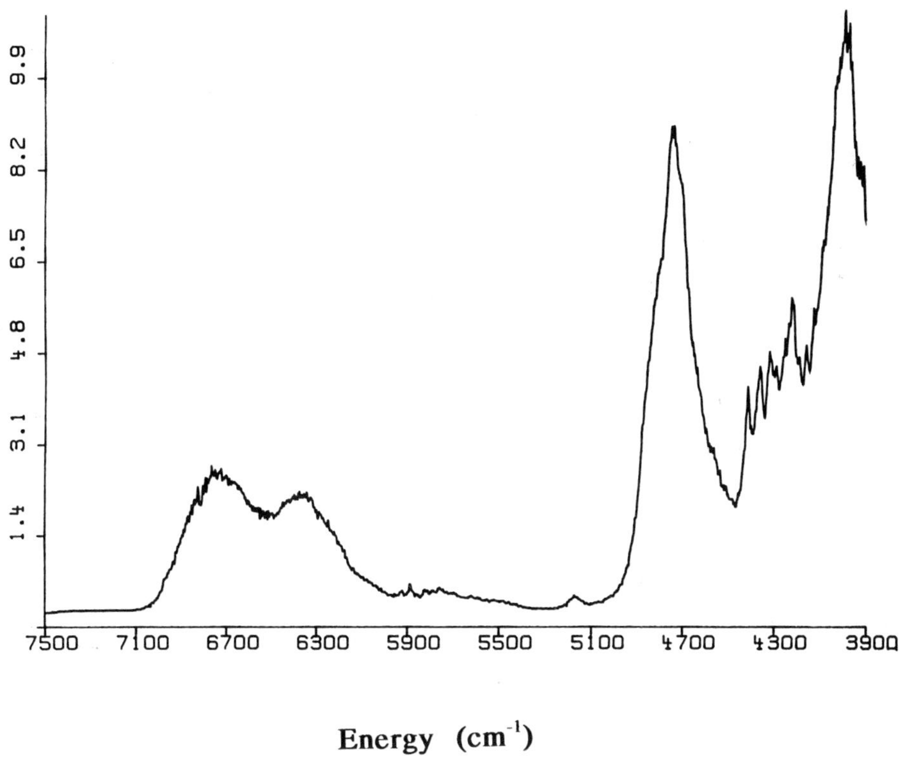

Figure 5. Near-infrared absorption spectrum of anhydrous lactose (Sheffield DT grade), obtained using diffuse reflectance. The intensity scale is presented in Kubelka Munk units.

overtone absorptions of fundamental vibrational modes, they are therefore not very useful for identity purposes. Near-IR spectral features are of greatest utility in the detection and determination of functional groups that contain unique hydrogen atoms. For instance, the hydrous lactose water band (5168 cm^{-1}) is only barely detected for anhydrous lactose. The water detected in the anhydrous lactose material probably was absorbed during sample preparation.

The solid state ^{13}C CP/MAS NMR spectrum for anhydrous lactose is shown in Figure 6. Analogous to the mid-IR spectra, the solid state NMR spectra display differing resonant bands for differing crystal structures. The differences in crystallinity between the hydrated and anhydrous modifications of lactose are evident in their non-equivalent NMR spectra. The resonances in the anhydrous lactose spectrum that do not correlate with the hydrous sample are due to the magnetically nonequivalent nuclei of the anhydrous crystal structure.

4. Methods of Analysis

4.1 Compendial Tests

According to the United States Pharmacopeia [13], lactose is tested according to its color and clarity of solution, identification, specific rotation, microbial limits, acidity or alkalinity, water content, residue on ignition, alcohol-soluble residue, and heavy metal content. The details of these tests are as follows:

Clarity and Color of solution: A solution consisting of 3 g of lactose in 10 mL of boiling water must be clear, colorless or nearly colorless, and odorless.

Identification: 5 mL of 1 N sodium hydroxide is added to 5 mL of a hot, saturated solution of lactose, and the mixture is further warmed. The liquid should become yellow, and finally brownish-red. The solution is cooled to room temperature, and a few drops of alkaline cupric tartrate test solution (Fehling's

Figure 6. Solid state ^{13}C nuclear magnetic resonance spectrum of anhydrous lactose (Sheffield DT grade), obtained using cross polarization and magic angle spinning. The intensity scale is presented in arbitrary units.

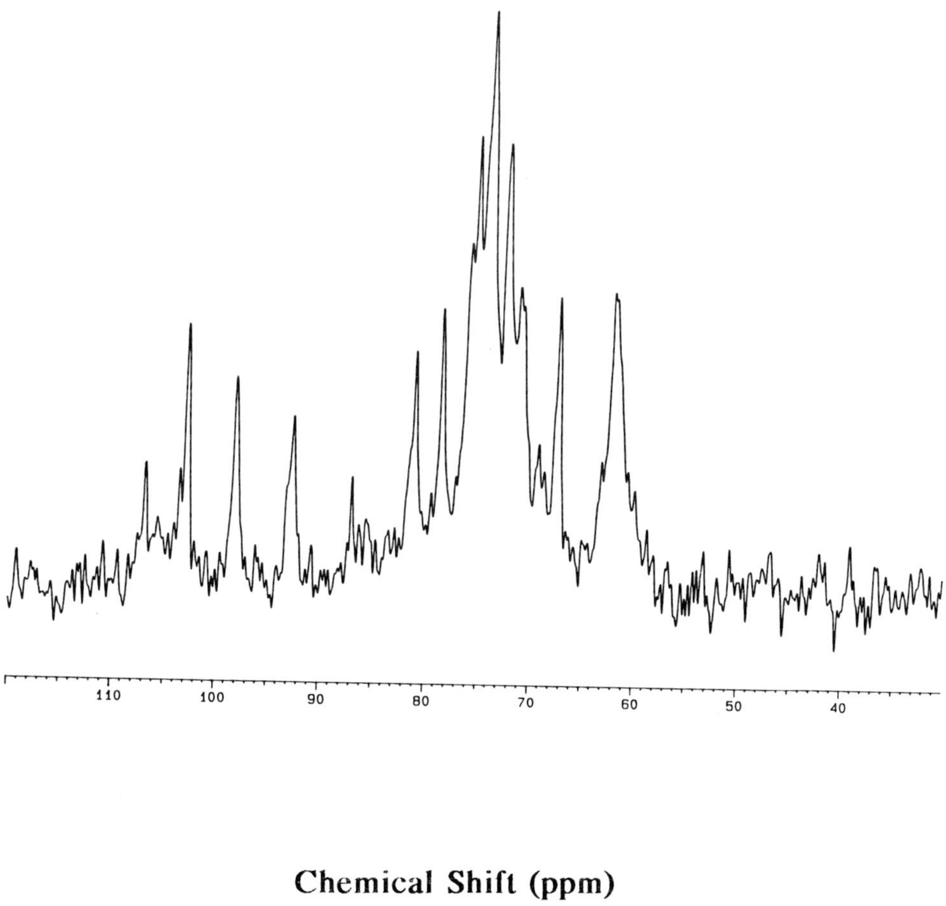

Solution) is added. Positive identification consists of the observation of a red precipitate of cuprous oxide.

Specific rotation: 10 g of lactose and 200 μL of 6 N ammonium hydroxide are dissolved in 100 mL water. The specific rotation of this solution (measured at 20°C) should be between +54.8° and +55.5°, calculated on an anhydrous basis.

The final value measured for the specific rotation represents the equilibrium value obtained after mutarotation has taken place. Since the specific rotation of α-D-lactose is +90°, and the specific rotation of ß-D-lactose is +35°, it follows that the equilibrium composition of lactose in aqueous solution is 35% α-lactose and 65% ß-lactose.

Microbial limits: The total bacteria count cannot exceed 100 per gram, and the tests for *Salmonella* species and *Escherichia coli* must be negative.

Acidity or alkalinity: 30 g of lactose are dissolved in 100 mL of carbon dioxide-free water with heating. 10 drops of phenolphthalein test solution are added, and the solution should remain colorless. The solution is titrated with 0.1 N sodium hydroxide, and not more than 1.5 mL of titrant should be required to produce the red color of the endpoint.

Water content: When tested using the Karl Fischer titration method, anhydrous lactose cannot contain more than 1.0% of water.

Residue on ignition: Approximately 1-2 g of lactose is thoroughly charred in a crucible, and then cooled. 1 mL of sulfuric acid is added, heated gently until the evolution of white fumes ends, and then ignited at 800°C. The amount of residue should not exceed 0.1%.

Alcohol-soluble residue: 10 g of very finely powdered lactose is added to 40 mL of alcohol, and the suspension shaken for 10 minutes. The supernatant is filtered from the solids, 10 mL of which is evaporated to dryness, and any residue dried at 100°C for 10 minutes. The amount of residue cannot exceed 20 mg (0.2%).

Heavy metals: 4 g of lactose is dissolved in 20 mL of warm water, 1 mL of 0.1 N hydrochloric acid is added, and the mixture diluted to 25 mL with water. Using the procedure described in the Pharmacopeia under Heavy Metals Method I (method <231>), not more than 5 ppm heavy metals can be present.

4.2 Identification

Anhydrous lactose melts with decomposition around 235°C, and this transition may be observed using hot stage microscopy. When thusly observed, lactose will melt with a yellow discoloration and the formation of gas bubbles. This melt can be solidified into a glass [14].

Positive identification for anhydrous lactose can also be made on the basis of its x-ray powder diffraction, and chemical identification through its infrared absorption and nuclear magnetic resonance spectra.

Lactose is classified as a reducing disaccharide, and is therefore capable of reacting with Fehling's or Benedict's solution. Both reagents consist of the cupric ion, kept in solution by the formation of a complex salt with either tartrate (Fehling's reagent) of citrate (Benedict's reagent). In alkaline solution, these reagents react with the aldehyde group present on the glucose portion of lactose to yield a precipitate of red cuprous oxide. This reaction can be generalized as:

$$\text{sugar-CHO} + 2\,Cu(OH)_2 \rightarrow \text{sugar-COOH} + Cu_2O + 2\,H_2O$$

Although the use of Fehling's solution is more widespread, the use of Benedict's solution should be preferred, owing to its greater stability and greater specificity.

Lactose is known to react with phenylhydrazine, forming from hot aqueous solutions a yellow crystalline phenylosazone:

$$
\begin{array}{l}
CHO \\
| \\
CHOH \\
| \\
(CHOH)_3 \\
| \\
CH_2OH
\end{array}
\quad + \quad 3\ C_6H_5NHNH_2 \quad \rightarrow \quad
\begin{array}{l}
CH{=}N\text{-}NHC_6H_5 \\
| \\
C{=}N\text{-}NHC_6H_5 \\
| \\
(CHOH)_3 \\
| \\
CH_2OH
\end{array}
\quad
\begin{array}{l}
+\ C_6H_5NH_2 \\
\\
+\ NH_3 \\
\\
+\ 2\ H_2O
\end{array}
$$

Unlike many other osazones, the phenylosazone of lactose can be solubilized in hot water and precipitated from cold water [15].

An aqueous solution of lactose mixed with an ethanolic α-naphthol solution will form a violet ring when layered above concentrated sulfuric acid.

4.3 Elemental Analysis

No special problems are encountered with the elemental analysis of anhydrous lactose. The compound analyzes as $C_{12}H_{22}O_{11}$, having a molecular weight of 342.30 daltons.

The theoretical elemental composition for anhydrous lactose works out as:

C	42.10%
H	6.48%
O	51.42%.

4.4 Spectrophotometric Methods of Analysis

The colorimetric analysis of lactose and other sugars has received extensive attention in the past, with numerous methods and assay procedures being developed [16].

A number of general reactions have been developed which are suitable for the spectrophotometric determination of sugars [17]. These methods first require the hydrolysis of the disaccharide into the monomer constituents, and then a subsequent reaction at the aldehyde portion of the resulting sugars. In this general approach, an aqueous solution of lactose would be initially treated with concentrated sulfuric acid. The reaction products are then reacted with either α-naphthol, tryptophan, or indole to produce colored products. The absorbance of these derivatives is then used to evaluate the sugar content in the original material. The α-naphthol derivative is determined at 570 nm, the tryptophan derivative is determined at 500 nm, and the indole derivative is determined at 470 nm.

The reaction of hydrolyzed sugars with anthrone has been found to be a superior spectrophotometric method [18]. The concentration of the blue analyte derivative is obtained by measurement at 625 nm.

An intensely colored compound is formed by the reaction of sugars with 3-methyl-2-benzothiazolinone hydrazone [19]. After completion of the reaction, the absorbance at 670 nm is used for the determination. Considerable variety in the detection wavelengths and detection limits can be achieved through the use of other structurally related reagents [20].

Other useful colorimetric reagents which have been proposed include ethyl malonate [21], 3,6-dinitrophthallic acid [22], methylamine [23], and o-toluidine [24].

Although near-IR analysis is widely used to quantitate the active components in a formulation, this spectroscopic technique has not been actively applied to the study of excipient materials. Lactose in milk powders has been determined using near-IR spectroscopy [25], with quite reasonable precision being obtained in the results. Similarly, near-IR spectroscopy has been used to determine lactose in a variety of artificial mixtures [26].

4.5 Thin Layer Chromatographic Methods of Analysis

Prior to the introduction of HPLC methods, the use of thin layer chromatography as a separation and detection assay method for lactose and other sugars was widespread, and the applications numerous [27].

TLC methods have been used for the separation of sugars (including lactose) on cellulose [28], silica gel [29], or silica cell impregnated with phosphate [30]. A very useful and general method involves the use of silica gel G buffered with boric acid [31]. Two-dimensional thin layer chromatography has found extensive use in the separation of complicated mixtures of mono-, di-, and trisaccharides [32].

Most recently, high-performance thin-layer chromatography has been used for the quantitative determination of sugars [33]. In these works, various stationary phases and derivatization techniques were evaluated, and applications for determinations in different matrices were made.

4.6 Gas-Liquid Chromatographic Methods of Analysis

A wide variety of silation reactions have been used in the determination of lactose and other carbohydrates, with these derivitizations permitting the use of gas chromatography as an analytical method [34].

Permethylated sugar alcohols and permethylated methyl glycosides have been used as derivatives of sugars to facilitate the GC analysis of carbohydrates [35].

The use of appropriate (and inert) internal standards has been found to greatly simplify the GC analysis of carbohydrates [36].

Since the ratio of α-lactose and ß-lactose can change within a variety of applied conditions, the determination of compound type is important. After reaction with trimethylsilylimidazole, the anomeric composition of lactose can be determined [37].

4.7 High Performance Liquid Chromatographic Methods of Analysis

With the advent of HPLC, the majority of analytical work concerning lactose and other carbohydrates has centered around application of this method [38,39]. A comparison of HPLC and GC methods has been made, with recommendations being made on the method of choice for a given problem [40].

Since sugars do not exhibit characteristic absorbance in useful regions of the UV spectrum, the use of derivatives in chromatography is essential. Recently, dinitrophenylphenyl-hydrozones [41] and 4-N,N-dimethylamino-4-azobenzene sulfonyl hydrazide [42] have been used for this purpose.

General methods for the separation of reducing sugars have been developed which use detection methods applied after formation of the borate complexes. These complexes are separated on a trimethylammonium-bonded resin column, using anion exchange. The analytes are detected using either postcolumn fluorescence derivitization with 1,2-bis(4-methoxyphenyl)ethylenediamine [43], or potentiometrically with hexacyanoferrate(III)-hexacyanoferrate(II) [44].

Pulsed amperometric detection after post-column alkalization of the eluent has been used to determine lactose and other carbohydrates. Columns packed with cation exchange materials were used to effect the separation [45]. The use of post-column reactors and chemically modified electrodes has also been described for the determination of lactose [46]. Complexes of carbohydrates with ethylenediamine are electrochemically oxidizable, and may be chromatographed and detected as such [47].

After separation of sugars on diol bonded silica, they may be detected using a cuprammonium reagent [48]. The detection is based on formation of a complex which contains two molecules of the carbohydrate and one molecule of $Cu(NH_3)_4(H_2O)_2^{2+}$. The complexation process alters the absorption of the copper reagent, and yields a shift toward longer wavelengths.

5. Stability

5.1 Stability

When stored in adequately sealed containers, anhydrous lactose is a very stable material. Its normal high stability and purity have led to the widespread use of lactose as one of the most important pharmaceutical excipients. In the solid state, exposure to high relative humidities can lead to formation of the α-lactose monohydrate form, and to the growth of mold.

Acidic hydrolysis of lactose will convert lactose into equal amounts of glucose and galactose. Lactose may also be oxidized with bromine water, yielding an acid which hydrolyses to galactose and gluconic acid.

In the solution phase, the anomeric composition of lactose changes with time, ultimately ending up as 35% α-lactose and 65% ß-lactose. Equilibrium can be reached instantly through the addition of dilute ammonia to the lactose solution.

5.2 Incompatibilities with functional groups

Many workers have commented on the incompatibility of lactose with primary amines, where condensation of the two yields brown colored products. The reaction is more facile with amorphous lactose than with crystalline lactose, thus making spray-dried material more prone toward discoloration [1]. It has been claimed that elimination of a trace lactose impurity (hydroxylmethyl furfural) from the bulk minimizes the problem, since this compound acts as a catalyst in the reaction [49]. However, the reaction is also catalyzed by base, and therefore it is prudent to maintain the neutrality of lactose-based products.

Lactose is known to readily absorb odors and volatile compounds, and should therefore be protected from the atmosphere. In a study of the retention of aliphatic alcohols by anhydrous lactose, it was found that the retention order decreased in the order methanol > ethanol > 1-propanol ≈ 1-butanol [50]. This behavior could lead to retention of alcohols by lactose-based formulations, if the formulated products were incompletely dried.

6. References

1. Handbook of Pharmaceutical Excipients, American Pharmaceutical Association, Washington, D.C., 1986, pp. 153-162.

2. B.B. Sheth, F.J. Bandelin, and R.F. Shangraw, in Pharmaceutical Dosage Forms: Tablets, volume 1, H.A. Lieberman and L. Lachman, eds., Marcel Dekker, New York, 1988.

3. C.A. Beevers and H.N. Hansen, **Acta Cryst.**, B27, 1323 (1971).

4. T. Itoh, M. Katoh, and S. Adachi, **J. Dairy Res.**, 45, 363 (1978).

5. E. Berlin, P.G. Kliman, B.A. Anderson, and M.J. Pallansch, **Thermochim. Acta**, 2, 143 (1971).

6. S.G Lim and T.A. Nickerson, **J. Dairy Sci.**, 56, 843 (1973).

7. T. Itoh, M. Katoh, and S. Adachi, **J. Dairy Sci.**, 60, 1230 (1977).

8. H.G. Brittain, S.J. Bogdanowich, D.E. Bugay, J. DeVincentis, G. Lewen, and A.W. Newman, **Pharm. Res.**, 8, in press (1991).

9. S. Brunauer, The Adsorption of Gases and Vapors, volume 1, Princeton University Press, Princeton, NJ, 1945.

10. R.L. Carr, **Chemical Engineering**, 72, 163 (January issue, 1965); **ibid.**, 69 (February issue, 1975).

11. E. Berlin, P.G. Kliman, B.A. Anderson, and M.J. Pallansch, **Thermochim. Acta**, 2, 143 (1971).

12. United States Pharmacopeia, XXII, p. 8.

13. United States Pharmacopeia, XXII, p. 1942.

14. M. Kuhnert-Brandstätter, Thermomicroscopy in the Analysis of Pharmaceuticals, Pergamon Press, Oxford, 1971, p. 275.

15. R.L. Shriner and R.C. Fuson, The Systematic Identification of Organic Compounds, 2nd edition, John Wiley & Sons, New York, 1940.

16. G. Ashwell, in Methods in Enzymology, volume III, Academic Press, New York, 1957, pp. 73-105.

17. Z. Dische, Methods of Biochemical Analysis, volume II, Interscience Pub., New York, 1955, pp. 313-358.

18. W.B. Tuemmler and B.S. Wildi, **J. Am. Chem. Soc.**, 80, 3772 (1958).

19. E. Sawicki, R. Schumacher, and C.R. Engel, **Microchem. J.**, 12, 377 (1967).

20. S. Hunig, H. Balli, and H. Quast, **Angew. Chem.**, 74, 28 (1962).

21. T. Harada, **Biochim. Biophys. Acta**, 63, 334 (1962).

22. T. Momose and Y. Mukai, **J. Pharm. Soc. Jap.**, 81, 227 (1961).

23. S. Adachi, **Anal. Chem.**, 37, 896 (1965).

24. A.-M. Desmaison, J.-C. Brenton, M. Tixier, and J. Chatelut, **Bull. Soc. Pharm. Bordeaux**, 109, 118 (1970)].

25. R.J. Baer, J.F. Frank, and M. Loewenstein, **J. Assoc. Off. Anal. Chem.**, 66, 858 (1983).

26. G.Z. Wetherill, I. Murray, and C.A. Glasbey, **J. Agric. Sci.**, 114, 253 (1990).

27. M. Ghebregzabher, S. Rufini, B. Monaldi, and M. Lato, **J. Chromatogr.**, 127, 133 (1976).

28. D.W. Vomhof and T.G. Tucker, **J. Chromatogr.**, 17, 300 (1965).

29. P.G Pifferi, **Anal. Chem.**, 37, 925 (1965).

30. K. Kringstad, **Acta Chem. Scand**, 18, 2399 (1964).

31. E.J. Shellard and G.H. Jolliffe, **J. Chromatogr.**, 24, 76 (1966).

32. M. Lato, B. Brunelli, G. Ciuffini, and T. Mezzetti, **J. Chromatogr.**, 34, 26 (1968).

33. K. Patzsch, S. Netz, and W. Funk, **J. Planar Chromatogr.**, 1, 39, 177 (1988).

34. G. Rumpf, **J. Chromatogr.**, 43, 247 (1969).

35. Y.S. Ovodov and E.V. Evtushenko, **J. Chromatogr.**, 31, 527 (1967).

36. Y. Halpern, Y. Houminer, and S. Patai, **Analyst**, 92, 714 (1967).

37. S.K. Dwivedi and A.G. Mitchell, **J. Pharm. Sci.**, 78, 1055 (1989).

38. A. Heyraud and M.J. Rinaudo, **J. Liq. Chromatogr.**, 4, 175 (1981).

39. S. Inoue, S. Hoshi, and M. Matsubara, **Talanta**, 33, 611 (1986).

40. F.G.R Reyes, R.E Wrolstad, and C.J. Cornwell, **J. Assoc. Off. Anal. Chem.**, 65, 126 (1982)

41. N.K. Karamanos, T. Tsegenidis, and C.A. Antonopoulos, **J. Chromatogr.**, 405, 221 (1987).

42. K. Muramoto, R. Goto, and H. Kamiya, **Anal. Biochem.**, 162, 435 (1987).

43. Y. Umegae, H. Nohta, and Y. Ohkura, **Anal. Sci.**, 5, 675 (1989).

44. H. Ohura, T. Imato, S. Yamasaki, and N. Ishibashi, **Anal. Sci.**, 6, 777 (1990).

45. J. Haginaka and T. Nomura, **J. Chromatogr.**, 447, 268 (1988).

46. G. Marko-Varga, **J. Chromatogr.**, 408, 157 (1987)].

47. S. Honda, K. Enami, T. Konishi, S. Suzuki, and K. Kakehi, **J. Chromatogr.**, 361, 321 (1986).

48. M. Abbou and A.M. Siouffi, **J. Liq. Chromatogr.**, 10, 95 (1987).

49. Sheffield Products, technical bulletin.

50. F.W. Parrish, P.E. Pfeffer, K.D. Ross, D.P. Schwartz, and K.M. Valentine, **J. Agric. Food Chem.**, 27, 56 (1979).

Analytical Profile of Methoxamine Hydrochloride

Abdulrahman M. Al-Obaid and Maher M. El-Domiaty***

**Pharmaceutical Chemistry Department, College of Pharmacy, King Saud University, P.O. Box 2457, Riyadh-11451 (Saudi Arabia)*

***Pharmacognosy Department, College of Pharmacy, P.O. Box 2457, Riyadh-11451 (Saudi Arabia)*

METHOXAMINE HYDROCHLORIDE

1. Introduction

Overview and Therapeutic Category

Methoxamine is a synthetic aminobenzyl alcohol derivative, active as a directly acting sympathomimetic amine with a prompt and prolonged pressor action which results almost exclusively from α-adrenergic receptor agonist activity to increase peripheral resistance. The drug is indicated for the treatment of hypotensive states when it is desired to raise blood pressure without cardiac stimulation. It is mainly used to support blood pressure during anesthesia, including anesthesia by heart-sensitizing anesthetic agents. The drug has also been useful in terminating episodes of paroxysmal supraventricular tachycardia. Because of its relatively pure α-receptor agonist activity, methoxamine belongs to the therapeutic category of the adrenergic vasopressor agents and it is similar to phenylephrine in pharmacological actions.

2. Description (1, 2, 3)

2.1 Nomenclature

2.1.1 Chemical Names

Benzene methanol, α-(1-aminoethyl)-2,5-dimethoxy-hydrochloride.

(±)-α-(1-Aminoethyl)-2,5-dimethoxybenzyl alcohol hydrochloride.

2-Amino-1-(2,5-dimethoxyphenyl)-1-propanol hydrochloride.

β-Hydroxy-β-(2,5-dimethoxyphenyl)-isopropylamine hydrochloride.

β-(2,5-Dimethoxyphenyl)-β-hydroxyisopropylamine hydrochloride.

2,5-Dimethoxynorephedrine hydrochloride.

2.1.2 Generic Name

Methoxamine hydrochloride.

2.1.3 Trade Names

Vasoxyl HCl, Vasoxine HCl, Vasylox HCl and Pressomin HCl.

2.2 Formulae

2.2.1 Empirical

$C_{11}H_{17}NO_3$ (Methoxamine).

$C_{11}H_{18}ClNO_3$ (Methoxamine hydrochloride).

2.2.2 Structural

Methoxamine HCl is an aminoethyl benzenemethanol derivative which has the chemical structure shown below (Fig. 1).

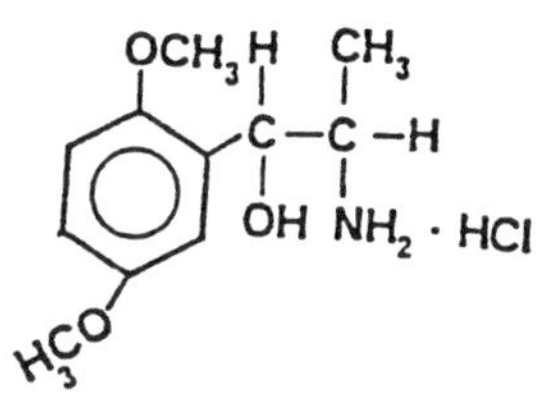

Fig. 1: Chemical structure of methoxamine hydrochloride.

2.2.3 CAS Registry Numbers

Methoxamine [390-28-3]
Methoxamine hydrochloride [61-16-5]

2.3 Molecular Weight

211.26 (Methoxamine).
247.72 (Methoxamine hydrochloride).

2.4 Elemental Composition

C, 62.54%; H, 8.11%; N, 6.63%; O, 22.72% (methoxamine).
C, 53.33%; H, 7.32%; Cl, 14.31; N, 5.65%; O, 19.38% (methoxamine hydrochloride).

2.5 Appearance, Color and Odor

Methoxamine HCl may occur as colorless or white, plate-like crystals, or a white, crystalline powder; odorless or has only a slight odor.

3. Dose and Pharmaceutical Dosage Forms (4)

3.1 Dose

Intramuscular: adults, pre- and post-operative: 5 to 10 mg; with spinal anesthesia or to correct hypotension: 10 to 15 mg; children, 250 µg/kg.

Intravenous: adults, for hypotensive emergencies, 3 to 5 mg slowly; for paroxysmal supraventricular tachycardia, 10 mg slowly, children, 80 µg/kg given slowly in divided doses.

3.2 Pharmaceutical Dosage Forms

Methoxamine hydrochloride injection USP 20 mg/ml Vasoxyl® with citric acid anhydrous 0.3%, potassium metabisulfite 0.1% and sodium citrate 0.3%) Burroghs Wellcome (5).

4. Physiochemical Properties

4.1 Melting Range

216° Methoxamine hydrochloride.

4.2 Solubility (5)

One gram of methoxamine HCl is soluble in 2.5 ml of water, in 12 ml of ethanol; very slightly soluble in chloroform and ether.

4.3 Stability and Storage (6)

Although methoxamine hydrochloride and injections containing the drugs are thermostable, the drug is sensitive to light. Commercially available methoxamine injections should be stored in the original airtight container at 15-30°C and protected from light. These injections have an expiration date of 3 years following the date of manufacture. In a related study (7) methoxamine was found to decompose in aqueous solution under air, yielding the major breakdown product 2,5-dimethoxybenzaldehyde.

4.4 Dissociation Constant

The pKa of methoxamine HCl is 9.2 (25°).

4.5 pH Range

Solutions of methoxamine HCl are acidic to litmus, having a pH between 3 to 5.

4.6 Differential Scanning Calorimetry (DSC)

The differential scanning calorimetry (DSC) thermal curve for methoxamine hydrochloride is given in Fig. 2. The scanning has been run at a rate of 10°C min^{-1} from 185 to 250°C. The hydrochloride salt of methoxamine melts at 216°C, the ΔH-value is 36.4 K.J. $mole^{-1}$ for 99.8 mole % purity. A DuPont TA-9900 Thermal Analyzer attached to a DuPont Data Unit were used for the DSC-scanning (8).

4.7 X-Ray Powder Diffraction

The X-ray diffraction pattern of methoxamine HCl was determined using Philip's full automated X-ray diffraction spectrogoniometer equipped with W/PW1730/10 generator (8). Radiation was provided by a copper target (Cu anode 2000W, λ = 1.5418 A) high intensity X-ray tube operated at 40 KV and 35 MA. The mono-chromator was a curved single crystal one (PW1752/100). Divergance slit and the receiving slit were 1 and 0.1, respectively. The scanning speed of the goniometer used was 0.02, 2θ per second. The instrument is combined with Phillips PM8210 printing

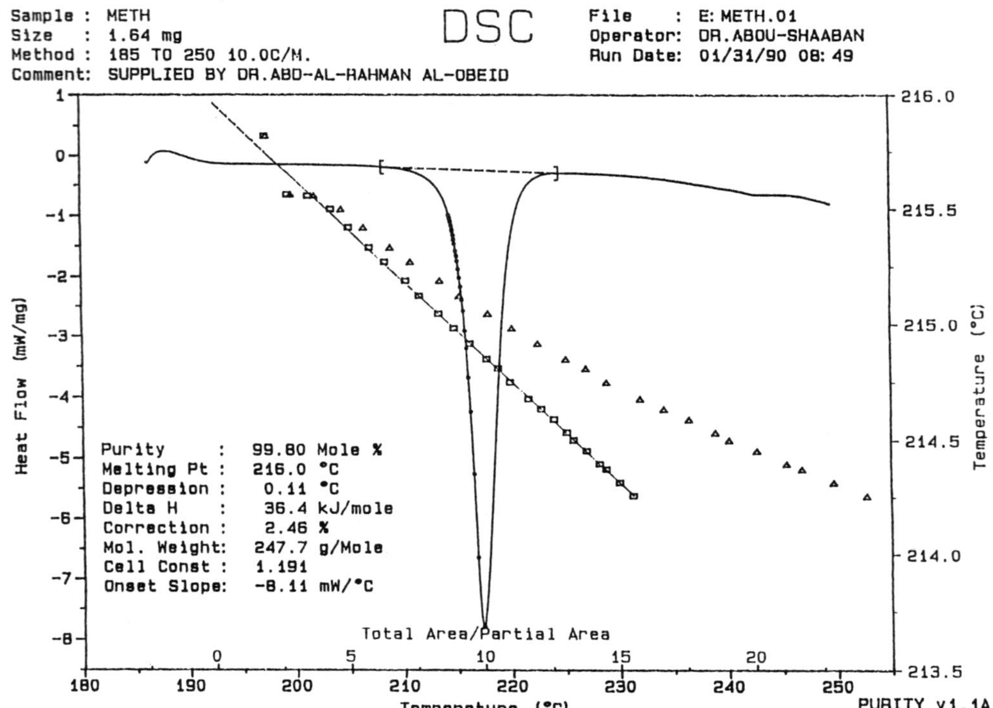

Fig. 2: Differential Scanning Curve of Methoxamine HCl.

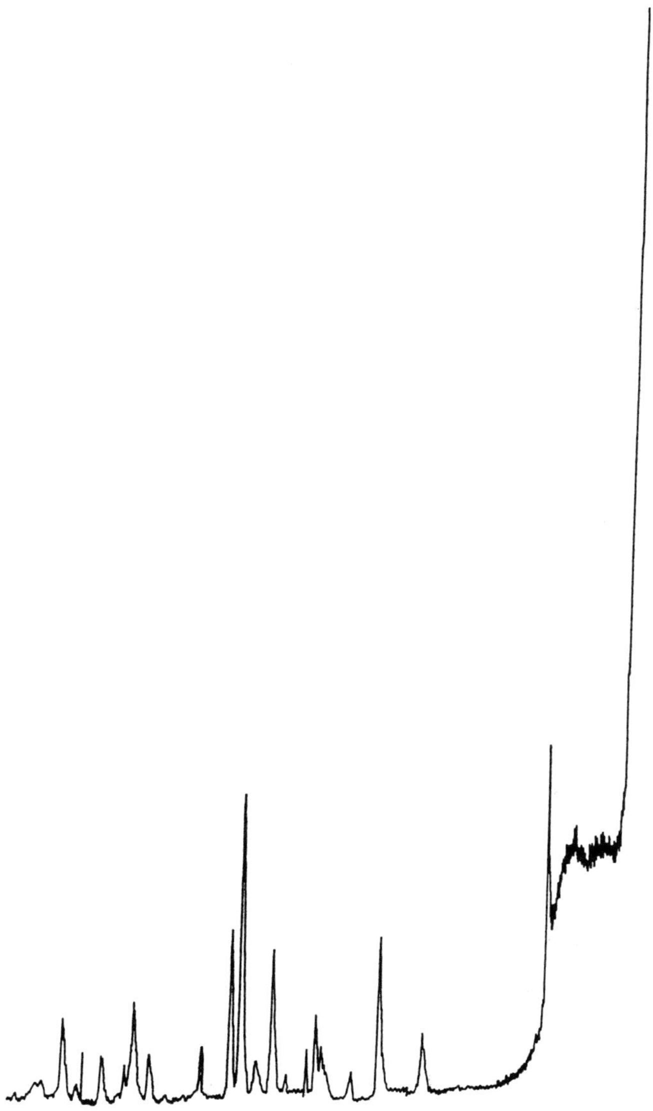

Fig. 3: X-ray Diffraction Curve of Methoxamine HCl.

recorder with both analogue recorder and digital printer.

The X-ray pattern of methoxamine HCl is presented in Fig. 3. Interplannar distance (dÅ), scattering angle (2 θ), and % relative intensity (I/I_o x 100) are placed in table 1.

Table 1: Characteristic Lines of the X-ray Diffraction of Methoxamine HCl

2θ	d(Å)	I/I_o %	2θ	d(Å)	I/I_o %
4.767	18.5373	88.55	25.773	3.4566	22.27
6.224	14.2000	99.41	26.317	3.3864	9.81
12.107	7.3103	8.89	27.090	3.2915	11.04
12.486	7.0889	12.79	27.786	3.2106	6.78
13.052	6.7826	23.36	28.787	3.1012	19.17
15.137	5.8530	18.99	29.722	3.0057	33.53
15.535	5.7040	52.53	30.209	2.9584	16.00
17.028	5.2070	38.35	30.537	2.9274	6.63
18.507	4.7941	18.15	31.559	2.8348	18.77
18.792	4.7220	21.50	32.662	2.7449	4.41
19.130	4.6394	31.68	32.955	2.7179	10.35
19.627	4.5230	20.34	33.803	2.6517	28.98
20.618	4.3077	42.32	34.424	2.6052	7.72
21.666	4.1016	52.93	34.934	2.5683	10.90
22.527	3.9467	17.79	35.360	2.5383	10.31
23.475	3.7896	100.00	35.688	2.5157	6.96
24.141	3.6865	54.17	36.056	2.4909	6.01
25.100	3.5478	9.84	36.511	2.4609	6.74

4.8 Spectral Properties

4.8.1 Ultraviolet Spectra (UV)

The UV spectra of methoxamine hydrochloride in water and in methanol were scanned from 200 to 375 nm (Fig. 4 and Fig. 5, respectively) using LKB Ultrospec Plus 4054 UV/Visible spectrophotometer. The spectra in water and in methanol show almost identical absorption

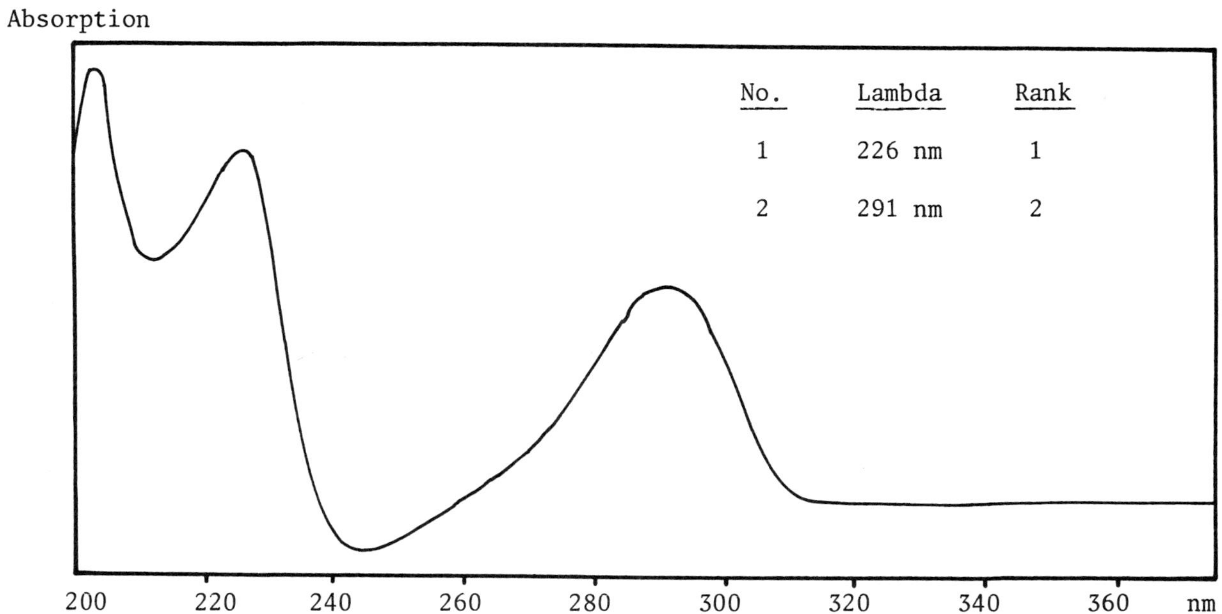

Fig. 4: U.V. Spectrum of Methoxamine HCl in dist. H_2O.

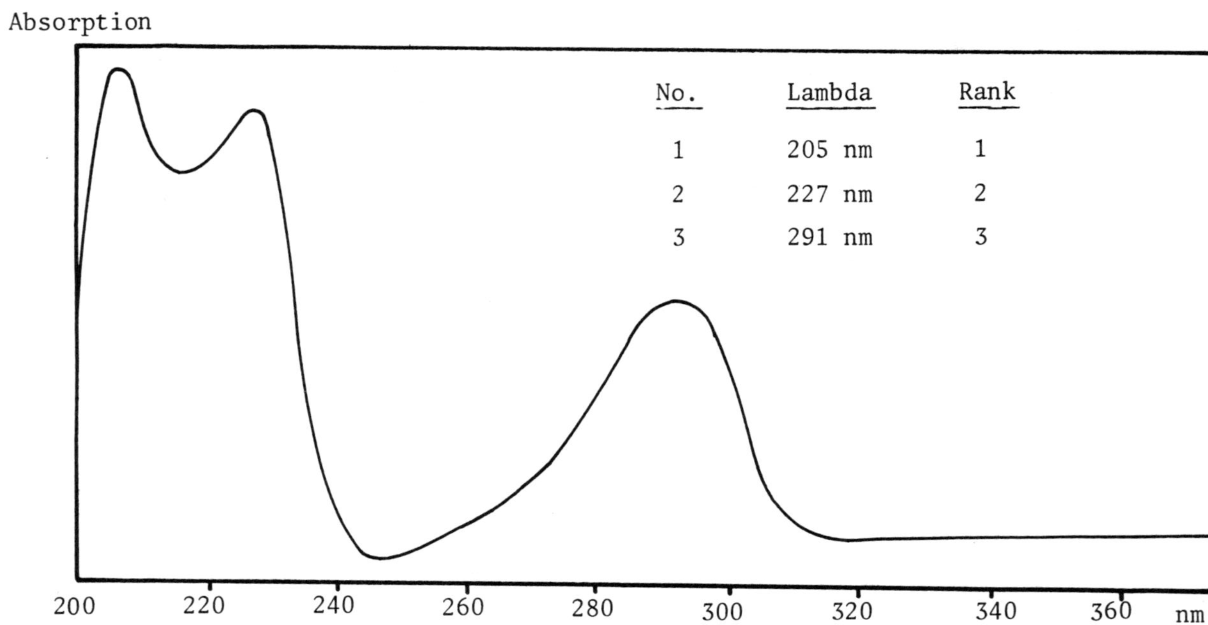

Fig. 5: UV Spectrum of Methoxamine HCl in methanol.

maxima (water-226 and 291 nm; methanol-205, 227, and 291 nm). The molar absorptivity for methoxamine HCl in water was 3467.261. The recorded values are in general agreement with those reported in the literature (3).

4.8.2 Infrared Spectrum (IR)

The infrared spectrum of methoxamine HCl is shown in Fig. 6. The spectrum was obtained with a Perkin-Elmer 1310 infrared spectrophotometer from a compressed KBr disc. The structural assignments have been correlated with the following frequencies (Table 2).

Table 2: IR Characteristics of Methoxamine HCl.

Frequency cm^{-1}	Assignment
3260	O-H stretching
3040-2900	$^{+}NH_3$ stretching
2700-2400 and 1980	$^{+}NH_3$
1590 & 1495	C = C aromatic stretching
1275 & 1220	C-O-C stretch in OCH_3
1040 & 1020	alkyl C-O stretching
820 & 795	$^{+}NH_3$ rocking
710	C-H out of plane deformation.

IR data for methoxamine HCl have also been reported (3, 9).

4.8.3 Nuclear Magnetic Resonance (NMR) Spectra

4.8.3.1 Proton Magnetic Resonance (PMR) Spectrum

The ^1H-NMR spectrum of methoxamine hydrochloride in $CDCl_3$ (Fig. 7) was recorded on a Varian XL 200 MHz NMR spectrometer using TMS as an internal reference. The interpretation of the spectrum is given in Fig. 8.

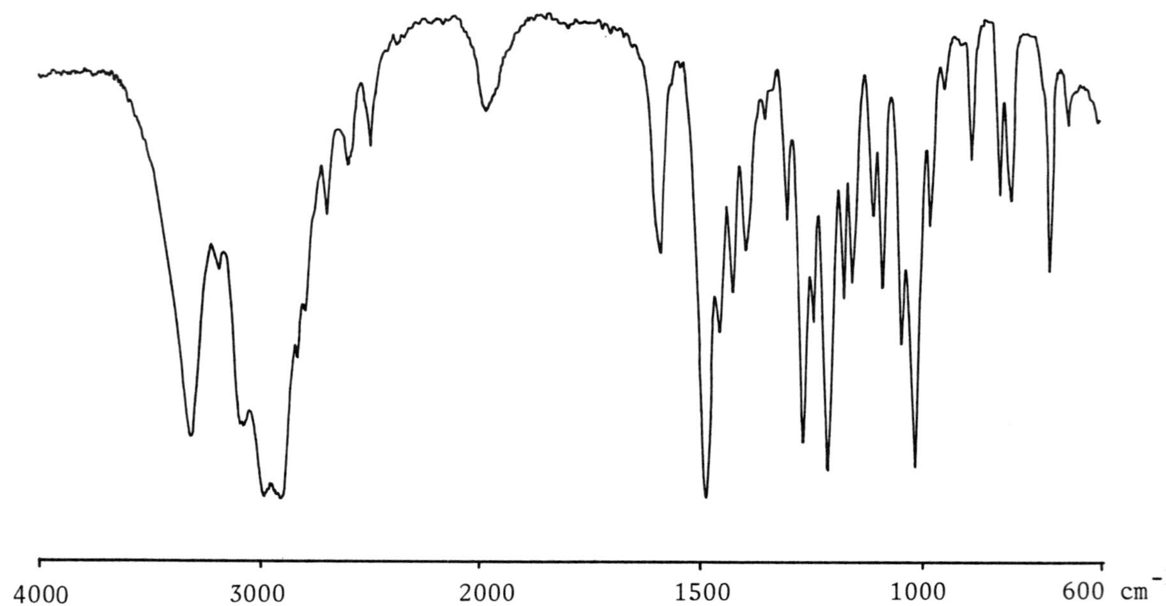

Fig. 6: I.R. Spectrum of Methoxamine HCl as KBr disc.

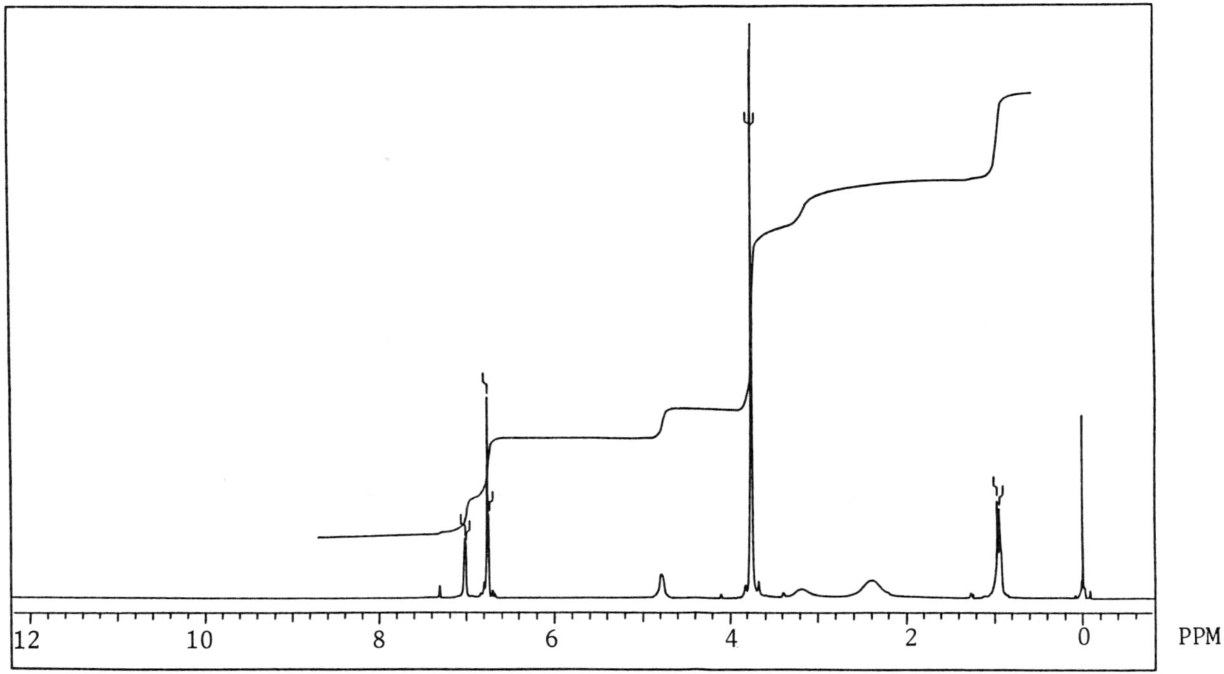

Fig. 7: PMR Spectrum of Methoxamine HCl in $CDCl_3$.

An inspection of the D_2O exchange spectrum shows disappearance of the resonance at δ 2.5 ppm. This corresponds to three protons, one hydroxyl proton and two amine protons.

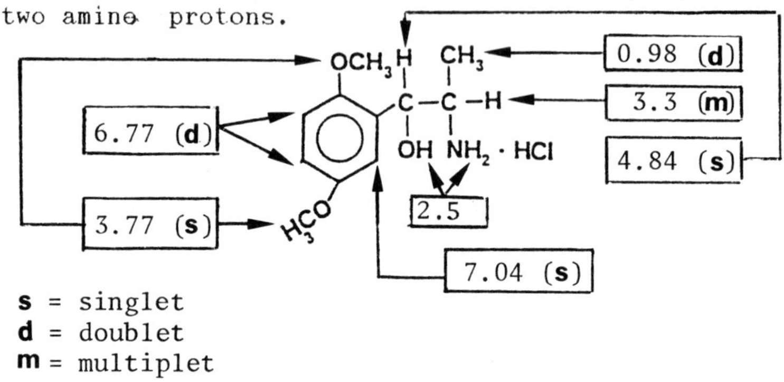

Fig. 8: Interpretation of PMR Spectrum of Methoxamine HCl.

4.8.3.2 Carbon-13 Nuclear Magnetic Resonance Spectrum (^{13}C-NMR)

The ^{13}C-NMR spectrum of methoxamine hydrochloride in $CDCl_3$ (Fig. 9) was recorded on a Varian XL 200 MHz NMR spectrometer using TMS as internal reference. The assignments are given in Fig. 10.

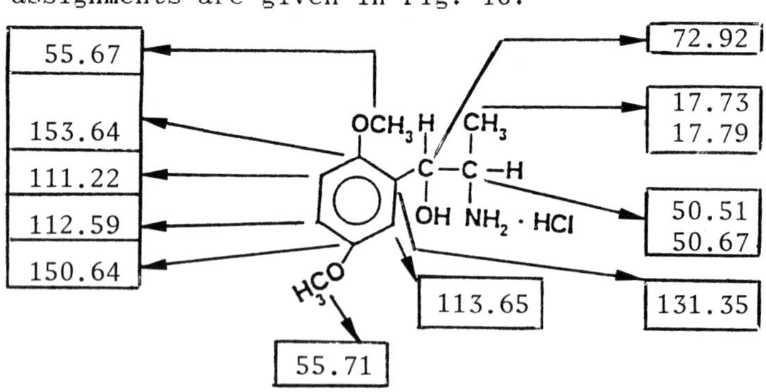

Fig. 10: Interpretation of ^{13}C-NMR Spectrum of Methoxamine HCl.

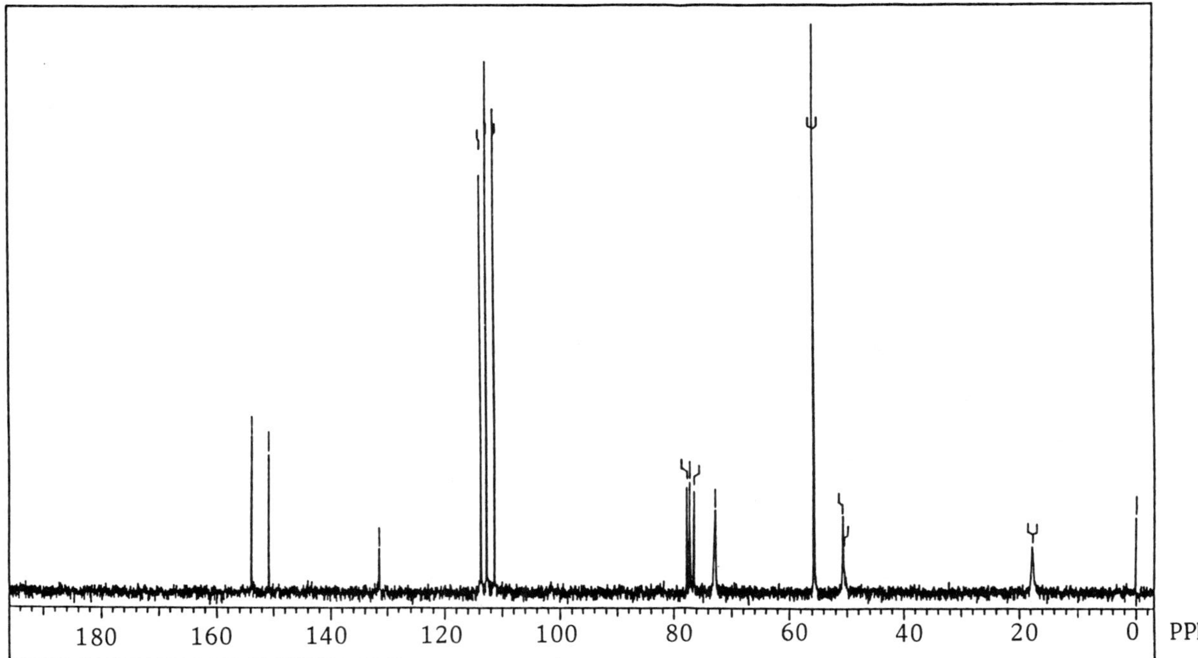

Fig. 9: ^{13}C-NMR spectrum of Methoxamine HCl in $CDCl_3$.

4.8.4 Mass Spectrometry (MS)

The mass spectrum of methoxamine HCl is presented in Fig. 11. This was obtained by electron impact ionization on a Finnigan-Mat 5100 GC/MS spectrometer. The spectrum was scanned to 250 amu. It shows a molecular ion peak M^+ at m/z 247. The base peak is at m/z 168. The most prominent ions, their relative intensities and corresponding fragments are presented in Table 3.

The mass spectrum and fragmentation pattern of methoxamine HCl were previously reported (3,7) with fragment ions of M/z 168, 137, 44, 139, 43, 152, 124, 167 in descending order of relative intensity.

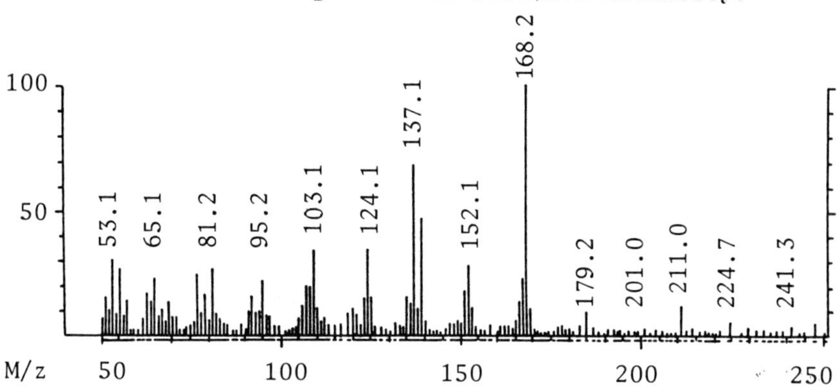

5. Synthesis (10)

Methoxamine HCl was synthesized in Burroughs Wellcome & Co. Laboratories. Among several ways, from 2',5'-dimethoxypropiophenone dissolved in absolute ether through reaction with CH_3NO_2 and HCl to give 2,5-dimethoxy-α-isonitrosopropiophenone HCl. This isonitroso ketone was dissolved in absolute alcohol containing an excess of HCl and was then hydrogenated with palladized charcoal to yield ß-(2,5-dimethoxyphenyl)-ß-ketoisopropylamine HCl. Further hydrogenation of the ß-ketoisopropylamine HCl with PtO_2-Pt black afforded the ß-(2,5-dimethoxyphenyl)-ß-hydroxyisopropylamine HCl (methoxamine HCl). The

Table 3: Mass Spectral (EI) Data for Methoxamine

M/z	Relative intensity %	Fragment ion
211	12.9	M^+
168	100	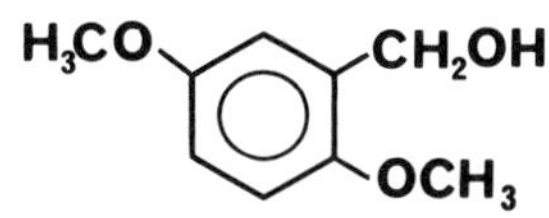
152	30	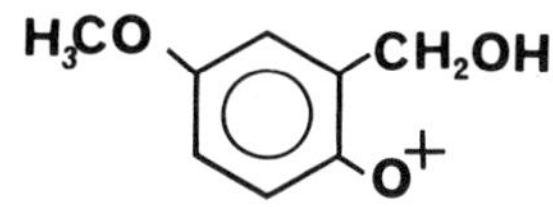
137	70	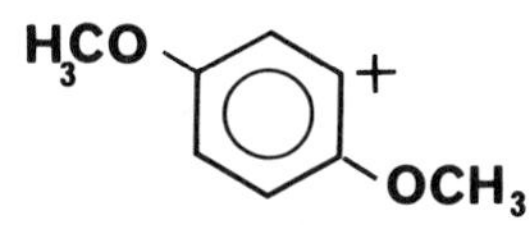
124	35.8	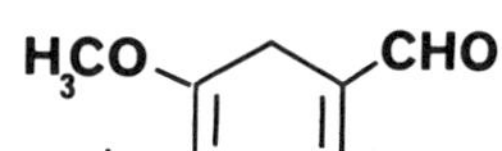
109	35	
95	22.9	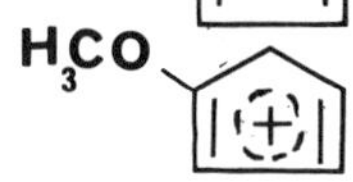
81	28.6	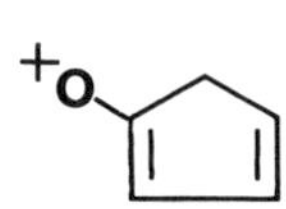
65	24.3	
53	32.9	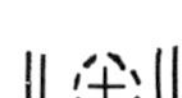

synthetic route of methoxamine HCl is presented in Fig. 12.

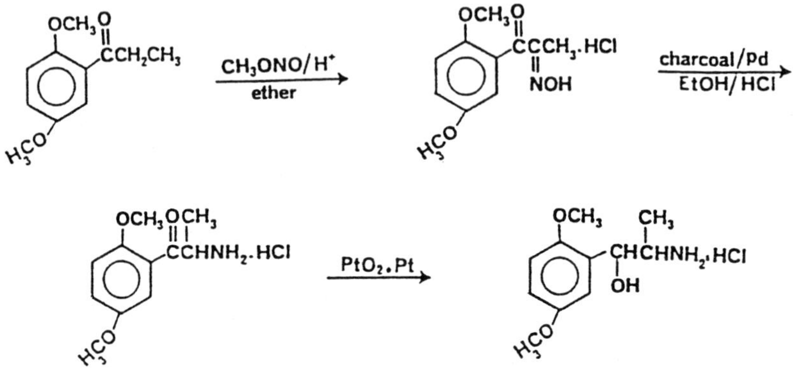

(Fig. 12)

6. Pharmacokinetics, Metabolism and Interactions

6.1 Onset and Duration

Onset of pressor effects following IV administration is 1 to 2 miutes, as compared to 15 to 20 minutes following IM injection (6). Other reports (11,12) indicated onset of 15 seconds following IV injection. Duration of action following IV and IM administration is 60 and 90 minutes, respectively (13).

6.2 Absorption, Distribution and Elimination

There are no data on the pharmacokinetics of methoxamine in the literature.

6.3 Metabolism (14)

Methoxamine and the O-dealkylated metabolite of methoxamine (2-hydoxymethoxamine) were detected in urine samples from dogs treated with methoxamine HCl. N-isopropyl methoxamine (IMA), a structurally related compound, has been shown to be O-dealkylated and N-dealkylated when given to rats and dogs. Fig. 14 shows metabolic transformation of methoxamine.

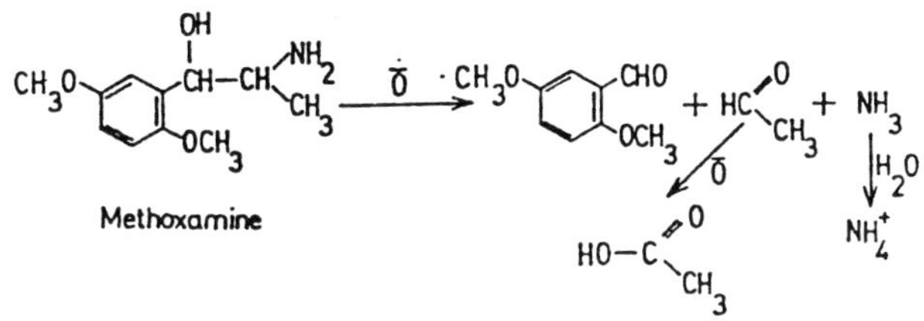

(Fig. 13)

6.4 Interactions (15)

The following drug interactions of methoxamine have been observed.

6.4.1 α-Adrenergic Blocking Agents

Prior administration of α-adrenergic blocking agents, such as phentolamine, may block the pressor response to methoxamine, possibly resulting in severe hypotension. Interactions with medications with α-adrenergic blocking actions such as phenothiazines may result in decreasing the pressor effect and shortening the duration of action of methoxamine.

6.4.2 Anesthetics, Parenteral-local

Use of methoxamine in combination with local anesthetics to prolong their action at local sites is not recommended, since methoxamines extended effect may cause excessive restriction of circulation and lead to sloughing of tissue.

6.4.3 Antidepressants

Concurrent use of antidepressants with methoxamine may potentiate the cardiovascular effects of the latter, thus resulting in arrhythmias, tachycardia, or severe hypertension, or hyperpyrexia.

6.4.4 Antihypertensive and Diuretics

Concurrent use of either one of these agents with methoxamine may reduce their antihypertensive effects.

6.4.5 Digitalis Glycosides

Concurrent use with methoxamine may increase the risk of cardiac arrhythmias; caution and electrocardiographic monitoring are very important in this case.

Other drug-interactions and/or related medical problems were mentioned under the same reference.

7. Methods of Analysis

7.1 Elemental Analysis

	Theory (%)	
Element	Hydrochloride	Base
C	53.33	62.54
H	7.32	8.11
N	5.56	6.63
O	19.38	22.72
Cl	14.31	-

7.2 Identification Tests (16)

7.2.1 Mandelin's Test

Add one drop of Mandelin's reagent to a small amount of methoxamine on a white tile; a green-yellow colour is produced.

7.2.2 Marquis Test

Add one drop of Marquis reagent to a small amount of methoxamine on a white tile; a green colour is developed.

7.3 Microcrystal Formation

The microcrystals of methoxamine HCl were performed on a solution of methoxamine HCl in water. One to two drops of this solution were treated with an equal quantity of the specific reagent on a microscopical

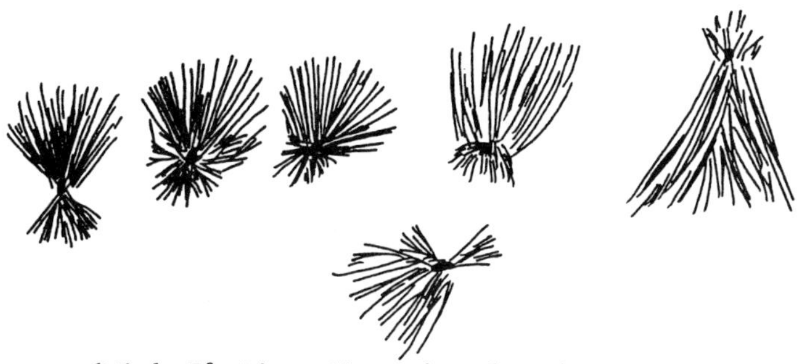

(With 5% Di-sodium phosphate)

(With Mercuric chloride)

(With Pot. dichromate)

Fig. 14: Microcrystals of Methoxamine HCl with different reagents.

slide. After a specific time, the crystals so formed were microscopically examined. Shapes of the crystals upon treatment with each reagent are shown in Fig. 14.

7.4 Chromatographic Analysis

7.4.1 Thin Layer Chromatography (TLC)

Several TLC procedures have been employed for the evaluation of methoxamine and its metabolic transofrmation products in biological fluids and standard solutions. After extraction of methoxamine and related compounds from the urine it can be chromatographed on aluminium oxide (14) or silica gel G (16) layers. TLC was also employed for the identification of methoxamine decomposition products after storing of methoxamine in aqueous solution for various periods of time (7).

Other quantitative TLC methods involved the conversion of methoxamine into a highly fluoresceing derivative by means of 4-(6-methylbenzothiazol-2-yl)phenyl isocyanate followed by TLC separation and measuring the fluorescence (17). o-Phthalaldehyde can also rapidly form fluorescent derivative with methoxamine in presence of mercaptoethanol at room temperature. This reaction was carried out as a dipping procedure for the quantitative determination of methoxamine using HPTLC plates (18). TLC systems used in the analysis of methoxamine are compiled in table 4.

7.4.2 Paper Chromatography

An ascending paper chromatographic technique was used for the analysis of methoxamine and its metabolites after being extracted from urine (14,19). This method involved the use of Whatman No. 1 paper and a solvent system consisted of chloroform - methanol - water and acetic acid (2:1:1:0.1, v/v). The paper were allowed to equilibrate overnight with the upper phase and then developed with the lower layer, Pauly reagent (20) is sprayed for detection.

Table 4: TLC Systems for Methoxamine

Adsorbent	Solvent system	Detection	Rf	Ref.No.
- Silica gel G (250 μm, thick), impregnated with 0.1 M KOH/MeOH	MeOH-strong ammonia solution (100:5)	Sprayed with acidified $KMnO_4$	0.55	16
- Aluminium oxide G	CF:EtOH:H_2O:HAc. (20:10:35:2)	Sprayed with diazotized benzidine (yellow spot[a]).	0.28	14
- Silica gel G	n-but.-HAC-H_2O (5:1:4)	Sprayed with 0.2% ninhydrin in benzidrin (intense red spot[a])	0.84	7
- Silica gel G (Macherey-Nagel)	CF:benz.:n-but. (35:5:5)	Spectrofluorometric[b]	0.25	17
- Silica gel 60 HPTLC layers 10 x 10 (Merck).	EtOAc:MeOH:HF (69:30:1)	Spectrofluorometric[c]	0.18	18
- Silica gel G (0.25 mm)	Hept:MeOH (60:40) (69:30:1)	Spectrofluorometric[d]	0.3	19
- Silica gel G (0.25 mm)	Benz.:MeOH (83:17)	"	0.4	19
- Silica gel G (0.25 mm)	Benz.:Acet.:MeOH: diox. (40:40:4:5).	"	0.3	19

Notes: Solvents: acet. = Acetone; Benz. = benzene, n-but. = n-butanol, CF = Chloroform, Diox. = dioxane, EtOAc = ethylacetate, EtOH = ethanol, HAc = acetic acid, Hept. = heptane. HF = formic acid, MeOH = methanol.

a) Although methoxamine does not fluoresce (UV ligth, long wave), some of its decomposition product could be detected by this mean.

b) After reaction with 4-(6-methylbenzothiazol-2-yl)phenyl isocyanate.

c) After dipping the developed plates in o-Phthalaldehyde solution and then in polyethylene glycol.

d) In this case fluorescamine is spotted over the sample spot before development.

7.4.3 Gas Liquid Chromatography (GLC)

Gas liquid chromatography has been exploited in few reports for routine screening, separation or identification (16,21,22) of methoxamine and related compounds. The following are the reported operating conditions:

a) Column: a glass column (2 m X 4 mm i.d.), packed with 2.5% SE-30 on 80-100 mesh Chromosorb G, Temperature 172°, nitrogen flow rate 45 ml/min. R I of methoxamine was 1726 (16,21).

b) Column: a glass column (2 m X 3 mm i.d.), packed with 10% Apiezon L and 2% KOH on 80-100 mesh Chromosorb W HP. Column temperature was programed at 16° per minute from 170° to 260° and held for 8 min; nitrogen flow rate 30 ml./min. RI value of methoxamine was 1735 (16).

c) Column: Capillary 15 m, packed with SE-30 oven temp. was programmed from 100° to 295° at 5°/min. Helium carrier gas was used at 45 cm/s (100°C). Methoxamine RI value was 1700 (22).

7.4.4 Combined Gas Liquid Chromatography-Mass Spectrometry (GC-Mass) (7)

GC-mass spectrometer has been used for the identification of the decomposition products of methoxamine HCl in aqueous solution. The column consisted of 2% SE-52 on Embacel (M8 B Kieselguhr) packed in a 5 ft stainless steel column (¼ inch o.d.), column temperature 150°, a helium flow rate was 50 ml/min the molecular separator was contained within the GLC oven and the inlet line to the mass spectrometer source was kept at 200°. Results indicated that the major breakdown product of methoxamine HCl was 2,5-dimethoxybenzaldehyde.

7.4.5 High Performance Liquid Chromatography (HPLC)

Only two methods are available for the analysis of methoxamine by HPLC.

a) Method I (23)

This method is suitable for the qualitative HPLC analysis of methoxamine HCl employing a non-aqueous ionic eluent under the following operating conditions: Column was stainless steel (12.5 cm X 4.9 mm i.d.) packed with silica (Spherisorb S5W, 5 µm) and a mobile phase was a solution containing 1.175 g (0.01 M) of ammonium percolate in 1000 ml of methanol and the pH was adjusted to 6.7 by the addition of 1 ml of 0.1 M NaOH in methanol, UV-spectrophotometer (254 nm) or electrochemical detector was used. (K of methoxaine HCl was 0.9).

b) Method II (24)

This method was applied for the quantiative and qualitative determination of methoxamine HCl in both biological fluids and pharmacueitcal preparations. It involves the use of an internal standard (ℓ-norephedrine), under the following chromatographic operating conditions:

Column: a stainless steel (30 cm X 3.9 mm i.d.) packed with µ-Porasil-10; mobile phase: a mixture of 1,2-dichloroethane-n-propanol- acetic acid - diethylamine - water (840:160:10:5:5); flow rate 3 ml./min.; UV-spectrophotometric detector set at 290 nm and l-norephedrine was used as internal standard (K for methoxamine HCl and l-norephedrine were 2.8 and 3.9, respectively.

7.4.6 Miscellaneous

Method I: Characterization of Some Adrenergic Amines as Toluene-p-sulphonate (25)

The toluene-p-sulphonate of some adrenergic amines including methoxamine were prepared by two different methods. The m.p., elementary analyses, U.V., and I.R. absorption spectra and Rf values in T.L.C. were determined for these derivatives. The main reaction products were N-tosylates, but four ditosylates were obtained. The derivatives had very strong absorption max. at a longer wavelength. Ditosylation greatly increased the ε values. Tosylation had potential value in the characterization and differentiation of structurally related amines.

Method II: Spectrometric Determination of Methoxamine by Hydroxylation (26)

Methoxamine was determined after monohydroxylation with H_2O_2 as oxidizing agent, and with Fe^{3+} and catechol as catalysts, glucose being used as stabilizer when required. The product was coupled with 4-amino antipyrine, and the absorbance was measured at 490 nm. Calibration graph was rectilinear for 0.25 to 1.75 mM. Tablet excipients did not interfere.

Method III: Resolution of Enantiomeric Drugs of Some ß-Amino Alcohols (27)

A number of enantiomeric amino alcohols such as ephedrine, methoxamine, propranolol, timolol, etc., were well resolved by conversion to their urea derivatives with α-naphthyl isocyanate and HPLC on γ-aminopropyl silica modified with (R)-N-(3,5-dinitrobenzoyl)phenylglycine chiral stationary phase and a mobile phase of n-hexane-iso-PrOH-MeCN (90:10:2).

Method IV: Reactions of 2,4-Dinitrofluorobenzenes with Some Arylalylamines and Determination of Their Derivatives (28)

The 2,4-dinitrophenyl derivatives of primary and secondary sympathomimetic arylalkylamines were prepared and the m.p., elemental analyses, Rf value, and UV and IR absorption spectra were characterized. Spectrometric and polarographic analysis of these derivatives can be used to detect the presence of these sympathomimetic amines in pharmaceutical preparations.

8. Determination of Methoxamine HCl in Biological Fluids and Pharmaceutical Formulations

8.1 Combined TLC-Spectrophotometric Method (14)

After extraction and purification of methoxamine and its metabolic product from urine. The sample to be analysed was applied to a doubly layered thin layer plates and chromatographed along with a sample of authentic methoxamine using a mixture of chloroform-ethanol-water-acetic acid (20:10:35:2) as a developing solvent system. The chromatogram was then visualized by I_2 reagent and the corresponding methoxamine could be eluted by methanol from the adsorbent after scraping off the adosrbent and then measured spectrophotometrically.

8.2 Combined TLC-Spectrofluorimetric Method (19)

This method can be used for determination of methoxamine HCl in pharmaceutical preparation. The method involved spotting the sample on TLC then fluorescamine solution was spotted over the sample spots. The plates were air-dried, developed and wrapped in aluminium foil to protect the samples from light for 30 min. Fluorescence measurements were made with a spectrofluorometer with TLC accessory (see Table 4).

8.3 HPLC Method (24)

This method has been published very recently for the qualitative and quantitative estimation of methoxamine HCl in plasma and pharmaceutical formulations (e.g.: Vasoxine(R) ampules) with high selectivity and sensitivity (detection limits were as low as 15 ng per injection, in less than 6 min.). The operating condition are listed in section 7.4.5b.

Procedure:

To 500 µl of rabbit plasma containing 10 mg/ml of internal standard (l-norephedrine) a volume of 30 µl of concentrated ammonia solution is added. The mixture is immediately extracted with 5 ml ethylacetate vortexed for one minut and centrifuged for 5 minuts at 4000 r.p.m. The organic layer is filtered and a 4 ml aliquot of the filtrate is evaporated to dryness at 50°C under a stream of nitrogen. The residue is reconstituted in 400 µl of methanol and an appropriate volume (usually 30 µl) is injected in triplicate into the HPLC system working at the operating conditions mentioned in section 7.4.5.b. The concentration of methoxamine HCl is then calculated from a previously constructed calibration graph.

For determination of methoxamine HCl in Vasoxine(R) ampules, the content of 5 ampules are pooled and a 2 ml aliquot (40 mg methoxamine HCl) was diluted to a final concentration of 20 µg/ml. A volume of 1 ml of this solution is mixed with 1 ml of stock solution of the internal standard and the mixture diluted to yield a theoretical final concentration of 2 µg and 10 mg/ml of methoxamine HCl and internal standard, respectively. An appropriate volume (30 µl) of this solution is injected to the HPLC system and the concentration of methoxamine HCl is calculated as mentioned above.

9. Place in Theraphy and Cautions (29)

Methoxamine is used almost solely to prevent or treat hypotension occuring during anesthesia, particularly spinal anesthesia. The drug has also been used to terminate paroxysmal supraventricular tachycardia. The drug can be administered for the treatment of hypotension occurring during surgery and can be used with anesthetics since it has minimal, if any, arrhythmogenic activity. Methoxamine has produced mixed results in the treatment of shock (30).

Methoxamine is contraindicated in severe hypertension and coronary artery diseases. Concomitant administration of methoxamine with local anesthetics that prolong the duration of action is also contraindicated. Methoxamine should be used with caution in patients with hyperthyrodism, congestive heart failure and after parentral injection of ergot alkaloids.

10. Pharmacology and Mechanism of Action (31)

Methoxamine is a relatively pure α-receptor agonist similar to phenylephrine. The drug produces peripheral vasoconstriction, resulting in increase in blood pressure, with virtually no stimulant effects on the heart. Methoxamine produces a predictable rise in systolic and diastolic pressures for 60 to 90 minutes following IV or IM administration secondary to an increase in peripheral resistance. It is believed that the α-adrenergic effects of methoxamine result from the inhibition of the production of c-AMP by inhibition of the enzyme adenyl cyclase. The main effect of therapeutic doses of methoxamine is vasoconstriction.

11. Manufacturer Information

Burroughs - Wellcome
3030 Cornwallis Road,
Research Triangle Park, NC 27709
U. S. A.

12. Acknowledgements

The authors would like to thank Mr. Mohammad Hafiz for microcrystal formation and Mr. Tanvir A. Butt for his efforts and valuable secretarial assistance in typing of this manuscript.

13. References

1. The United States Pharmacopia (USP XIX), "The United States Pharmacopeial Convention Inc.", Md. (1985), p. 316.

2. The Merck Index, 10th Edn., M. Windholz (Editor), Merck & Co., Inc., Rahway-NJ, 1983, p. 860-No. 5863.

3. Clarke's Isolation and Identification of Drugs in Pharmaceuticals, Body Fluids, and Post-Mortem Material, 2nd Edn., A.C. Moffat (Editor), The Pharmaceutical Press, London, 1986, pp. 758.

4. Remington's Pharmaceutical Sciences, 17th Edn., A.R. Gennaro (Editor), Mack Publ. Co., Easton, Penn. (1985), p. 888.

5. Product Information: Vasoxyl®, Burroughs-Wellcome, 1987.

6. Martindale's The Extrapharmacopeia, 27th Edn., Wade A. (Editor), The Pharmaceutical Press, London, 1977.

7. B.J. Millard, D.J. Priaulx, and E. Shottoni, J. Pharm. Pharmac., 23, 369 (1971).

8. Al-Obaid, A.M.; Al-Domiaty, M. & Abou-Shaaban, R. Unpublished data (1990).

9. W.E. Thompson, R.J. Warren, I.B. Eisdorfer, and J.E. Zarembo; J. Pharm. Sci., 54(12), 1819 (1965).

10. R. Bltzly, E. Beer and J. Buck, U.S. Patents: 2,359,707, Oct. 3, 1944. Chem. Abstr., 39 1511 (1945).

11. N.T. Smith and C. Witcher; "Acute Hemodynamic: Effects of Methoxamine in Man", Anesthesiology, 28, 735 (1967).

12. N.T. Smith and C. Witcher; "A Comparison of the Acute Hemodynamic Effects of Methoxamine and Mephentermine in Man", Fed. Proc., 26, 287 (1967a).

13. A.G. Gilman, L.S. Goodman and A. Gilman "The Pharmacological Basis of Therapeutics, 6th Edn. MacMillan Publishing, NY (1980).

14. A. Kluthch and M. Bordun; J. Med. Chem., 10(5), 860 (1967).

15. Drug Information for the Health Care Professional, 9th Ed. 1989, Vol. IB. USPDI. p. 1618.

16. Clarke's Isolation and Identification of Drugs in Pharmaceuticals, Body Fluids, and Post-mortem Material, 2nd Edn., A.C. Moffat (Editor), The Pharmaceutical Press, London, 1986, pp. 137, 139 and 758.

17. R. Wintersteiger, G. Gamse, and W. Pacha; Fresenius Z. Anal. Chem., 312, 455 (1982).

18. G. Giibitz; Chromatographia, 12(12), 779 (1979).

19. R. Wintersteiger, G. Giibitz, and A. Martinger; Chromatographia, 13(5), 291 (1980).

20. I. Smith, "Chromatographic and Electrophoretic Techniques", Vol. I, Interscience Publishers, Inc., New York, N.Y., 296 (1960).

21. R.E. Ardery and A.C. Moffat; J. Chromatography, 220, 195 (1981).

22. R.H. Rohrbangh and Pol. Jurs; Anal. Chem., 60, 2249 (1988).

23. I. Jane, A. McKinnon, and R.J. Flanagon; J. of Chromatogr., 323, 191 (1985).

24. I.A. Al-Meshal, M.M. El-Domiaty and A.M. Al-Obaid; J. Liq. Chromatogr., 12(9), 1589 (1989).

25. J.H. Halmekoski and J. Taipale; Farm. Aikak., 82 (5-6), 81 (1973).

26. J.M. Halmekoski and A. Karki; Farm. Aikak., 85(4), 125 (1975).

27. Q. Yang, Z. Sun, and D. Ling; J. Chromatogr., 447(1), 208 (1988).

28. S. Tammilehto and T. Astala; Acta. Pharm. Fenn., 87(3), 115 (1978).

29. AHFS Drug Information; G.K. McEvoy (Editor), p. 625 (1989).

30. Brown Brown "Comparative Evaluation of Sympathomimetic Amines in Clinical Shock", Circulation, 34, 260 (1966).

31. Comprehensive Online Search on Methoxamine HCl. King Abdulaziz City for Science and Technology (KACST), private communication.

ANALYTICAL PROFILE OF MEXILETINE HYDROCHLORIDE

MOHAMMAD A. ABOUNASSIF

MOHAMMAD SALEEM MIAN

NEELOFUR ABDUL AZIZ MIAN*

Pharmaceutical Chemistry Department,
College of Pharmacy, King Saud University,
P.O. Box 2457, Riyadh-11451,
Saudi Arabia.

* Clinical Laboratory Sciences Dept.,
College of Applied Medical Sciences,
King Saud University, P.O. Box 10219,
Riyadh-11433, Saudi Arabia.

Contents

1. Description

1.1 Nomenclature

1.1.1 Chemical Names

a) 1-(2,6-Dimethylphenoxy)-2-propanamine. (1)
b) 1-(2,6-Xylyloxy)-2-propylamine (1).
c) 1-(2 , 6 -Dimethylphenoxy)-2-aminopropane (1).
d) 1-Methyl-2-(2,6-xylyloxy)ethylamine (1,2).

1.1.2 Generic Names

Katen, Mexitil, Mexitilen, Ritalmex.

1.2 Formulae

1.2.1 Empirical

$C_{11}H_{17}NO$ (Mexiletine).
$C_{11}H_{17}NO$, HCl (Mexiletine hydrochloride).

1.2.2 Structural

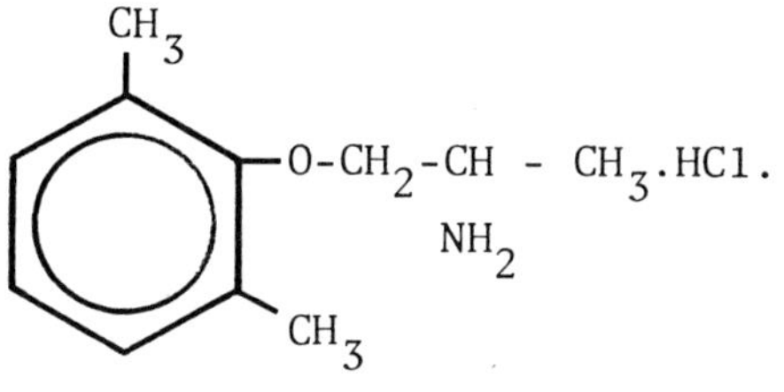

1.2.3 CAS (Chemical Abstract Service Registry Number)

31828-71-4 mexiletine (2, 3).
5370-01-4 mexiletine hydrochloride (2,3).

1.3 Molecular Weight

Mexiletine = 179.3 (2)
Mexiletine. HCl = 215.7 (2)

1.4 Elemental Composition

For mexiletine
C 73.70%, H 9.56%, N 7.81%, O 8.93%
For mexiletine hydrochloride
C 61.22%, H 8.36%, N 6.51%, O 7.43%, Cl 16.48%

1.5 Appearance, Color, Odour and Taste

A white or almost white, odourless, crystalline powder (3).

2. Physical Properties

2.1 Melting Range

203-205° (1).

2.2 Solubility

Freely soluble in water and in methyl alcohol, sparingly soluble in chloroform, insoluble in ether (3).

2.3 pH

A 10% solution in water has a pH of 4.0 to 5.5 (3,4).

2.4 Dissociation Constant (2)

The drug has a dissociation constant (pKa = 9.0).

2.5 Clarity and Color of Solution

A 5.0% w/v solution is clear and colourless (4).

2.6 Sulphated Ash

Not more than 0.1% (4).

2.7 Water

Not more than 0.5% w/v. (4)

2.8 X-Ray Powder Diffraction

The X-ray diffraction pattern of mexiletine hydrochloride was determined by a Philips full automated X-ray diffraction spectrogoniometer equipped with PW 1730/10 generator (5). Radiation was provided by a copper target (Cu anode 2000 w, γ = 1.5480 Å and high intensity X-ray tube operated at 40 KV and 35 MA. The monochromator was a curved single crystal one (PW 1752/100). Divergance slit and the receiving slit were 1 and 0.1° respectively. The scanning speed of the goniometer (PW 1050/81) used was 0.02-2 θ per second. The instrument is combined with Philips PM 8210 printing recorder with both analogue recorder and digital printer. The goniometer was aligned using silicone sample before use. The X-ray pattern of mexiletine HCl is presented in Fig. (1). The scattering angle 2θ, interplanner distance dÅ and relative intensity I/I_o are shown in table (1).

Table (1). X-ray diffraction patterns of mexiletine HCl.

2 θ	dÅ	I/I_o %	2 θ	dÅ	I/I_o %
5.561	15.8913	78.075	28.249	3.1590	44.790
6.309	14.0100	61.143	28.727	3.1075	18.668
8.949	9.8818	20.405	29.456	3.0322	12.662
10.183	8.6861	88.350	30.395	2.9407	20.477
11.686	7.5722	47.829	31.501	2.8400	14.327
12.686	7.5722	29.739	31.988	2.7978	10.564
14.406	6.1481	16.570	32.517	2.7535	8.972
14.995	5.9082	8.465	33.106	2.7058	16.570
16.130	5.4948	10.709	33.703	2.6592	17.438
16.532	5.3620	18.885	34.577	2.5940	7.235
17.392	5.0989	87.771	34.934	2.5683	8.972
18.777	4.7256	9.334	37.855	2.3766	13.024
19.922	4.4567	100.	38.123	2.3605	13.458
21.425	4.1473	13.386	38.807	2.3205	10.347
22.255	3.9944	14.254	40.248	2.2407	12.156
22.914	3.8810	16.859	41.373	2.1823	9.189
23.635	3.7642	76.121	42.038	2.1493	8.393
24.585	3.6209	12.300	42.573	2.1235	11.722
25.056	3.5539	51.953	44.136	2.0519	7.018
25.527	3.4894	28.871	44.619	2.0308	9.189
25.717	3.4640	30.318	45.382	1.9984	8.104
26.381	3.3784	13.675	46.105	1.9687	5.282
27.018	3.3001	30.028	46.893	1.9340	7.235
27.542	3.2385	14.471	48.964	1.8603	5.788

I/I_o = relative intensity (based on highest intensity as 100)
dÅ = interplanner distance.
2θ = scattering angle.

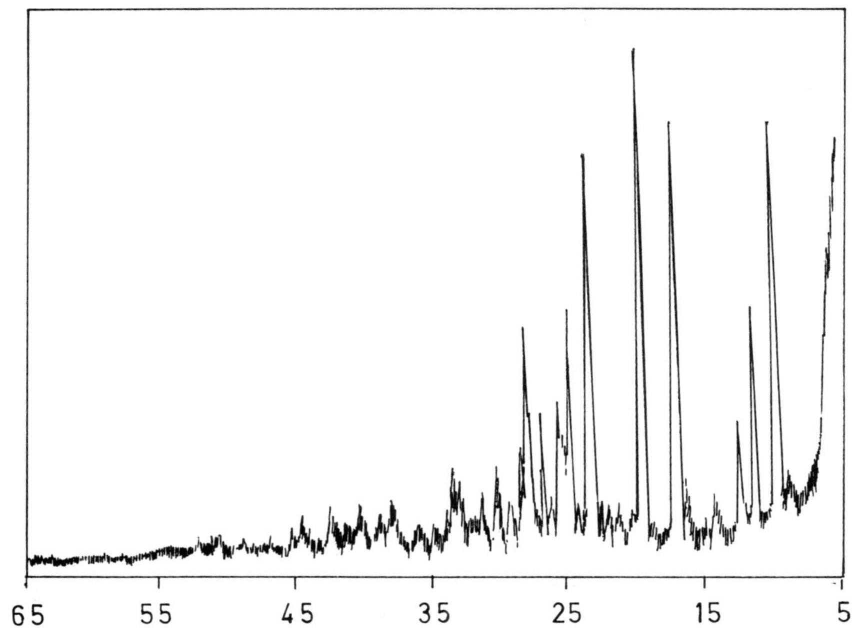

Fig. 1: X-Ray diffraction pattern of mexiletine hydrochloride.

2.9 Spectral Properties

2.9.1 Ultraviolet Spectrum (UV)

The UV spectrum of mexiletine hydrochloride in dist. H_2O Fig. (2) was scanned from 200 to 360 mm using LKB 4054 UV/Vis spectrophotometer (5). The exhibited UV data is presented in table (2).

Table 2. UV data of mexiletine HCl.

λ_{max} nm	Molar absorptivity (ϵ)
212	5980.603
262	23.708

J. Kracmar and J. Kracmarova reported the following UV data (6).

Solvent	min nm	λ_{max}	A(1%, 1 cm)	Optimal conc. µg/ml
CH_3OH	235	211.5 (257) 263 (267) (271.5)	11.3	50-1600
$CHCl_3$		(258) 264 (268) (273)	12.0	50-1600
H_2O	236.5	(210) (255) 262 (267) (271)	11.9	50-1600
HCl (0.01 mol/L in H_2O)	236	(210) (254) 262 (267) (271)	12.0	50-1600

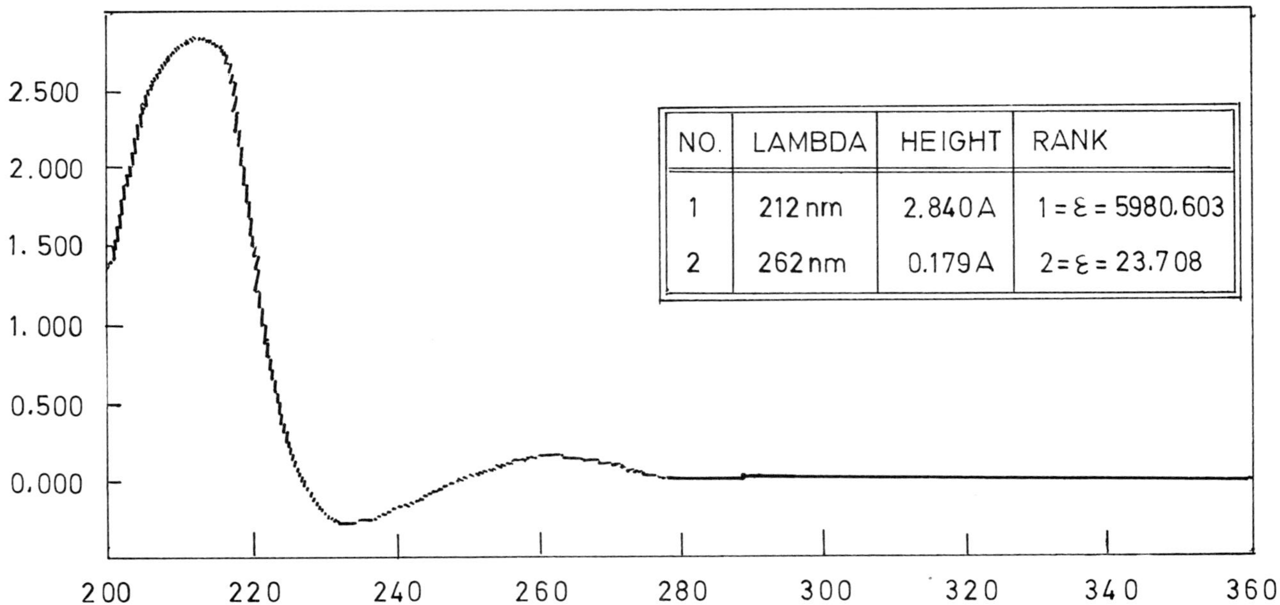

Fig. 2: U.V. spectrum of mexiletine HCl.

2.9.2 Infrared Spectrum (IR) (5)

The IR spectrum of mexiletine hydrochloride as KBr disc was recorded on a Perkin Elmer 580 B infrared spectrometer to which an infrared data station is attached (Fig. 3). The structural assighments have been correlated with various band frequencies and are given in table 3.

Table 3: I.R. spectral assignments for mexiletine HCl

Frequency cm^{-1}	Assignments
2850-3000	C-H Aliphatic stretch.
3200	$-NH_2$
1560, 1600	C = C aromatic stretch.
1040	O-C stretch.

The principal peaks as reported by Clark (2) are 1200, 1040, 770, 760, 1395 and 1090.

2.9.3 Nuclear Magnetic Resonance Spectra (5)

2.9.3.1 Proton Spectrum (PMR)

The PMR spectrum of mexiletine HCl in DMSO-d_6 (Fig. 4) was recorded on a Varian FT80-A, 80 MHz NMR spectrometer using TMS as an internal reference. The structural assignments are presented in table (4).

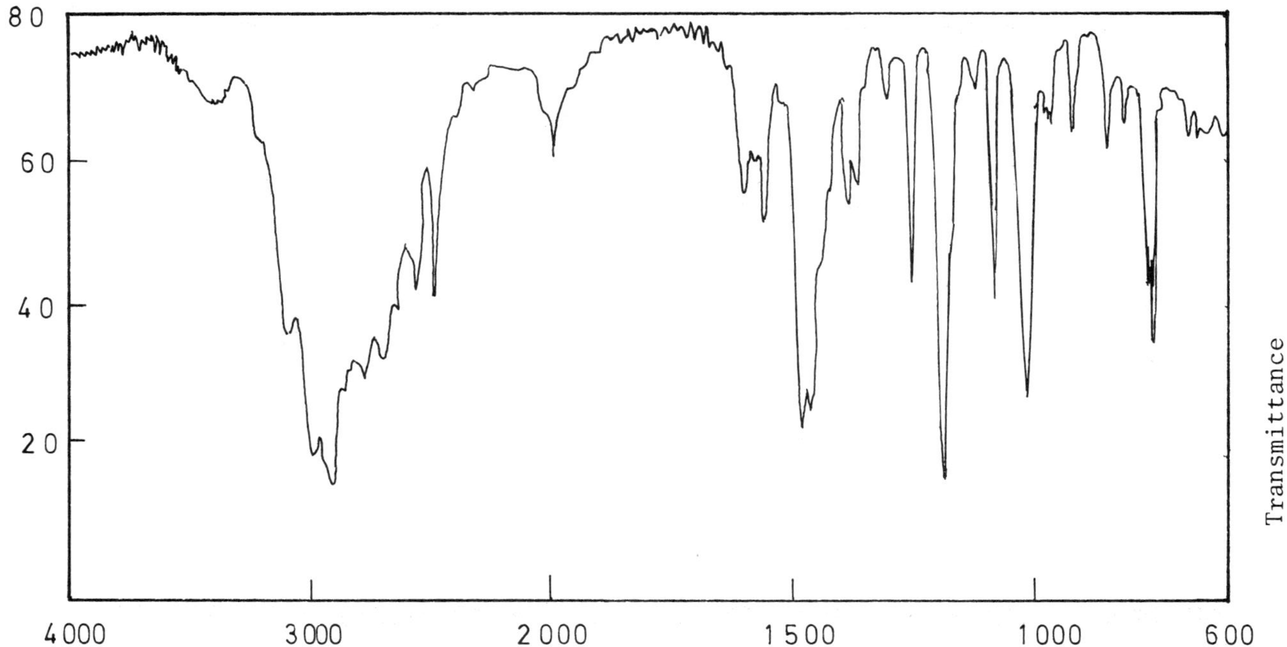

Fig. 3: I.R. spectrum of mexiletine HCl (KBr disc).

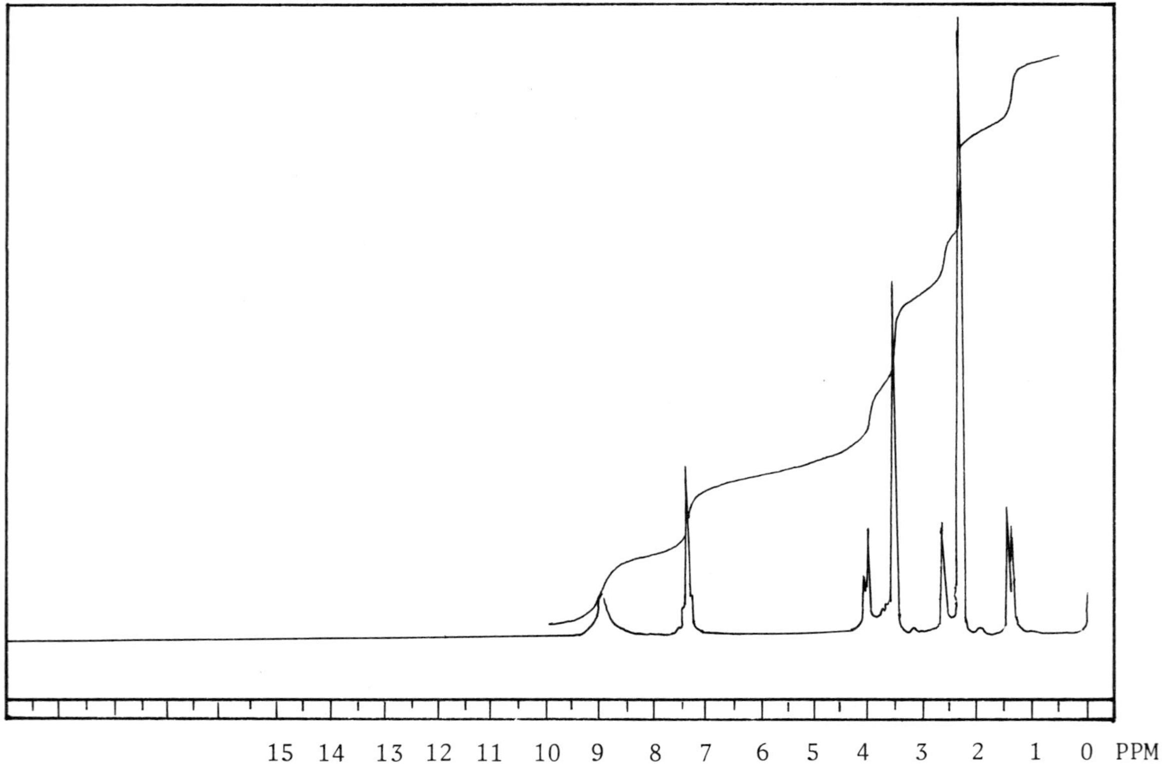

Fig. 4: PMR spectrum of mexiletine HCl.

Table (4): ^{1}H-NMR assignments for mexiletine HCl in $DMSO-d_6$.

Proton assignment	Chemical shift	Multiplicity
$CH-C\underline{H}_3$	1.37	d
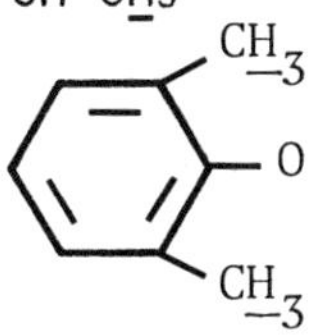	2.268	s
$CH_2-C\underline{H}-CH_3$	3.43	m
$O-C\underline{H}_2$	3.86	d
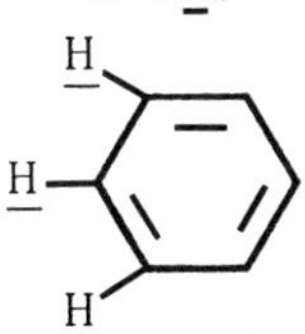	6.94-7.07	m
$-CH-N\underline{H}_2$	8.49	bs

s = singlet, d = doublet, m = multiplet,
bs = broad singlet.

2.9.3.2 ^{13}C-NMR spectra

The ^{13}C-NMR noise decoupled and off resonance spectra of mexiletine HCl in $DMSO-d_6$ are recorded over 6024 Hz range on a Joel FX-100 NMR spectrometer and is presented in Fig. (5 and 6) respectively. The carbon chemical shifts are assigned on the basis of the additivity principals and off resonance splitting pattern and are presented in (Table 5).

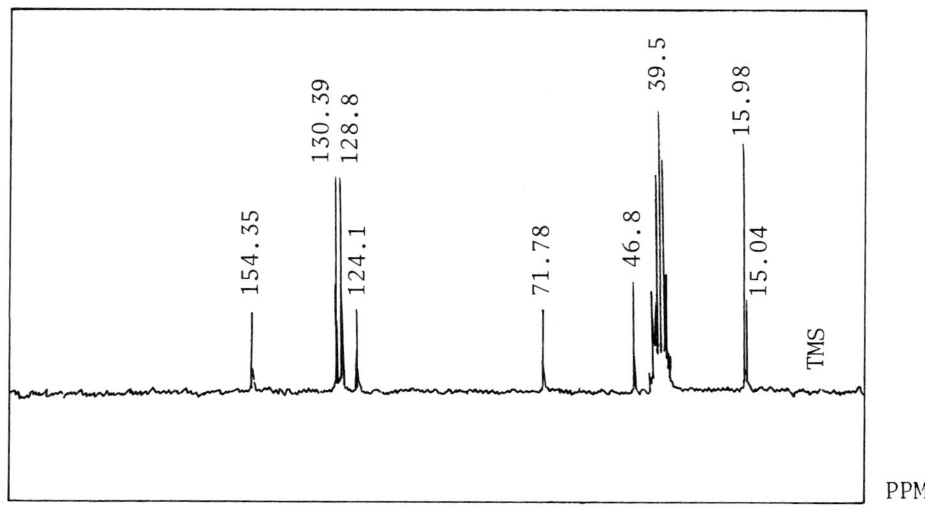

Fig. 5: Noise decoupled ^{13}C-NMR spectrum of Mexiletine HCl in DMSO-d_6.

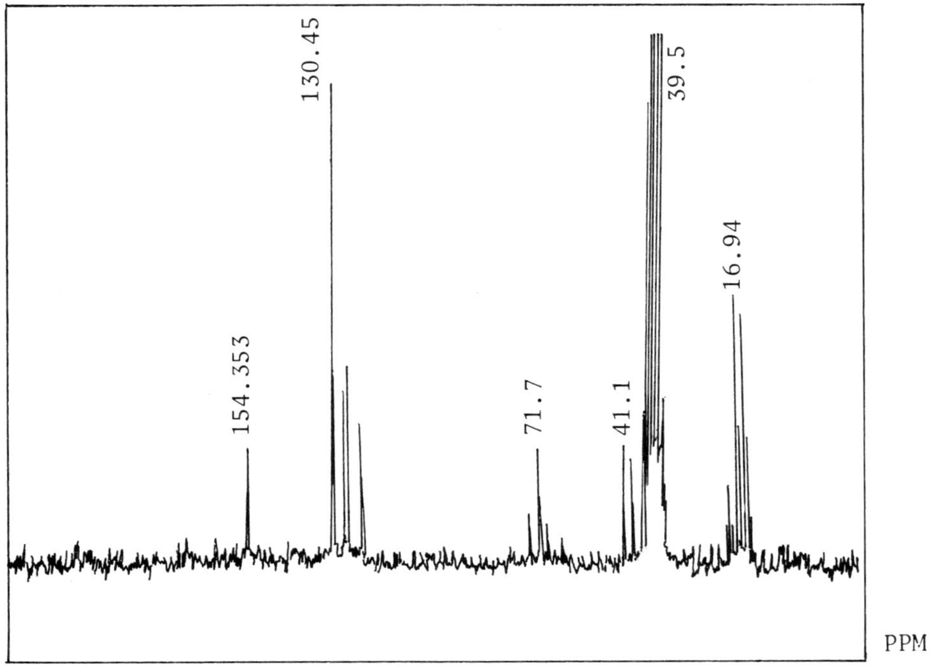

Fig. 6: Off Resonance ^{13}C-NMR spectrum of mexiletine HCl in $DMSO-d_6$.

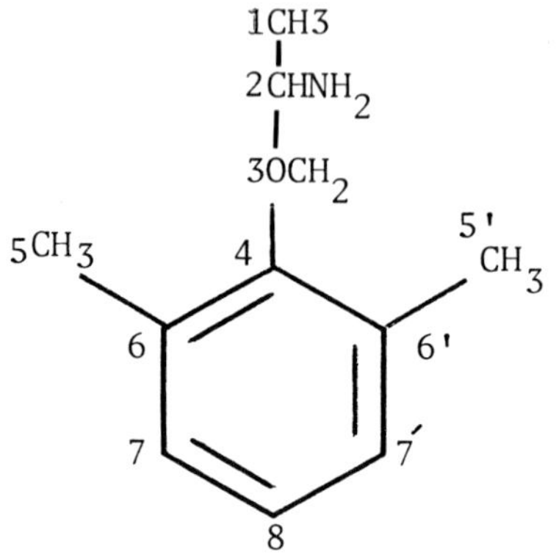

Table (5). Carbon chemical shifts of mexiletine HCl.

Carbon No.	Chemical shifts (ppm)
C_4	154.353 (s)
C_8	130.389 (s)
C_7, C_7	128.102 (d)
C_6, C_6	124.102 (d)
C_3	71.772 (t)
C_2	46.869 (d)
C_5, C_5	15.98 (t)
C_5, C_1	15.04 (t)

s = singlet, d = doublet, t = triplet

2.9.4 Mass Spectrum

The mass spectrum of mexiletine hydrochloride is presented in Fig. (7). This was obtained by electron impact ionization on a Finigan-MAT 1020 GC/Mass spectrometer. The spectrum was scanned to mass 450 amu. The spectrum shows a molecular ion peak M^+ at m/z 179 with a relative intensity of 12%. The base peak is 44 with a relative intensity of 100%. The most prominent fragments, their relative intensities and

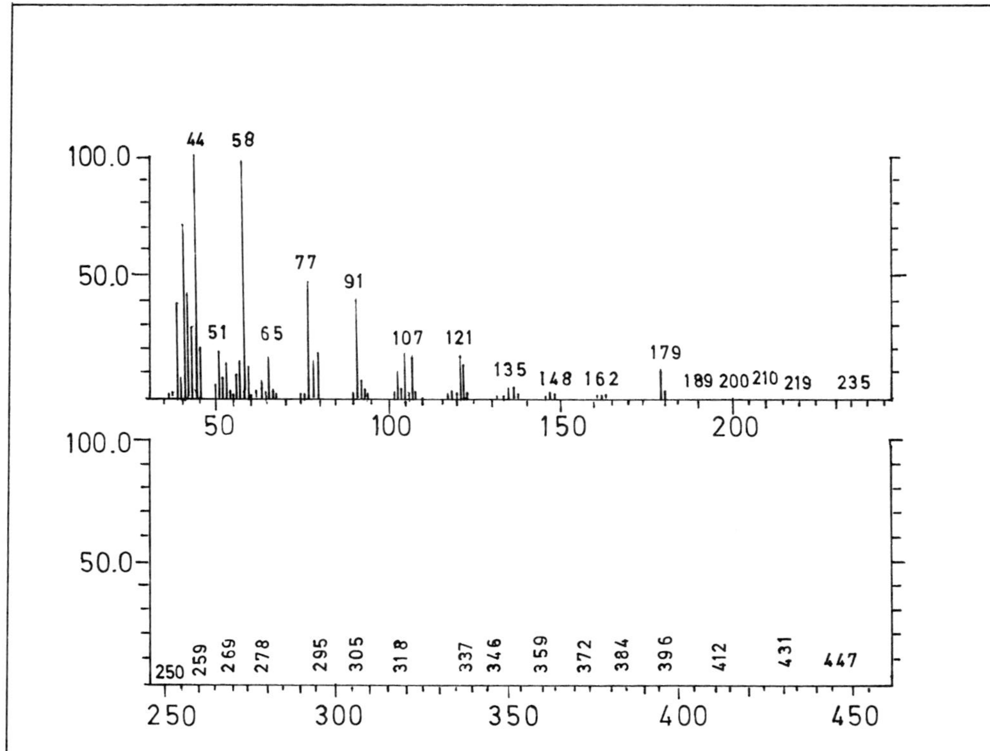

Fig. 7: Mass spectrum of mexiletine HCl.

some proposed ion fragments are presented in table (6).

Table (6). Mass fragments of mexiletine HCl.

m/z	Relative intensity %	Fragments
179	12	M^+ (mexiletine)
121	19	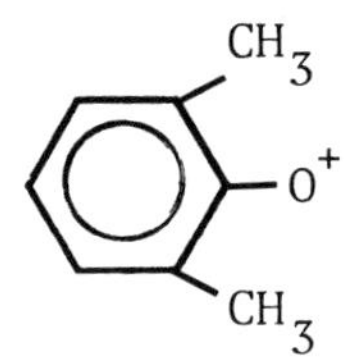
107	19	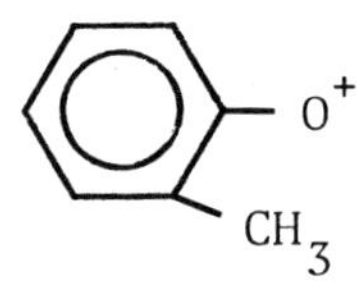
105	19	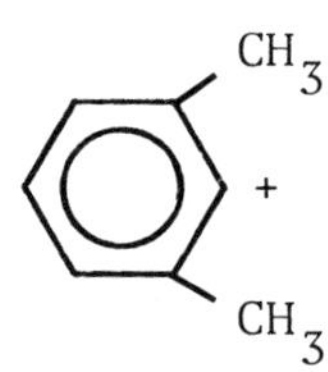
91	42	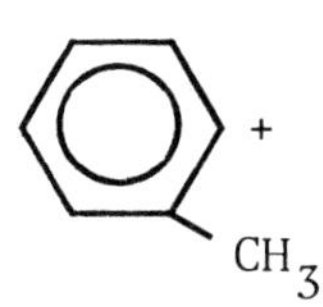
58	98	+ CH_2-CH - CH_3 | NH_2
44	100	+ CH-CH_3 | NH_2

3. <u>Synthesis</u>

<u>Scheme I</u>

The drug is prepared from the condensation of sodium salt of 2.6-dimethylphenol with chloroacetone, with subsequent conversion to oxime followed by hydrogenation to yield mexiletine (7).

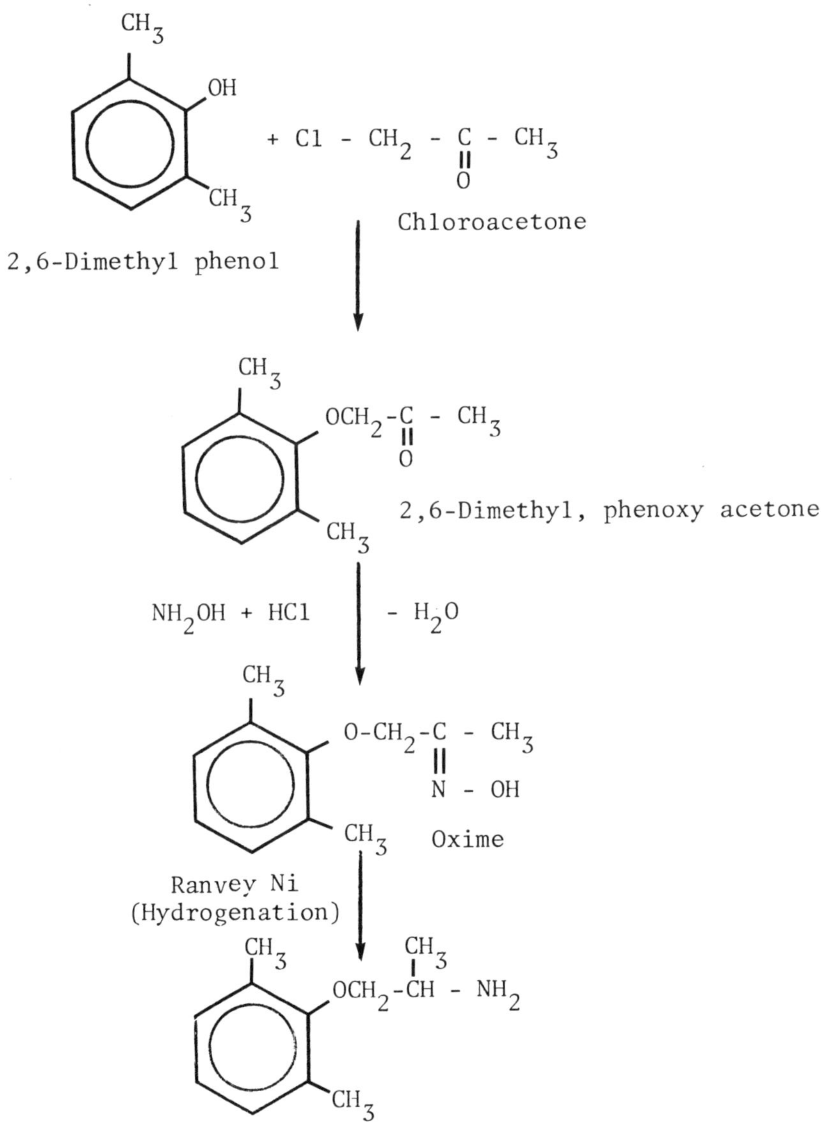

Mexiletine

Scheme II

Mexiletine can be prepared by heating 36 gm of 2,6-$(CH_3)_2$ $C_6H_3OCH_2COCH_3$ and 90 ml of 23% NH_3 solution in CH_3OH over 1.2 gm activated charcoal and 0.1 % platinum at 85-92°/1.5-3 MPa H followed by dilution with $CH_3C_6H_5$-$(CH_3)_2CHOH$ and treatment with HCl gave 84% mexiletine hydrochloride (8).

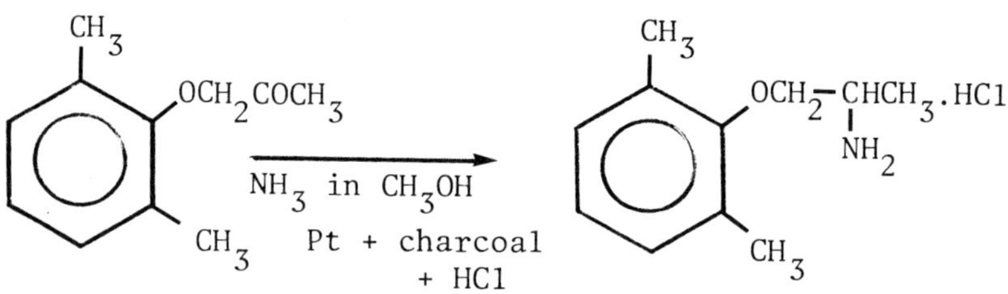

4. Purification of Mexiletine Hydrochloride (9)

The antiarrhythmic 1-(2,6-dimethylphenoxy)-2-aminopropane hydrochloride was purified by liberation of the base mexiletine. Extraction of mexiletine into a non polar organic solvent, evaporation, of solvent and distillation of mexiletine, then salification of mexiletine with HCl and final recrystallization. A mixture of 0.21 mol. of crude mexiletine hydrochloride containing 1.5% 2,6-dimethylphenol, 90 ml of H_2O, 50 ml toluene, and 0.25 mol NaOH (predissolved in 20 ml additional H_2O) was stirred at 15-20° for 0.5 h. After separation of the aqueous layer, recrystallizing the aqueous layer and aqueous washing of the organic layer, concentration of the latter and distillation at 125-128° and 1.8 KPa gave pure mexiletine. Salification of mexiletine in 143 ml 10:1 toluene/ethanol with HCl and stirring for one hour at 15-20° and 3 h at 2-10° gave recrystalized mexiletine hydrochloride which was decolorized with charcoal in EtOH solution (40 ml). Dilution with hot (60-70°) PhMe and stirring hot for 1 h and at 0-15° for 4 h gave pure mexiletine hydrochloride which was washed with acetone and dried to weigh 39 gm (87%) containing only 3 ppm 2,6-dimethyl phenol.

5. Pharmacokinetic Studies

Pharmacokinetic studies with simultaneous determination of mexiletine and one of its metabolites are being performed in a cardiological unit for drug monitoring (10). In order to adapt the posology to the patient, pharmacokinetic parameters were measured after the administration of a single dose of mexiletine by oral route with chronic ventricular ectopic beats. Pharmacokinetic parameters of mexiletine were determined according to one compartment model; and were found similar to those quoted by other authors (11, 12). In order to control the range of therapeutic plasma levels of mexiletine at a steady rate, determinations were performed during a long term oral maintenance therapy, the results are presented in table (7). Plasma concentrations of mexiletine were in the therapeutic range and found similar to those reported by Campbell et al. (12), i.e. varying from 0.75 to 2 μg/ml with little risk of toxicity whereas those of the metabolite reach the level of 200 ng/ml (presented in table 8).

Table (7): Mexiletine pharmacokinetics data after single administration.

Posology	Oral (200 mg)
T ½ (h)	11
Cl (1/h)	38
V_d (1)	613

T ½ = half life

Cl = total body clearance of mexiletine calculated by dividing the dose by the area under the plasma concentration - time curve extrapolated to infinity.

V_d = total volume of distribution.

Table (8): Plasma concentration of mexiletine and its metabolite during maintenance oral therapy of mexiletine (200 mg every 8 h).

Time (h)	0	4
Mexiletine (ng/ml)	1190	1567
Metabolite (ng/ml)	187	217

Effect of Cigarette Smoking on the Pharmacokinetics

The effect of cigarette smoking on the pharmacokinetics of a single dose of mexiletine was studied by Grech-Belanger et al. (13) 0.2 g, was investigated in 14 healthy white adults, mean age 22.5 ± 2.7 years.

Cigarette smoking had no effect on absorption or distribution of the drug but it significantly reduced the elimination $t\frac{1}{2}$ from 11.1 ± 3.4 to 7.2 ± 1.8 h; the effect of clearance was less significant. Determination of the urinary concentrations of the 3 major metabolites of mexiletine glucuronide conjugate, hydroxymethylmexiletine, and p-hydroxymexiletine indicated that cigarette smoking selectively induced conjugation of drug with glucuronic acid as well as aliphatic hydroxylation to yield hydroxymethylmexiletine, but that had no effect on the formation of p-hydroxymexiletine.

Mexiletine eliminated (14) primarily by hepatic metabolism. The influence of age on the plasma pharmacokinetics of mexiletine was assessed in seven young and ten elderly healthy subjects. Mexiletine (50 mg) was administered orally three times daily for 10 days. The use of low doses of drug was possible owing to the development of a new fluorescence HPLC method. No differences in the pharmacokinetic parameters of mexiletine related to age were observed.

6. Metabolism of Mexiletine

6.1 Absorption and Fate

Mexiletine is readily and almost completely absorbed from the gastrointestinal tract.

Mexiletine is metabolized in the liver to a number of metabolites. It is excreted in the urine, mainly in the form of its metabolites with a small proportion of unchanged mexiletine, the clearance of drug is increased in acid urine, mexiletine is widely distributed throughout the body and is about 60-70% bound to plasma proteins. It has a plasma half-life of 10 hours in healthy subjects but this may be prolonged in patients with heart disease. Its therapeutic effect has been correlated with plasma concentrations of 0.75-2 µg/ml but the margin between therapeutic and toxic concentrations is narrow, and severe toxicity may occur within this range (3).

Mexiletine absorbed after oral administration; bioavailability 80-90%. It is extensively metabolised, the major metabolites are the 4'-hydroxy and 2'-hydroxy methyl derivatives which may be further metabolised by deamination to the corresponding alcohols. The rate and extent of renal excretion is dependent on urinary pH under normal conditions, up to about 20% of a dose may be excreted in the urine as unchanged drug in 24 hours, but this may be increased to about 50% if the urinary pH is maintained at about 5. (2).

6.2 Therapeutic Concentration

In plasma, usually in the range of 0.5 to 2 µg/ml. Following a single oral dose of 400 mg to 6 subjects, peak plasma concentrations of 0.65 to 0.94 µg/ml (mean 0.77) were attained in about 2 hours (15).

In another study, after a daily oral doses of 450 mg to 10 subjects, steady state plasma concentrations of 0.75 to 2.18 µg/ml (mean 1.6) were reported (16).

6.3 Use and Administration (3)

Mexiletine is used for the prevention and treatment of ventiricular arrhythmias particularly after myocardial infarction. It is given by mouth as the hydrochloride in initial doses of 400 mg followed by 200-250 mg three to four times daily starting 2 hours after the loading dose. The usual maintenance dosage of mexiletine is 600-800 mg daily in divided doses. Higher doses may be necessary in patients after myocardial infarction to overcome delayed absorption. Use of a controlled-release preparation allows 12-hourly administration.

Mexiletine may be given by slow intravenous injection of the hydrochloride in doses of 100-250 mg at a rate of 25 mg per minute, followed by an infusion at a rate of 250 mg over 1 hour, 250 mg over the next 2 hours and then at about 0.5 mg per minute for maintenance, according to the patients's response when the appropriate patient may be transferred to oral therapy with doses of 200-250 mg of the hydrochloride 3-4 times daily. Alternatively an initial intravenous dose of 200 mg at a rate of 25 mg per minute, may be followed by an oral dose of 400 mg on completion of the injection, with subsequent oral therapy as before.

7. Toxicology

7.1 LD_{50}

Oral Route (17)

350 mg/kg (male rats).
400 mg/kg (female rats).
310 mg/kg (male mice).
400 mg/kg (female mice).
180 mg/kg (male rabbits).
160 mg/kg (female rabbits).

I.V. Route (17)

27 mg/kg (male rats).
30 mg/kg (female rats).
43 mg/kg (male mice).
50 mg/kg (female mice).

Other LD_{50} determination test performed by Boehringer Ingelheim (18) are as follows:

Species	Route	24 hours
Mouse	p.o.	430 mg/kg
	s.c.	170 mg/kg
	i.m.	150 mg/kg
	i.v.	34 mg/kg
Rat	p.o.	530 mg/kg
	s.c.	390 mg/kg
	i.m.	380 mg/kg
Rabbit	i.v.	10 mg/kg
Dog	p.o.	150 mg/kg
	s.c.	240 mg/kg
	i.v.	60 mg/kg
Cat	p.o.	240 mg/kg
	s.c.	80 mg/kg

7.2 Toxicity and Adverse Effects

Mexiletine has a narrow therapeutic ratio many adverse effects of mexiletine are dose-related and will respond to dosage reduction but may be severe enough to necessitate discontinuation of therapy. Toxicity is commonly seen with oral or parental loading doses when plasma concentrations are high (3).

The most common adverse effects are nausea and vomiting. Effects on the nervous system include tremor, confusion, light-headedness, dizziness, blurred vision and other visual disturbances. The most frequent cardiovascular effects are hypotension, sinus bradycardia, atrioventricular dissociation, and arterial fibrillation. Other adverse effects which have been reported include skin rashes, abnormal liver function tests and overdose convulsions (3).

Thrombocytopenia was reported in a patient four days after starting treatment with mexiletine. Platelet count had returned to normal within 10 days of discontinuing the drug. There was no evidence of platelet antibody formation. (19)

A 22 year old man who had ingested 4.4 g of mexiletine died (20). Symtoms include paraesthesia of the tongue, nausea, convulsions, cyanosis, rigididy, bradycardia and an unrecordable blood pressure. The ECG showed complete heart block, ventiricular asystole was unresponsive to isoprenaline, adrenaline, electrical pacing, and cardiac massage. The blood-mexiletine concentration post mortem was 34-37 μg/ml compared with the therapeutic range of 1-2 μg per ml.

In another fatality due to mexiletine overdose, postmortem blood and liver concentrations of 44.3 μg/ml and 636 μg/g, respectively, were reported (21).

Adverse cardiovascular effects of mexiletine will call for symtomatic therapy. Hypotension associated with bradycardia may be reversed with atropine. It has been reported that animal studies indicate the toxic neurological effects of mexiletine may respond to diazepam (3).

7.3 Precautions

Mexiletine should be used with special caution in patients with sinus node dysfunction, conduction effect, bradycardia, hypotension cardiogenic shock, or cardia or hepatic failure. ECG and blood pressure monitoring should be carried out during treatment.

Absorption of mexiletine may be delayed by concurrent administration of opiates, and after myocardial infarction. Plasma concentrations of mexiletine may be reduced by the concurrent administration of drugs which induce hepatic enzymes such as phenytoin and rifampicin (3).

Evidence that metaclopramide enhances and atropine decreases the rate of mexiletine absorption without altering the relative oral bioavailabilty (22).

Several beneficial interactions of mexiletine with other anti-arrhythmic agents are of clinical importance since the efficiency of mexiletine used alone is limited. However, combinations of mexiletine with quinidine or beta-blockers can increase it's efficiency and reduce adverse effects. It is probably necessary to use reduced doses of mexiletine in these combinations (23).

Administration of rifampicin was shown markedly to enhance the elimination of mexiletine in a study in 8 healthy subjects. The suggested mechanism was increased metabolism due to induction of hepatic enzymes by rifampicin (24).

8. Methods of Analysis

8.1 Identification

a) Dissolve 0.1 g in 3 ml of 0.02 M hydrochloric acid and add a few crystals of sodium nitrite. Nitrogen is evolved and a yellow colour may be produced slowly (4).

b) A 1% w/v solution yields reaction a characteristic of chlorides (4).

c) The light absorption, in the range 230-350 nm of a 0.04% w/v solution in 0.01 M hydrochloric acid exhibits a maximum at 260 nm. The absorbance at 260 nm is about 0.46 (4).

8.2 Potentiometric Determinations

I. Normal potentiometric method (25)

Weigh accurately a specified quantity (100-200 mg) of authentic mexiletine hydrochloride in a small beaker. Add about 30 ml glacial acetic acid and 15 ml mercuric acetate. Stir over a magnetic stirrer for 10 minutes and record the titration curve using 0.1 N acetous perchloric acid. Locate the equivalence point graphically by the method of parallel tangents (22) and compute the content of the active ingredient using the expression:

$$(\text{mg})\ \text{mexiletine HCl} = V \times f \times F$$

where V, is the titrant volume chemically equivalent to the drug; f, is the titrant standardisation factor and F, is the titrant equivalence factor (= 21.57 mg mexiletine HCl).

II. D_1-differential potentiometric method (25)

For authentic mexiletine HCl. — Dissolve about 15-20 mg mexiletine hydrochloride accurately weighed, in 20 ml glacial acetic acid and 15 ml mercuric acetate with continuous stirring for 10 minutes. Titrate against 0.025 N perchloric acid while recording the D_1-differential curve (_E/_V, versus V). The end point coincides with the sharp peak at the maximum D_1-value. Calculate the amount of drug using the same expression as in I, noting that the titrant equivalence factor, F, equals 5.39 mg mexiletine hydrochloride.

For mexiletine HCl capsules. — Weigh accurately 20 capsules, empty them, and collect the material. Clean the empty shells with a current of air and subtract the weight of empty capsules from the gross weight to obtain the net weight of active ingredient from which weight per capsule can be computed. Weigh and dissolve a specified quantity equivalent to (15-25 mg) of the active ingredient in 20 ml glacial acetic acid and proceed as previously described for authentic mexiletine hydrochloride.

For mexiletine HCl injections. — Transfer 5 ml of the injection volume into a separating funnel, add 2 ml 1 N NaOH and extract with 4 X 10 ml chloroform, washing each $CHCl_3$ extract with the same 5 ml of water. Filter through a filter paper containing anhydrous sodium sulphate and collect the chloroform extracts into a 100 ml volumetric flask. Complete to volume and evaporate the chloroform gently to about 5 ml. Add 30 ml glacial acetic acid and proceed as described for authentic mexiletine hydrochloride.

8.3 Nuclear Magnetic Resonance Method (25)

For authentic mexiletine HCl. — Dissolve about 100 mg accurately weighed mexiletine hydrochloride and 65 mg maleic acid in 2 ml DMSO-d_6. Transfer about 0.5 ml of the clear solution to a NMR tube and record the spectrum, adjusting the spin rate to reduce the spinning side bands. Integrate the peaks of interest (either the CH_3 doublet at 1.37 ppm or the aromatic protons at 6.94-7.07 ppm for mexiletine hydrochloride and the CH = CH protons of maleic acid at 6.25 ppm) and record the mean of three integrations for each peak. The amount of the drug is then calculated as follows:

$$\text{(mg) mexiletine HCl} = \frac{A_1}{A_2} \times 1.24 \times \text{mg, maleic acid}$$

where A_1 is the integral value of the aromatic protons of mexiletine hydrochloride, A_2 that of the CH = CH singlet of maleic acid, and 1.24 is a factor obtained by dividing a third of the molecular weight of mexiletine hydrochloride by half the molecular weight of maleic acid.

For mexiletine HCl capsules. — To a quantity of the mixed contents of 20 capsules equivalent to about 100 mg mexiletine hydrochloride accurately weighed, add 65 mg maleic acid and 2 ml DMSO-d_6. Shake for 10 minutes and centrifuge at about 3000 rpm for 3 minutes. Transfer about 0.5 ml of the clear supernatent into a NMR tube and proceed as previously described for authentic mexiletine hydrochloride.

For mexiletine HCl injections. — Evaporate 4 ml portion of the injection to dryness, add 65 mg maleic acid accurately weighed, dissolve in 2 ml DMSO-d_6, and proceed as previously described for authentic mexiletine hydrochloride.

8.4 Spectrofluorometric Method

Scheja et al. (26) described the simple quantitative spectrofluorometric method for the determination of mexitil in plasma of patients treated orally or intravenous with therapeutic doses of drug. Following alkalinization and ether extraction of the blood sample, drug was obtained by reverse-extraction into a HCl-solution. The relative flourescence maximum of mexiletine is 218 nm. The recovery rate of the drug is 95% and the precision is 4.3%. The limit of detectibility is 0.06 μg mexiletine/ml. This method was used to determine the pharmacokinetics of mxiletine in humans.

8.5 Chromatographic Methods

8.5.1 Gas Chromatography

Quantitative analysis of mexiletine in urine was done by Beckett et al. (27) in which 4 ml of urine samples collected at consecutive intervals after dosing, or blank urine to which the drug had been added, were placed in glass centrifuge tubes and 1 ml aqueous solution of n-butylamphetamine hydrochloride (10 µg/ml of abse) was added as an internal standard. After making alkaline to pH 12.9 wtih 1 ml 4N NaOH, the samples were extracted with freshly re-distilled ether (3 X ml), shaking the tubes for 30 minutes on a mechanical shaker during each extraction. The combined ethereal extracts of each sample, in evaporating glass tube with tapered bases, were concentrated on water bath at 43° to about 20 µl. The concentrates were analysed by g.l.c. using a glass column (1 m X 0.64 cm o.d.) with Carbowax 20 M (7.5%) on Chromosorb W, 80-100 mesh, AW-DMCS-treated and operated at 125-135° using N_2 as carrier gas. Summary of conditions used for some of the GLC determinations of mexiletine is given in the table (8).

8.5.2 Thin Layer Chromatography (TLC)

Mexiletine may be analysed by TLC, glass plates (20 X 20 cm) were spread to a thickness of 0.25 mm with a mixture of silica gel G and H_2O (1:2). The plates were pre-dried at room temperature for 30 minutes and then oven dried at 110° for 1 hour before use. The solvent system used was benzene:methanol (80:20). The detection was done by using the spray Dragenorff's (27).

8.5.3 High Pressure Liquid Chromatography (HPLC)

High pressure liquid chromatography (HPLC) method has wide applications for the estimation of mexiletine HCl. A summary of variable parameters in a few cases is given in table (9).

Table (8) Summary of conditions used, for GLC of mexiletine.

Column support	Mesh	Temperature	Flow rate and Detection	Retention time	Sample	Ref.
(1.5 m X 4 mm) containing 3% of OV-17 on gas Chrom Q.	80-100	150°	N_2 carrier gas 30 ml/min flame-ionization detection	-	Plasma or urine	28
(1.8 m X 2 mm) of 1.5% of Carbowax 20M plus 5% of KOH on supelcoport.	80-100	-	N.P flame ionisation detector.	3.5 min	Blood or plasma	29
(1.1 m X 2 mm) packed with 10% of SP-1000 on supelcoport.	80-100	180°	20 ml/min N_2 carrier gas. N.P flame ionization detector (F.I.D.)	-	Serum	30
(2 m X 3 mm) packed with 3% of OV-225 on acid-washed and silanised chromosorb W.	-	-	Thermionic detection	-	Plasma or aq. solution.	31
Glass column (3 Ft X 2 mm) containing 2.8% of OV-210 and 3.2% of OV-1 on Chromosorb WHP.	80-100	120°	25.5 ml/min N_2 carrier gas flame ionization detection	-	Plasma	32

Continued/...

Continued (Table 8)

Glass column (3 m X 4 mm) packed with 1% of HI-EFF-8BP on Gas Chrom Q.	100-120	180	45 ml/min methane: Ar (1:9) as carrier gas and detection by electron capture.	-	Plasma bile and urine	33
Silanised glass column (2 m X 2 mm) packed with 5% of carbowax 20M plus 5% of KOH on Chromosorb WHP.	100-120	155	40 ml/min N2 gas flame ionization detection.	-	Plasma	34
(2 m X 2.5 mm) packed with 3% of OV-17 on Gas-Chrom Q.	80-120	150	25 ml/min Ar-CH4 (9:1) as carrier gas detection by electron capture.	-	Plasma	35
Column 3% of OV-17 on Gas Chrom Q.	10-100	-	N-sensitive detection.	-	Plasma	36
Silanized glass column packded with 5% of Carbowax 20M-5% of KOH or 5% of OV-101 on Chromosorb WHP, or with 10% of Apiezon L-2% KOH on Chromosorb WAW.	-	-	-	-	-	37

Continued/...

Continued (Table 8)

(1.2 m X 4 mm) of Gas. Chrom Q coated with 3% of OV-16.	-	180°	N2 carrier gas and flame ionization detection	-	Serum	38
(1.2 m X 4 mm) packed with 10% of Apiezon L plus 2% of KOH on Chromosorb WAW.	80-100	195°	60 ml/min N2 carrier gas flame ionization detection	-	Plasma or serum	39
Glass column (100 cm X 0.64 cm o.d.) containing 7.5% of Carbowax 20M on Chromosorb WAW-DMCS.	80-100	125°-135°	100 ml/min N2 carrier gas flame ionization detection	-	Urine	27
Glass column (4 ft X 0.25 in o.d.) containing 3% of HI-EFF 8BP on Gas Chrom Q.	100-120	220°	70 ml/min He carrier gas Rb Br flame ionization detector.	-	Plasma or urine	40
Glass column (200 cm X 2.5 mm) packed with 3% OV-17 on Gas Chrom Q.	80-100	280°	25 ml/min argon:methane (90:10) as carrier gas.	-	Plasma	10

Table (9) Summary of HPLC conditions for the determination of mexiletine

Column	Mobile phase	Flow rate	Sample	Detection	Ref.
(30 cm X 4 mm) Spherisorb ODS (5 µm).	Heptane-1-sulphonic acid: acetonitrile: tetrahydrofuran (21:24:5).	2.4 ml/min	Plasma, urine or cerebro-spinal fluid.	352 nm	41
(25 cm X 4.6 mm) Zorboax C8.	Methanol: 0.04M amm. acetate (80:20)	2.5 ml/min	Serum	Flourescence	42
(12.5 cm X 5 mm) Spherisorb S5W silica	Methanol: 2,2,4 trimethylpentane (4:1) containing 1 mM -(+)camphor-10-sulphonic acid.	2 ml/min	Plasma or serum	Flourescence	43
(30 cm X 3.9 mm) µ Bondapak C18	0.1 M potassium phosphate (pH 4.0): acetonitrile (3:1)	3 ml/min	Plasma	210 nm	44
(30 cm X 4 mm) µ Bondapak C18 (10 µm)	Methanol 10 mM: Phosphate buffer of pH 5.5 (13:7).	2 ml/min	Plasma	250 nm	45
(30 cm X 4 mm) µ Bondapak C18	20% Acetonitrile in 30 mM - phosphate buffer (pH 3.7).	2 ml/min	Plasma	214 nm	46
(30 cm X 4 mm) µ Bondapak C18	Aq. 65% acetonitrile containing 0.4% of pentane sulphonic acid and 1% of acetic acid.	1 ml/min	Serum	475 nm	47

Continued/...

Continued (Table 9)

(30 cm X 4 mm) µ Bondapak C18	23-29% solution of acetonitrile in 0.196M acetate buffer of pH 4.80.	2 ml/min	Plasma	254 nm	48
(25 cm X 4.5 mm) 5 µ Hypersil	7 g Na_2HPO_4 + 380 acetonitrile up to 1 L pH 5.0	1.5 ml/min	Serum	214 nm	49
Clin. Elut columns	Et_2O-EtOAC (1:1)	-	Plasma	-	50
Nucleosil 10 C18	0.05 M KH_2PO_4 (pH 2.3): MeCN	-	Plasma or serum	UV	51
µ Bondapak columns	-	-	Blood serum	-	52

8.6 Extraction from Biological Fluids (10)

Scheme III represents a diagram of the extraction procedure. 1 ml of plasma was introduced into a glass-stoppered 20-ml centrifuge tube; 100 µl of the aqueous solutions of both internal standards were successively added to plasma alkalinized (pH = 11.7) with 0.5 ml of a NH_4OH (32%) concentrated solution.

The mixture was shaken with 5 ml of ethylic ether for 15 min and centrifuged at 2400 g for 10 min. The upper organic phase was carefully transferred to another centrifuge tube and acidified (pH = 1) by 1 ml of a H_2SO_4 aqueous solution (0.2 N). The mixture was shaken for 10 min and centrifuged at 2400 g for 3 min. The upper organic phase was removed and the aqueous phase kept was alkalinized again (pH = 11) by 0.3 ml of a NH_4OH concentrated solution (32%) and then extracted by 1.5 ml of diethyl ether. The mixture was shaken for 20 min, then centrifuged at 2400 g for 5 min and the ther layer poured into a glass tube (microvial) was evaporated off under a stream of dry nitrogen at room temperature.

8.7 Thermal Analysis (DSC) (5)

A differential scanning calorimetry curve (Fig. 8) was obtained on a Perkin-Elmer DSC-2C differential calorimeter. Nitrogen was used as the purge gas. Scan was performed at a rate of 25°C/min from 170-230°C. The DSC curve revealed an endothermic melting peak at 203.18°C.

9. Acknowledgements

The authors are highly thankful to Mr. Tanvir A. Butt, Pharmaceutical Chemistry Department, College of Pharmacy and Mr. Babkir Awad Mustafa from College of Applied Medical Sciences, King Saud University for their great help in typing and extending the technical services, respectively, in preparing the manuscript.

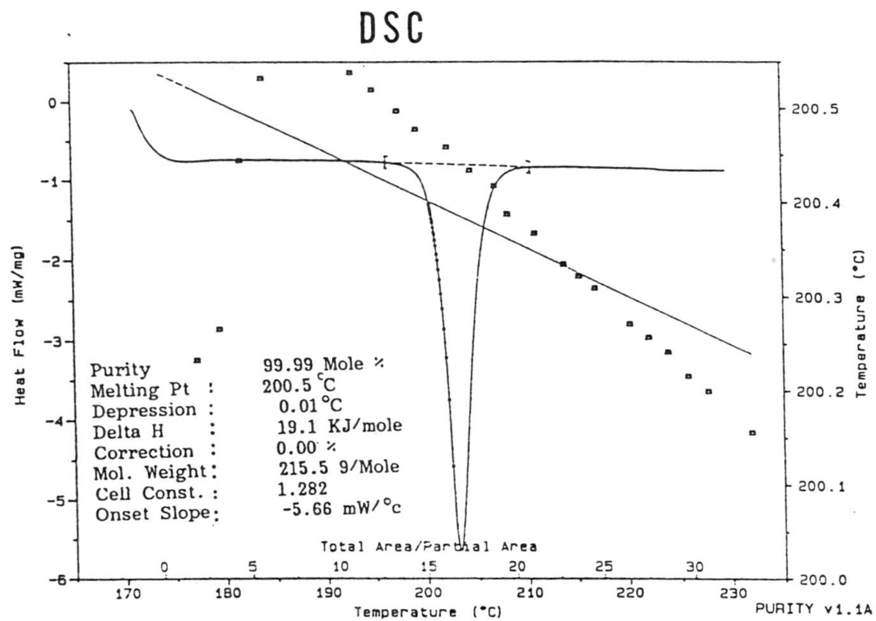

Fig. 8: DSC thermal curve of mexiletine HCl.

Plasma 1 ml

↓

add 500 µl of conc. NH_4OH

↓

extract with 5 ml of diethylether

↓

Organic layer

↓

back extract into 1 ml of 0.2 N sulfuric acid

↓

acid/aqueous layer

↓

and 300 µl of concentrated NH_4OH

↓

extract with 1.5 ml of diethylether

↓

Organic layer

↓

evaporate to dryness

↓

residue

dissolve in 70 µl of cyclohexane

↓

heat at 70°C for 10 min

↓

add 15 µl of HFBA mix 30 s and evaporate immediately to dryness

↓

residue

dissolve in 50 µl of methanol

↓

Ready for injection to column

Scheme III: Diagram of the plasma extraction procedure.

10. References

1. "The Merck Index" 11th Edition, p. 971 (1989).

2. "Clarks' Isolation and Identification of Drugs" 2nd ed. p. 782. The Pharmaceutical Press, London (1986).

3. "Martindale" The Extra Pharmacopeia" 29th Ed., p. 80, The Pharmaceutical Press, London (1989).

4. "British Pharmacopeia" p. 375, Her Majestys' Stationary Office, London (1988).

5. Mohammad A. Abounassif and Mohammad Saleem Mian, Unpublished Data, College of Pharmacy, King Saud University, 1991.

6. J. Kracmar and J. Kracmarova, Pharmazie, 38(8), 524-527 (1983).

7. Pharmaceutical Chemistry, Vol. I (drug synthesis) p. 122. Edited by H.J. Roth and A. Kleeman, John Wiley and Sons, New York.

8. Chemical Abstract. 109, 92467c (1988).

9. Chemical Abstracts 110, 94677r (1989).

10. A. Pachecus, Y. Santoni, M. Fornairs, S. Magnan, C. Aubert, A. Ragon and J.P. Cano. Arzeim. Forsch. Drug Research. 32(1), 688-693 (1982).

11. Mc Comish, M. Robionson, C., Kitson, D., Jewitt, D.E., Postrad. Med. J., 53 (Suppl. I) 85, (1977).

12. Campbell, N.P.S., Kelly, J.G., Adgey, A.A.J., Shanks, R.G. Brit. J. Clin. Pharmacol., 6, 372 (1978).

13. Grech-Belanger, O.; Gilbert, M.; Turgeon, J., LeBlank, P.P. Clin. Pharmacol. Ther., 37, 638-643 (1985).

14. El-Allaf, D.; Carlier, J., Dresse, A. Int. J. Clin. Pharmacol. Res., 6(4) 303-308 (1986).

15. V. Haselbarth et al., Clin. Pharmac. Ther., 29, 729-36 (1981).

16. K. Ohashi et al., Arzeimittel-Forsch., 34, 503-507 (1984).

17. Kast, Alexander et al., Iyakuhim Kenkyu, 13(4), 922-956 (1982) Japan.

18. Mexitil, "for treating ventricular arrhythmias" Boehringer Ingelheim, p. 9 (Germany).

19. G.P. Fasola et al. Ann. Intern. Med., 100, 162 (1984).

20. P. Jequier et al. Lancet, 1, 429 (1976).

21. R.C. Blackmore and M.D. Ossleton, Bull. Int. Ass. Forens. Toxicol., 16(3), 7-8 (1983).

22. L.M.H. Wing et al., Br. J. Clin. Pharmac., 9, 505 (1980).

23. J.T. Bigger, Am. Heart J., 107, 1079 (1984).

24. P.J. Pentikainen et al., Eur. J. Clin. Pharmacol., 23, 261 (1982).

25. M.A. Abounassif. Il Farmaco, Societa' Chimica Italiana, accepted for publication (1991).

26. Scheja, W.; Scheja, J.W.; Bruester, H. GIT Labor-Med., 6(1), 14-15 (1983).

27. A.H. Beckett and E.C. Chidomere, J. Pharm. Pharmac., 29, 281-285 (1977).

28. Chan, K.; Latham, Susan M.; and Tse, J., J. Pharm. Pharmacol., 32, 98P (1980).

29. Smith, K.J.; and Meffin, P.J., J. Chromatogr., 181(3-4); Biomed. App. 7(3-4), 469-472 (1980).

30. Elfving, Sirkka, M.; Sven, Eivor, H.; and Leskinen, E.E.A., J. Clin. Chem. Clin. Biochem., 19(12), 1189-1191 (1981).

31. Sharman, J.R. J. Anal. Toxicol., 5(4), 153-156 (1981).

32. Perchalski, Robert, J.; Wilder, B.J.; and Hammer, R.H. J. Pharm. Sci., 63(9), 1489-91 (1974).

33. Willox, S.; and Singh, B.N. J. Chromatogr., 128(1), 196-198 (1976).

34. Kaeprowicz, Arther T. Clin. Chem., 28(1), 245-246 (1982).

35. Pachecus, A.; Santoni, Y.; Fornasis, M.; Magnan, S.; Aubert, C.; Ragon, A.; and Cano, J.P., Arzneim-Forsch., 32(I) (6), 688-693 (1982).

36. Pilling, Mark; Tse, John; and Chan Kelvin. Methods Find. Exp. Clin. Pharmacol., 4(4), 243-247 (1982).

37. Kacprosicz, A.T. J. Chromatogr., 269(2), 61-70 (1983).

38. Grech Benager, O. J. Chromatogr., Biomed. Appl., 34(1) (J. Chromatogr. 309), 165-169 (1984).

39. Holt, David W.; Flanagan, Robert, J.; Hayler, Angela M.; and Loizou, Mary. J. Chromatogr., 169, 295-301 (1979).

40. Kelly, J.G.; Nimmo, J.; Rae, R.; Shanks, R.G.; and Prescott, L.F. J. Pharm. Pharmacol., 25(7), 550-553 (1973).

41. Bresthaupt, Henning; and Wilfling, Michael. J. Chromatogr., 230(1), 97-105 (1982).

42. Farid, N.A.; White, S.M. J. Chromatogr., 275(2), 458-462 (1983).

43. Bhamra, R.K.; Flanagan, R.J.; Holt, D.W. J. Chromatogr., 307, 439-444 (1984).

44. Mastropaolo, W.; Holmes, D.R.; Osborn, M.J.; Rook, J.; Moyer, T.P. Clin. Chem., 30(2), 319-322 (1984).

45. Grech-Belanger, O.; Turgeon, J.; Gilbert, M. J. Chromatogr., 337, 172-177 (1985).

46. Dusci, L.J.; Hackett, L.P. J. Anal. Toxicol., 9(2), 67-70 (1985).

47. Grech-Belanger, O.; Turgeon, J.; Gilbert, M. J. Chromatogr. Sci., 22(11), 490-492 (1984).

48. Kelly, R.; Christmore, D.; Smith, R.; Doshier, L.; and Jacobs, S.L. Ther. Drug Monit., 3(3), 279-286 (1981).

49. Caroline Stokl and John C. Crawhall. Analytical Letters, 22(7), 1693-1701 (1989).

50. Susanto, F.; Humfeld, S.; Reinaur, H. Chromatographia, 21(1), 41-3 (1986).

51. Erdweg, M.; Hoffmann, M.; Weinand, H.A. Med. Welt., 39(37), 1082-6 (1988).

52. Wilczynska-Wojtulewicz, Irena, Sadlej. Biol. Inf. Inst. Lekow (Polish), 35(1), 82-81 (1988).

NICOTINAMIDE

Ezzat M. Abdel Moety*, Mohammad Tariq** and
Abdullah A. Al-Badr*

* Department of Pharmaceutical Chemistry, College of Pharmacy, King Saud University, P.O. Box 2457, Riyadh-11451, Saudi Arabia.

** Department of Pharmacology, College of Pharmacy, King Saud University, P.O. Box 2457, Riyadh-11451, Saudi Arabia.

CONTENTS

NICOTINAMIDE

1. INTRODUCTORY

Nicotinamide, a substituted pyridine derivative, is an antipellagra agent and is called Vitamin PP (pellagra-preventing). It is used for treating vitamin deficiency in hypertensive patients (1) because it lacks the vasodilator action. Warburg and Associates (2) obtained nicotinic acid amide (nicotinamide) from a coenzyme isolated from the red blood cells of the horse. This finding stimulated further studies on the nutritional value of nicotinic acid. Liver extracts were known to be highly effective in curing human pellagra and canine black tongue. Elvehjem and Associates (3) prepared concentrates of liver that were highly effective in the treatment of canine black tongue, and identified the active substance as nicotinamide (3). It was first prepared in 1894 by reacting ethyl nicotinate with concentrated ammonia (4). The drug decomposes by hydrolysis and an obvious and effective means of stabilization is to limit access of the drug to water. In such cases solid dosage forms are normally recommended (5). As the amide is laiable to hydrolytic conversion to the prototype nicotinic acid; the acid can mainly metabolized to the corresponding amide. Nicotinamide can also be synthetically obtained from nicotinic acid by several modifications. Most of the published analytical procedures deal with the solution of the coexistance of the two drug forms in bulk material, dosage formulations and/or in biological specimens. This demonstrates how it is important to include both the drug substances in one profile.

2. DESCRIPTION

2.1 Nomenclature

2.11 Chemical Names

Nicotinamide: Pyridine-3-carboxamide; Nicotinic acid amide; 3-Pyridine carboxylic acid amide; 3-Pyridine carboxamide (6).

Nicotinic acid: Pyridine-3-carboxylic acid; 3-Pyridinecarboxylic acid; Pyridine-ß-carboxylic acid.

2.12 Generic Names

Nicotinamide: Niacinamide; Nicotylamide; Nicotinamidum; Vitamin PP; Vitamin B_3; Aminocotin; Dipegyl; Pilonin amide; Amide PP.

Nicotinic acid: Niacin; PP-(Pellagra Preventive) factor; antipellagra vitamin; Acidum Nicotinicum.

2.13 Trade Names

Nicotinamide: Nicamina; Nicobion; Nicosedine; Nicovel; Nicovit; Nicamindon; Nicotamide; Nicofort; Niozymin; Pelmine; Benicot; Vinieotyl; Amisyn; Farmabion; Nicotinamide; Pelo; Apisate (in combination) and Effico (6,7).

Nicotinic acid: Nico-400; Nicobid; Nicotinex; Nico-Span; Nicyl; Equivert, Pernivit, Nicolar, Niconacid; Nicacid; Nicagin; Nicotene; Nicotinipca; Akotin; Daskil; Nicolar; Tinic; and Wampocap. (6,7).

2.14 Chemical Abstract Registry Number [CAS]

Nicotinamide: [98-92-0]
Nicotinic acid: [59-67-6]

2.2 Formulae

2.21 Empirical

Nicotinamide: $C_6H_6N_2O$
Nicotinic acid: $C_6H_5NO_2$

2.22 Structural

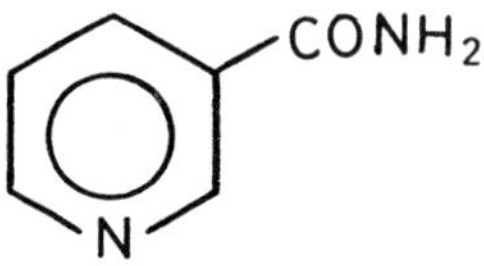

Nicotinamide

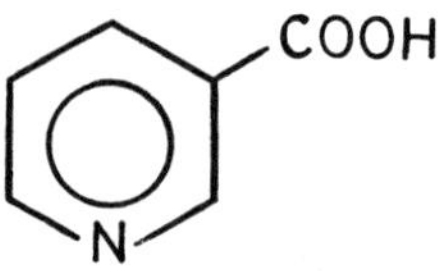

Nicotinic acid

2.23 Wiswesser Line Notation

Nicotinamide: T6NJ CVZ (8)
Nicotinic acid: T6NJ CVQ (8)

2.3 Molecular Weight

Nicotinamide: 122.13
Nicotinic acid: 123.11

2.4 Appearance, Color, Odor and Taste

Nicotinamide exists as colorless crystals or white crystalline powder; faint characteristic odor, salty and bitter taste (9); while nicotinic acid appears as needles or white to creamy-white crystals or crystalline powder (7).

3. PHYSICAL CHARACTERISTICS

3.1 Melting Point

Nicotinamide: 128°-131°C (7,9,10).
Nicotinic acid: 236.6°C (6).
234-237°C (7).

3.2 Solubility

Nicotinamide is highly soluble in water and alcohols; 1 g dissolves in about 1 ml of water, in about 1.5 ml alcohol and in 10 ml glycerol. It is slightly soluble in ether and chloroform, while it is soluble in solutions of alkali (9,12,13). Nicotinic acid, 1 g, is soluble in 55 to 60 ml of water and in 100 ml of ethanol, very slightly soluble in chloroform; practically insoluble in ether; soluble in alkaline solutions of hydroxides and carbonates (7,11).

3.3 pH-Range

A 1.3% (w/v) aqueous solution of nicotinamide gives a pH 3-3.5; its 5% solution in water has a pH of 6 to 7.5 and the 10% solution in water is neutral to litmus (6,11,12). The pH of saturated solution of nicotinic acid is about 2.7 (6).

3.4 Optical Activity

Both the drug substances are optically inactive because of the absence of chemical asymmetry.

3.5 Moisture Content and Hygroscopicity

Nicotinamide absorbs insignificant amounts of moisture at 25°C at relative humidities up to about 90% (4).

3.6 Dissociation Constant

The dissociation constant, pK, of nicotinamide: 3.3 [20°C] (4); while the pK of nicotinic acid = 4.85 [25°C] (6), pKa 2.00 (-N=), 4.80 (-COOH) [25°C] (7).

3.7 Stability

Nicotinamide is stable in air, light and at certain pH-changes (11).

3.8 Storage

Both drug substances should be stored in air-tight containers (4).

3.9 Identification

3.91 Color Test

To 2 ml of 0.1% solution of nicotinamide or nicotinic acid add 6 ml of cyanogen bromide solution and 1 ml of a 2.5% v/v solution of aniline golden yellow; i.e.

König's reaction. Nicotinamide gives brown-orange color with Nessler's reagent (7).

3.92 Microcrystal Test

Nicotinamide yields with gold chloride solution dense rosettes (sensitivity 1 in 400) and with platinic iodide solution a mass of hair - like neeldes (sensitivity: 1 in 400) (7).

3.10 Spectral Properties

3.101 Ultraviolet Spectrum

The ultraviolet spectrum of nicotinamide in ethanol (using a Varian UV-vis spectrophotometer model DMS 90) is shown in Figure 1. The spectrum of the drug exhibits a maximum at 261 nm [A(1%, 1cm) = 451, ϵ = 5.507 x 10^3]. The UV-running of methanolic solutions of nicotinic acid exhibits a characteristic maximum at about 263 nm [A(1%, 1cm)) = 229, ϵ = 2.819 x 10^3].

3.102 Infrared Spectrum

The infrared spectrum of nicotinamide, in KBr disc, is presented in Figure 2, (recorded on Pye Unicam SP 1025 infrared spectrophotometer). The frequencies and structural assignments of nicotinamide and nicotinic acid are shown below:

Frequency (cm^{-1})	Assignment
1600-1630	C ═ C (stretch, amide and acid)
1700	C ═ O (amide)
1710	C ═ O (acid)
3200	N — H (stretch; amide)
3100	O — H (stretch, acid)

3.103 Proton Nuclear Spectrum [^{1}H-NMR]

The ^{1}H-NMR runnings were undertaken for each drug substance. The ^{1}H-NMR spectrum of nicotinamide is shown in Figure 3. Each drug substance was dissolved

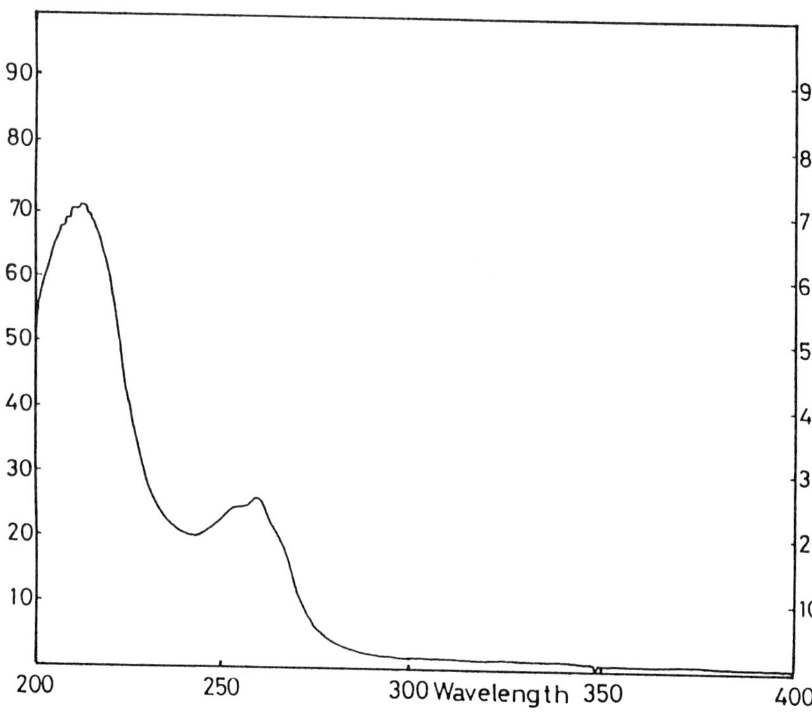

Fig. 1: The ultraviolet (UV) scanning of ethanolic solution of nicotinamide.

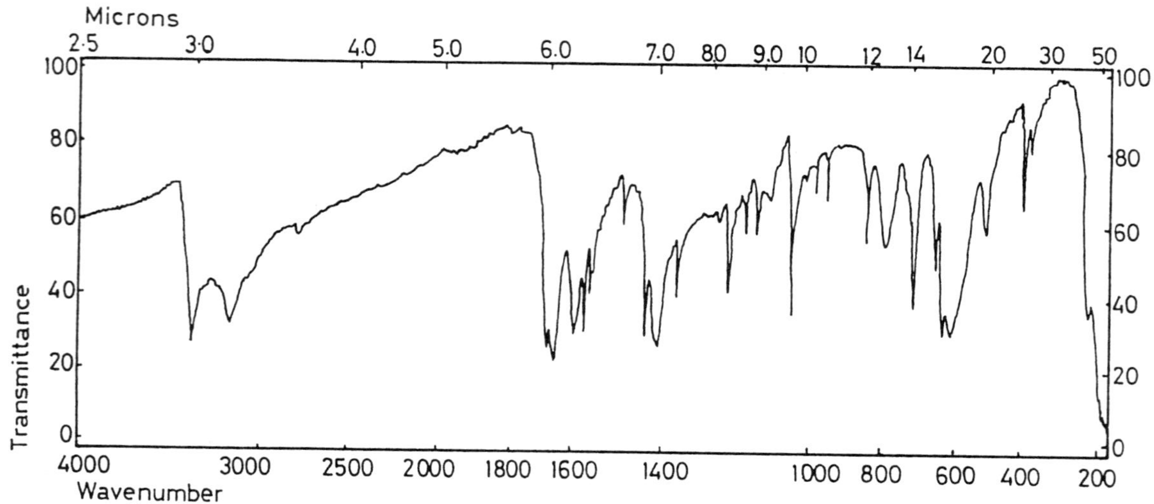

Fig. 2: The infrared (IR) spectrum of nicotinamide.

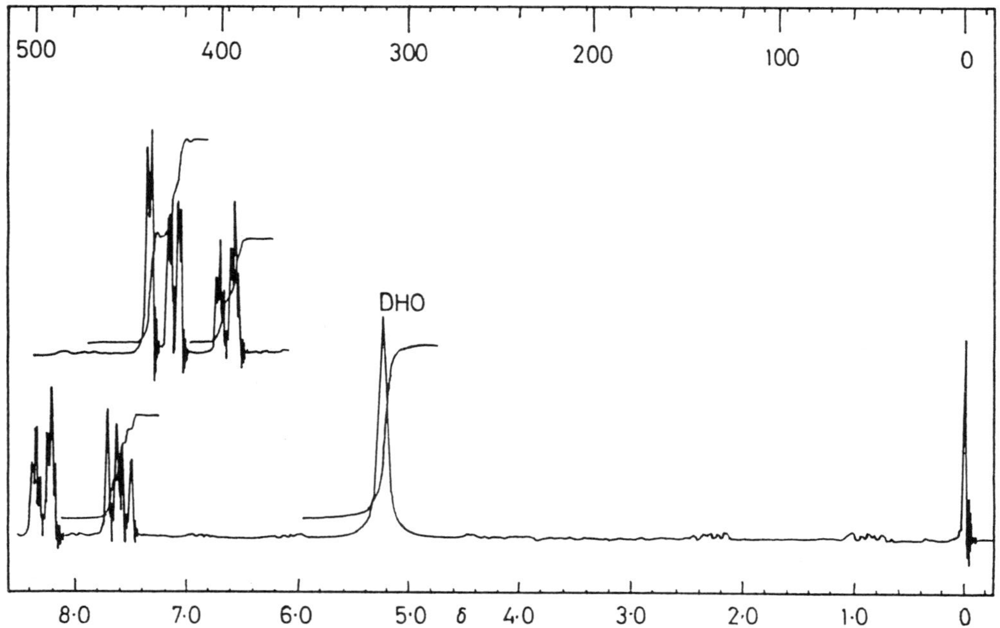

Fig. 3: The proton nuclear magnetic resonance (^1H-NMR) spectrum of nicotinamide.

in deuterium oxide (D_2O) and its spectrum determined on a Varian T6A NMR spectrometer using DSS (sodium-2,2-dimethyl-2-silapentane-5-sulfonate) as the internal standard. The structural assignments are shown below:

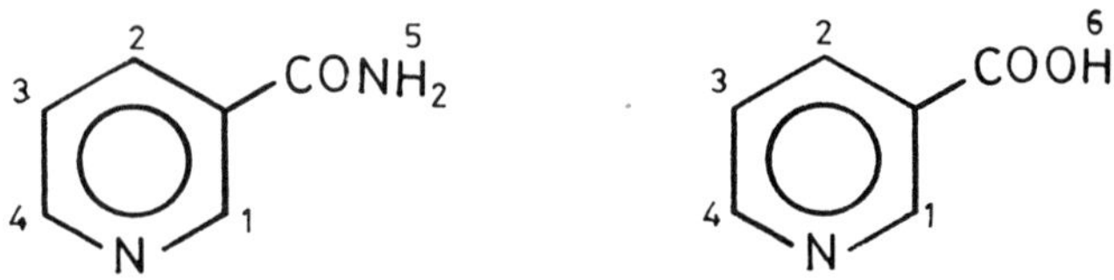

Nicotinamide Nicotinic acid

Chemical Shift (δ, ppm)	Multiplicity	Proton Assignment
7.60	multiplet	H(3)
8.25	multiplet	H(2)
8.72	multiplet	H(4)
8.93	doublet	H(1)
9.16	singlet	H(5) (amide)
9.14	singlet	H(6) (acid)

3.104 Carbon-13 NMR Spectrum [^{13}C-NMR]

The ^{13}C-NMR spectra of nicotinamide in deuterium oxide (D_2O) using DSS as the internal reference were obtained using Jeol FX 100 MHz spectrometer at an ambient temperature. Figure 4 represents the ^1H-decoupled spectrum of nicotinamide.

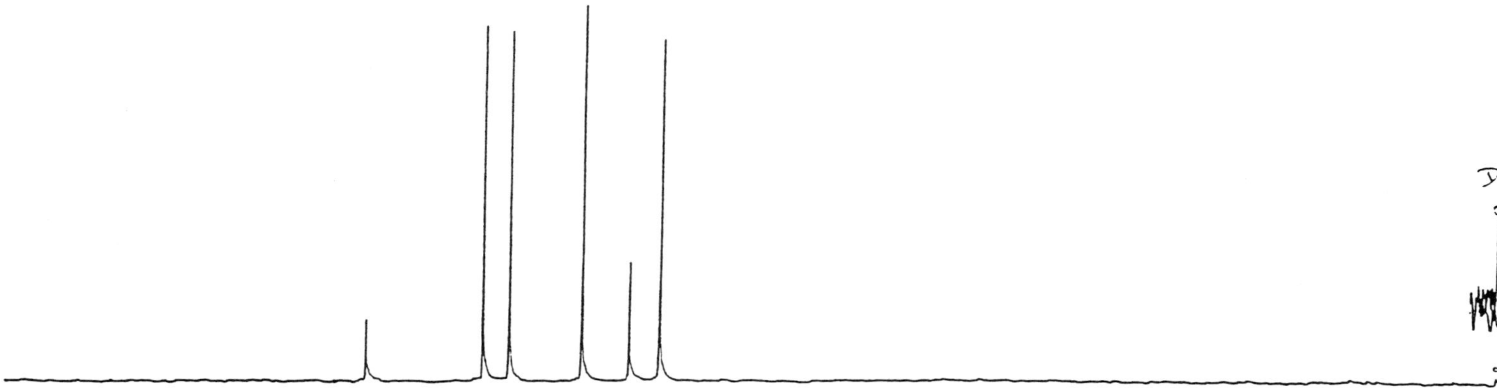

Fig. 4: The carbon^{13} nuclear magnetic resonance (^{13}C-NMR) of nicotinamide (carbon assignment see text).

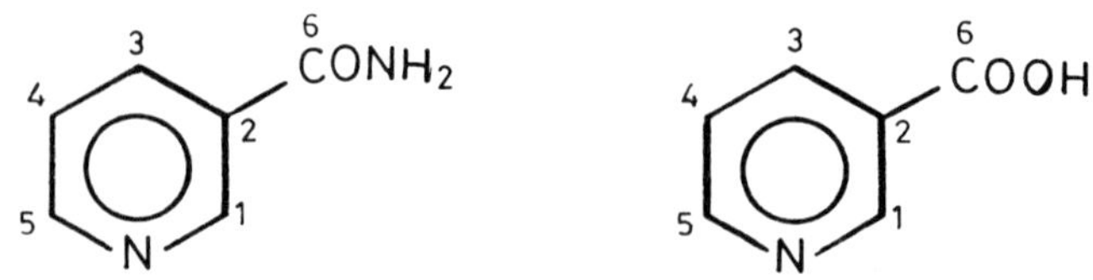

Chemical shift (δ, ppm) Acid[+]	Amide	Multiplicity	Carbon Assignment
	Nicotinamide		**Nicotinic acid**
150.3	150.2	doublet	C(1)
133.5	131.3	singlet	C(2)
138.6	138.7	doublet	C(3)
124.8	126.8	doublet	C(4)
151.5	154.3	doublet	C(5)
173.8	172.0	singlet	C(6)

[+]E. Lecte; *Bioorg. Chem.*: 6, 273 (1977); cf. M. Shamna and D.M. Hindenlang (Eds.) in: Carbon-13 NMR Slight Assignments of Amines and Alkaloids, Plenum Press, New York (1979) p. 24.

3.105 Mass Spectrum

The electron impact (EI) mass spectrum of nicotinamide at 70 eV recorded on Varian Mat 311 mass spectrometer and the methane derived chemical ionization (CI) mass spectrum obtained with Finnigan 4000 mass spectrometer are shown in Figures 5 & 6, respectively. The EI spectrum shows a molecular ion peak at m/e 122 (base peak). Other fragments are at m/e 106 and 78 which correspond to $M-NH_2$ and $M-CONH_2$ respectively. The CI spectrum, is simple, the base peak (MH) is the only

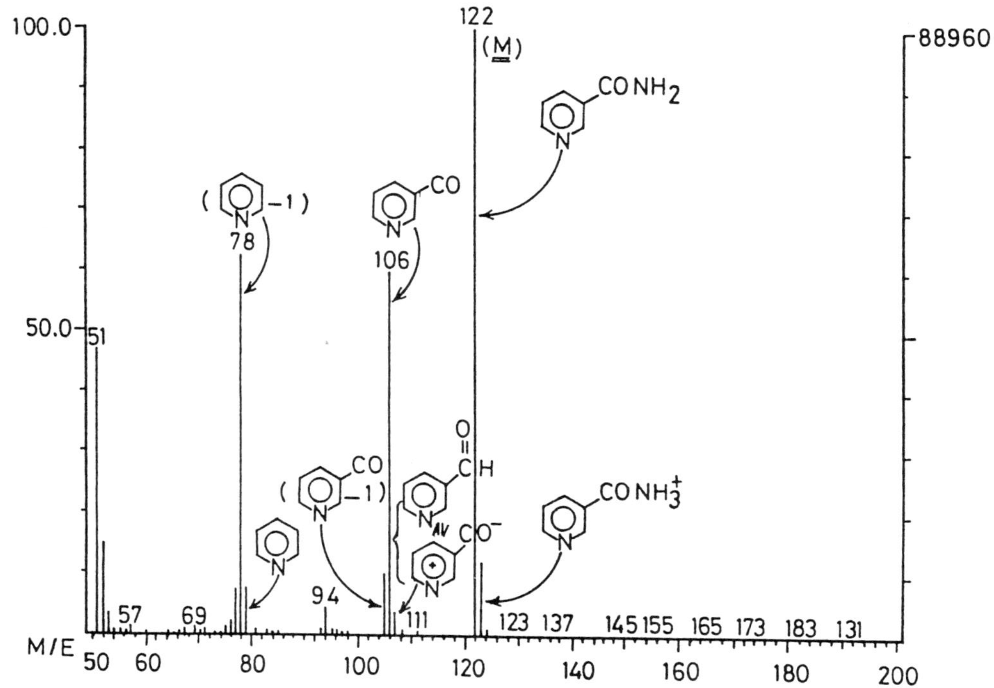

Fig. 5: The electron impact (EI) spectrum of nicotinamide.

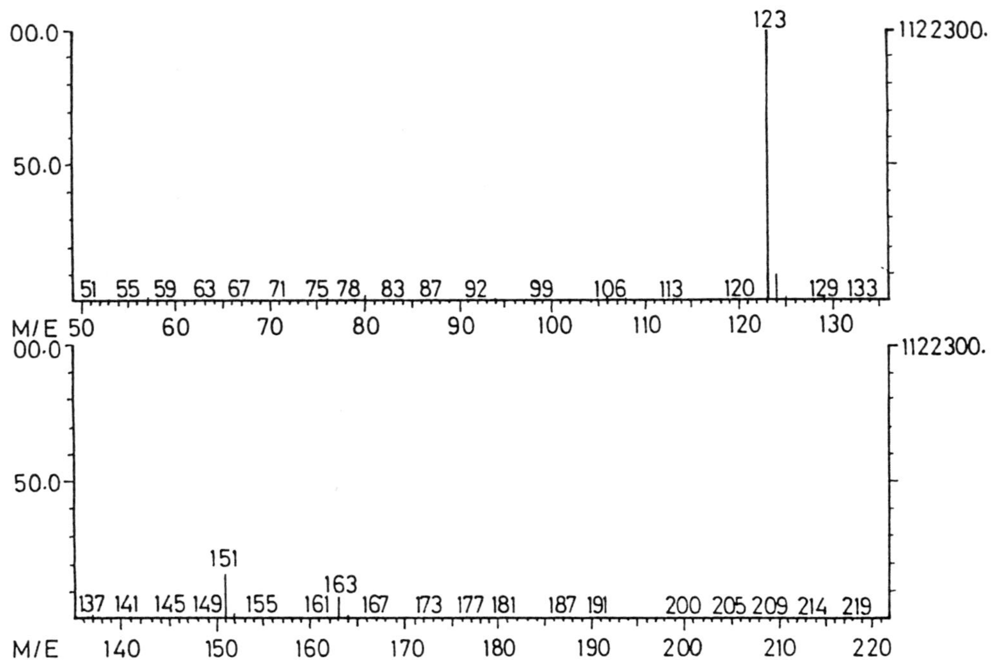

Fig. 6: The chemical ionization (CI) spectrum of nicotinamide.

SCHEME 1

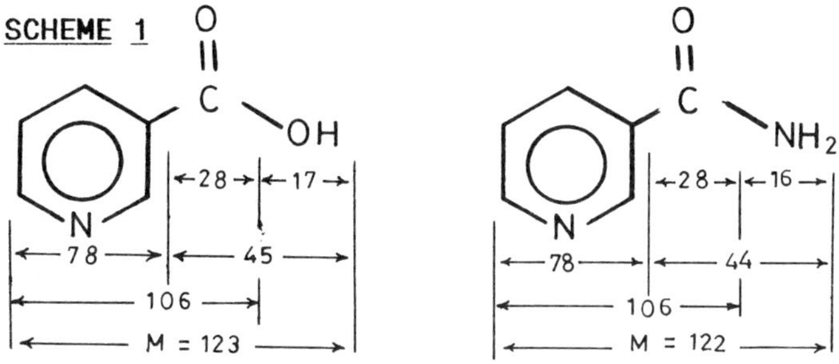

Nicotinic acid   Nicotinamide

peak and is at m/e 123. Scheme 1 illustrates the possible fragmentation patterns of nicotinamide and nicotinic acid.

3.11 Crystallographic Characteristics

3.111 X-Ray Powder Diffraction Patterns

The X-ray powder diffraction pattern of nicotinamide were determined using Philips fully automated X-ray diffraction spectrogoniometer equipped with PW1730/10 generator. Radiation was provided by a copper target (Cu anode 2000W, τ = 1.5480 A) high intensity X-ray tube operated at 40 KV and 35 MA. The monochromator was a curved single crystal one (PW1752/00). Divergance slit and the receiving slit were 1 and 0.1 nm, respectively. The scanning speed of the goniometer (PW1050/81) used as 0.02-2θ per second. The instrument was combined with Philips PM8210 printing recorder and connected with both analogue recorder and digital printer. The goniometer was aligned using silicon sample before use (14). The X-ray pattern is demonstrated in Fig. 7.

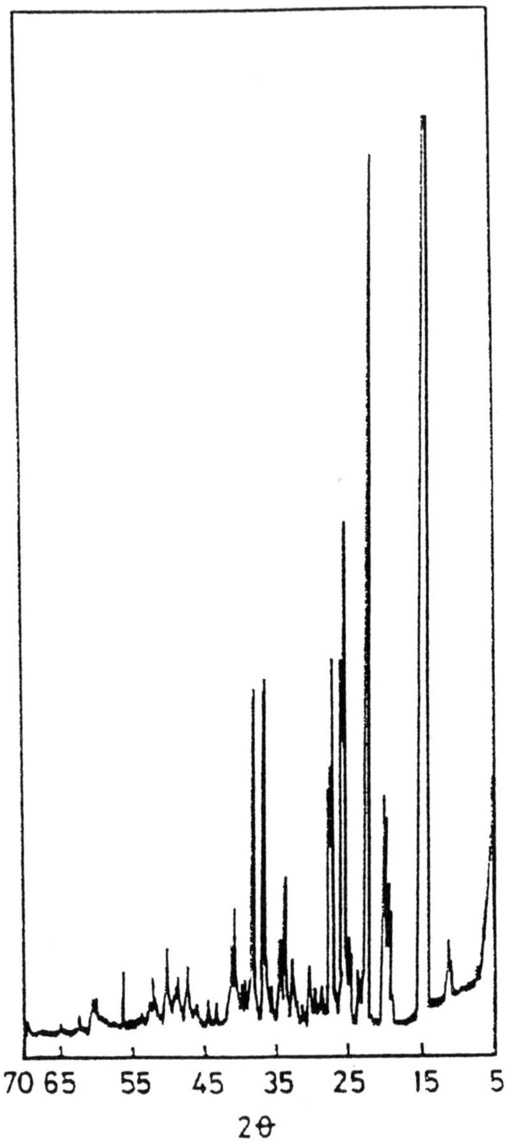

Fig. 7: The X-ray powder diffraction pattern of nicotinamide.

Table 1: The X-ray diffractional principal lines of nicotinamide

2θ	d(Å)	I/Io X100	2θ	d(Å)	I/Io X100
5.351	16.5148	10.62	32.478	2.7565	3.61
11.052	8.0057	3.33	33.586	2.6682	6.80
11.215	7.8892	4.10	34.122	2.6275	2.58
14.731	6.0133	100.00	34.459	2.6026	4.28
19.045	4.6598	6.35	35.716	2.5139	2.30
19.521	4.5473	8.50	36.947	2.4329	14.68
19.867	4.4689	9.85	38.625	2.3310	11.92
22.197	4.0048	35.04	39.097	2.3039	2.08
22.653	3.9252	3.32	39.618	2.2748	2.60
23.203	3.8334	2.77	40.134	2.2467	2.42
24.659	3.6101	4.10	40.955	2.2036	5.25
25.286	3.5221	8.22	41.372	2.1823	3.84
25.664	3.4711	21.02	47.554	1.9121	2.84
27.138	3.2857	15.50	48.740	1.8683	2.82
27.364	3.2592	9.70	49.188	1.8523	2.18
29.331	3.0450	2.25	50.346	1.8124	3.90
30.093	2.9695	3.20	52.250	1.7507	2.71

3.112 Differential Thermal Scanning (DSC)

The DSC-curves for nicotinamide and nicotinic acid were obtained separately on a DuPont TA-9900 Thermal Analyzer attached to a data processing unit. Figure 8 shows the DSC-curve of nicotinamide. The runnings are between 120-270°C at a heating rate of 10°C.min^{-1}. The heat of activation and the purity of the samples were determined using purity program. The ΔH-values are 22.7 and 25.6 KJ.mole^{-1} for nicotinamide and nicotinic acid, in order.

4. SYNTHESIS

Various synthetic pathways have been demonstrated for preparation of the nicotinic acid and/or its amide utilizing various starting materials and procedures. The prototype acid, i.e. nicotinic acid, can be obtained from quinoline *via* oxidation followed by decarboxylation. Nicotinamide can be prepared by conversion of nicotinic acid through the treatment either with urea (15-18) or ammonia (19-23). The nicotinic acid esters can be converted into the amide

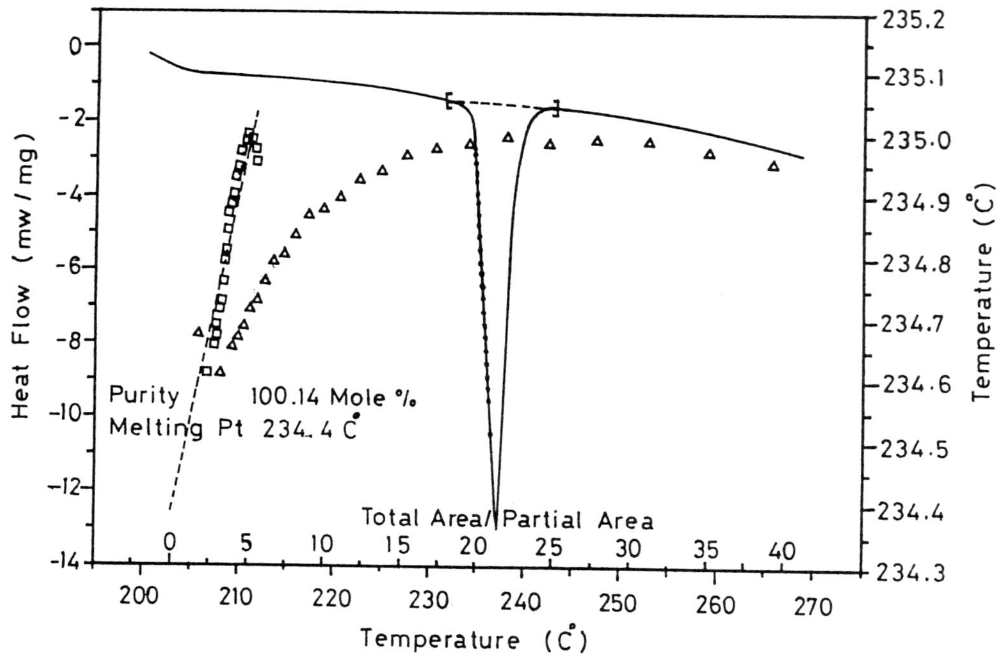

Fig. 8: The DSC-scanning of nicotinamide.

by reaction with ammonia (4, 24). Nicotinamide can also be synthesized from the ammonium nicotinate (25) or acetyl pyridine (26) by the treatment with ammonia. Reaction of asparagine with glutamic acid can yield nicotinamide (27). The vitamin amide can be obtained from nicotinonitrile by adopting various hydrolyzing reagents and conditions. Among the reported reagents are water (28), aqueous NaOH (29-31), alkaline H_2O_2 (32-37), ammonia (38-40), dilute acid (41 & 42), concentrated acids (43-46), and on moistened anion-exchanger; *viz* ionite AB-17, OH-form (47). Nicotinic acid and its amide can be synthetically prepared by vapor-phase oxidative ammonolysis of mixtures of 3-cyanopyridine and quinoline (48). Different pyridine derivatives including nicotinic acid and nicotinamide can be obtained starting from methane, ammonia and water by high frequency electric discharge followed by TLC-separation on silica gel G using AcOEt-MeOH-HCO_2H, 75:20:5, by volumes (49). Some other miscellaneous procedures have been described for the preparation of antipellagra acid/or amide (50-54).

5. PHARMACOKINETICS

5.1 Absorption and Distribution

Nicotinamide is readily absorbed from all portions of gastrointestinal tract (2). Nicotinic acid is also absorbed from the food contents in the gastrointestinal tract, which affects the bioavailability of orally administered nicotinamide (55). In untreated human subjects nicotinic acid plasma levels generally range from about 0.1-0.3 mg% (56). Continuous biochemical interactions at varying rates between nicotinamide, nicotinic acid, their complexes and their metabolites may predominate over the rate and completeness of gastrointestinal (GI) absorption of nicotinamide (57). The peak level is achieved in about 30 min following oral administration, and the blood half life ($T\frac{1}{2}$) is 5.3 hours and the drug disappeared from the liver with a half-life of 4.4 hours (4). Nicotinamide level has been reported to be highest in the kidney and in decreasing order, the liver, muscle, intestine, heart, spleen and lungs (58).

SCHEME 2

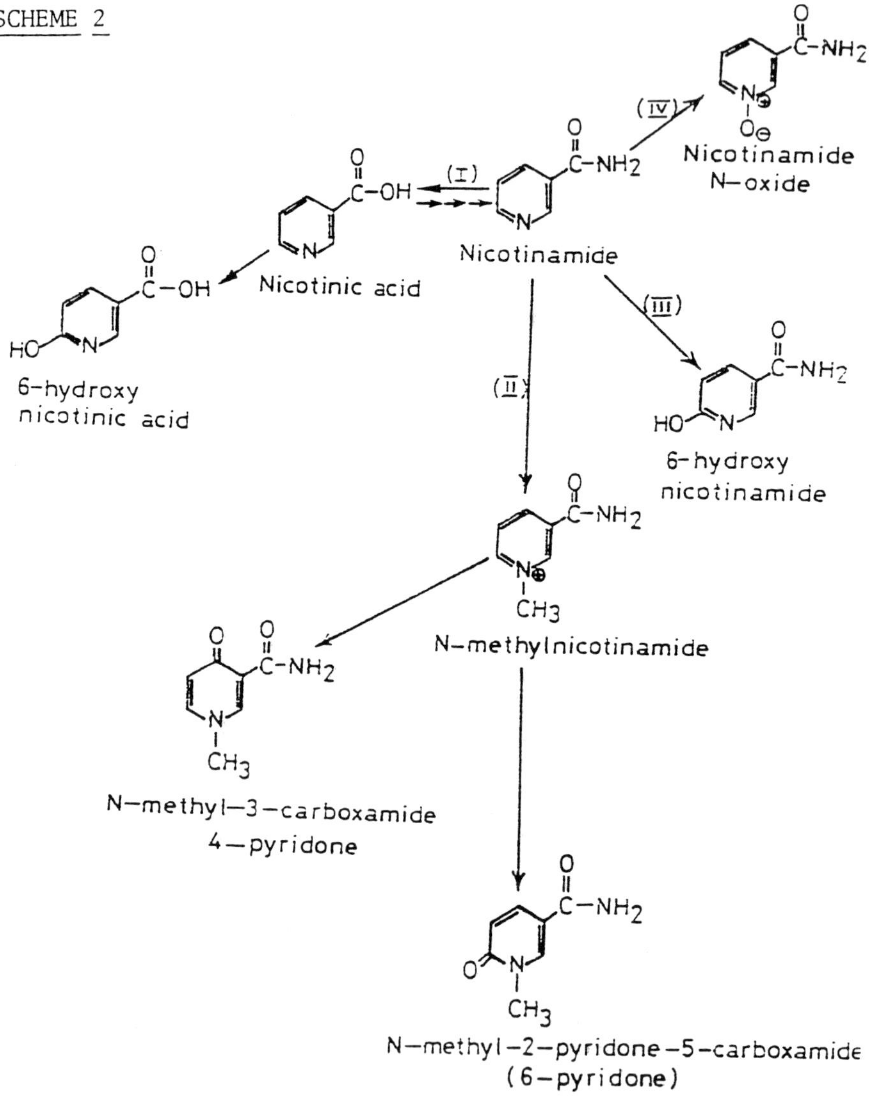

O
C–NH2
N⊕
O⊖
Nicotinamide
N–oxide
(IV)
O
C–NH2
(I)
O
C–OH
N
Nicotinamide
O
C–OH
N
Nicotinic acid
HO
N
6–hydroxy
nicotinic acid
(III)
O
C–NH2
HO
N
6–hydroxy
nicotinamide
(II)
O
C–NH2
N⊕
CH3
N–methylnicotinamide
O
O
C–NH2
N
CH3
N–methyl–3–carboxamide
4–pyridone
O
C–NH2
O
N
CH3
N–methyl–2–pyridone–5–carboxamide
(6–pyridone)

5.2 Metabolism

The metabolism of nicotinamide in human body takes place into four pathways as mentioned in Scheme 2. (I) Formation of nicotinic acid and 6-hydroxy nicotinic acid, (II) Formation of N-methyl nicotinamide and subsequent conversion of this compound to N-methyl-3-carboxamide 4-pyridone and N-methyl-2-pyridone-5-carboxamide 6-pyridone, (III) Formation of 6-hydroxy nicotinamide and (IV) Nicotinamide N-oxide. In animals most of the nicotinamide is metabolized to nicotinamide N-oxide, in addition to N-methylnicotinamide, 6-pyridone and 4-pyridone. The level of nicotinamide in liver and tissue cells are under feedback control and mainly depends on dietary intake and metabolic utilization (55). Nicotinamide levels in serum are buffered by the liver by being converted into storage form of NAD. In cell, the turn over of nicotinamide is proportional to the formation of poly-ADPR as shown in scheme 3. There is indeed deamination, since free 6-hydroxynicotinic acid and nicotinic acid are seen in the mouse and rat after injection of nicotinamide (59-63).

5.3 Excretion

When therapeutic doses are administered only small amounts of the unchanged nicotinamide appear in the urine (61 & 62). When extremely high doses of this vitamin is given, the unchanged vitamin represent the major component in the urine (64). Chaykin *et al* (65) have reported that a minimum of 98% of radioactivity was present in mouse urine after the administration of nicotinamide-7- ^{14}C, it was found in the form of the following metabolites: N^1-methylnico- tinamide, nicotinic acid, nicotinamide-N-oxide, N^1-methyl-4-pyridine-3-carboxamide, N^1-methyl-4-pyridine-3-carboxamide, N^1-methyl-2-pyridone-5-carboxamide and nicotinamide.

6. TOXICITY AND SIDE EFFECTS

Nicotinamide has a LD_{50} of 1.68 g/kg body weight for rats (66-68) and 1.75 g/kg for mice. The toxicity does not vary greatly with the route of administration. It has a stimulant action on the central nervous system and does not increase the work output of the perfused

SCHEME 3

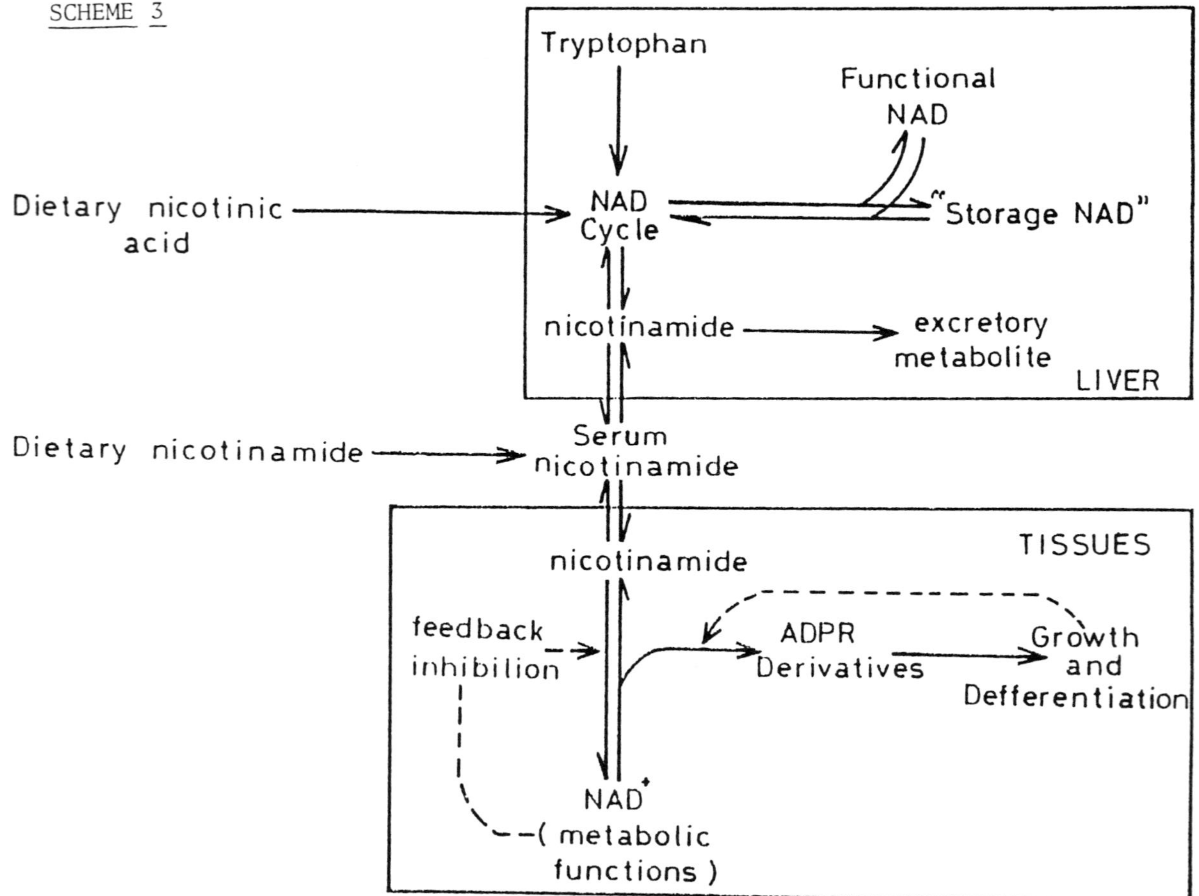
Tryptophan
Functional NAD
Dietary nicotinic acid
NAD Cycle
"Storage NAD"
nicotinamide
excretory metabolite
LIVER
Dietary nicotinamide
Serum nicotinamide
TISSUES
nicotinamide
feedback inhibition
ADPR Derivatives
Growth and Defferentiation
NAD+
(metabolic functions)

frog gastroinemius muscle (69). Lethal doses cause convulsions (67) and still higher doses (0.5-1.0 mg/kg) injected intravenously or intraperitoneally cause a fall of blood pressure and a rise in the heart rate and increase the ratio and amplitude of breathing and sometimes in pulse rate (70). The parentral administration of nicotinamide in large doses may cause furunculosis and other skin lesions, malaise, activation of peptic ulcer, jaundice and impairment of liver function (64). Nicotinamide is suspected of being teratogenic if taken in excess by pregnant women in the first trimester (1).

7. METHODS OF ANALYSIS

7.1 Qualitative Analysis (Identification)

7.11 Microchemical Tests

Gautier (71) investigated a precipitating spot test using mercuric chloride to differentiate nicotinamide from the prototype nicotinic acid. The precipitated nicotinic acid or its amide can be easily regained by microsublimation. Sandri (72) recommended bromothallic acid as a precipitating reagent to differentiate between several bases including nicotinamide. Lüdy-Tenger (73) described what he called "Bile 1.5" reagent for characterization of nicotinamide and nicotinic acid. Nicotinamide gives characteristic and reproducible reactions with different aldehydes, particularly vanillin and benzaldehyde. Comparative recommendation of bromothallic acid, bromoplatinic acid, bromoauric acid and sulfocyanides of some metal ions was discussed for microchemical identification of nicotinamide (74). König's reaction, involving the reaction of pyridine derivatives with cyanobromide and aniline, was studied extensively on the example of nicotinic acid and its amide by Ciusa and Barbiroli (75). High absorption data were shown in aqueous media in case of nicotinic acid and nicotinamide. Quantification of the König's reaction for determination of nicotinamide in urine (76) and pharmaceutical formulations (77-79) was evaluated too. Reaction of nitric acid with nicotinamide provides rapid precipitation of the nitrate salt (m.p. 223-5°C) characteristic for the drug (80). Zivanov *et al* (81) reported that on heating nicotinamide with methyl iodide (55-60°C, 40 min.) or ethyl iodide (80-85°C, 2 hrs) it gives the quaternary salts. The precipitated nicotinamide-MeI (yellow, m.p. 201-6°C) and nicotinamide-EtI (yellow-green, m.p. 146-8°C) can be used for gravimetric quantification of the drug (81). Van der Wegen (82) described several microchemical spot precipitating reagents for some basic drug substances including nicotinamide. Dobry (83) suggested Roussin's salt, [$KFe_4(NO)_7S_2$)] (iron nitrosyl sulphide), to form characteristic precipitations with many amines of biological interests including nicotinamide. Inorganic precipitation can be used for gravimetric determinations and as the precipitates are deeply colored, such amines can be colorimetrically determined; as little as 106 µg of nicotinamide can be determined by the proposed method (83).

7.12 Elemental Composition

	C (%)	H (%)	N (%)	O (%)
Nicotinic acid (C6H5NO2, 123.11):	58.53	4.09	11.38	25.99
Nicotinamide (C6H6N2O, 122.13):	59.00	4.95	22.94	13.10

7.13 Pharmacopoeial Specifications

The USP XXI (10) specifies the comparison of the IR- & UV-spectra of nicotinic acid and nicotinamide samples with the USP-reference standard of the acid or its amide, in addition to the identity test with 2,4-dinitrochlorobenzene and alc. KOH (deep red to wine red coloration) and test with copper sulphate on the neutral solution (blue precipitation) in case of nicotinic acid.

The BP-1988 (84) also recommends the concordance of the sample of nicotinic acid or its amide with the corresponding IR- and UV-spectra of the BP-reference drug substance. Color reaction with cyanogen bromide and aniline, i.e. König's reaction, (gives yellow coloration) is also described for both drug substances, while the evolution of ammonia vapours on hot reaction with sodium hydroxide is mentioned for the amide only.

7.2 Quantitative Analysis (Determination)

7.21 Volumetry

7.211 Determination in bulk material

Krewson (85) described an acid-base residual titrimetric method by involving acid splitting of the amide bonding then followed by alkalinization and distilling the equivalent NH_3 into a known excess of standard acid. The unreacted acid could be acidimetrically determined using methyl red as a suitable indicator. Sreeram and Naidu (86) investigated the non-aqueous titration of nicotinamide and nicotinic acid with chlorosulphuric acid in dioxan using crystal violet and photometric titration at 610 nm. Bianchi (87) described a similar non-aqueous photometric titration utilizing 0.2% malachite green or nile blue in acetic acid as indicators.

Titrimetric quantification of nicotinamide in non-aqueous medium has been investigated by using $HClO_4$ as the standard and Fettblau as the suitable indicator (88). Helmstaedter (89) suggested a titrimetric procedure for quantitative determination of nicotinamide *via* its precipitation from aqueous solutions by excess $HgCl_2$. By dissolving the precipitated mercury salt of the drug in HCl, the titration was then carried out indirectly with EDTA after rendering ammonical and using standard $ZnSO_4$ to determine the residual amount of the chelating standard. To quantify nicotinic acid by following the suggested procedure, $CuSO_4$ was advised for precipitation instead of $HgCl_2$ in case of the amide. Neutralimetry with 0.1 M NaOH was described by the BP-1988 (84) for nicotinic acid in bulk material (using phenolphthalein) and in tablets (using phenol red).

7.212 Determination in Pharmaceutical Preparations

All the volumetric methods described for the quantification of nicotinamide in dosage formulations are carried out in non-aqueous media utilizing standard $HClO_4$ but different non-aqueous solvents and indicators. Table 2 summarizes the solvents (for

Table 2: Non-aqueous determination of nicotinamide

Solvent				
Standard	Drug	Indicator (solvent)	Dosage form	Reference
Glacial acetic acid		Methyl violet (chlorobenzene)	Tablets & injections	90
Dioxane $CHCl_3$ (hot)		Methyl red	Dragées	91
Glacial acetic acid		Crystal violet (acetic acid)	Injections	92
" " "		" " " "	Tablets	84

standard and drug) and the indicators in each non-aqueous procedure.

The use of 1,3-dibromo-5,5-dimethylhydantoin for direct determination of nicotinamide in some dosage formulations by using methyl red, methyl orange, indigo (C.I. Vat Blue 1) or amaranth (C.I. Acid Red 27) was compared with indirect iodimetry and those of N-bromosuccinimide (NBS) as well as an official methods (93).

7.22 Electrochemical Methods

7.221 Oscillometry

Vajgand and Pastor (94) recommended an oscillometric titration procedure for quantification of some organic amines including nicotinamide in glacial acetic acid. Reproducible results could be obtained in case of the HCl-salt of the drug.

7.222 Potentiometry

Nicotinamide can be potentiometrically titrated using $HClO_4$ solution in acetonitrile containing 1-30% water as standard and glass electrode, i.e. indicator electrode, with a reference electrode containing $Ag^+/AgCl(s)/Cl^-$. The equilibrium potential of the cell reached in 10-15 min. showed higher accuracy with low pH-deviation than when the time was in 1 min. (95). Grzybowska (96) investigated a potentiometric method for determination of nicotinamide after conversion to the prototype nicotinic acid by refluxing with NaOH then acidifying with HCl. The liberated acid is dissolved in acetone and titrated against standard ethanolic KOH-solution. Potentiometric acidimetry, using standard HCl, was suggested by Subert *et al* (97) for determination of nicotinamide in medium of $MgCl_2$ or NaCl. Tungsten rod could be used as indicator electrode in the potentiometric titration of small amounts of nicotinamide in a mixture containing acetic acid and acetic anhydride (1 + 6, v/v) (98). Membrane graphite electrodes, pretreated with acidified $KMnO_4$ to attain maximal improvement, was described for quantification of nicotinamide in acetic acid medium using standard $HClO_4$ as the titrant (99). The utility of the half-neutralization point in the potentiometry was investigated for the quantification of nicotinamide in a medium of constant ionic strength, 1M aqueous solution in NaCl, where the ionic strength of the titrant was to be 0.01M - 0.5M-HCl (100).

Abou-Ouf *et al* (93) showed the applicability of using the halogenating agent 1,3-dibromo-5,5-dimethylhydantoin in direct potentiometric determination of nicotinamide in some pharmaceutical preparations.

7.223 Polarography

Shikata and Tachi (101) first investigated the behaviour of nicotinic acid at the dropping mercury electrode and obtained a reduction step with a $E_{\frac{1}{2}}$ value of - 1.6 V *vs* saturated calomel electrode (S.C.E.) from a 0.1 M sodium bicarbonate solution of pH 8.4. More positive potentials were observed as the nicotinic acid concentration was increased. Tompkins and Schmidt (102) investigated the behaviour of nicotinic acid from several base electrolytes, but these authors recommended a borate buffer of pH 8 to 9 also; however, the diffusion current was not proportional to the nicotinic acid concentration unless the buffer concentration was at least 0.6 M. From a 0.6 M sodium borate solution the nicotinic acid step has a half-wave potential of - 1.66 V *vs* S.C.E.

Nicotinamide was reported by Wenig and Kopecky (103) to reduce at the dropping mercury electrode from a 0.1 M sodium hydroxide solution producing a step with an $E_{\frac{1}{2}}$ value of - 1.82 V *vs* S.C.E. Later Tompkins and Schmidt (104) observed a step with an $E_{\frac{1}{2}}$ of - 1.6 V *vs* the mercury-pool anode from a 0.1 M tetramethylammonium bromide base solution and, in general, they obtained well-defined steps for nicotinamide from solutions with pH values in the region of 8, whereas at lower pH values they encountered considerable interference from the hydrogen step. The $E_{\frac{1}{2}}$ values in V *vs* S.C.E. reported for the nicotinamide steps from various base solutions are as follows: 0.1 M sodium acetate (pH 7.5) - 1.65 V; 0.1 M sodium citrate (pH 7.9) - 1.67 V; 0.1 M sodium bicarbonate (pH 9.2) - 1.40 V; 0.1 M sodium borate (pH 8.7) - 1.56 V. These half-wave potential values clearly vary with the pH and also slightly with the nature of the buffer constituents. In all cases, however, the step-height proved to be proportional to the nicotinamide concentration and independent of the buffer capacity of the solution.

Knobloch (105) observed a considerably larger step following the nicotinamide step from neutral and acidic buffer solutions, due to the catalytic evolution of hydrogen at the dropping mercury electrode. Using a 5 X 10^{-5} M solution of the amide the catalyzed step with a very pronounced maximum first appeared at about pH 7.24 and the height of this step increased as the pH of the final solution was decreased. At pH 4.78 the maximum on the step disappeared completely, resulting in a well-defined catalytic step, but at lower pH values it coalesced with the step owing to the normal reduction of hydrogen and it consequently proved impossible to measure the height of the catalytic step under these circumstances (106).

Cernatescu *et al* (107) described the polarographic determination of nicotinamide ($E\frac{1}{2}$ = 1.6 V) alone and in the presence of nicotinic acid ($E\frac{1}{2}$ = ~ 1.2 V) in 0.1M NaCl or KCl solutions as the supporting electrolytes. In alkaline solutions the wave due to the acid will disappear allowing the determination of the amide in the mixture. Oscillographic polarography with an electronic polaroscope and a streaming- or dropping-Hg electrode differentiate between nicotinamide, nicotinic acid, picolinic, dipicolinic, dinicotinic, isocinchomeronic acids, nicotine, trigonelline and isoniazide, in individual form or in their mixtures (108). Lindquist and Farroha (109) investigated the applicability of differential pulse (D.P.P.) polarography for determination of some vitamins including nicotinamide and nicotinic acid with low detection limits (0.01-1 p.p.m.; i.e. $\mu g.L^{-1}$). There are several other polarographic methods for identification and quantification of nicotinamide and nicotinic acid, the following table 3 collects such procedures.

Table 3: Polarographic characterization and determination of nicotinamide and nicotinic acid

Analyte form	Electrode	(pH) Electrolyte	Scanning range ($E_{\frac{1}{2}}$, V)	Reference cell	Conc. range	Reference
Raw materials						
Nicotinamide Nicotinic acid	Dropping-Hg	0.1M NaOH containing geltin (> 7-10)	-			110
Nicotinamide Nicotinic acid		Buffer (12, amide and 8.7, acid)	- -			111
Nicotinamide Nicotinic acid		0.1M KOH 0.1M HCl	(- 1.8) (- 1.1)			112
nicotinamide		0.1M LiOH	(- 1.5)	$Ag/AgCl(s)/Cl^-$	0.1-14 mg%	113
Pharmaceutical preparations						
Nicotinamide	d.c. polarography with H-cell.	0.1M NaOH with 0.002% gelatin	(- 1.5 to - 2.1)		- 10 mg%	114
Nicotinamide[+]	Cathode-ray polarography	0.2M KOH & 0.05M KCl (5.7-6)	(- 1.65)		- 10 mg%	115

Continued /...

Table 3 Contined ...

Nicotinamide+	dropping-Hg	0.1M NaOH	(-1.5 to - 1.9)		0.06-12 mg%	116
Nicotinamide+	dropping-Hg with H-cell	KCl-solution	(-1.4 to 1.9)	S.C.E.		117
Nicotinamide+		alk. Na-citrate (12)	- 0.05 to - 1.5)	S.C.E.		118
Nicotinamide+	diff. puls polaro-graphy (DPP)	Phosphate buffer pH 6.8 and 2.5M NaOH (12-13)	(- 0.8 to - 1.5)	$Ag/AgCl(s)/Cl^-$ (sat)		119
Nicotinamide+		3% KOH, Kolthoff buffer pH 8.6	(- 1.80 - 1.55)			120
Nicotic acid		Kolthoff buffer pH 8.6	(- 1.55)			

+in multivitamin preparations.

7.23 Spectrophotometry

7.231 Colorimetry, UV-Spectrophotometry and Spectrofluorimetry

7.2311 Colorimetric Methods

Different chromogens have been investigated for colorimetric quantification of nicotinamide and/or nicotinic acid. Although the color development in most of the quantitative procedures can be recommended for identification of the drug substance, all the color spot tests are not suitable in most cases for quantification. The most adopted color reaction for qualitative and quantitative analyses of nicotinamide and nicotinic acid in pure forms, in dosage formulations and in biological fluids is the König's reaction. The color formation in the König's reaction depends on oxidation of the pyridine-based drugs including nicotinamide and its prototype acid by cyanogen bromide (CNBr) to the corresponding aldehyde followed by coupling with a base like aniline to give the colored products which can be estimated colorimetrically. There are different modifications of König's reaction by changing the oxidizing agent and/or the coupling base-species. Abdoh and Täufel (121) were the first, who applied König's color reaction with CNBr and aniline for the quantitative determination of nicotinamide and nicotinic acid in pure form and in plant or animal tissues extracts containing either of the two drugs. Since nicotinamide gives less color intensity than nicotinic acid if they where treated according to the König's principle utilizing CNBr and aniline, the amide should be first hydrolyzed by heating with 2.5 M HCl for 1.5 hrs at 90-95°C before developing the color. Such modification enables accurate quantification of the drug in urine, tissues and blood (122).

The colored hydrazides formed by adding benzidine to the oxidation products of nicotinamide or nicotinic acid with CNBr were studied for quantitation of these drugs (123). Spectrophotometric studies on the action of CNBr on the pyridine nucleus in nicotinamide and nicotinic acid in the presence of quinoline showed the presence of typical functional grouping (124). Quantitative modification of König's colorimetric reaction has been investigated by coupling barbituric acid for determination of nicotinamide in the presence of nicotinic acid (125). Ciusa and Barbiroli (75) compared the extinction coefficients (ϵ) at 460-465 nm, of the König's reaction products of nicotinamide and nicotinic acid in aqueous and ethanolic media. The ϵ-values obtained in ethanol are depressed with respect to those in water for both drug substances. Mesnard (126) discussed the applicability of coupling 2-thiobarbituric acid to the oxidation products of nicotinamide and nicotinic acid with CNBr for the drug assay. The König's reaction could be applied for determination of nicotinamide in urine after chromatographic separation from nicotinic acid and nicotinuric acid by ion-exchanging utilizing Amberlite IRA-400 (OH^-form) and elution with acetone-H_2O (2:1, v/v) (127). Automation of König's color reaction has been undertaken for vitamin analysis of nicotinamide at a flow-cell measuring at 420 nm (128). Nicotinamide contents in tablets could be accurately (~ 102%) determined by following the König's reaction (75). A similar method to that of Jeffus and Kenner (129) was applied for direct determination of nicotinamide in multivitamin preparations (130). Modified König's reaction was carried out by using an excess of yeast nicotinamide deaminase to yield the nicotinic acid giving rise to relatively consistent quantitative results (131). Dessouky *et al* (132) suggested the use of 3-methyl-1-phenylpyrazolin-5-one as coupling reagent for the quantitative determination of nicotinamide and nicotinic acid in tablets. Nicotinamide and nicotinic acid have been quantified in pharmaceutical formulations by automating the König's reaction using barbituric acid and sulphanilic acid as the coupling reagents (133). The manual AOAC-colorimetric method No. 43.044, involving König's reaction with CNBr and sulphanilic acid, was automated for determination of nicotinamide and nicotinic acid in cereal products (134). Farsam and Mahmoudian (135) discussed the mechanistic possibility of the König's reaction using sodium 1-phenyl-2,3-dimethyl-5-pyrazolone-4-methylaminomethansulphonate (dipyrone) in case of nicotinamide. The possibility of coupling TLC-separation with spectrodensitometric determination of the color formed according to the König's reaction gives good analytical way for quantitative separation and determination of nicotinamide in some multivitamin preparations (77,78). Kringstad and Naess (136) advised the phosphate buffer, pH 6.1, for best stabilization of the color maximally formed after 7 min. for colorimetric determination of nicotinamide and nicotinic acid according to the König's procedure. There are some other colorimetric methods, rather than the König's reaction, adopted for quantification of nicotinamide and nicotinic acid. Khattab (137) recommended the bromination of nicotinamide by using aqueous bromine, where the formed N-bromoderivative was determined by reaction with starch-CdI_2 at 600 nm. Pauer (138) determined the cyanide contents in nicotinamide and nicotinic acid by separation following microdiffusion and then determination with 2% tartaric acid, 0.5% chloramine-T solution followed by pyridine-pyrazolone reagent at 620 nm. Dimethylglyoxine (DMG)-Fe^{2+} chelate was recommended as a colorimetric reagent for quantification of nicotinamide (139). Reaction of nicotinamide with 1-chloro-2,4-dinitrobenzene yields pyridinium derivative which yields on alkaline decomposition a measurable yellowish red color of glutaconaldehyde (140). Nudelman and Nudelman (141) recommended a nearly identical method to that of Karrer and Keller (140) by utilizing 1-chloro,2,4,-dinitrobenzene as the chromogen in the presence of alcoholic solution of NaOH without interferences that may be due to nicotinic acid and some other vitamins. Nicotinamide is the principal metabolite of nicotinic acid in urine, which could be colorimetrically assayed after conversion to the amino-compound by heating with hypobromite, followed by diazotization and coupling with N-(1-naphthyl)ethylenediamine.2HCl to give the azodye (142). The method could be successfully applied to determine the urinary excretion of nicotinamide in three adult volunteers. Walksmundzki and Romanowski (143) showed the specificity of epichlorohydrin as an analytical reagent for detection and determination of nicotinamide in tablets. Indemans and Rademakers (144) described a colorimetric procedure by treating with hypobromite-phenol, sulphamat and N-(1-naphthyl)-ethylenediamine.2HCl to yield the diazonium salt measurable at 500 nm. Domnas (145) recommended a mixture of hypochlorite-phenol reagent for determination of nicotinamide in raw materials and in uptake studies of the drug substance during metabolism in yeast. Nicotinamide, R, could be colorimetrically

Table 4: Colorimetric Determination of nicotinamide and nicotinic acid in pure form, dosage forms and in biological fluids

Substance	Specimen	Chromogen (Reagent)	Special treatment	max (nm)	Sensitivity range	Reference
Nicotinamide Nicotinic acid	pure form, plant &	CNBr + aniline	16°C/23 min. using PO_4-buffer pH 6.1-6.3			121
Nicotinamide Nicotinic acid	bulk form	CNBr + benzidine				123
Nicotinamide	raw materials	CNBr + barbituric acid	20-22°C/30 min. using PO_4-buffer (M/15, pH = 7.2)	513	- 3 $\mu g.ml^{-1}$	125
Nicotinamide Nicotinic acid	raw materials	CNBR + 2-thiobarbituric acid				126
Nicotinamide	biological fluids (urine)	CNBr + aniline	20°C/20 min., pH 6.5-7	460	Recovery from urine - 91 ± 8.9%	127
Nicotinamide	vitamins mixtures	CNBr + N,N-diethyl-p-phenylenediamine	ambient, 2% NaOH and ammonia buffer	420		128

Continued /...

Table 4 Continued ...

Nicotinamide	tablets	CNBr + barbituric acid	room temperature 30 min, 0.3% KH_2PO_4	550	Recovery = ~ 102%	129
Nicotinamide	multivitamin preparation	CNBr + satd. barbituric acid in 3% KH_2PO_4	room temp./25-30 min, 3% KH_2PO_4	550	~ 5 $\mu g.ml^{-1}$; recovery 100.2%	130
Nicotinamide nicotinic acid	tablets & capsules	CNBr + barbituric acid & sulphanilic acid	single discrete-sample automated analysis		~ 1 $mg.ml^{-1}$ ~ 5 $\mu g.ml^{-1}$	133
Nicotinamide	raw materials	CNBr + sodium 1-phenyl-2,3-dimethyl-5-pyrazolone-4-methylaminomethanesulphonate	ambient/30 min.	390	8-33 $\mu g.ml^{-1}$	135
Nicotinamide	raw materials	Br_2/starch-CdI_2	1M NaOH – H_2SO_4/HCO_2H	600	80-400 µg	137
Nicotinamide Nicotinic acid	raw materials	N – HCN + 2% tartaric acid + 0.5% chloramine-T + pyridine pyrazolone reagent	microdiffusion (20 hr), reaction 30 min.	620		138
Nicotinamide	raw materials	Dimethylglyoxime (DMG)-Fe^{2+}	pH 8.4-11, color stability > 30 min.	516	10^{-3}-10^{-4}M	139

Continued /...

Table 4 Continued ...

Nicotinamide	raw materials	1-chloro-2,4-dinitrobenzene + 20% KOH	90°C/1 hr.-ethanol		5-10 $\mu g.ml^{-1}$	140
Nicotinamide	tablets, capsules, syrups & elexir	1-chloro-2,3-dinitrobenzene + 1M NaOH	80°C/2 hrs.-ethanol	504	- 4 $\mu g.ml^{-1}$	141
Nicotinamide	Urine	– NH_2, diazotization & coupling with N-(1-naphthyl)-ethylenediamine			1-50 μg	142
Nicotinamide	tablets & solutions	epichlorohydrin	60°C/5 hrs.		˜ 10 $pg.ml^{-1}$	143
Nicotinamide	multivitamin preparations	NaOBr, C_6H_5OH, diazotization & coupling with N-(1-naphthyl)-ethylenediamine	heating	500	- 10 $\mu g.ml^{-1}$	144
Nicotinamide	raw materials & uptake studies	NaOCl - C_6H_5OH		625	10-50 $pg.ml^{-1}$	145
Nicotinamide	raw materials	cobalt thiocyanate complex, $H_2[Co(SCN)_4]$	room temp./2 hrs., $CHCl_3$, or 24 hrs. nitroso-R salt solution	625 330	8-40 $\mu g.ml^{-1}$	140

Contined /....

Table 4 Continued ...

Nicotinamide	pharmaceutical preparations	H^+-hydrolysis (reflux, 1 hr.) + Na 1,2-naphthaquinone-4-sulphonate + $Na_2S_2O_3$.	80°C/1 hr, acetone	540	- 15 $\mu g.ml^{-1}$	147
Nicotinamide	pharmaceutical preparations	4-dimethylaminobenzaldehyde (PDAB)		445	80-200 $\mu g.ml^{-1}$	148

quantified by using cobalt thiocyanate complex, $H_2[Co(SCN)_4]$ to give $(RH)_2-[Co(SCN)_4]$ which exerts the color (146). Acid hydrolysis of nicotinamide followed by reaction with sodium 1,2-naphtha-quinone-4-sulphonate at 80°C for 1 hr. gives purple coloration on adding thiosulphate (147). The absorbance of the complex formed between 3-aminopyridine, the product of Hoffmann's reaction between nicotinamide and 4-dimethylaminobenzaldehyde was used for quantitative determination of the drug in pharmaceutical preparations (148). Table 4 collects and compares in a summarized form the different colorimetric procedures for identification and quantification of the mentioned drug substances.

Marshal and Kenner (149) compared the direct spectrophotometry at 262 nm of acid extracts of nicotinamide separated from capsules, syrups and elixirs, with other two colorimetric methods by using either bromothymol blue, measured at 430 nm, or with cyanogen bromide and barbituric acid solutions, measured at 550 nm. Table 4 collects the various colorimetric methods described for determination of nicotinamide and/or nicotinic acid in dosage formulations and in biological specimens.

7.2312 <u>UV-Spectrophotometry</u>

The USP XXI (10) recommends the UV-measurement at 262 nm for nicotinamide and nicotinic acid in bulk forms or in injections and tablets, while the BP 1988 (84) specifies the same wavelength for the quantification of only nicotinamide in its tablets. Spectrophotometric determination of nicotinamide in admixtures with thiamine, riboflavin and pyrodixine, is undertaken by measuring A at 267 nm for solutions at pH 2 & 7 (150). Estimation of nicotinamide in admixtures with thiamine and pyridoxine by direct measurement at 247 nm, 260 nm and 290 nm was only possible after removal of any riboflavin by the use of ion-exchange resin (Decolorite anion-exchanger) (151 & 152), presence of panthenol does not interfere because of its null absorptions at the wavelengths mentioned above (153). Indirect spectrophotometric analysis of nicotinamide and ascorbic acid in ethanol containing cysteine.HCl to protect the acid could be carried out at 263 nm and 245 nm, in order (154). Similar method was described for quantification of nicotinamide at 262 nm in the presence of thiamine and pyridoxine (155). Nicotinamide in a quaternary mixture with thiamine-HCl, riboflavin phosphate and pyridoxine-HCl and chloramphenicol could be investigated by adopting indirect spectrophotometry (156). Resolved nicotinic acid and its amide on silica gel G, using acetic acid-acetone-methanol-benzene (5:5:20:70), could be determined in the presence of thiamine.HCl, pyridoxine.HCl and p-aminobenzoic acid by means of UV-reflectance spectrometry (157). Nicotinic acid could be quantified at 262 nm after its separation from dosage formulations on activated polyamide powder in a microcolumn (158). Among other 30 different compounds with pyridine chromophores, the effect of substitution and solvent on the UV-spectrophotometric behaviour of nicotinamide was studied for drug control (159). Direct UV-measurement of nicotinamide at 262 nm could be undertaken in admixtures with inosine and uridine 5-phosphate after TLC-separation on silica gel GF_{254} using butanol-acetic acid-5% ammonia solution (7:2:2) (160). Partition chromatography, between a diatomaceous earth-water column and chloroform, was a preliminary step for separation of nicotinamide from pharmaceutical preparation before its UV-measurement at 262 nm (149). Paczek and Wardynska (161) also adopted the chromatographic separation of nicotinamide from some other water-soluble vitamins in multivitamin drageés before its quantification *via* direct UV-measurement at 262 nm. Direct UV-measurement at 246 nm & 262 nm at pH = 2 has been described for determination of nicotinamide with some other water-soluble vitamins (162).

7.2313 Spectrofluorimetry

Scudi (163) described the addition of an alkali to the solution of nicotinamide and cyanogen bromide to quantify the drug spectrofluorimetrically. A fluorimetric method for quantification of 1-methylnicotinamide and nicotinamide in serum was recommended by Clark *et al* (164) by developing the fluorescence in 9% formic acid; EX: 370 nm, EM: 450 nm. Modified Hankder spectrofluorimetric procedure has been recommended by Mestres (165) to determine nicotinamide; the fluorescence formation takes place by KCN-buffered solution and chloramine-T, EM: 462 nm. Nicotinamide can be determined in acid extracts of homogenized food products by measuring the fluorescence at 426 nm (166). Mohran (167) described a spectrofluorimetric procedure for estimating nicotinamide intake, based on the determination of plasma levels of one to its metabolites, namely N'-methyl-4- pyridone-3-carboxamide, which causes fluorescence on UV-irradiation. Nikolic *et al* (168) demonstrated fluorescence reactions for determination of a variety of natural and synthetic drugs including nicotinamide. *In situ* spectrofluorimetry after TLC-separation was adopted by Frei *et al* (169) for quantification of traces of nicotinic acid and its amide. Spectrofluorimetric quantification of nicotinamide and its methyl derivative in biological fluids was recommended by Williams (170). Spectrofluorimetric measurement, EX: 436 nm, of the Schiff's base formed by reacting nicotinic acid or its amide with cyanogen bromide, acetone and 4-aminobenzoic acid has been investigated by using 3,5-diacetyl-1,4-dihydrolutidine as comparing standard for quantification of each drug substance (171).

7.24 Chromatography

7.241 Paper Chromatography (PC)

PC-technique using different paper types in various solvent systems has been investigated for resolution, identification or determination of nicotinic acid and its amide. The separation of the antipellagra drugs can be achieved in one direction, two-dimensions, or circular diffusion utilizing different visualizing regents (172-193). Table 5 summarizes collectively the different PC-possibilities for identification and quantification of nicotinic acid and/or nicotinamide.

7.242 Thin Layer Chromatography (TLC)

Different adsorbents and developing solvents have been described for fractionation of nicotinic acid and/or nicotinamide from admixtures and in multivitamin preparations. Color development or UV-visualization are among the various methods recommended for location (194-213). The following table 6 collects the various

Table 5: Paper chromatography (PC) of nicotinamide and/or nicotinic acid

Drug substance	Dimensions one (1), two (2)	Developing solvent (v/v-ratios)	Visualization, reagent	Rf-value (color)	Reference
Nicotinic acid Nicotinamide n-methylnicotinamide	(1)	BuOH contg. urea	Dragendorff's reagent	0.29 (red) 0.68 (") 0.08 (")	176
Nicotinic acid Nicotinaide	(1)	PrOH-0.1M HCl (2:1)	spot coloration	--	177
Nicotinamide tabac	(1, ring)	BuOH-AcOH-H_2 (6:1:3)	0.25% benzidine in 50% EtOH	0.67	178
Nicotinic acid Nicotinamide	(1)	--	a- CNBr vapors (2 min.)/10% NaOH (100°C) b- 2% aq. o-$C_6H_4(NH_2)$/CNBr vapors c- p-$MeNHC_6H_4CHO$ in gl. AcOH/(6%)	a: dark blue b: carmine yellow to carmine c: yellow yellow	179
Nicotinic acid Nicotinamide	(1, circular)	BuOH-H_2O	growth zones of *L. plantarum* on agar incubated at 30-37°C/4.5 hrs.	0.27 0.67	180
Nicotinamide (in multivitamins preparation)	(1, descending)	80% EtOH	--	0.68-0.72	181

Continued /...

Table 5 Continued ...

Nicotinic acid Nicotinamide (tabac)	(1)	Na citrate buffer, pH 6.3	p-aminobenzoic acid/CNBr vapors	-- --	182
Nicotinic acid Nicotinamide	(1)	Partridge's mixture ($BuOH-AcOH-H_2O$) (4:1:5)	Diazotization (0.5% sulphanilic acid in 1% HCl of 0.5% aq. $NaNO_2$) and Ehrlich's reagent (2% p-aminobenzaldehyde in 5% HCl) or 1% meth. 4-chloro-1,3-dinitrobenzoic acid/3M NaOH.	-- (orange) -- (wine red)	183
Nicotinamide	(1, descending)	BuOH satd. H_2O	UV-absorption for quantification (up to 100 µg per sample).	--	184
Nicotinamide	(1)	$BuOH-AcOH-H_2O$ (4:1:5)	1-chloro-2,4-dinitrobenzene/ethan. KOH	-- (brown)	185
Nicotinamide	(1, descending)	$BuOH-AcOH-H_2O$ (40:5:55)	ninhydrin (1% in citrate buffer)/70°C-30'.	-- (violet)	186
Nicotinamide	(1)	$AmOH-AcOH-H_2O$ (6:3:3)	1% 2,4-dinitrobenzyl chloride in EtOH/ 1% ethan. NaOH.		187
Nicotinic acid Nicotinamide	chroMax paper column	BuOH-1.5M NH_4OH (25:4), PrOH-0.1M PO_4-buffer pH 6.8 (11:9) successive runnings.	--	--	188

Continued /...

Table 5 Continued ...

Nicotinic acid Nicotinamide	(1, descending)	$CHCl_3$-EtOH-MeOH-HCO_2H (49.5:0.5:5:0.2:4.8) or EtMeCO-AcOH-H_2O (4.9:0.1:5) or PrOH-H_2O (4:1).	1.5% ethan. picryl chloride	--	189
Nicotinic acid Nicotinamide (tobacco)		EtOH-100 PrNH-H_2O (16:2:1)	aniline phosphate buffer. pH 6.1/CNBr vapors or photometrically (quantitative)	-- (yellow)	190
Nicotinic acid Nicotinamide	(1, descending)	80% aqu. $(CH_3)_2CO$/8 hrs. 20°C	UV (250-260 nm): colorimetrically or microbiologically	0.60 0.80	191
Nicotinic acid Nicotinamide	(1)	a) BuOH satd. with H_2O; b) EtOH (95%)-1M NH_4OAc (7:3)	1% Dinitrofluorobenzene in 96% EtOH/110°C, pH 5. 15 min- and 5% aq. NaOH	0.2 (a) 0.7 (b) 0.69 (a) 0.75 (b)	192
Nicotinic acid Nicotinamide	(1)	a) C_6H_6-$(CH_3)_2CO$-AcOH (15:4:1) + 0.83% H_2O, b) BuAc-AcOH (5:1) satd. with H_2O.	König's reaction (CNBr + amine)		193

Table 6: TLC-separation, identification and quantification of nicotinic acid and nicotinamide

Substance (specimen)	Adsorbent	Developing solvent (v/v) (temperature)	Visualization	Rf-value (color)	Reference
Nicotinamide (in vitamin preparations)	silica gel G containing 2% Leuchtstoff ZS-super	AcOH-$(CH_3)_2CO$-MeOH-C_6H_6 (1:1:4:14), in dark	Fluorescence at 254 nm	0.65	194
Nicotinamide (in vitamin preparations)	Al_2O_3-G	AcOH-$(CH_3)_2CO$-MeOH-C			
Nicotinamide	silica gel G	a- BuOH-AcOH-H_2O (20:4:3). b- PrOH-10% aq. NH_3-glcerol (4:1:1)	10% $CuSO_4$-2% aq. NH_3 (5:1)	0.63 (blue) 0.76 (blue)	196
Nicotinamide	silica gel F_{254}	BuOH-MeOH-AcOH-H_2O (4:4:1:2)	Quantification by fluorescence quenching at 254 nm.	--	197
Nicotinic acid Nicotinamide	polyamide-silica gel G (2:1)	a- 10% NaCl, or b- 10% NaOH	0.07% Rhodamine B in alc. and under UV (254 nm).	a- 0.74 b. 0.73 a- 0.53 b. 0.52	198
Nicotinic acid Nicotinamide	purified cellulose	BuOH satd. with 3% aq. NH_3	CNBr (vapors) and 2% 4-aminobenzoic acid in 0.75 M HCl-96% EtOH (3:1)	0.30	199
Nicotinic acid	Kieselgel G	PrOH-10% aq. NH_3 (19:1)	2% 1-chloro-2,4-dinitrobenzene in EtOH, heating 180°C/30-45 min.	- (violet-red- - orange)	200

Continued /...

Table 6 Continued ...

			then with 3% NaOH in MeoH and NH_3-vapors.		
Nicotinic acid Nicotinamide	precoated UV_{254} sheets	PrOH-aq. NH_3 (19:1) or $C_7H_{16}(CH_3)CO$ (3:2)	2% 4-aminobenzoic acid in 0.75 M HCl-EtOH (3:1), then vapors of CNBr/1 hr.	(yellow-orange)	201
Nicotinamide (in injections)	silica gel G	EtOH (98%) - $CHCl_3$ (5:12)	Quantification by measuring $A_{261\ nm}$	--	202
Nicotinamide (in multivitamin preparations)	silica gel (fluores-cent. 254 nm)	H_2O	Cl_2-atmosphere (20 min.), then spraying with o-tolidine (0.03% in aq. AcOH)	--	203
Nicotinamide acid, nicotinamide (dosage forms)	silufol & silufol UV 254	different solvent systems	UV-light and colorimetrically	--	204
Nicotinamide	cellulose MN300 HR	BuOH-AcOH-H_2O (8:1:11)	iodoplatinic acid	--	205
Nicotinic acid, Nicotinamide	silica gel	C_6H_6-MeOH-$(CH_3)_2$CO-AcOH (14:4:1:1)	different spraying reagents	--	206
Nicotinic acid, Nicotinamide	silufol UV_{254} or celluiose	0.1 M HCl	UV-light (254 nm); quantification at 261 nm was possible.	--	207

Continued /....

Table 6 Continued ...

Nicotinamide (coffee extract)	$Al_2O_3F_{254}$	$MeOH-H_2O$ (23:2)	UV-light (254 nm)	--	208
Nicotinic acid Nicotinamide	silica gel GF_{254}	BuOH-EtOH (1:1)	UV-light (254 nm)	--	209
Nicotinamide (in multivitamin preparation)	silica gel G	$CHCl_3$-EtOH (12:5)	aniline and 4% aq. CNBr vapors, quantification by densitometry	0.51 (yellow)	210
Nicotinamide (in multivitamin preparation)	silica gel 60 F_{254}	$(CH_3)_2CO-CHCl_3$-BuOH-conc. NH_4OH (30:30:40:5 & 30:30:40:2)	quantification by remission measurement at 262 nm.	--	212
Nicotinamide (in Terramycin capsules)	silica gel HF_{254}	tert.BuOH-dioxane-iso. Pr ether-H_2O (2:5:2:1); PrOH-10% NH_4OH (95:5) or H_2O.	UV-light, identification & quantification	--	213

Table 7: GLC-Procedures for identification and determination of nicotinic acid and its amide

Substance (specimen)	Column; internal standard	Temperature (°C)	Rt (min.)	Detection	Carrier gas (flow rate)	Reference
Nicotinic acid, Nicotinamide (vitamin mixture)	3% SE-30 (after silylation with N,O-bis (trimethyl-silyl)acetamide) (BSA)	100-270°C (Progr.)	--	FID	--	218
Nicotinic acid Nicotinamide	2.5% 2,2-dimethylpropane-1,3-diol succinate and 10% SE-30 on 100-120 mesh silanized Gas-Chrom P (after alkylation; ethylation).	180°C (isoth.)	--	FID or ArID	Ar(50 ml.min^{-1})	214
Nicotinic acid, Nicotinamide	10% silicone oil (2 m) ethyl ester of the acid or the nitrile derivative of the amide)	220°C (isoth.)	--	FID	He (140 ml.min^{-1})	219
Nicotinamide (with other vitamins)	15% silicone grease on 60-80 mesh Gas-Chrom Q (silanized) (after silylation with BSA: phenanthrene).	175°C (isoth.)	--	Therm. conduct.	He (8 ml.min^{-1})	220
Nicotinamide (in vitamin prepara-	4% silicone gum nitrile XE-60 on acid washed	210°C (isoth.)	(5)	FID	N (30 ml.min^{-1})	215

Continued /....

Table 7 Continued ...

tions)	chromosorb G (silanized); phenacetin.					
Nicotinamide (in dosage formulations)	1% carbowax 20 M on Gas-Chrom P; p-dimethylaminobenzaldehyde.	190°C (isoth.)	--	FID	N (100 ml.min^{-1})	221
Nicotinic acid, Nicotinamide	15% polymethylphenylsiloxane on chromaton N-AW DMCS (selanized) (after silylation with hexamethyl disilazane); eiphenyl ether.	100-170°C (progr., 8°C.min^{-1})	--	Therm. conduct.	He (30 ml.min^{-1})	222
Nicotinamide (in multivitamin form)	20% Carbowax 20 M-tetraphthalic acid on 100-120 mesh Gas-Chrom P AW-DMCS (silanized) (after nitrilation).	156°C (isoth.)	--	FID	N (30 ml.min^{-1})	223
Nicotinic acid (in dosage formulations)	3% SE-30 on 60-70 mesh Diatomite C-AW DMCS (silanized); without derivatization.	180°C (isoth.) (1.2)	--	FID	N (45 ml.min^{-1})	224
Nicotinic acid	20% Marlophen 87 on Chromosorb W, (after methylation); N,N-dimethylaniline.	165°C (isoth.) (3.2)	--	FID	--	225

TLC-methods for fractionation, identification and/or quantification of nicotinic acid and nicotinamide.

7.243 Gas-Liquid Chromatography (GLC)

Some GLC-procedures (214-216) have often been recommended, but do not take into account the likelihood of encountering nicotinic acid and its relatives in a samples (217). The methods are ranging from the isothermal to the programming technique at different temperature ranges. Different polarities have been specified for identification of nicotinic acid and its amide in raw materials, biological fluids and pharmaceutical formations (218-225). The following table 7 compares the various GLC-procedures.

7.244 High Performance (Pressure) Liquid Chromatography (HPLC)

The determination of nicotinic acid and/or nicotinamide has been performed using HPLC on various normal and reversed-phase columns. The HPLC technique has advantage over the GLC one in requiring almost no derivatization, such as silylation or conversion to the corresponding alkyl esters, prior to quantification (226-241). The following table 8 summarizes the different HPLC-methods recommended for the qualitative and quantitative analyses of nicotinic acid and its amide. Guilleman *et al* (242) describe a high performance, at moderate pressure (10-30 bar), liquid chromatographic procedure for separation of nicotinamide and some other organic compounds on columns packed with high specific area and consequently short columns. Two grades of Spherosil, *viz*, XOA 600 and XOA 800 (with specific areas 600 & 800 $m^2.g^{-1}$, respectively) with mean particle diameters of 5-7 µm. Efficiencies up to 800 theoretical plates cm^{-1} with analysis times of 5-15 min. could be obtained. Ion-pair, reverse-phase HPLC for separation of nicotinamide in admixtures with other water soluble vitamins has been investigated by Coleman (243). The effects of pentane, hexane and heptane sulfonates, as ion-pair reagents at various water-methanol ratios on the retention times on a RP-column were investigated.

Table 8: HPLC-Determination of nicotinic acid and nicotinamide

Substance (specimen)	Column/temp. (°C)	Developing solvent (mobile phase)/flow rate-Int. Std.	Retention time (min.)	Detection ( ), nm	Sensitivity	Reference
Nicotinic acid nicotinamide	1% SCX (sulfonated fluoro-carbon) Zipax glass bead/ (27)	0.1 M $NaNO_3$-0.1 NH_3PO_4/1.7 $ml.min^{-1}$-picolinamide.	8.40	UV (254)	> 0.01% (50 ng-5 µg)	226
Nicotinamide (multivitamin preparations)	HS Pellionex SCX	0.025 M HNO_3/10-20 $ml.h^{-1}$	--	UV (254 or 280)		227
Nicotinic acid (with other vitamins)	Aminex A-5 resin	0.03 M KCl in 0.1 M PO_4-buffer (pH 8) linear gradient of pH and ionic strength at acid	--	--	100-250 ng	228
Nicotinamide	Aminex A-14 resin	pH-value isocratically pH 2.5.	--	--		
Nicotinic acid, nicotinamide (dosage formula-tions)	µ-Bondapak C18	0.02 M Na dioctyl sulphasuc-cinate pH 2.5 - CH_3OH (1:1)/ 2.5 $ml.min^{-1}$ - acetophenone	3.10 3.60	UV (280)	25 µg.ml 7.5 $µg.ml^{-1}$	229,230
Nicotinic acid. nicotinamide (food)	µ-Bondapak C18	Detto	--	UV (280)	> 0.1 $µg.g^{-1}$	231

Continued /.....

Table 8 Continued ...

Nicotinamide (with levoglutamide)	Nucleosil 10 SB	Phosphate buffer pH 5	--	UV (210)	--	231
Nicotinamide (plasma)	μ-Bondapak C18 or LiChrosorb RP-18 (10 μm)	0.31% Na dioctylsulfosuccinate, pH 2.5, MeOH (1.4 = 1)/2 ml.min^{-1}-isonicotinamide.	--	UV (254)	0.1-10 μg.ml^{-1}	232
Nicotinamide (multivitamin preparations)	μ-Bondapak Phenyl (10 μm)/(30°C)	MeOH-H_2O (0 to 80% linear gradient)/2 ml.min^{-1}-0.025 M $C_6H_{11}SO_3H$ as ion-pair reagent	--	UV (280)	--	233
Nicotinic acid nicotinamide	μ-Bondapak C18 (10 μm)	0.05 M $C_5H_9SO_3H$-0.025 M KH_2PO_4 (pH 2.5).	--	UV (254)	1-10 ng	234
Nicotinic acid nicotinamide (plasma)	Zipax 8 CX (25-37 μm)/(45°C)	0.02 M PO_4-buffer, pH 2.6-guinaldic acid	--	UV (260)	0.1-2 μg ml^{-1} 1-10 μg ml^{-1}	235
Nicotinic acid Nicotinamide (urine)	Ultrasphere ODS/24.5°C (pH 3.30)	(MeCN-0.01 M $C_5H_{11}SO_3H$, & $(Et)_3NHCl$)-(0.01 M $C_5H_{11}SO_3H$ & KH_2PO_4) (0 to 6% linear gradient).	--	UV (254)	0.1-1 μg (0.01 μg)	236

Continued /....

Table 8 Continued ...

Nicotinamide (with other vitamins)	C18-bonded phase	MeOH - NH_4PO_4 buffer (gradient, 0 to 70%)	--	--	--	237
Nicotinamide (with other vitamins in dosage forms)	FAST-LC-8 (5 µm)	Tetrahydrofuran-Na $C_7H_{13}SO_3^-$-KH_2PO_4-H_3PO_4-$(Et)_3N$ (pH 2.5)/ 2 or 4 $ml.min^{-1}$.	--	UV (272)	--	238
Nicotinamide (with other vitamins)	Silica gel silanized with $Me_2C_8H_{17}$ silane.	0.3% $(Bu)_4NOH$ in 0.015 M PO_4^{3-} buffer, pH 3 - MeOH (17 = 8)/1 $ml.min^{-1}$-C_6H_5OOH	--	UV (280)	--	239
Nicotinamide (with other vitamins, dosage forms)	C8-HPLC column/(25°C)	MeOH-0.04 M $NaC_7H_{15}SO_3^-$ in 2% (AcOH (1:9).	--	UV (280)	--	240
Nicotinamide (multivitamin-mineral preparations)	RP-C18	MeOH & H_2O (3:17)-0.005 M $C_7H_{15}SO_3H$ & 0.5% $(Et)_3N$, pH 3.6/2 $ml.min^{-1}$	--	UV (280)	--	241

7.245 Ion Exchange and Electrophoresis

7.2451 Ion-Exchange Chromatography

Klotz and Poethke have separated the vitamin B complex components and determined the niacinamide with ion-exchange resin in a column of Wofatit CP300 (H^+) and Wofatit L150 (OH^-) (244). These authors (245) used Amberlite IR-120, IRA-401, IRC-50 and XE-100 and Wofatit CP300 and L150 for the isolation of nicotinamide. Nicotinamide was retained on the weakly acidic cation exchangers at pH < 4, but can be eluted with N-HCl or with NaOH solution. Sweeney and Hall (246) used the ion-exchange resin for the quantitative differentiation between nicotinic acid and nicotinamide. The sample solution at pH 5 is passed over the anion-exchange resin IRA-400. The nicotinamide is eluted with water and the nicotinic acid with hot 1N HCl. Both substances are then determined by means of König's reaction with CNBr and sulphanilic acid. Table 9 summarizes the ion-exchange chromatography reported in the literature (244-252).

7.2452 Electrophoresis

Nicotinamide, in pharmaceutical preparations, have been separated by gel filtration on Sephadex G-50 and eluted from the column by H_2O and saturated butanol. The eluate after concentration were subjected to paper electrophoresis (110 V; 8-10 mA) with pyridine-formic acid-water (1:1:98) buffer solution (pH 3.35). Zones were located under U.V. light (253). In an earlier method (254) nicotinamide in a mixture of vitamins was separated by paper electrophoresis in acetate buffer (pH 5.1) with a current of 3.5 mA. Spots are developed with diazotized *p*-aminoacetophenone. Electrophoresis on Whatmann No. 1 filter paper to separate nicotinamide from a number of derivative at a potential difference of 280 V in 0.03 M-borate buffer has been reported by Sundaram *et al* (255). Fukuda and Mimura (256) have investigated a paper electrophoresis method to separate and identify nicotinamide in a mixture of vitamins by using 0.5% sodium acetate, 1% sodium borate and 0.1% ammonium chloride as electrolyte. Paper electrophoresis technique has also been employed for the separation and determination of nicotinamide in pharmaceutical preparations of

Table 9: Ion exchange chromatography of nicotinamide and/or nicotinic acid

Mobile phase	Column	R_f	Reference
Acidified water (pH 4)	Wofatit CP300 (H^+) Wofatit L150 (H^-)	-	244
1N-HCl or NaOH solution	Amberlite IR-120, IRA-401, IRC-50, XE-100, Wofatit CP300 and L150	-	245
Water	IRA-400	-	246
AcO form Cl form (buffered at pH 1.6).	Amberlite SB-2 (AcO form) Amberlite SB-2 (Cl form) Amberlite WA-2 (buffered at pH 1.6) Amberlite WB-2	0.5 0.52 0.18 0.02	247
Acetate buffer solution pH 4.62.	Amberlite WA-2 and carboxylic acid resin	0.18	248
0.1N-$NaNO_3$-0.1N-H_3PO_4	1% of SCX (a 1% sulphonated fluorocarbon	-	249
Water	Strong basic anion exchangers (OH^-)	-	250
-	Amberlite IRC, XE-50, IR-120 and LRA-401 and Wofatit CP-300 and KPS-200 resins	-	251
0.005 N HCl	Acid-precipitated alginic acid	-	252

vitamins (257). Table 10 collects the different electrophoretic separations of the drug substances (258-276).

7.25 Gravimetry

Nicotinamide can be precipitated from its solutions in amyl acetate or ether (10-100 mg) by adding excess of 0.025 M chloranilic acid solution in the same solvent. After standing for 30 min. (amyl acetate) or 5 min. (ether) filtration through a preweighed Gooch crucible is advised. Weighing of the precipitate can be done after its washing and drying till constant weight (277). Colorimetric quantification of nicotinamide can be carried out by dissolving an aliquot of the colored precipitate in water. The crystallized nicotinamide-chloranilic acid salts sublime at 180°C and decompose at 240-245°C, its aqueous solution has pH-value = 3, at 23°C with conductivity of 6 X 10^{-4} mho.cm^{-1}. Nicotinamide can also be determined gravimetrically by precipitation *via* heating with methyl iodide (40 min, 55-60°C) or ethyl iodide (2 hrs, 80-85°C) to give the corresponding quaternary salts (278). The nicotinamide-EtI, yellow-green crystals, melt at 146-148°C, while the MeI-salt, yellow crystals, melts at 201-206°C.

7.26 Automated Analysis

Continuous, non-segmented, flow injection system with colorimetric detection at 420 nm, has been investigated for quantification of nicotinamide admixed with some other water-soluble vitamins. The yellow color formation could be obtained by reacting nicotinamide with polymethine dye mixed with NaOH and ammonia buffer containing CNBr (279). Nicotinic acid and nicotinamide in one-component and multivitamin tablets and capsules can be quantified by adopting autoanalysis with colorimetry by using CNBr-sulfanilic acid for nicotinic acid and CNBr-barbituric acid for nicotinamide (280). Another automated method for determination of nicotinic acid and nicotinamide in cereal products has been collaboratively studied by Gross (281). The AOAC microbiological method 43.121 were compared in such study. Coverly (282) described the possibility of automating an HPLC-procedure for quantification of nicotinamide in admixtures with some

Table 10: Electrophoretic separation of nicotinamide and/or nicotinic acid

Mobile phase	Column	Detector	Reference
Methanol-water (3:17), 0.005M heptane sulfonic acid, and containing 0.5% triethylamine at pH 3.6.	C_{18} (reversed phase)	UV 280 nm	258
0.3% Tetrabutyl ammonium hydroxide solution in 0.015M phosphate buffer (pH 3) - methanol (17:8)	Silica gel treated with dimethyl(octyl)silane.	UV 280 nm	259
Methanol-4 mM sodium heptanesulfonate in aqueous 2% acetic acid (1:9).	Silica gel C_{18}	UV 280 nm	260
Tetrahydrofuran sodium heptanesulfonate-KH_2PO_4-H_3PO_4-triethylamine (pH 2.5).	FAST-LC-8	UV 272 nm	261
[Acetonitrile with 10 mM pentane sulfonic acid - 10 mM triethylammonium chloride (9:1)] in 10 mM pentae sulphonic acid - 10 mM - triethylammonium chloride and KH_2PO_4 in water.	Ultrasphere ODS	UV 254 nm	262

Continued /...

Table 10 Continued ...

Ammonium phosphate buffer solution, then aq. 70% methanol.	C_{18}-bonded phase	-	263
0.02 M - Phosphate buffer (pH 2.8)	Zipax 8CX	UV 260 nm	264
0-80% Methanol in water and (2.5 mM hexane sulfonic acid as ion-pair reagent).	μ Bondapak-phenyl	UV 280 nm	265
5 mM Pentanesulfonic acid in 25 mM KH_2PO_4 (pH 2.5)	μ Bondapak C_{18}	UV 254 nm	266
1050 ml Methanol-4.446 gm Na dioctyl sulfosuccinate in 1450 ml of water adjusted to pH 2.5 with formic acid.	μ Bondapak C_{18} or LiChrosorb RP-18	UV 254 nm	267
Methanol-water-hexanesulfonic acid	-	-	268
Phosphate buffer (pH 5)	Nucleosil 10 SB	UV 210 nm	269
--	Spherosil XOA 600 and	-	270
--	Spherosil XOA 800	-	

Continued /...

Table 10 Continued ...

2 mM-Na (dioctyl sulfosuccinate-methanol (1:1) adjusted to pH 25 with 10% form acid.	μ Bondapak C_{18}	UV 254 nm	271
0.03 M-KCl in 0.1 M K-phosphate buffer (pH 8).	Aminex A-5 resin	-	272
0.025 M Nitric acid	HS Pellionex SCX	UV 254 or 280 nm	273
--	Spherosil	-	274
0.1 N Sodium nitrate-0.1 N H_3PO_4	1% CSX Zipax glass bead	UV	275
0.3% Tetrabutyl ammonium hydroxide solution in 0.015 M - phosphate buffer (pH 3) - methanol (17:8).	Silica gel treated with dimethyl (octyl) silane	UV 280 nm	276

other water-soluble vitamins in tablets by continuous-flow sample preparation with UV-detection at 272 nm.

ACKNOWLEDGEMENTS

The authors would like to thank Mr. Syed Rafatullah, Assistant Researcher, College of Pharmacy, King Saud University for his technical assistance and Mr. Tanvir A. Butt, for typing the manuscript.

References

1. W.C. Bowman and M.J. Rand, "Text Book of Pharmacology", 2nd Ed. Blackwell Scientific Publications, Oxford, (1980) p. 43.

2. Goodman and Gilman's, The Pharmacological Basis of Therapeutics, 6th Ed., MacMillan Publishing Co. Inc., New York, (1985) p. 1557.

3. F.A. Robinson, "The Vitamin B Complex", John Wiley & Sonc Inc., New York, (1951) p. 216.

4. The Pharmaceutical Codex, 11th edn., The Pharmaceutical Press, London, (1979) p. 592.

5. K.A. Connors, G.L. Amidan, and L. Kennon (Eds.), Chemical Stability of Pharmaceuticals, A Handbook for Pharmacist, J. Wiley & Sons, New York (1979) p. 77.

6. The Merck Index, 10th edn., Merck and Co. Inc., Rahway-NJ, (1983) No. 6366-pp. 930 & 935.

7. E.C.G. Clarke, "Isolation and Identification of Drugs" edn. The Pharmaceutical Press, London, A.C. Moffat (Ed.) (1986) p. 807 & 808.

8. Smith, E.G. and Baker, P.A. (Eds.), The Wiswesser Line-Formula Chemical Notation (WLN), 3rd edn., Chem. Inf. Mang. Inc. (CIMI), Cherry Hill, New Jersey (1975) p. 60-62.

9. European Pharmacopoeia, Vol. II, Maisonneuve, S.A. France, (1971) p. 314-315.

10. United States Pharmacopoeia XXI, Pharmacopoeial Convention, Inc., Rockville-Md., (1985) pp. 731 & 732.

11. Martindale, The Extra Pharmacopoeia, 29th edn., The Pharmaceutical Press, London, (1989) p. 1268.

12. "The British Pharmacopeia 1980", Vol. I, Her Majesty's Stationary Office, London, (1980) p. 303.

13. "The United States Dispensatory", 27th edn., J.B. Lippincott (Ed.), Philadelphia, (1973) p. 782-783.

14. R.R. Abou-Shabaan; unpublished data (1989).

15. P.W. Garbo, U.S. Patent, 2,427,400; cf. Ref. 3.

16. P.W. Garbo; U.S. Patent, 2,419,831; April 29 (1947).

17. P.W. Garbo, U.S. Patent 2,460,002, Jan. 25 (1949).

18. M.E. Haltquist; Brit. Patent 561,019, May 2 (1944).

19. P.W. Garbo, U.S. Patent 2,453,706, Nov. 16 (1948).

20. L.J. Wissow, U.S. Patent 2,617,805, Nov. 11 (1952).

21. H.B. Thomasen; Dan. Patent 87,228, July 27 (1959); cf. C.A.: 54, 7742F (1960).

22. Aries Associates, Inc.; Brit. Patent 822,519, October 28 (1959).

23. E.F. Pike and R.S. Shane; U.S. Patent 2,412, December 17 (1946).

24. H.F. Seibert, L.J. Szabo, Wyeno and E.R. Eckhardt, U.S. Patent 2,280,040, Apr. 14, 1941; cf. C.A.: 36, 51876 (1942).

25. Ciba Ltd., Brit. Patent, 561,019; May 2 (1944).

26. Ciba Ltd., Brit. Patent, 558,774; cf. Ref. 3.

27. M.R. Bovarnick; J. Biol. Chem.: 149, 301, (1943) & 153, 1 (1944).

28. H.R. Rosenberg; U.S. Patent 2,446,957, Aug. 10 (1948).

29. R.G. Benner and I. Hochhauser; U.S. Patent 2,685,586, Aug. 3 (1954).

30. B.F. Deusel and H.L. Friedman; U.S. Patent 2,471,518, May 31 (1949).

31. Pyridium Corp.; Brit. Patent 563,184, Aug 2, 1944, c.f. C.A.: 40, 24734 (1946).

32. J.F. Couch and C.F. Krewson; U.S. Patent 2,453,496, Nov. 9 (1948).

33. R. Jujo; J. Pharm. Soc. Japan: 66, 21 (1946); cf. C.A.: 45, 620i (1951).

34. E.C.,Jr. Taylor and A.J. Crovetti; J. Org. Chem.: 19, 1633 (1954).

35. P.A. George and P. Bachmann; Helv. Chim. Acta: 26, 358 (1943).

36. D. Gold and S.S.V. Roessler; Ger. Patent 828,247, July 2, (1953).

37. J. Van de Kamp and M. Tishler; U.S. Patent 2,435,809, Feb. 10 (1948).

38. Z.J. Vejdelek, Chem. Listy: 46, 770 (1952); cf. C.A.: 47, 11191 (1953).

39. C.F. Krewson and J.F. Couch; J. Am. Chem. Soc.: 65, 2256 (1943).

40. J.F. Couch and C.F. Krewson, U.S. Pat. (Applied) 63, 432, Official Gazette 644, 307 (1951).

41. J. Schmidt-Thome and H. Goebel; Z. Physiol. Chem.: 288, 237 (1951).

42. Heyl; J. Am. Chem. Soc.: 70, 3434 (1948).

43. Jones; J. Am. Chem. Soc.: 74, 1489 (1952).

44. A.H. Berrie, G.T. Newbold and F.S. Spring; J. Chem. Soc.: 2042 (1952).

45. I.A. Balaban and F. Manchester; Brit. Patent 575, 119, Feb. 7 (1946).

46. Wm. White and R.A. Glen; U.S. Patent, 2,409,806, Oct. 22 (1946).

47. E.F. Kozlova, M.I. Kustanovich, M.M. Yanina and I.B. Chekmareva; Khim. Farm. Zh.: 2, 28 (1968).

48. V.I. Trubnikov, V.V. Petrov, E.S. Zhdanovich and N.A. Preobrazhenskii; Khim-Farm. Zh.: 3, 49-52 (1969).

49. D. Stefanescu, D. Moraree, I. Predescu, P. Barza, T. Stanciu, Fl. Coman.; Farmacia (Bucharest): 17(2): 115-20 (1969).

50. B.V. Suvorov, S.R.Rafikov, B.A. Zhubanov, and M.I. Khmura; U.S.S.R. Patent 119878, May 21,(1959).

51. S.M. McElvain and M.A. Goese; U.S. Patent 2,418,977, Apr. 15 (1947).

52. V. Mikkelsen; Arch. Pharm. Chem.: 46, 479-497 (1939).

53. Merck & Co. Inc.; Brit. Patent 687,312, Feb. 11 (1951).

54. J.R. Berg; U.S. Patent 2,496,114, Jan 31 (1950).

55. M. Weiner and J. Van Eys (Eds.); in: Nicotinic Acid, Nutrient-Cofactor-Drug, Vol. 1, Marcel Dekker, Inc., New York & Basel (1983), p. 232.

56. J.L.S. Hollman, C.G. Davis and Leeper; J. Am. Geriatr. Soc.: 10, 903 (1962).

57. A.M. Diab; Arzneimittel-Forsch./Drug Research: 27, 2314 (1977).

58. C.J. Walters, R.R. Brown, M. Kaihara and J.M. Price; J. Biol. Chem.: 217, 489 (1955).

59. K.K. Reddi and E. Kodicek; Biochemistry: 53, 286 (1953).

60. M. Weiner; J. Clin. Pharmacol.: 16, 550 (1976).

61. B. Petrack, P. Greengaard and H. Kalinsky; J. Biol. Chem.: 241, 2367 (1966).

62. P. Greengaard, B. Petrack and H. Kalinsky; J. Biol. Chem.: 242, 152 (1967).

63. B.B. Brodie, M. Weiner, J.J. Burns, G. Simson, and E. Yale; J. Pharmacol. Exp. Ther.: 106, 543 (1952).

64. R.S. Satoskar and S.D. Bhandarkar, "Pharmacology and Pharmacotherapeutics", Revised 9th edn., Popular Parkasham, Bombay, (1983) p. 851.

65. S. Chaykin, M. Dagani, L. Johnson and M. Samli; J. Biol. Chem.: 240, 932 (1965).

66. F.G. Brazda, R.A. Coulson, Proc. Soc. Exp. Biol. Med.: 62, 19 (1946).

67. P. Ellinger, G. Fraenkel and M.M. Abdel-Kader; Biochem. J.: 41, 559 (1947).

68. K. Unna; J. Pharmacol.: 65, 95-103 (1939).

69. N.W. Shock and W.H. Sebrell; Am. J. Physiol.: 46, 52 (1946).

70. F. Bergman and L. Wislicki; Brit. J. Pharmacol: 8, 49 (1953).

71. J.A. Gautier; Ann. Chim. Anal.: 26, 89 (1944); cf. C.A.: 40, 15496 (1946).

72. G. Sandri; Mikrochim. Acta: 2, 253 (1958); cf. Anal. Abstr.: 5, 3587 (1958).

73. F. Lüdy-Tenger; Pharm. Acta Helv.: 28, 22 (1953); cf. C.A.: 47, 7376 f (1953).

74. G. Sandri; Atti Accad. Sci. Ferrara: 35, (1957-58); cf. C.A.: 55, 13777b (1961).

75. W. Cuisa and G. Barbiroli; Ann. Chim. (Rome): 53, 1248 (1963); cf. Anal. Abstr.: 12, 1447 (March 1965).

76. A. Wenusch; Österr. Chem.-Ztg.: 45, 87 (1943); cf. C.A.: 37, 44156 (1943).

77. S.A. Ismaiel, W.G. Haney, and E.M. Abdel-Moety; Pharmazie: 40, 804 (1985).

78. D. Ropte and K. Zieloff; Pharmazie: 40, 793 (1985).

79. Y.M. Dessouky, F.T. Hussein, and S.A. Ismaiel; J. Pharm. Belg.: 29, 43 (1974).

80. M.M.L. Inacio, M.A. Andrade, and A.M. Leite; Rev. Port. Farm.: 14, 17 (1964); cf. Anal. Abstr.: 12, 4172 (1965).

81. D. Zivanov, Z. Blagojevic, and M. Mladenovic; Arhiv. Farm.: 14, 273 (1964); cf. C.A.: 62, 2671h (1965).

82. T.P.A. van der Wegen; Pharm. Weekbl.: 94, 717-733 (1959); cf. Anal. Abstr.: 7, 3951 (1960).

83. A. Dobry; Trav. Membres Soc. Chim. Biol.: 23, 1438-44 (1971); cf. C.A.: 39, 1431 (1940/41).

84. The British Pharmacopoeia 1988, H.M.S. Office, London, (1988) pp. 388, 973 & 974.

85. C.F. Krewson; Am. J. Pharm.: 115, 122 (1943); cf. C.A.: 37, 45307 (1943).

86. S.V. Sreeram and P.R. Naidu; Indian J. Chem.: 20, 1046 (1981).

87. A. Bianchi; Farmaco, Ed. Prat.: 16, 476 (1961); cf. Anal. Abstr.: 9, 2065 (May 1962).

88. M. Rink, R. Lux, and E. Franken; Dtsch. Apoth Ztg: 98, 660 (1958); cf. Anal. Abstr.: 6, 1885 (May 1959).

89. G. Helmstaedter; Arch. Pharm. (Berlin): 292, 91 (1959); cf. Anal. Abstr.: 7, 1549 (April 1960).

90. H.H. Kavarana; Indian J. Pharm.: 20, 360 (1958); cf. Anal. Abstr.: 6, 3738 (Sept. 1959).

91. Z. Klellin and K. Sumanovie; Creat. Chem. Acta: 30, 181 (1958); cf. Anal. Abstr.: 6, 1512 (April 1959).

92. V.A. Devyatnin and M. Ya Moizhes; Med. Prom. (SSSR): 17, (1963); cf. Anal. Abstr.: 12, 1362 (March 1965).

93. A.A. Abou-Ouf, M.I. Walash, M. Rizk, and F. Belal; J. Drug Res.: 11, 81(1981); Anal. Abstr.: 41, 5E8 (1981).

94. V.J. Vajgand and T.J. Pastor; Glas. Hem. Drus (Beograd): 31, 9 (1966); cf. C.A.: 67, 39955y (1967).

95. G. Velinov, D. Ivanov, and O. Budevski; J. Electroanalyt. Chem.: 57, 97 (1974); cf. Anal. Abstr.: 28, 6E11 (1975).

96. M. Grzybowska; Chemia Analit.: 19, 801 (1974); cf. Anal. Abstr.: 28, 3C39 (1975).

97. J. Subert, M. Bachrata, L. Knazko, and A. Stemberova; Cslka Farm.: 24, 68 (1975); cf. Anal. Abstr.: 30, 1E1 (1976).

98. T.J. Pastor and V.J. Vajgand; Mikrochim. Acta: 2, 85 (1976); cf. Anal. Abstr.: 32, 1J104 (1977).

99. T.J. Pastor, V.J. Vajgand, and Z. Simonovic; Anal. Chim. Acta: 98, 233 (1978); cf. Anal. Abstr.: 35, 5E26 (1978).

100. G. Petho and K. Burger; Acta Chim. Acad. Sci. Hung.: 108, 161 (1981); cf. Anal. Abstr.: 42, 6A14 (1982).

101. M. Shikata and I. Tachi; Mem. Coll. Agr. Kyoto Imp. Univ.: 4, 35 (1927).

102. P.C. Tompkins and C.L.A. Schmidt; J. Biol. Chem.: 143, 643 (1942).

103. K. Wenig and M. Kopecky, Cas. Cesk. Lek.: 56, 49 (1943); cf. ref. 40.

104. P.C. Tompkins and C.L.A. Schmidt; Univ. Calif. Pub. Physiol., 8, 247 (1943).

105. K. Knobloch, Coll. Czech. Chem. Commun., 12 407 (1947); cf. Ref. 40.

106. G.W.C. Milner, "The Principles and Applications of Polarography and Other Electroanalytical Processes", Longmans 4th edn. part 3, 1966, p. 598.

107. R. Cernatescu, M. Poni, and R. Ralea; Studii Cercetarri Stiint: 4, 117 (1953); cf. C.A.: 50, 13663e (1956).

108. J. Volke and V. Volkova; Chem. Lasty: 48, 1031 (1954); cf. Anal. Abstr.: 2, 2477 (1955).

109. J. Lindquist and S. Farroha; Analyst (London): 100, 377 (1975).

110. P.C. Tompkins and C.L.A. Schmidt; Univ. California Pub. Physiol.: 8, 221 (1943); cf. C.A.: 38, 36796 (1944).

111. W. Kemula and Y. Chodowski; Roczn. Chem.: 29, 839 (1955); cf. Anal. Abstr.: 3, 1791 (1956).

112. V.A. Serazeolinova, A.D. Kagarlitskif, D.Kh. Sembaev, K.U. Umarova, F.A. Ivanovskaya and B.V. Suvorov; Vest. Akad. Nauk. kazakh. (SSR): 2, 68 (1968); cf. Anal. Abstr.: 17, 931 (Aug. 1969).

113. S.M. Farroha and J. Lindquist; Bull. Coll. Sci.-Univ. Baghdad (Iraq): 16, 287 (1975); cf. Anal. Abstr.: 33, 4E23 (Oct. 1977).

114. J.M. Moore; J. Pharm. Sci.: 58, 1117 (1969); cf. Anal. Abstr.: 19, 3354 (Oct. 1970).

115. M.E. Schertel and A.J. Sheppared; J. Pharm. Sci.: 60, 1070 (1971); cf. Anal. Abstr.: 22, 3488 (1972).

116. E. Jacobsen and K.B. Thorgersen; Analytica Chim. Acta: 71, 175 (1974); cf. Anal. Abstr.: 28, 2E21 (1975).

117. A.Y. Taira; J. Ass. Off. Anal. Chem.: 57, 910 (1974); cf. Anal. Chem: 28, 3E29 (1975).

118. M. Jozan, G. Szasz and K. Szemeredy; Acta Pharm. Hung.: 50, 153 (1980); cf. Anal. Abstr.: 40, 3E36 (1981).

119. E. Jacobsen and T.M. Tommelstd; Anal. Chim. Acta: 162, 379 (1984); cf. Anal. Abstr.: 47, 6E44 (1985).

120. V.A. Deyatnin and L.A. Kuznetsova; Med. Prom. (SSSR): 18, 58 (1964); cf. C.A.: 61, 2908f (1964).

121. J. Abdoh and K. Täufel; Z. Untersuch.-Lebensm.: 84, 385 (1942); cf. C.A.: 37, 6594 (1943).

122. K. Ritsert; Klin. Wochschr.: 18, 934 (1939); cf. C.A.: 33, 86594 (1939).

123. J. Deltombe; J. Pharm. Belg.: 8, 59 (1953); cf. Anal. Abstr.: 1, 1054 (1954).

124. P. Douzou and A.M. Le Clerc; Anal. Chim. Acta: 12, 239 (1955); cf. Anal. Abstr.: 2, 2134 (1955).

125. B.P.Lisboa; Naturwissenschaften: 44, 617 (1957); cf. Anal. Abstr.: 5, 3172 (1958).

126. P. Mesnard; Arch. Pharm. (Athens): 22, 105 (1966); cf. C.A.: 66, 118877g (1967).

127. T. Wronska-Nafer and J. Sokal; Acta Pol. Pharm.: 24, 89 (1967); cf. Anal. Abstr.: 15, 2186 (1968).

128. B.E. Albright, E.F. Degner; Automat. Anal. Chem., Technicon Symp., 1967, 3rd, 1, 461 (publ. 1968); cf. C.A.: 71, 33460U (1969).

129. M.J. Jeffus and C.T. Kenner; J. Pharm. Sci.: 58, 749 (1969); cf. Anal. Abstr.: 19, 1685 (Aug. 1970).

130. O. Pelletier and J.A. Campbell; J. Pharm. Sci.: 50, 926 (1961); cf. Anal. Abstr.: 9, 2498 (June 1962).

131. L. Fuller and L.S. Dietrich; Analyt. Biochem.: 39, 538 (1971); cf. Anal. Abstr.: 22, 350 (1972); C.A.: 74, 83911z (1971).

132. Y.M. Dessouky, F.T. Hussein and S.A. Ismaiel; J. Pharm. Belg.: 29, 43 (1974); cf. Anal. Abstr.: 28, 2E20 (1975).

133. J. Ramirez-Munoz; Analyt. Chim. Acta: 71, 321 (1974); cf. Anal. Abstr.: 28, 4E28 (1975).

134. A.F. Gross; J. Ass. Off. Anal. Chem.: 58, 799 (1975); cf. Anal. Abstr.: 30, 2F 49 (1976).

135. H. Farsam and M. Mahmoudian; J. Pharm. Belg.: 32, 587 (1977); cf. Anal. Abstr.: 36, 3E22 (1979).

136. H. Kringstad and T. Naess; Nature wissenschaften: 26, 709 (1938); cf. C.A.: 33, 2071z (1939).

137. F.I. Khattab; Anal. Lett.: 16 (B14), 1121 (1983); cf. Anal. Abstr.: 46, 5E6 (1984).

138. K. Pauer: Chemia Analit: 18, 1069 (1973); cf. Anal. Abstr.: 26, 2856 (May 1974).

139. W.-K. Lee; Yakhak Hoeji (Korea): 13, 22 (1969); cf. C.A.: 73, 69915b (1970).

140. P. Karrer and H. Keller; Helv. Chim. Acta: 21, 463 (1938); cf. C.A.: 33, 5727 (1938).

141. N.S. Nudelman and O. Nudelman; J. Pharm. Sci.: 65, 65 (1976) cf. Anal. Abstr.: 31, 2E33 (1976).

142. W.I.M. Holman; Biochem. J.: 56, 513 (1954); cf. Anal. Abstr.: 1, 1024 (1954).

143. A. Walksmundzki and H. Romanowski; Acta Polon. Pharm.: 11, 243 (1954); cf. C.A.: 49, 8565 (1955).

144. A.W.M. Indemans and H.E.J. Rademakers; Pharm. Weekbl.: 95, 413 (1960); cf. Anal. Abstr.: 8, 1246 (March, 1961).

145. A. Domnas; J. Biochem. (Tokyo): 50, 1246 (March 1961); cf. C.A.: 56, 7634g (1962).

146. J. Bosly; J. Pharm. Belg.: 19, 513 (1964); cf. C.A.: 62, 15993b (19065); cf. Anal. Abstr.: 13, 1480 (March 1966).

147. J.P.S. Sarin, R.B. Chakravarty and G.K. Ray; Indian J. Pharm.: 26, 165 (1964); cf. Anal. Abstr.: 12, 5408 (Oct. 1965).

148. F. Namigohar, A. Chanbarpour and D. Beltran; Ann. Pharm. Fr.: 39, 445 (1981); cf. Anal. Abstr.: 42, 5E40 (1982).

149. T.J. Marshal and C.T. Kenner; J. Pharm. Sci.: 58, 479 (1969).

150. K. Capele and R. Holch; Pharm. Acta Helv.: 33, 163 (1958).

151. D.K. Bardhain, R.N. Bhattacharya and S.K. Dutta; Sci. & Cult.: 24, 85 (1958); cf. Anal. Abstr.: 7, 759 (Feb. 1960).

152. G. Scapini, P. Condorelli and F. Innocenti; Boll. Sedute Accad. Gioenia Sci. Nature. Catanica: 8, 626 (1966); cf. C.A.: 70, 14451e (1969).

153. D.K. Bardhain, R.N. Bhattacharya and S.K. Dutta,; J. Instn. Chem. India: 32, 86 (1960); cf. Anal. Abstr.: 8, 817 (Feb. 1961).

154. G. Machek and F. Lorentz; Sci. Pharm: 30, 25 (1962); cf. C.A.: 57, 4761d (1962).

155. *Idem*; Sci. Pharm.: 32, 185 (1964); cf. Anal. Abstr.: 13, 357 (Jan. 1966).

156. *Idem, ibid*: 34, 213 (1966); cf. C.A.: 66, 22258e (1967).

157. H.M. Frodyma and V.T. Lieu; cf. Anal. Chem.: 39, 814 (1967).

158. G. Lehmann, H. Hahn and M. Schirra; Fresenius'Z. Anal. Chem.: 243, 554 (1968).

159. J. Kracmar, J. Kracmarova and J. Zyka; Pharmazie: 23, 567 (1968).

160. G. Cosi; Farmaco, Ed. prat.: 23, 331 (1968).

161. K. Paczek and H. Wardynska; Acta Pol. Pharm.: 28, 71 (1971); cf. Anal. Abstr.: 22, 4407 (June 1972).

162. A. Jan, H. Ali and M. Ashraf; J. Pharm. (Lahore): 1, 12 (1979); cf. Anal. Abstr.: 40, 2E33, (1981).

163. J.V. Scudi; Science: 103, 567 (1946).

164. B.R. Clark, R.M. Halpern, and R.A. Smith; Analyt. Biochem: 68, 54 (1957); cf. Anal. Abstr.: 30, 4070 (1976).

165. R. Mestres; Trav. Soc. Pharm. Montpellier: 22, 46 (1962); cf. C.A;: 58, 14431g (1963).

166. A. Sebesta and L. Dalma; Sci. Aliment.: 10, 77 (1964); cf. Anal. Abstr.: 12, 6128 (Nov. 1965).

167. Y.Z. Mohran; Proc. Int. Congr. Nutr., 7th; 5, 573 (1967); cf. C.A.: 70, 35551 (1969).

168. K. Nikolic, D. Malesev, and K. Velasevic; Acta Pharm. Jugosl.: 18, 209 (1968); cf. Anal. Abstr.: 19, 4330 (Nov. 1970).

169. R.W. Frei, A. Kunz, G. Pataki, T. Prims, and H. Zuercher; Anal. Chim. Acta: 49, 527 (1970).

170. A.T.R. Williams; Lab. Equip. Dig.: 14, 67-77 (1976); cf. Anal. Abstr.: 32, 4D113 (1977).

171. J. Bartos; Annls Pharm. Fr.: 29, 71 (1971); cf. Anal. Abstr.: 21, 3502 (Nov. 1971).

172. G. Schoen; Acta Vitaminol.: 7, 151 (1953); cf. C.A.: 48, 13851h (1954).

173. C. Bergamini; Sperimentale, Sez. chim. biol.: 4, 38 (1953); cf. C.A.: 48, 13851h (1954).

174. R.C.R. Barreto, H.S.R. Barreto, and I.P. Pinto; J. Chromatogr.: 6, 416 (1961); cf. C.A.: 56, 12295e (1962).

175. R. Allouf and R. Munier; Bull. Soc. Chim. Biol.: 34, 196 (1952); cf. C.A.: 47, 447e (1953).

176. M. Kato; Kagaku (Science): 23, 478 (1953); c.f. C.A.: 48, 1632i (1954).

177. A. Heyndrickx; J. Am. Pharm. Assoc.: 42, 680 (1953); cf. C.A.: 48, 1632i (1954).

178. S.N. Tewari; Naturwissenschaften: 41, 217 (1954); cf. C.A.: 49, 4332g (1955).

179. F. Leuschner; Naturwissenschaften: 40, 554 (1954); cf. C.A.: 48, 8098b (1954).

180. A. Jones, M.P. Taylor, and D.N. Gore; Chem. & Ind.: 461 (1954); cf. C.A.: 49, 14080h (1955).

181. L.D. Rodrigues and J.A. deSilva; Rev. Port. Farm.: 5, 75 (1955); cf. C.A.: 50, 1264d (1956).

182. D.A. Buyske, J.M. Flowers, jun., P. Wilder, Jun., and M.E. Hobbs; Science, 124, 1080 (1956); cf. Anal. Abstr.: 4, 1958 (1957).
183. L. Maiwald and H. Maske; Hoppe Seyl. Z.: 306, 143 (1956); cf. Anal. Abstr.: 4, 2811 (1957).

184. K.M. Jones; Biochem. J.: 73, 719 (1959); cf. Anal. Abstr.: 7, 3896 (Sep. 1960).

185. D. Blitek, K. Strzyga and H. Wardynska; Acta Polon. Algm.: 16, 389 (1959); cf. Anal. Abstr.: 7, 3605 (July 1960).

186. P.S.M. Simatoepang; Suara Pharm. Madjalah: 4, 67 (1959); cf. C.A.: 58, 13724g (1963).

187. M. Danilovic, B. Lavica, and Z. Savic; Ach. Farm.: 16, 13 (1966); cf. C.A.: 66, 68982q (1967).

188. C. Ovenlors; Sci. Tools: 7, 1 (1966); cf. Anal. Abstr.: 8, 2586 (June 1961).

189. I. Gutorska; Chem. Anal. (Warsaw): 9, 597 (1964); cf. Anal. Abstr.: 12, 4634 (Sep. 1965).

190. R. Wahl; Z. Anal. Chem.: 204, 25 (1964); cf. Anal. Abstr.: 12, 4501 (Oct. 1965).

191. V. Colombatti; Farmaco, Ed. prat.: 20, 593 (1965); cf. Anal. Abstr.: 14, 2800 (May 1967).

192. V. Pallini, C. Vasconetto, and C. Ricci; Boll. Soc. Ital. Biol. Sper.: 41, 673 (1965); cf. C.A.: 64, 3401e (1966).

193. C. Hoehne and R. Patsch; Pharmazie: 26, 440 (1971); cf. Anal. Abstr.: 22, 3506 (May 1972).

194. H. Ganshirt and A. Malzacher: Naturwissenschaften: 47, 279 (1960); cf. Anal. Abstr.: 8, 332 (Jan. 1961).

195. S. Ishikawa and G. Katsui; Bitamin (Kyoto): 29, 203 (1964); cf. C.A.: 61, 15932b (1964).

196. K.C. Güven and Ö. Pekin; Eczaciha Bülteni: 8, 119 (1966); cf. Anal. Abstr.: 14, 5689 (Sept. 1967).

197. V. Massa; Frar. Soc. Pharm. Montpellier: 29, 35 (1969); cf. C.A.: 71, 105248f (1969).

198. H.-C. Chiang, Y. Lin and Y.-C. Wu; J. Chromatogr.: 45, 161 (1969).

199. R.W. Frei, A. Kunz, G. Paraki, T. Prims and H. Zuercher; Analytica Chim. Acta: 49, 527 (1970).

200. J. Washuettl; Mikrochi. Acta: 3, 621 (1970); cf. Anal. Abstr.: 21, 468 (July 1971).

201. H. Thielemann and M. Paepke; Sci. Pharm.: 40, 210 (1972); cf. Anal. Abstr.: 24, 2433 (April (1973).

202. S.A. Ismaiel and D.A. Yassa; Analyst (London): 98, 816 (1973).

203. H. Thielemann; Sci. Pharm.: 42, 145 (1974); cf. Anal. Abstr.: 28, 5E18 (1975).

204. M. Blesova and M. Zahradnicek; Cslka Farm.: 23, 303 (1974); cf. Anal. Abstr.: 28, 6E44 (1975).

205. S. Baczyk and L. Szczesniak; Acta Pol. Pharm.: 32, 347 (1975); cf. Anal. Abstr.: 30, 2E22 (1976).

206. H. Thielemann; Pharmazie: 32, 51 (1977); cf. Anal. Abstr.: 33, 2E29 (1977).

207. M. Blesova, M. Zahradnicek, J. Vanerkova, and M. Hlavacova; Cesk. Farm.: 27, 204 (1978); cf. Anal. Abstr.: 36, 4E24 (1979).

208. G. Lehmann, I. Haug and R. Schlosser; Z. Lebensm. Untersu.-Forsch.: 172, 87 (1981); cf. Anal. Abstr.: 41, 3F37 (1981).

209. J. Zhang and X. Du; Yaowu Fenxi Zazhi: 2, 85 (1982); cf. Anal. Abstr.: 45, 1E72 (1983).

210. I. Sunshine; Am. J. Clin. Pathol.: 40, 576 (1963).

211. S.A. Ismaiel, W.G. Haney and E.M. Abdel-Moety; Bull. Pharm. Sci. (Assuit Univ.): 8, 1 (1985).

212. D. Ropte and K. Zieloff; Pharmazie: 40, 793 (1985).

213. A. Nancovska; Bilt. Sojuzot Zruzenijata Farm. Farm. Teh. SR Maked: 29, 3 (1982); cf. C.A.: 102, 124575 (1985).

214. A.R. Prosser and A.J. Sheppard; J. Pharm. Sci.: 57, 1004 (1968).

215. J.D. Ashby and J.C. Deavin; J. Pharm. Pharmacol.: 21, 525 (1969).

216. J. Vessman and S. Tromberg; J. Pharm. Sci.: 64, 311 (1975).

217. D.P. Wittmer and W.G. Haney; "Water-soluble vitamins" in GLC & HPLC determination of Therapeutic Agents, Part 3, K. Tsuji (Ed.), Chromatographic Science Series, Vol. 9, Marcel Dekker, Inc., New York, (1979) p. 1243.

218. H. Janecke and H. Voege; Naturwissenschaften: 55, 447 (1968); cf. Anal. Abstr.: 18, 2731 (April 1970).

219. V.I. Trubnikov, L.M. Malakhova and E.S. Zhdanovich; Zh. Anal. Khim.: 23, 1546 (1968); cf. C.A.: 70, 43889j (1969).

220. L.T. Sennello and C.J. Argoudelis; Anal. Chem.: 41, 171 (1969).

221. P. Tarli, S. Benocci and P. Neri; Boll. Chim. Farm.: 108, 554 (1969); cf. C.A.: 73, 7303s (1970).

222. A.M. Poromarev, O.A. Uselyubavskaya, and M.Ts. Yanotovskii; Zh. analit. Khim.: 28, 2456 (1973); cf. Anal. Abstr.: 29, 1E28 (1975).

223. J. Vessman and S. Stromberg; J. Pharm. Sci.: 64, 311 (1975).

224. E.M. Abdel-Moety, M.G. El-Bardicy and S.A. Ismaiel; Acta Pharm. Jugosl.: 35, 255 (1985).

225. H. Liliedahl; Acta Chem. Scand.: 20, 95 (1966).

226. C.P. Talley; Anal. Chem.: 43, 1512 (1971); cf. C.A.: 75, 147598h (1971).

227. K. Callmer and L. Davis; Chromatographia: 7, 644 (1974).

228. A. Floridi, C. Fini, C.A. Palmerini and A. Rossi; Riv. Sci. Tecnol. Alimenti Nutr. Um.: 6, 197 (1976); cf. Anal. Abstr.: 33, 6D132 (1977).

229. D.P. Wittmer and W.G. Haney: Water soluble vitamins, in "GLC & HPLC determination of therapeutic agents", K. Tsuji (Ed.), Chromatographic Science Series, Vol. 9 - Chapter 36, Marcel Dekkern Inc., New York, 1979, p. 1245.

230. S.P. Sood, D.P. Wittmer, S.A. Ismaiel and W.G. Haney; J. Pharm. Sci.: 66, 40 (1977).

231. N. Nashimoto, Y. Mitani and S. Hayashi; J. Chromatogr.: 176, 448 (1979).

232. J.X. DeVries, W. Guenthert and R. Ding; J. Chromatogr.: 221 (Biomed. Appl.) 10: 161 (1980).

233. R.P. Kwok, W.P. Rose, R. Tabor and T.S. Pattison; J. Pharm. Sci.: 70, 1014 (1981).

234. J.T. Heard and G.J. Tritz; J. Liq. Chromatogr.: 4, 325 (1981).

235. K. Takikawa, K. Miyazaki and T. Arita; J. Chromatogr.: 233 (Biomed. Appl.) 22, 343 (1982).

236. R.W. McKee, Y.A. Kang-Lee, M. Panaqua and M.E. Swendseid; J. Chromatogr.: 230 (Biomed. Appl.) 19, 309 (1982).

237. R.M. Kothari and M.W. Taylor; J. Chromatogr.: 247, 187 (1982).

238. S.C. Coverly; J. Autom. Chem.: 5, 89 (1983); cf. Anal. Abstr.: 45, 4E29 (1983).

239. T. Cannella and G. Bichi; Boll. Chim. Farm.: 122, 205 (1983); cf. Anal. Abstr.: 46, 9E30 (1984).

240. G.D. Wachob; LC, Liq. Chromatogr. HPLC Mag.: 1, 110 (1983); cf. Anal. Abstr.: 46, 9E31 (1984).

241. F.-L. Lam, I.J. Holcomb and S.A. Fusari; J. Assoc. Off. Anal. Chem: 67, 1007 (1984).

242. C.L. Guilleman, J.P. Thomas, S. Thiault and J.P. Bounine; J. Chromatogr.: 142, 321 (1977).

243. G.L. Coleman; S. Afr. Pharm. J.: 46, 621 (1979).

244. L. Klotz and W. Poethke; Pharm. Zentralhalle: 103, 255 (1964).

245. L. Klotz and W. Poethke; Pharm. Zentralhalle: 103, 1 (1964).

246. J.P. Sweeney and L. Hall; J. Ass. Off. Agric. Chem.: 36, 1018 (1953).

247. L. Koltz and R. Huetten-rauch; Arch. Pharm.: 296, 721 (1963).

248. R. Huetlen-rauch and L. Klotz; Experientia: 19, 95 (1963).

249. C.P. Talley; Analyt. Chem.: 43, 1512 (1971).

250. Utsumi, M. Samejima and H. Kurita; J. Vitaminol. (Kyoto): 7, 209 (1961).

251. L. Koltz and W. Poethke; Pharm. Zentralh.: 103, 169 (1964).

252. J.S. Foster and J.W. Murfin; J. Pharm. Pharmacol.: 13, (Suppl.), 126T (1961).

253. M. Chakraborty and S.K. Dutta; J. Proc. Instu. Chem. India: 37, 114 (1965).

254. N. Siliprandi, D. Siliprandi and H. Lis; Biochim. Biophys. Acta: 14, 212 (1954).

255. T.K. Sundaram, K.V. Rajagopalan and P.S. Sarma; J. Chromatogr.: 2, 531 (1959).

256. T. Fukuda and K. Mimura; Bitamin: 41, 260 (1970).

257. O.R.P. da Cunha and A.P. da Cunha; Bol. Escola Farm., Univ. Coimhra, Ed. Cient, 25, 54 (1965).

258. F.L. Lam, I.J. Holcomb; and S.A. Fusari; J. Assoc. Off. Anal. Chem., 67, 1007 (1984).

259. T. Cannella, and G. Bichi; Boll. Chim. Farm.: 122, 205 (1983).

260. G.D. Wachob; LC, Liq. Chromatogr. HPLC Mag.: 1, 110 (1983).

261. S.C. Coverly; J. Autom. Chem.: 5, 89 (1983).

262. Ralph W. McKee, Y. Kang Lee, M. Panaqua and M.E. Swendseid; J. Chromatogr.: 230 (Biomed. Appl.) 19, 309 (1982).

263. R.M. Kothari, and M.W. Taylor; J. Chromatogr.: 247, 187 (1982).

264. K. Takikawa, K. Kiyazaki and T. Arita; J. Chromatogr.: 233 (Biomed. Appl.) 22, 343 (1982).

265. R.P. Kwok, W.P. Rose, R. Tabor and T.S. Pattison; J. Pharm. Sci.: 70, 1014 (1981).

266. J.T. Heard, and G.J. Tritz; J. Liq. Chromatogr.: 4, 325 (1981).

267. J.X. de Vries, W. Guenthert, and R. Ding; J. Chromatogr.: 221 (Biomed. Appl.) 10, 161 (1980).

268. G.L. Coleman; S. Afr. Pharm. J.: 46, 621 & 648 (1979).

269. N. Nishimoto, Y. Mitani and S. Hayashi; J. Chromatogr.: 176, 448 (1979).

270. C.L. Guileemin, J.P. Thomas, S. Thiault and J.P. Bounine; J. Chromatogr.; 142, 321 (1977).

271. S.A. Ismaile; Ph.D. Thesis [Pharm. Sci. (Anal. Chem.)], Faculty of Pharmacy, Cairo University (1982).

272. A. Floridi, C. Fini, C.A. Palmerini and A. Rossi; Riv. Sci. Tecnol. Alimenti Nutr. Um.: 6, 197 (1976).

273. K. Callmer and L. Davies; Chromatographia: 7, 644 (1974).

274. M. Caude and L.X. Phan; Chromatographia: 9, 20 (1976).

275. C.P. Talley; Anal. Chem.: 43, 1512 (1971).

276. T. Cannella, G. Bichi, Boll. Chim. Farm.: 122, 205 (1983).

277. A. Barreto, F.A. Gai and H.S.R. Baneto; Rev. Quim. Ind., Rio de Janeiro: 24, 12 (1955); cf. Anal. Abstr.: 3, 566 (1956).

278. D. Zivanov, Z. Blagojevic and M. Mladenovic; Archiv Farm. (Belgrade): 14, 273 (1964); cf. C.A.: 62, 2671h (1965).

279. B.E. Albright and E.T. Degner; Automat. Anal. Chem.; Technicon Symp., 3rd (1967): 1, 461 (publ. 1968); cf. C.A.: 71, 33460u (1969).

280. J. Ramirez-Munoz; Analytical Chim. Acta: 71, 321 (1974).

281. A.F. Gross; J. Assoc. Off. Anal. Chem.: 58, 799 (1975).

282. S.C. Coverly; J. Autom. Chem.: 5, 89 (1983); cf. Anal. Abstr.: 45, 4E29 (1983).

NORFLOXACIN

Claude Mazuel
Pharmaceutical Research & Development
Merck Sharp & Dohme-Chibret
Riom, France

1. HISTORY AND THERAPEUTIC CATEGORY

Norfloxacin is a synthetic antibacterial fluoroquinolone discovered in the late nineteen-seventies (1). It is related to nalidixic acid but its potency has been increased by a fluorine atom and by a piperazine moeity. In-vitro, it is active against a broad spectrum of gram-positive and gram-negative aerobic bacteria (2, 3). Norfloxacin inhibits bacterial deoxyribonucleic acid synthesis and is bactericidal. At a molecular level, three specific events are attributed to norfloxacin in E. Coli. cells (2) :

1. Inhibition of ATP-dependent DNA supercoiling reaction catalyzed by DNA gyrase
2. Inhibition of the relaxation of supercoiled DNA
3. Promotion of double-stranded DNA breakage

Chemistry, history, mechanisms of action, antimicrobial spectrum and pharmacokinetics have been reviewed in several articles (4-10).

2. DESCRIPTION

2.1 Nomenclature

2.1.1 Chemical Names

- 1-ethyl-6-fluoro-1,4-dihydro-4-oxo-7-(1-piperazinyl)-3-quinolinecarboxylic acid
- 3-Quinolinecarboxylic acid, 1-ethyl-6-fluoro-1,4-dihydro-4-oxo-7-(1- piperazinyl)
- 1-ethyl-6-fluoro-1,4-dihydro-4-oxo-7-piperazin-1-yl-quinoline-3-carboxylic acid (British Approved Name)

2.1.2 Generic Name

Norfloxacin

2.1.3 Laboratory Codes

- AM-715
- L-643,316-00J
- MK-366

2.1.4 Trademarks

- Noroxin
- Baccidal
- Chibroxin

2.1.5 CAS Registry Number

70458-96-7

2.2 Formula and Molecular Weight

$C_{16}H_{18}FN_3O_3$ Molecular Weight : 319.34

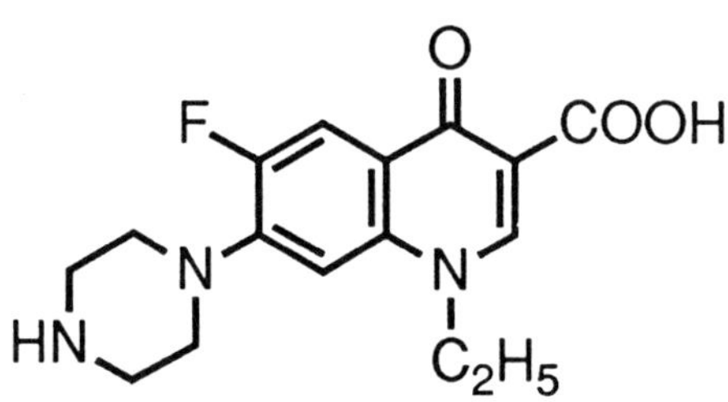

2.3 Appearance, Color, Odor, Taste

Norfloxacin is a white to light yellow crystalline, odorless powder with a bitter taste.

3. SYNTHESIS

Norfloxacin has been prepared by heating 1-ethyl-6-fluoro-7-chloro-1,4-dihydro-4-oxoquinoline-3-carboxylic acid with piperazine in a non-reactive solvent, as outlined in Figure 1 (1). It is obtained with an overall yield of 54 - 64%. Alternate synthetic routes have also been patented (11-17).

4. PHYSICAL PROPERTIES

4.1 Infrared Spectrum

The infrared spectrum of anhydrous norfloxacin shown in Figure 2 (18) was obtained in a potassium bromide pellet using a DIGILAB FTS/IMX spectrophotometer. Assignments of characteristic absorbance bands are given in Table I.

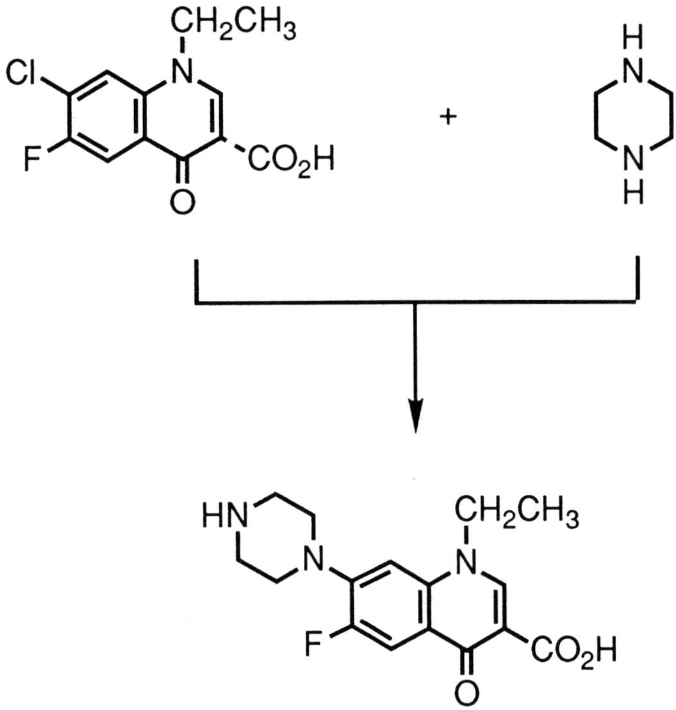

Figure 1. Synthesis of Norfloxacin

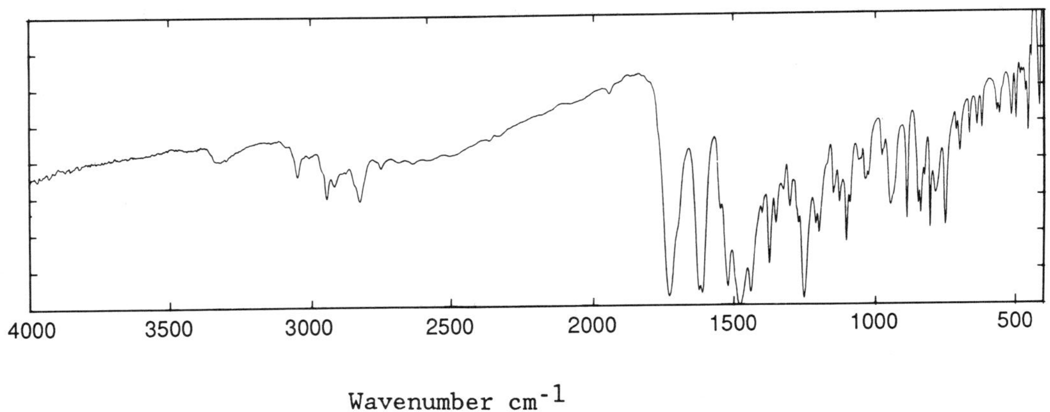

Figure 2. Infrared Absorbance Spectrum
KBr Pellet
Anhydrous Norfloxacin Reference Standard

TABLE I: Norfloxacin Infrared Band Assignment

Wavenumber (CM^{-1})	Assignment
3600 to 3250	N-H and O-H stretch
3050	Olefinic and aromatic C-H stretch
2950	Asymmetric C-H stretch in methyl and methylene of ethyl and piperazine groups
2830	Symmetric C-H stretch in methyl and ethylene of ethyl and piperazine groups
~2500	Hydrogen bonded O-H stretch
1725 - 1700	Carboxylic acid C=O stretch
1628	Pyridone C=O stretch
1619	Pyridone ring C=C stretch
1484	Quinoline ring C-C and C-N stretch
1250	C-F and carboxylic C-O stretch

The infrared spectrum of norfloxacin dihydrate obtained under similar conditions (Figure 3) does not exhibit a well-defined carboxylic C=O stretch.

4.2 Nuclear Magnetic Resonance Spectrum

4.2.1 Proton Magnetic Resonance

The proton magnetic resonance spectrum of norfloxacin shown in Figure 4 (19) was obtained using a Bruker Instruments Model WM-250 spectrometer and an approximately 8% w/v solution of the compound in deuteroacetic acid.

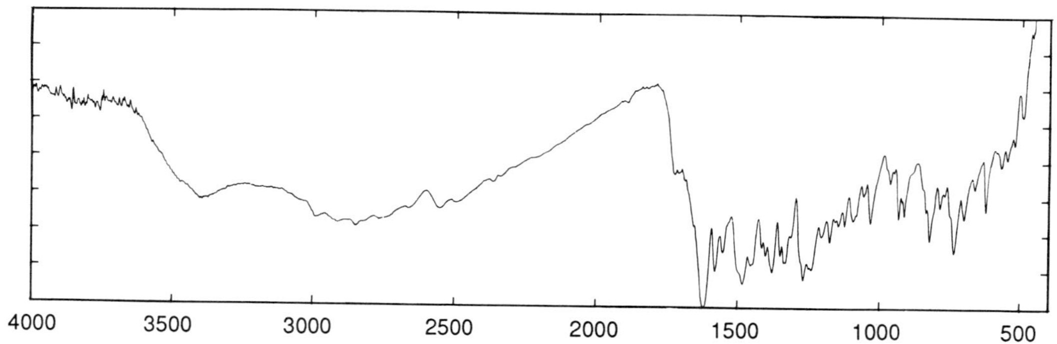

Figure 3. Infrared Absorbance Spectrum
KBr Pellet
Dihydrate Norfloxacin

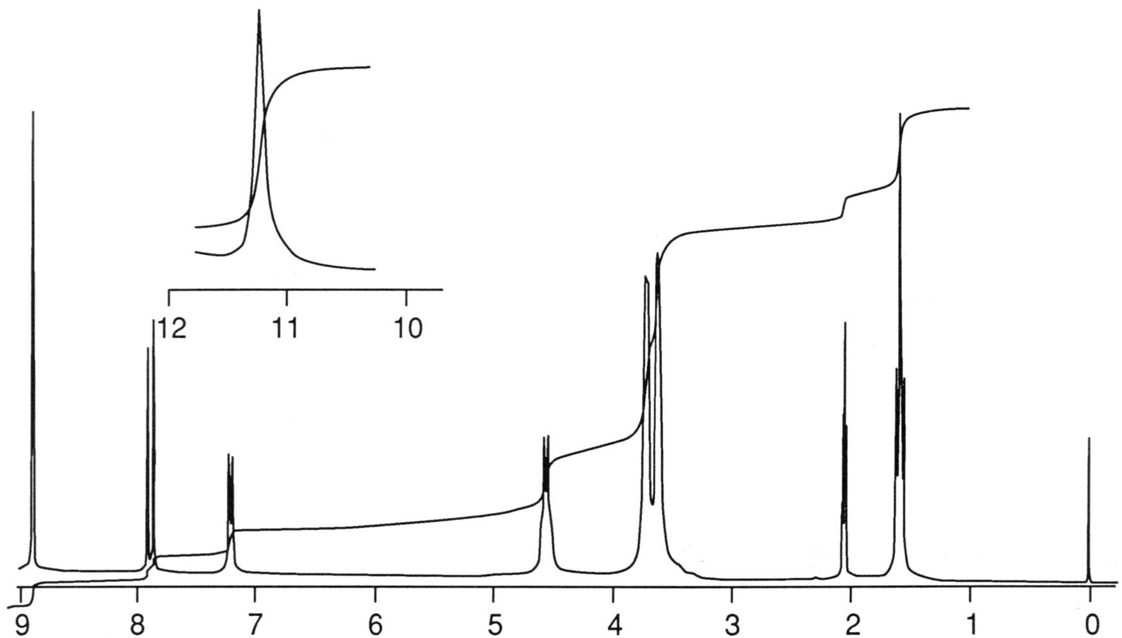

Figure 4. ^{1}H Magnetic Resonance Spectrum Norfloxacin Reference Standard

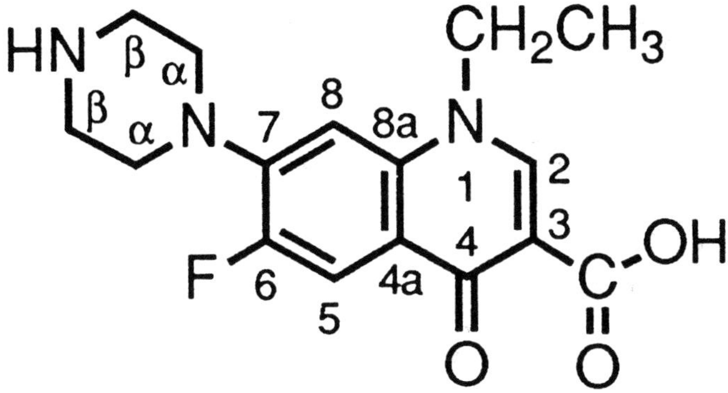

Norfloxacin Structural Formula

The reference compound was tetramethylsilane (internal). Assignments of the observed chemical shifts are tabulated (Table II) according to the following numbered norfloxacin structural formula.

TABLE II: Norfloxacin, Proton Magnetic Resonance Assignments

Chemical Shift (ppm)	Multiplicity/J	Relative No. of Protons	Assignment
1.58	Triplet	2.9	$>N-CH_2C\underline{H}_3$
3.5-3.8	Multiplet	8.2	Piperazine ring methylene protons
4.55	Quartet	2.0	$>N-C\underline{H}_2CH_3$
7.20	Doublet, J_{HF} = 7.0 Hz	1.0	C_8 proton
7.87	Doublet, J_{HF} = 13.0 Hz	1.0	C_5 proton
8.88	Singlet	1.0	C_2 proton

Note: Assignments refer to numbered structure above. Quintet at 2.04 ppm derives from an isotopic impurity in the NMR solvent CD_3COOD. Broad singlet at 11.2 is due to all active hydrogens in the sample

4.2.2 Carbon-13 Magnetic Resonance

The carbon-13 magnetic resonance spectrum of norfloxacin shown in Figure 5 (19) was obtained using a Bruker Instruments Model WM-250 NMR spectrometer and an approximately 8% w/v solution of the compound in deuteroacetic acid. Tetramethylsilane was used as an internal reference compound. Assignments of the observed chemical shifts are tabulated in Table III. Refer to the numbered norfloxacin structural formula shown above.

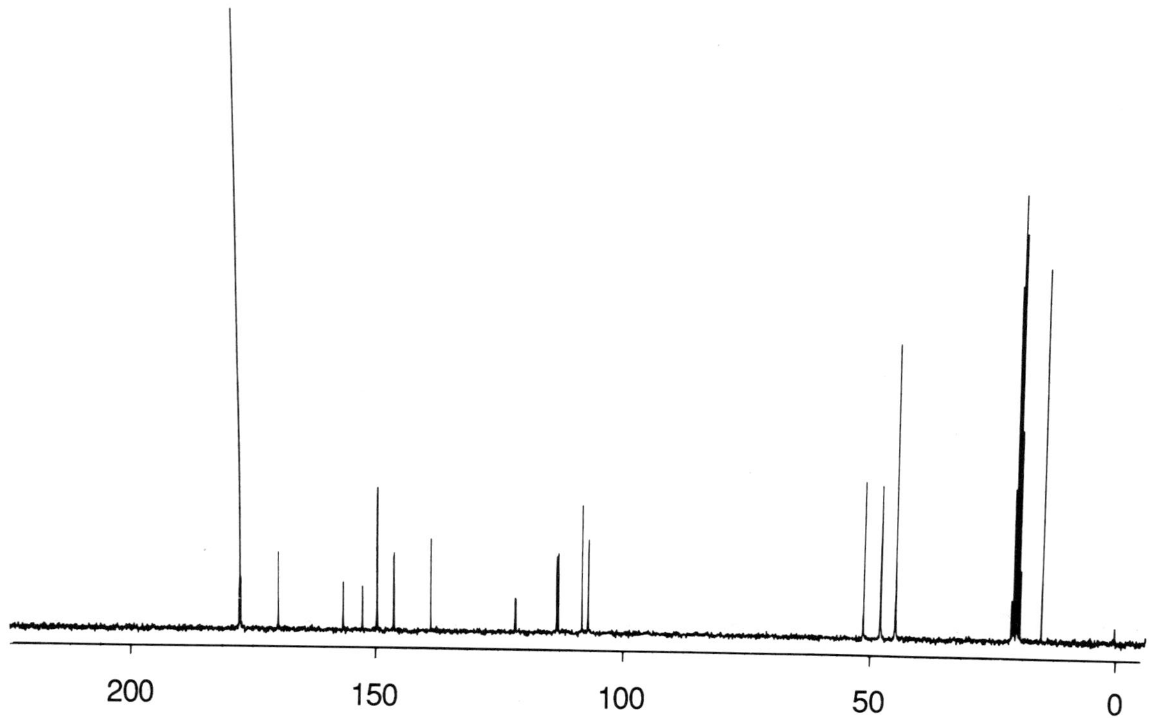

Figure 5. ^{13}C Magnetic Resonance Spectrum Norfloxacin Reference Standard

TABLE III: Norfloxacin, C-13 Magnetic Resonance Assignments

Chemical Shift δc (ppm)	J_{CF} (Hz)	Assignment
149.5	--	C_2
108.2	--	C_3
177.5	2.7	C_4
121.5	8.1	C_{4a}
113.1	23.5	C_5
154.6	250.2	C_6
146.1	10.8	C_7
106.9	2.6	C_8
138.6		C_{8a}
51.1		$CH_3\underline{C}H_2N<$
14.9		$\underline{C}H_3CH_2N<$
α = 47.67; 47.69 β = 44.7		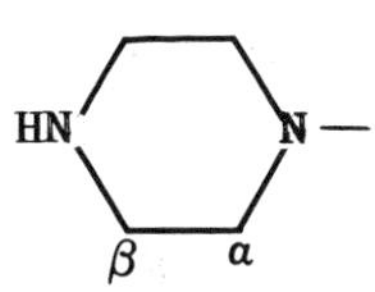
169.9		$-\underline{C}O_2H$

Note: Assignments refer to numbered structure above. The singlet at 177.7 ppm and the septet centered at 19.8 ppm derive from the NMR solvent CD_3CO_2D.

4.3 U.V. Spectrum

The ultraviolet absorbance spectrum of norfloxacin shown in Figure 6 (20) was obtained using a Perkin-Elmer Lambda 5 UV-VIS scanning spectrophotometer. The spectrum in 0.1N sodium hydroxide solution is characterized by maxima at ~274, 325 and 336 nm with respective A1%1cm values of ~1109, 437 and 425.

4.4 Mass Spectrum

The mass spectrum of norfloxacin shown in Figure 7 was obtained by the electron impact (70 eV) direct probe method using an LKB Model 9000 mass spectrometer. Fragment assignments are given in Table IV (21).

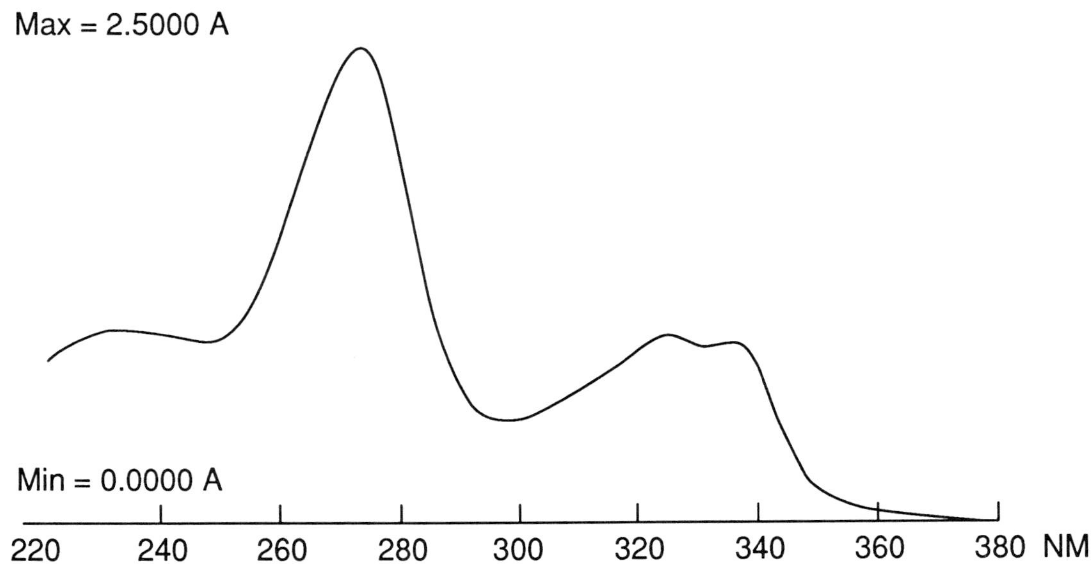

Figure 6. Ultraviolet Absorbance Spectrum
Norfloxacin Reference Standard
Solvent: 0.1N sodium hydroxide

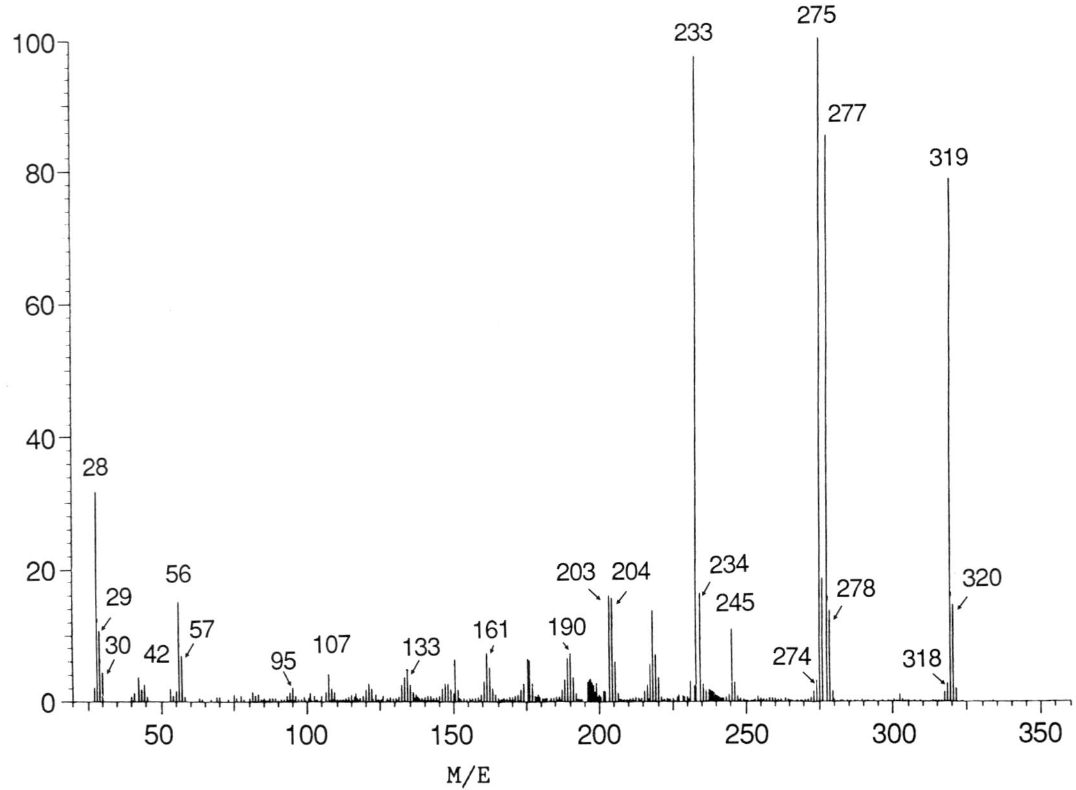

Figure 7. Mass Spectrum
Norfloxacin Reference Standard

Table IV: Norfloxacin Mass Spectrum Assignments

M/E	Assignment
319	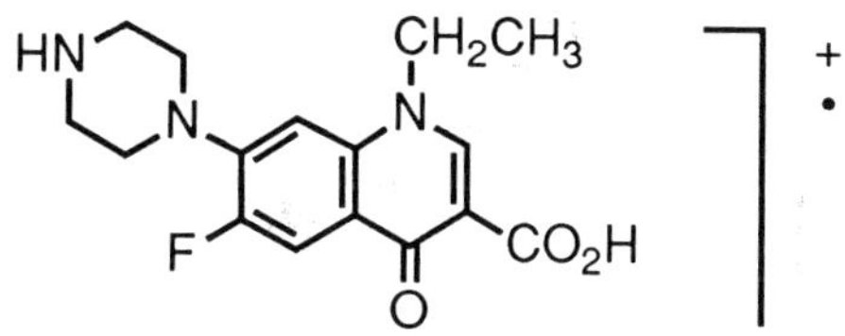 $C_{16}H_{18}FN_3O_3$ (Molecular ion, M^+)
277	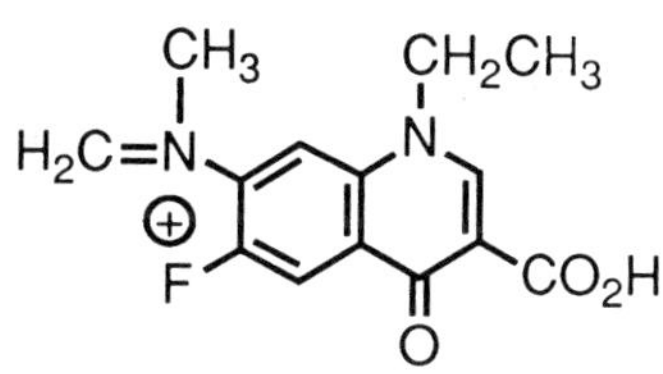 $C_{14}H_{14}FN_2O_3$ (M^+ minus C_2H_4N)
275	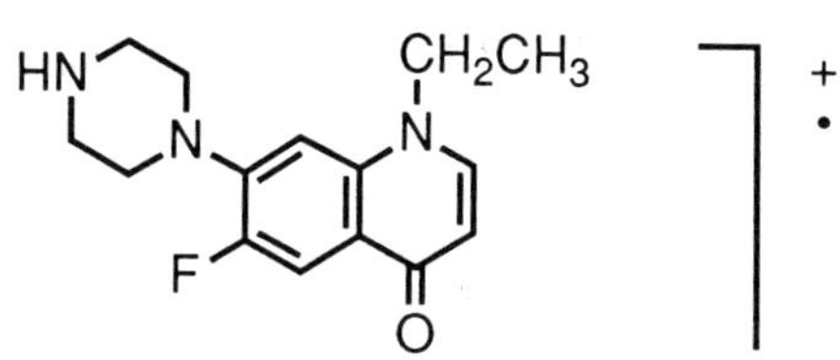 $C_{15}H_{18}FN_3O$ (M^+ minus CO_2)

(continued)

Table IV (continued)

M/E	Assignment
233	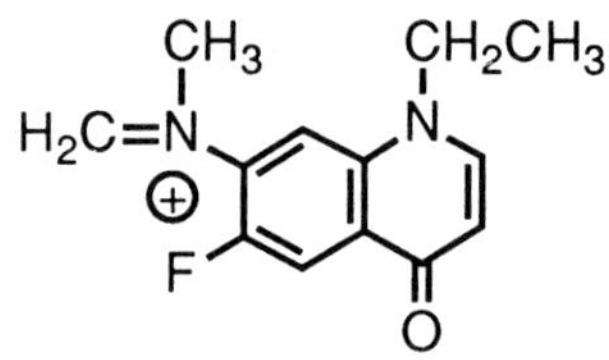 $C_{13}H_{14}FN_2O$ (M^+ minus CO_2 and C_2H_4N)
218	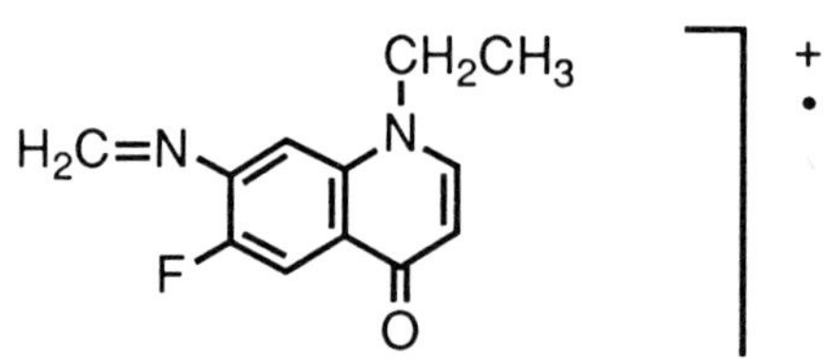 $C_{12}H_{11}FN_2O$ (M^+ minus CO_2 and C_3H_7N)
203/204	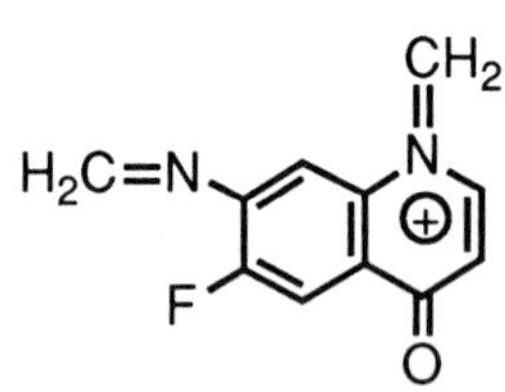 $C_{11}H_{8/9}FN_2O$ (m/e 218 minus CH_3)

4.5 Dissociation Constants

The pKa values for norfloxacin were determined by dissolution of the compound in dilute aqueous sodium hydroxide or hydrochloric acid solution and potentiometric titration of the solutions at 25°C with 0.1N hydrochloric acid or 0.1N sodium hydroxide. The pKa_1 (carboxylic acid) and pKa_2 (protonated piperazine nitrogen) are respectively 6.34 ± 0.06 and 8.75 ± 0.07 (21).

4.6 Solubility

The following approximate solubility values were determined at 25°C (21).

Table V: Norfloxacin 25°C Solubility

Solvent	Solubility (mg/ml)
Water	0.28*
Methanol	0.98
Ethanol	1.90
Acetone	5.10
Chloroform	5.50
Diethyl ether	0.01
Benzene	0.15
Ethyl acetate	0.94
Glacial acetic acid	340.00
Octyl alcohol	5.10

* Solubility in water is pH-dependent, increasing sharply either with decreasing pH below 5 or with increasing pH above ~10 as shown in Figure 8. The reduced solubility between pH 6 and 10 is consistent with the existence of a Zwitterionic form of norfloxacin expected from the 2 pKa values.

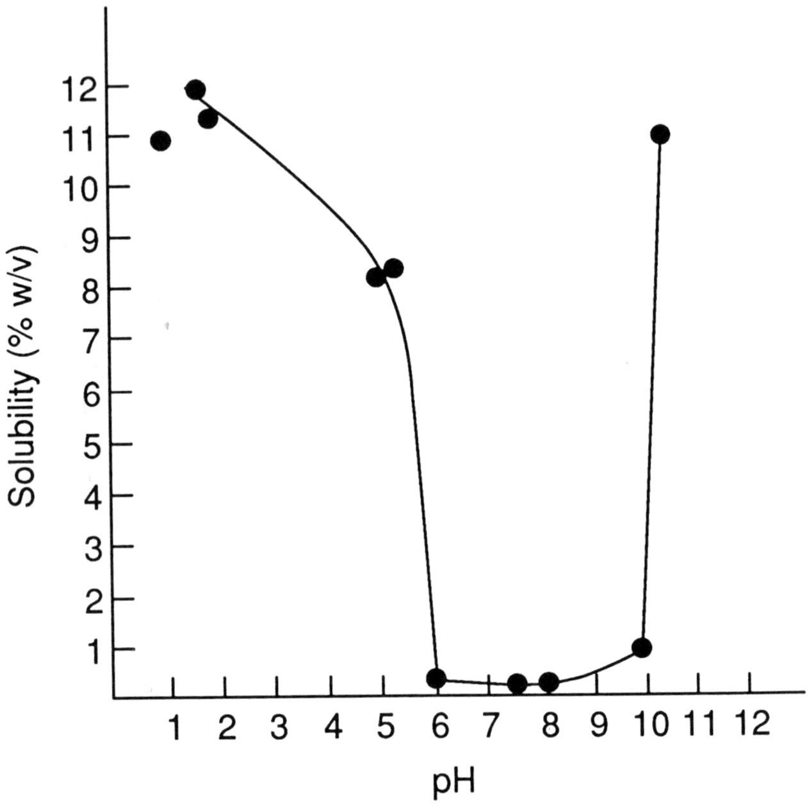

Figure 8. Norfloxacin Solubility vs pH Profile at 25°C

4.7 Partition Coefficients

Partition coefficients of norfloxacin in three aqueous solvent systems (21) are given in Table VI below.

TABLE VI. Partition Coefficients of Norfloxacin (25°C)

Water Phase (15 ml)	Organic Solvent Phase (15 ml)	Partition Coefficient*
Water	Octyl alcohol	0.46
0.1M pH 7.0 phosphate buffer	Chloroform	1.94
0.1M pH 7.0 phosphate buffer	Octyl alcohol	0.43

* The partition coefficient was calculated from the following expression :

$$\text{Partition coefficient (P)} = \frac{[Co]}{[Cw]} \times \left[1 + \frac{Ka_1}{Kw}[H^+] + \frac{Ka_2}{[H^+]}\right]$$

Where:
[Co] = Concentration in organic phase
[Cw] = Concentration in aqueous phase
$Kw = 10^{-14}$ (dissociation constant of water)
$Ka_1 = 10^{-6.34}$ (dissociation constant)
$Ka_2 = 10^{-8.75}$ (dissociation constant)
$[H^+]$ = H ion concentration

4.8 Metal Complexation

Norfloxacin forms complexes with different divalent metal ions in the molar ratio Norfloxacin to metal ion of 2 to 1 (18). Their dissociation constants determined in a pH 7.4 buffer by spectrophotometry are given in Table VII below (22). Formation of complexes with Ca^{++} and Pb^{++} was not evidenced.

TABLE VII: Norfloxacin Complexes Dissociation Constants

Ion	Dissociation Constant as pKD
Zn^{++}	3.9
Cu^{++}	5.7
Mg^{++}	3.6

The formation of a norfloxacin:ferric ion complex with a mole ratio 2:1 was reported (23); its formation constant is 4.0×10^8. This complex was used as the basis for a spectrophotometric determination of iron.

4.9 Hydrate Formation Crystal Properties

Anhydrous norfloxacin picks up moisture rapidly under ambient conditions to form different hydrates. Evidence of formation of a sesquihydrate (theoretical moisture content 7.8%) and of a dihydrate (theoretical moisture content 10.1%) was reported (24). The sesquihydrate and dihydrate were formed after 2 days of exposure of anhydrous norfloxacin to 40°C/75% RH and RT/75% RH respectively.

X-ray diffraction patterns of these 3 forms obtained by the same authors on a Philips Norelco X-ray diffractometer with the CuK_{α} radiation are shown in Figure 9. Reported 2θ values were 9.95, 7.85-11.55 and 8.0-10.0 respectively for anhydrous (Figure 9.a), sesquihydrate (Figure 9.b) and dihydrate (Figure 9.c). The dihydrate was the most thermodynamically stable of these 3 forms under ambient temperature and humidity.

Heat of solution at 30°C in 0.1N sodium hydroxide measured by microcalorimetry was 11.36 ± 0.02; 3.55 ± 0.04 and -1.92 ± 0.13 cal/g respectively for the anhydrous, sesquihydrate and dihydrate (24).

Hemipentahydrate (theoretical moisture content 12.4%), pentahydrate (theoretical moisture content 22.0%) (21) and a hemihydrate have also been reported (25).

4.10 Thermal Behavior

Norfloxacin melts at approximately 221°C (21).

Differential scanning calorimetric thermograms obtained under nitrogen purge are shown in Figure 10 (24). The thermogram of anhydrous norfloxacin (Figure 10.a) shows a single melting endotherm at approximately 220°C. The sesquihydrate (Figure 10.b) exhibits a peak at 120°C due to the loss of bound water. The dihydrate (Figure 10.c) shows a broad endotherm with twin peaks at 85 and 116°C.

5. METHODS OF ANALYSIS

5.1 Identification Tests

5.1.1 Elemental Analysis

Norfloxacin may be identified by its elemental composition:

C 60.18%, H 5.68%, F 5.95%, N 13.16%, O 15.03% (25).

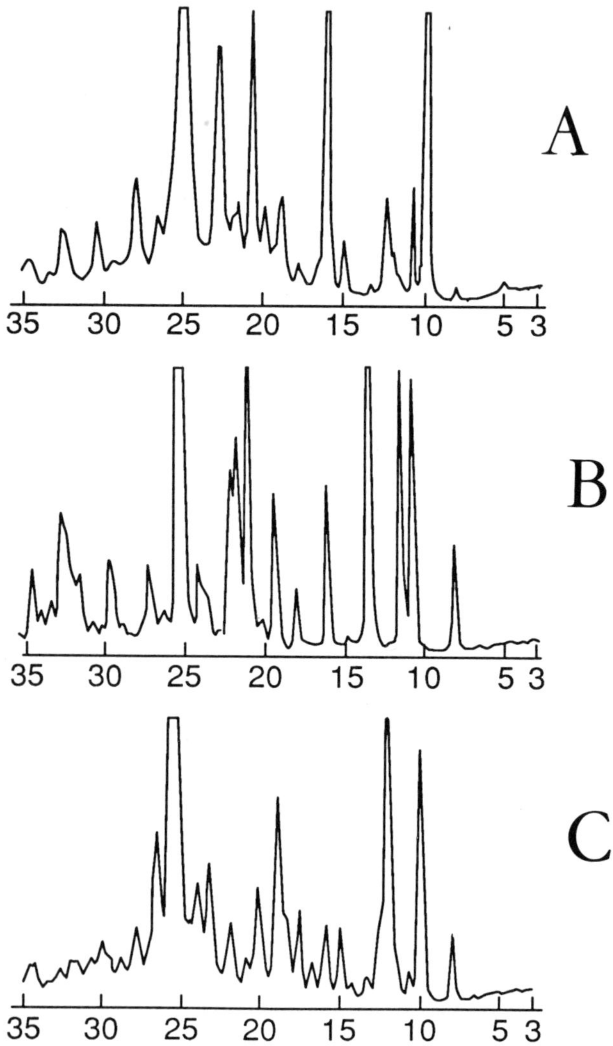

Figure 9. X-Ray Diffraction Pattern of Norfloxacin
A: Anhydrous, B: Sesquihydrate, C: Dihydrate

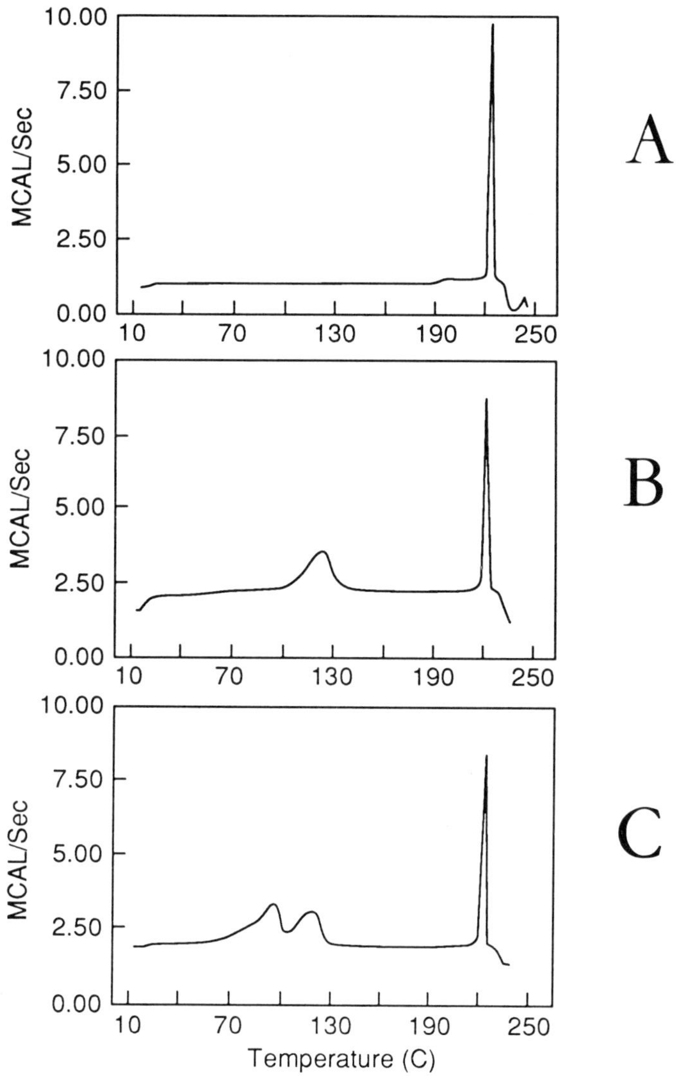

Figure 10. DSC Thermograms of Norfloxacin
A: Anhydrous, B: Sesquihydrate, C: Dihydrate

5.1.2 Infrared Spectroscopy (26)

Norfloxacin may be identified by its infrared absorption spectrum obtained as a mineral oil dispersion of material previously dried; it should exhibit maxima only at the same wavelengths as that of a similar preparation of a norfloxacin reference standard.

5.1.3 Ultraviolet Spectrophotometry (26)

(Note: Use low-actinic glassware in this procedure)

The ultraviolet absorption spectrum of a 1 in 200,000 solution of norfloxacin, previously dried, in a 0.1N sodium hydroxide exhibits maxima and minima at the same wavelengths as that of a similar preparation of a norfloxacin reference standard. The respective absorptivities calculated on a dried basis at the wavelength of maximum absorbance near 273 nm do not differ by more than 3.0%.

5.2 Titrimetric Analysis (26)

<u>Assay:</u> Dissolve about 460 mg of norfloxacin, accurately weighed, in 100 mL of glacial acetic acid. Titrate potentiometrically with standardized 0.1 N perchloric acid in glacial acetic acid using a suitable anhydrous electrode system. [Note: Remove any aqueous solution in the electrode(s), render anhydrous and fill with 0.1 N lithium perchlorate in acetic anhydride]. Perform a blank determination, and make any necessary correction. Each mL of 0.1 N perchloric acid is equivalent to 31.93 mg of $C_{16}H_{18}FN_3O_3$.

5.3 Chromatographic Analysis

5.3.1 Thin-layer Chromatography

Solutions of norfloxacin in methanol or in a mixture of methanol and methylene chloride (1:1) can be examined with the TLC systems summarized in Table VIII.

Table VIII: TLC Systems for Norfloxacin

Solvent System	Plate	Rf	Reference
Chloroform/methanol/ toluene/diethylamine/ water (40:40:20:14:8)	Silica gel 60 F_{254} (E. Merck)	0.35	20
Same as above	HPTLC Silica-gel	-	26
Chloroform/methanol/ Conc. ammonium hydroxide (10:10:3)	Silica gel 60 F_{254} (E. Merck)	0.37	21
Chloroform/glacial acetic acid/methanol/ water (20:10:3:3)	Silica gel 60 F254 (E. Merck)	0.43	21
Glacial acetic acid/ acetone/n-butanol/ toluene/water (1:1:1:1:1)	Silica gel GF (Analtech)	0.40	20

The spots are visualized under ultraviolet light or by exposure of the plate to iodine vapor.

5.3.2 High Performance Liquid Chromatography

Several reverse-phase HPLC systems have been reported for the determination of norfloxacin and impurities/degradates in raw material or in formulations. A summary of these systems is given in Table IX. Detection is done by U.V. at the absorbance maximum near 275 nm or by fluorescence with excitation and emission wavelengths of 285 and 440 nm respectively.

TABLE IX: HPLC Systems for Norfloxacin

Column	Mobile Phase	Temp.	Reference
Zorbax® C8 4.6 mm x 25 cm	75% A, 25% B A = 0.015M NaH_2PO_4, H_3PO_4 to pH 2.4 B = acetonitrile	Ambient	20
Zorbax® C8 4.6 mm x 25 cm	Gradient 100% A --> 40% A, 60% B in 60 min. A = 0.1% (v/v) aqueous H_3PO_4 B = 0.1% (v/v) H_3PO_4/ CH_3CN	30°C	20
Octadecyl Silane (L1) 3.9 mm x 30 cm	85% A, 15% B A = phosphoric acid 1 in 1000 B = Acetonitrile	40°C	26
Lichrosorb® RP18 4.6 mm x 25 cm	50% A, 50% B A = pH 10 borate buffer B = acetonitrile	25°C	27
Polystyrene divinylbenzene 4.6 mm x 25 cm	Acetonitrile/ methanol/THF/8% acetic acid in water (114:19.5:6.5:860)	-	28

5.4 Moisture Determination

Moisture content of norfloxacin raw materials can be determined by changes in the I.R. spectrum as a function of the hydration level. The ratio of the absorbance at 1730 cm^{-1} over absorbance at 1552 cm^{-1} (R) is directly related to the hydration level (18). R values for difference levels are shown in the table below.

TABLE X: Absorbance Ratio 1730/1552 as Function of Moisture Content

Hydration	R Value	% Water Content by TGA
Anhydrous	3.375	-
Dihydrate	0.445	10.8
Hemi-pentahydrate	0.306	13.1
Pentahydrate	0.183	22

6. STABILITY/DEGRADATION

6.1 Solid State Stability (21)

Norfloxacin raw material is stable when protected from light. No degradation was observed following storage at 100°C for 8 hours. Norfloxacin was found stable when stored 24 months at 40°C, and for 12 months at 50 and 60°C. The bulk drug was also stable when stored protected from light in open containers as follows : 24 months at 40°C/75% RH; 12 months at 40°C/96% RH; 12 months at 50°C/75% RH and 12 months at 60°C/75% RH. Exposure to high humidity levels resulted in an increased moisture content.

Norfloxacin is photosensitive. Prolonged exposure of bulk drug in sealed colorless glass ampules either under direct sunlight or under fluorescent light results in the formation of ethylenediamine, amino and formyl derivatives; their structure is shown in Figure 11. The same degradates were obtained by heating the raw material at 190°C.

The formation of a stearoyl derivative was also evidenced after prolonged storage at 60°C of tablets containing magnesium stearate (20).

6.2 Solution Stability

Norfloxacin solutions in 2N hydrochloric acid after heating at 100°C led to the formation of a decarboxylated degradate (see Figure 11), whereas no change was noted in 1N sodium hydroxide after 15 hours at 100°C (21). When protected from light, Norfloxacin solutions at pH 5.2 are stable for at least 36 months at 23 and 30°C and can be autoclaved (20 minutes at 121°C) without detectable degradation. Formation of the ethylene diamine degradate was evidenced in those solutions after prolonged exposure to direct sunlight (20).

Norfloxacin solutions can rapidly become discolored upon contact with lower grade stainless steel because of the formation of metal complexes as explained in Section 4.8.

Solution stability was found optimum at pH 5.5 for an ionic strength of 0.20 (29)

7. BIOPHARMACEUTICS AND METABOLISM

7.1 Pharmacokinetics and Bioavailability

Following single dose administration to fasting healthy volunteers, norfloxacin is readily absorbed to give peak levels in serum and urine between 1 and 2 hours (30, 31).

Single oral doses of 400 mg gave peak serum levels of 1.58 (32) and 1.45 μg/ml (33). About 30% of the administered dose was recovered from urine as unmetabolized norfloxacin.

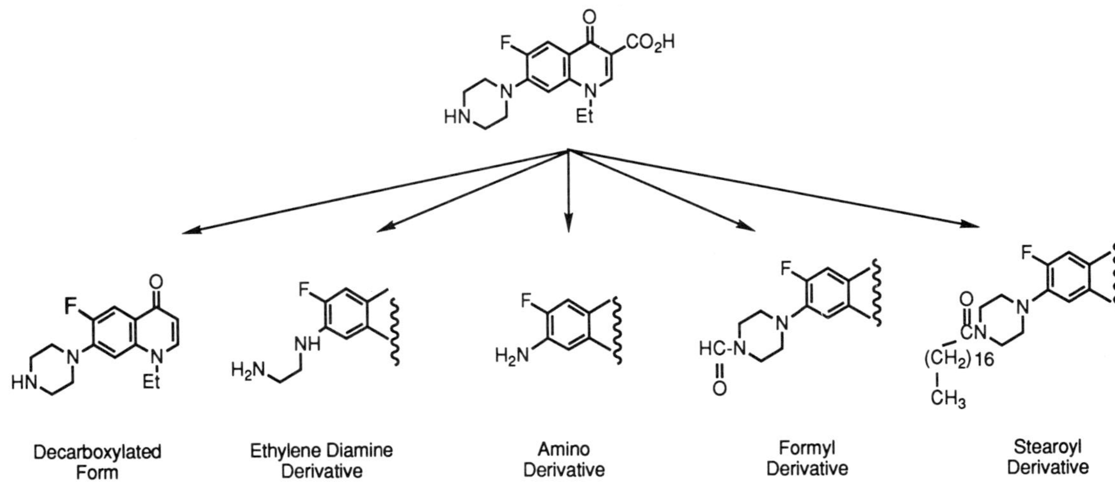

Figure 11. Known Degradates of Norfloxacin and Interaction Product with Magnesium Stearate.

The serum half-life reported by these different authors is approximately 3 hours. Urine concentrations are 100 to 300-times greater than those in serum (32). Bioavailability and pharmacokinetics were recently reviewed (34).

Norfloxacin precorneal clearance was studied following topical administration to the rabbit eye of a 0.3% solution (35). One and six hours after a single instillation of 2 drops, the drug concentration in the conjunctival sac was 89.0 ± 87.8 and 8.3 ± 9.9 μg/ml respectively (mean ± s.d., N = 6). Norfloxacin concentrations in various ocular tissues are also reported.

7.2 Metabolism

Norfloxacin is metabolized in man, to six active metabolites (30, 36). Their structures are shown in Figure 12. The urinary excretion of M1 and M2 concentrations represented 20 and 4% of the parent compound respectively. The other metabolites were present in quantities equal to or less than 1% of the intact drug.

8. DETERMINATION IN BIOLOGICAL FLUIDS

8.1 Liquid Chromatography

Norfloxacin has been assayed in biological fluids primarily by HPLC on reverse phase columns. U.V. or fluorescence detection provides enough sensitivity to obtain limits of quantitation as low as 10 nanograms per ml or per gram of tissue. A summary of the various HPLC systems published is shown in Table XI. Biological fluids have been chromatographed directly following simple dilutions with methanol, Soerensen buffer or diluted phosphoric acid. Extraction into halogenated solvents from a pH 7.5 buffered tissue homogenate followed by evaporation or back extraction either in basic or acid solutions was also used to separate norfloxacin and metabolites from interfering substances.

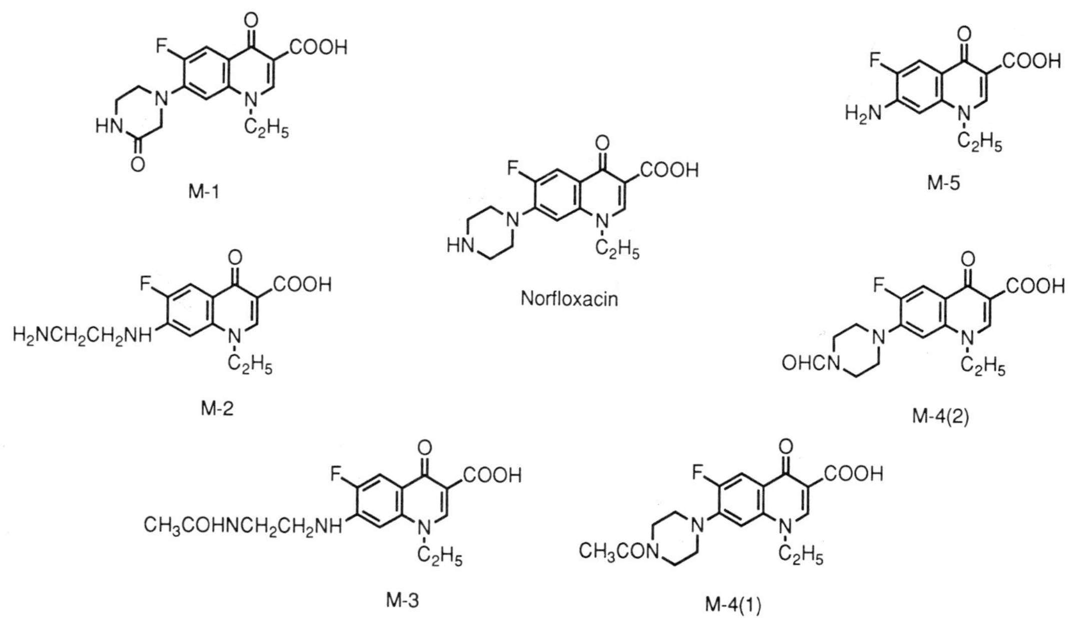

Figure 12. Norfloxacin and its Metabolites

TABLE XI. HPLC Analysis of Norfloxacin in Biological Fluids

Column	Mobile Phase	Detection	Reference
Vydac® 10 microns anion exchange 25 x 0.46 cm	20% A; 80% B A: Acetonitrile B: 0.05M pH 7.0 phosphate	U.V. 273 nm U.V. 280 nm	37 38
Lichrosorb® RP-18 10μ 10 x 0.46 cm	Gradient 20% B increased 2.5%/min for 10 min. A: Water B: Acetonitrile:Water 2:3, 0.4% $Na_2HPO_4.12H_2O$, 0.2% TBAI, pH 9 with Triethylamine	U.V. 270 nm	39
μBondapak® C18 30 x 0.4 cm	30% A; 70% B A: Methanol B: 20 mg/l sodium hesane sulfonate in pH 3.0 phosphate.	Fluorescence Ex: 280 nm Em: 445 nm	40
Nucleosil® C18 5μ 10 x 0.5 cm	Water 850 ml, Acetonitrile 150 ml, 2 g $CH_3COONa.3H_2O$, 2 g citric acid.H_2O, 1 ml Triethylamine	Fluorescence Ex: 330 nm Em: 440 nm	41

Code: Ex = Excitation wavelength. Em = Emission wavelength

(continued)

TABLE XI (continued)

Column	Mobile Phase	Detection	Reference
Nucleosil® C18 5μ 25 x 0.46 cm	Water 700 ml, 1.9 ml H_3PO_4 85% 1.4 g TBAI methanol 300 ml	U.V. 278 nm	42
μBondapak® C18 10μ 15 x 0.39 cm	Water 1000 ml, Acetonitrile 100 ml, TBAH 15 ml, 1.6 ml H_3PO_4 85%	Fluorescence Ex: 278 nm Em: 456 nm	43
Nucleosil® C18 5μ 20 x 0.46 cm	Water 900 ml, Acetonitrile 50 ml, Methanol 50 ml, H_3PO_4 1.5 g, TBAH qs pH 3.0	Fluorescence Ex: 278 nm Em: 446 nm	44
Partisil® PXS C8 10μ 25 x 0.46 cm	19% A; 3% B; 78% C A: Acetonitrile B: Methanol C: pH 2.5 phosphate	Fluorescence Ex: 300 nm Em: 420 nm	45 46
Nucleosil® C18 5μ 20 x 0.4 cm	11% A; 89% B A: Acetonitrile B: 0.025M H_3PO_4 TBAH qs pH 3.0	Fluorescence Ex: 278 nm Em: 445 nm	47

Code: Ex = Excitation wavelength. Em = Emission wavelength

(continued)

TABLE XI (continued)

Column	Mobile Phase	Detection	Reference
Lichrocart Select® B C18 7μ 25 x 0.4 cm	20% A; 80% B A: Acetonitrile B: Water 850 ml, CH_3COONa 1.6g, citric acid 1 g, TBAB 0.27 g, HCOOH qs pH 3.0	Fluorescence Ex: 282 nm Em: 418 nm	48
Nucleosil® C18 5μ 20 x 0.46 cm	80% A; 13% B; 7%C A: 0.015M pH 3.0 phosphate, TBAH B: Methanol C: Acetonitrile	Fluorescence Ex: 278 nm Em: 466 nm	49

Code: Ex = Excitation wavelength. Em = Emission wavelength

Norfloxacin was also separated from homogenized tissues prior to chromatographic analysis on Carbopak® B 80-120 columns after a preliminary wash with 0.02M methanoic acid in methanol and subsequent elution with chloroform/methanol/methanoic acid (45, 46).

8.2 Bioassays

Norfloxacin was assayed in body fluids by a bioassay using E. Coli NIHJ JC-2 and modified Müller-Hinton medium with a detection limit of 0.05 ug/ml (36) and with a disk-diffusion method in serum, tissue and urine with Klebsiella pneumoniae ATCC 10031 (50). The assay sensitivity of the latter is 0.2 μg/ml.

9. PHARMACEUTICAL DOSAGE FORMS

9.1 Potency Assay

A compendial assay for Norfloxacin Tablets (26), is outlined below :

Mobile phase - Filtered and degassed mixture of phosphoric acid solution (1 in 1000) and acetonitrile (850:150).

Standard preparation - An accurately weighed quantity of USP Norfloxacin RS is dissolved and diluted quantitatively in mobile phase to obtain a solution having a known concentration of about 0.2 mg per mL.

Assay preparation - Weigh and finely powder not less than 20 Norfloxacin Tablets. Transfer an accurately weighed portion of the powder, equivalent to about 100 mg of norfloxacin, to a 200 mL volumetric flask. Add 80 mL of Mobile phase, sonicate for 10 minutes, dilute with phosphoric acid solution (1 in 1000) to volume, and mix. Transfer 10.0 mL of this solution to a 25 mL volumetric flask, dilute with Mobile phase to volume, mix and filter through a filter having a porosity of 1 μm or less.

Chromatographic system - The liquid chromatograph is equipped with a 275 nm U.V. detector and a 3.9 mm x 30 cm column that contains chemically bonded octadeylsilane packing (L1), and is operated at 40° ± 1.0°. Precondition the column for several hours with degassed 0.1M monobasic sodium phosphate adjusted with phosphoric acid to a pH of 4.0. The flow rate is about 2 mL per minute; the injected volume about 10 1.

9.2 Dissolution Testing

Dissolution rate of Norfloxacin Tablets is evaluated with the USP XXII apparatus 2:50 RPM, 750 ml of a 0.05M acetate buffer at 37°C (26, 51, 52). Filtered aliquots are diluted with the dissolution medium and their U.V. absorbance determined at the maximum near 313 nm. USP tolerances are met if not less than 80% (Q) of the labeled amount of norfloxacin is dissolved after 30 minutes.

9.3 Moisture Adsorption/Desorption in Tablets

Norfloxacin Tablets adsorb or desorb water at rates which are dependent upon both temperature and relative humidity (53). A linear relationship was shown between relative humidity and moisture pick-up or loss at 25 and 40°C. Adsorption occurs at 25°C between 33 and 93% RH and at 40°C at 49 and 79% RH. The initial rates are affected by temperature. At a constant relative humidity of 75%, the moisture contents are directly proportional to temperature (51).

The effect of hydration on dissolution characteristics was investigated (51). Although breaking strength and disintegration time were not adversely affected, it is shown that moisture adsorption reduces disintegration time in a reversible way. The authors report that hydration of tablets at RT/75% RH improved their dissolution characteristics. Subsequent desorption of moisture upon exposure of these tablets to lower humidities did not revert the effect.

10. References

1. Japanese Patent, Kokai Tokkyo Koho J53141286, December 8 1978. Kyorin Pharmaceutical Co., Ltd., Tokyo, Japan.

2. Physicians' Desk Reference (PDR) Edition 43, 1989. Medical Economics Company Inc., New York

3. H.H. Gadebusch et al., Infection 10, 1, 41-44 (1982).

4. D.E. Nix & J.M. DeVito, Drug Review, 6, 105-117 (1987).

5. D.B. Jack, J. Clin. & Hosp. Pharm., 11, 75-93 (1986).

6. B. Rouveix & B. Pangon, Sem. Hosp. 62, 9, 553-564 (1986).

7. E.J. Goldstein, Am. J. Med., 82, 6B, 3-17 (1987).

8. B. Holmes et al., Drugs 30, 6, 482-513 (1985).

9. R.C. Rowen et al., Pharmacotherapy, 7, 4, 92-110 (1987).

10. J.S. Wolfson & D.C. Hooper, Ann. Intern. Med. 108, 2, 238-251 (19888).

11. Japanese Patent, Kokai Tokkyo Koho JP 59/122470 A2 [84/122470], 14 Jul 1984.

12. Spanish Patent, ES 540226 A1, 1 Dec 1985.

13. Spanish Patent, ES 537072 A1, 1 Jan 1986.

14. Spanish Patent ES 539111 A1, 1 Jan 1986.

15. Spanish Patent ES 548907 A1, 1 Mar 1986.

16. Spanish Patent ES 547912 A1, 1 Feb 1986.

17. Spanish Patent ES 548095 A1, 1 Feb 1986.

18. J. Ryan, Merck Sharp & Dohme Research Laboratories, West Point, Pa., personal communication.

19. L.Di Michele, Merck Sharp & Dohme Research Laboratoires, Rahway, N.J., internal communication.

20. Merck Sharp & Dohme Research Laboratories, West Point, Pa., unpublished data.

21. Kyorin Pharmaceutical Co., Ltd., Tokyo, Japan, unpublished data.

22. E. Oberholtzer, Merck Sharp & Dohme Research Laboratories, West Point, Pa., personal communication.

23. P.B. Issopoulos, Analyst 114, 627-630 (1989).

24. A.V. Katdare et al., Mikrochim. Acta (Wien) III, 1-12 (1986).

25. The Merck Index, 10th Edition, Merck & Co. Inc., Rahway, N.J. (1983).

26. U.S. Pharmacopeia, XXII, Mack Printing Company, Easton, Pa., (1990).

27. D. Marini et al., Rass. Chim., 37, 1, 19-22 (1985).

28. A. Rotar, P. Solmajer Lampic, Acta Pharm. Jugosl., 39, 123-128 (1989).

29. G.N. Singh et al., Pharmazie 43, 134 (1988).

30. T. Ozaki et al., Chemotherapy, 29, S-4, 128-135 (1981).

31. M. Eandi et al., Eur. J. Clin. Microbiol, 2, 3, 253-259 (1983).

32. B.N. Swanson et al., Antimicrob. Agents & Chemother., 23, 2, 284-288 (1983).

33. N. Adhami et al., J. Antimicrob. Chemother., 13: 87-92 (1984).

34. G. Stein, Am. J. Med. 82, suppl 6B, 18-21 (1987).

35. M. Ooishi et al., Acta Soc. Ophthalmol. Jpn., 91, 161-167 (1987).

36. Y. Oomori et al., Chemotherapy, 29, S-4, 91-97 (1981).

37. V.K. Boppana & B.N. Swanson, Antimicrobial Agents & Chemotherapy, 21, 5, 808-810 (1982).

38. C. Forchetti, J. Chromato. Biomed. Appl., 309, 177-182 (1984).

39. G. Montay et al., Chromatography Biomed. Appl., 272, 359-365 (1983).

40. L.T. Pauliukonis et al., J. Pharm. Sci., 73, 1, 99-102 (1984).

41. G. Montay & J.P. Tassel, J. Chromato. Biomed. Appl., 339, 214-218 (1985).

42. A.J.N. Groeneveld & J.R.B.J. Brouwers, Pharmaceutisch Weekblad Scientific Edition, 8, 79-84 (1986).

43. S.J. Morton et al., Antimicrob. Agents & Chemother. 30, 2, 325-327 (1986).

44. W. Schönfled et al., Zbl. Bakt. Hyg. A 261, 338-344 (1986).

45. A. Lagana et al., J. Chromato. Biomed. Appl., 417, 135-142 (1987).

46. A. Lagana *et al.*, J. Pharm. Biomed. Anal. 6, 3, 221-228 (1988).

47. I. Nilsson-Ehle, J. Chromato. Biomed. Appl., 416, 207-211 (1987).

48. J.P. McCann *et al.*, Brit. J. Clin. Pharmacol., 23, 5, 652P (1987).

49. J. Knöller *et al.*, J. Chromato. Biomed. Appl., 427, 257-267 (1988).

50. J. Bland *et al.*, Eur. J. Clin. Microbiol., 2, 3, 249-252 (1983).

51. A.V. Katdare & J.F. Bavitz, Drug Dev. & Ind. Pharm., 10, 5, 789-807 (1984).

52. A.V. Katdare & J.F. Bavitz, Drug Dev. & Ind. Pharm., 13, 6, 1047-1061 (1987).

53. A.V. Katdare & J.F. Bavitz, Drug Dev. & Ind. Pharm. 10, 7, 1041-1048 (1984).

Acknowledgements

The author wishes to thank Ms Florence Berg for performing the literature searches and Ms Marjorie Cremont for typing the manuscript.

OXAMNIQUINE

Iqbal Ahmad and Tauqir Ahmad

Department of Pharmaceutical Chemistry,
Faculty of Pharmacy,
University of Karachi, Karachi-75270, Pakistan

K. Usmanghani

Department of Pharmacognosy, Faculty of Pharmacy,
University of Karachi, Karachi-75270, Pakistan.

1. INTRODUCTION

Oxamniquine became available in 1975 for the treatment of the tropical disease schistosomiasis, a parasitic worm infection of man (1). Schistosomiasis (also known as bilharzia or "snail fever") is one of the most debilitating and widespread water- borne disease endemic in 74 countries, infecting 200 million people (2). Oxamniquine is a metabolite of the most active of a novel series of 2-aminomethyltetrahydroquinoline compounds that showed promising schistosomicidal activity and low toxicity in laboratory animals (3,4). It is highly active against *Schistosoma mansoni*—as a single dose against susceptible strains—but therapeutically useful activity has not been demonstrated against either *S. haematobium* or *japonicum* (1). The drug is effective in all stages of the infection and in patients with hepatosplenic involvement. It causes worms to shift from the mesenteric veins to the liver where they are destroyed (5). The primary mechanism of action of oxamniquine in unknown, even though the drug exhibits anticholinergic properties (6). Several reviews covering various aspects of the drug have been published (1,2,7-11).

2. DESCRIPTION

2.1 Name, Formula, Molecular Weight

Oxamniquine is 1,2,3,4-tetrahydro-2-[(isopropylamino) methyl]-7-nitro-6-quinolinemethanol (12). The generic name is derived from 6-hydroxymethyl-2-N-isopropylaminomethyl-7-nitro-1,2,3,4- tetrahydro-quinoline (10).

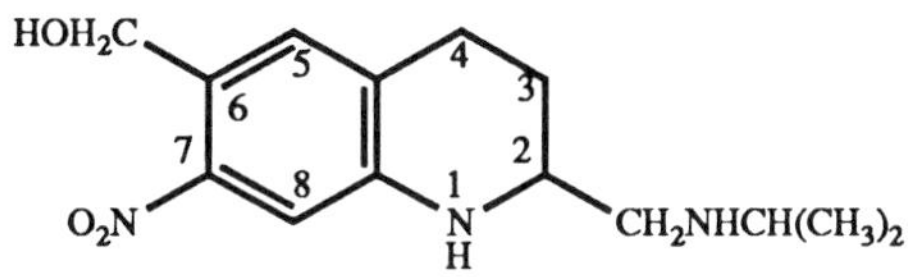

$C_{14}H_{21}N_3O_3$ = 279.3

The CAS registry No. is 21738-42-1
Official monographs for oxamniquine are given in Brazilian (1977) and United States (1990) Pharmacopeias.

2.2 Appearance, Colour

A light orange crystalline powder (13).

2.3 Proprietary Names

Mansil, Vansil (13,14).

3. SYNTHESIS

The synthesis of 2-aminomethyl-1,2,3,4-tetrahydroquinolines and derivatives with schistosomicidal activity has been described by Richards (15,16) and Baxter and Richards (17,18) respectively. Oxamniquine is prepared through a series of synthetic steps shown in Scheme I. 2,6-Dimethylquinoline (1) is converted via the 2- chloromethyl derivative (II) to the 2-isopropylamino compound (III) and reduced with Raney Ni to

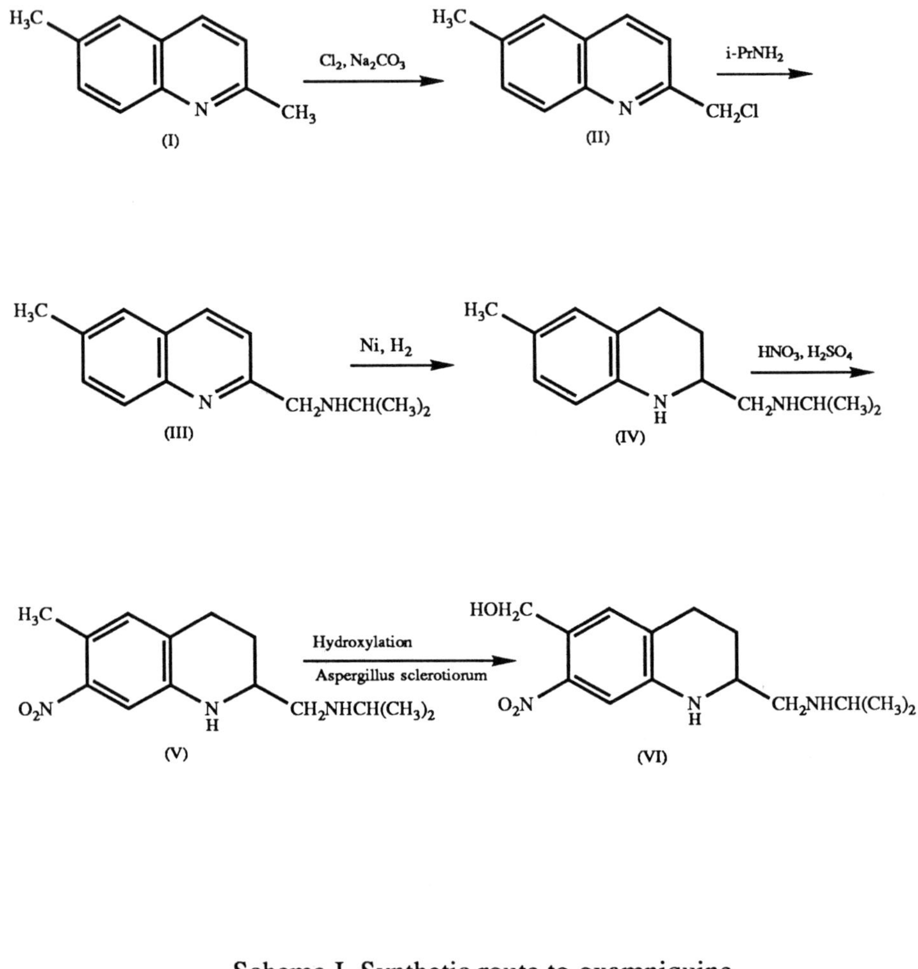

Scheme I. Synthetic route to oxamniquine

yield 2-N-isopropylaminomethyl-6-methyl-1,2,3,4-tetrahydroquinoline (IV), which is nitrated to give 2-N-isopropylaminomethyl-6-methyl-7-nitro-1,2,3,4- tetrahydroquinoline (V). This is transformed into the 6-hydroxymethyl analogue, oxamniquine (VI), by microbial hydroxylation using *Aspergillus sclerotiorum* Huber.

The preparation of embonate, hydroxynaphthoate, phthalate and salicylate salts of oxamniquine has been reported (19).

4. PHYSICAL PROPERTIES

4.1 Melting Point

It melts at 151o to 152° C (13,20).

4.2 Specific Rotation

Between - 4° and + 4° (2% solution in tetrahydrofuran) (12)

4.3 Crystallisation Solvent

Pale yellow crystals are obtained from isopropanol (14).

4.4 Polymorphic Forms

It occurs in two modifications which crystallise as orange spherulites. Modification I melts at 146-148°C and modification II at 117- 118°C. Modification II is slowly converted to modification I. Modification I is monotropic and is used in commercial products (21).

4.5 Solubility

It is soluble in about 3300 parts of water at 27°, soluble in acetone, chloroform and methyl alcohol (13).

4.6 Acidity

The pH of a suspension (1 in 100) is between 8.0 and 10.0 (12)

4.7 Water content

Not more than 1.0% (12).

4.8 Residue on Ignition

Not more than 0.2% (12).

4.9 Heavy Metals

The limit is 0.005% (12)

4.10 Hydrophobicity Parameter

The mean value of the hydrophobicity parameter, log P, in octanol/water system is 2.245 (22).

4.11 Ultraviolet (UV) Spectrum

The ultraviolet spectrum of oxamniquine (USP Reference Standard) in methanol was recorded on a Shimadzu 240 UV-Visible spectrophotometer and is shown in Figure 1. It exhibits sharp maxima at 205.5 and 249.5nm and a broad maximum at 389.5 nm. The spectral constants obtained are reported in Table I.

Table I

UV spectral Data for Oxamniquine

λ_{max},nm	A(1%,1cm)	Molar Absorptivity
205.5	486	13580
249.5	695	19410
389.5	62.5	1750

4.12 Infrared (IR) Spectrum

The infrared spectrum of oxamniquine is shown in Figure 2. It was obtained as KBr disc with a Jasco A-302 Infrared spectrophotometer. The assignments for characteristic bands are given in Table II.

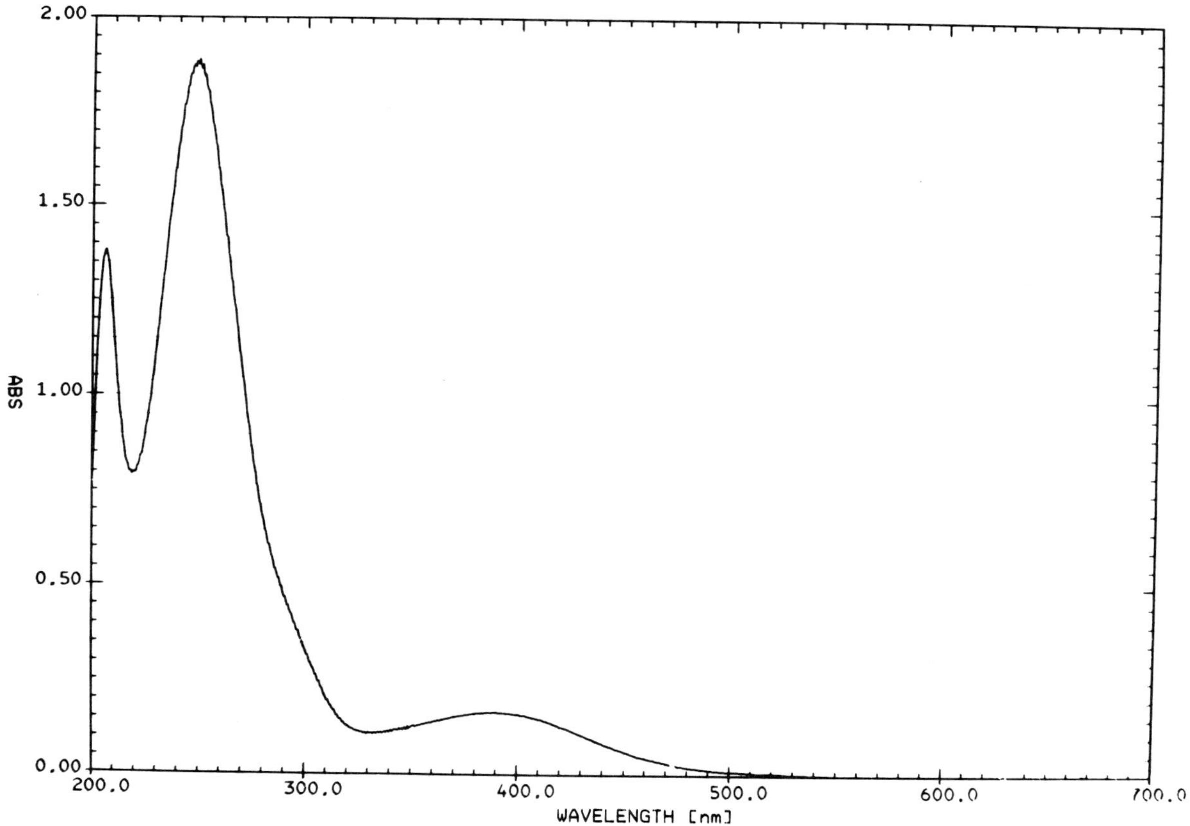

Figure 1. Ultraviolet Spectrum of Oxamniquine in methanol.

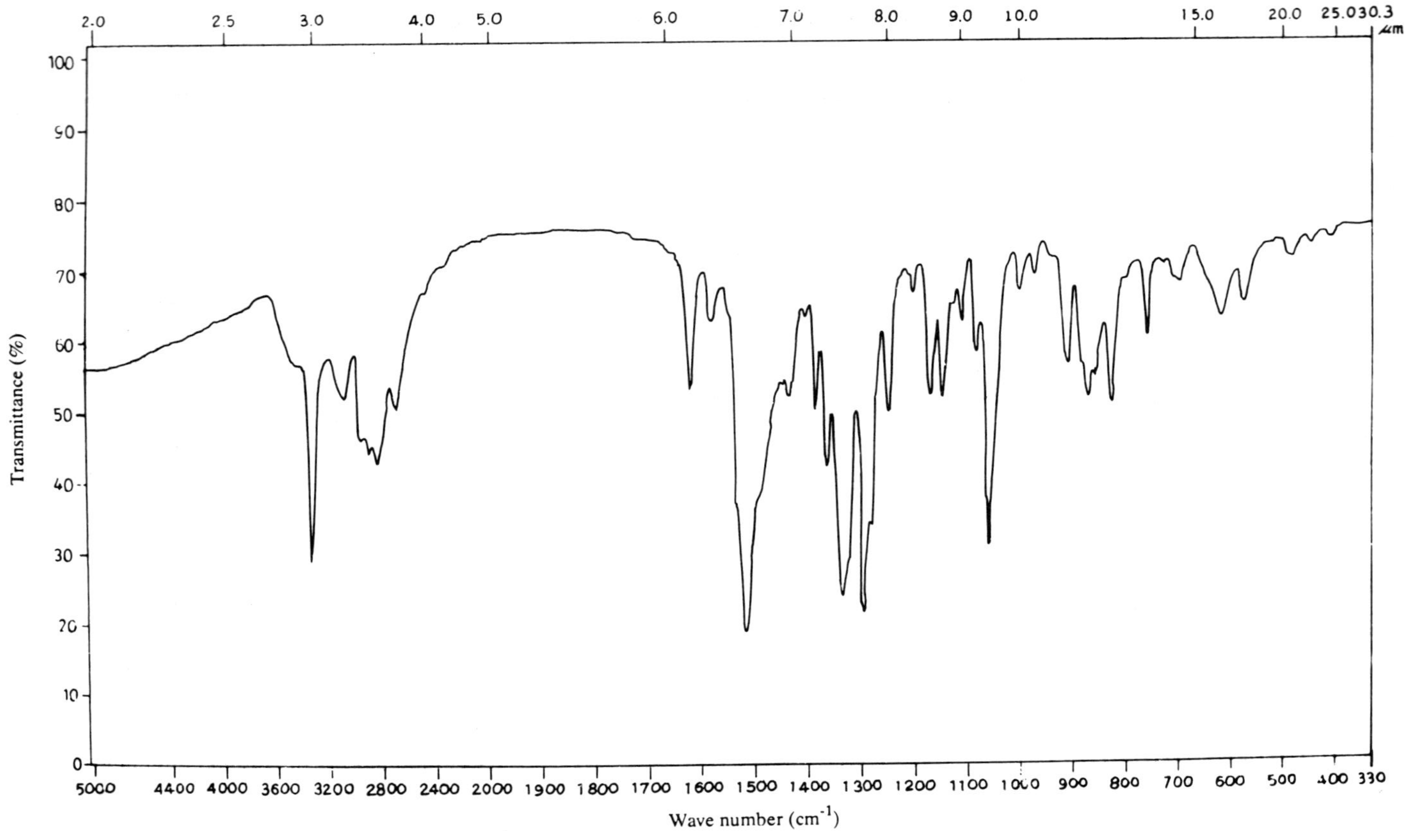

Figure 2. Infrared Spectrum of Oxamniquine (KBr disc).

Table II

IR Spectral Assignments for Oxamniquine

Frequency, cm^{-1}	Assignment
3340	- NH stretching
3100	- OH stretching
2850	- CH stretching
1620	C=C stretching
1520	- NO_2 symmetric stretching
1335	- NO_2 asymmetric stretching
1060	C-O stretching

4.13 Nuclear Magnetic Resonance (NMR) Spectrum

The ^{1}H-NMR and ^{13}C-NMR spectra of oxamniquine in $CDCl_3$ were determined at 300 MHz and 75.4 MHz respectively on a Bruker AM- 300 NMR spectrometer using tetramethylsilane as reference standard. The ^{1}H-NMR determinations included spin decoupling experiments, 2D J-resolved and COSY-45 measurements (Fig. 3-5). The ^{13}C-NMR spectra comprised DEPT and hetero-nuclear (C-H correlation) measurements (Fig. 6,7). The spectral assignments are listed in Table III.

Table III

^{1}H and ^{13}C Chemical Shifts Assignments and Coupling Constants for Oxamniquine

^{1}H-NMR Chemical shift ppm	Proton	Coupling constant	Hz	^{13}C-NMR Chemical shift ppm	Carbon
1.06	4'	$J_{(4',3')}$	6.22	51.39	2
1.08	5'	$J_{(5',3')}$	6.26	25.80	3
1.54	3b	$J_{1'a,1'b}$	11.68	26.57	4
1.94	3a	$J_{1'a,2}$	3.65	131.69	5
2.46	1'b	$J_{1'b,2}$	9.69	123.69	6
2.79	4a,4b,3'	$J_{2,1'a}$	3.69	127.71	7

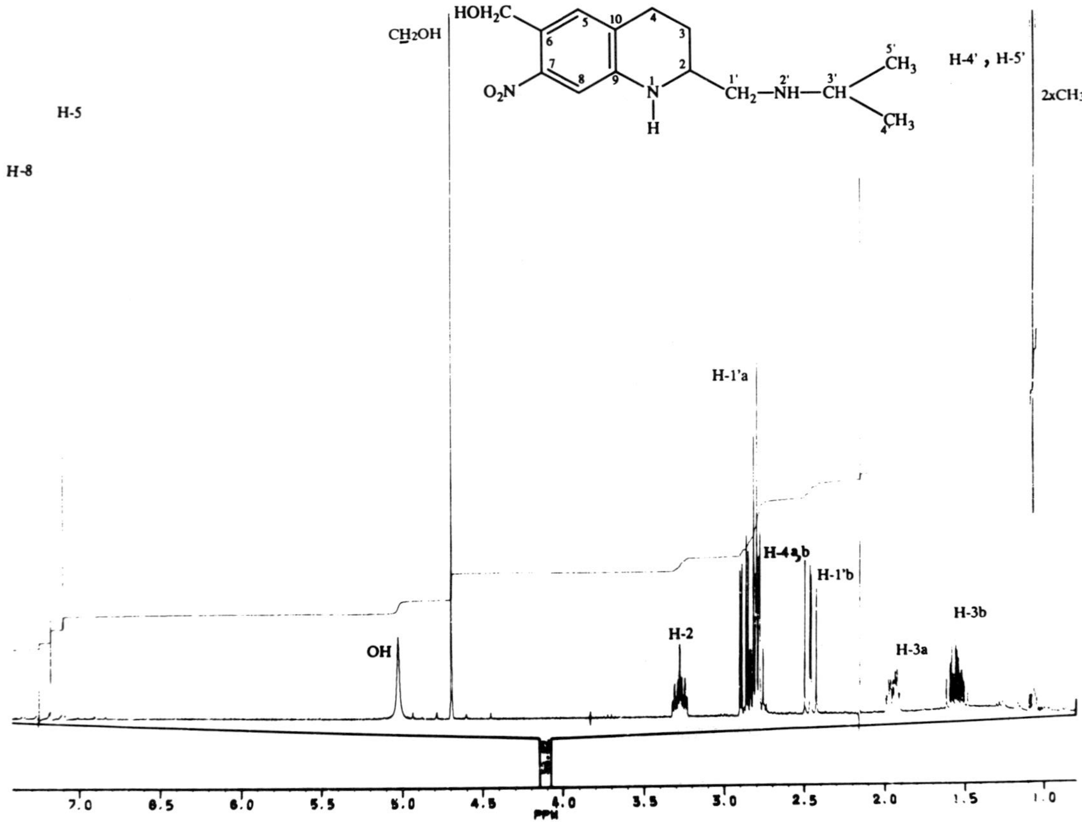

Figure 3. 300 MHz ^{1}H-NMR Spectrum of Oxamniquine (CDCl$_3$).

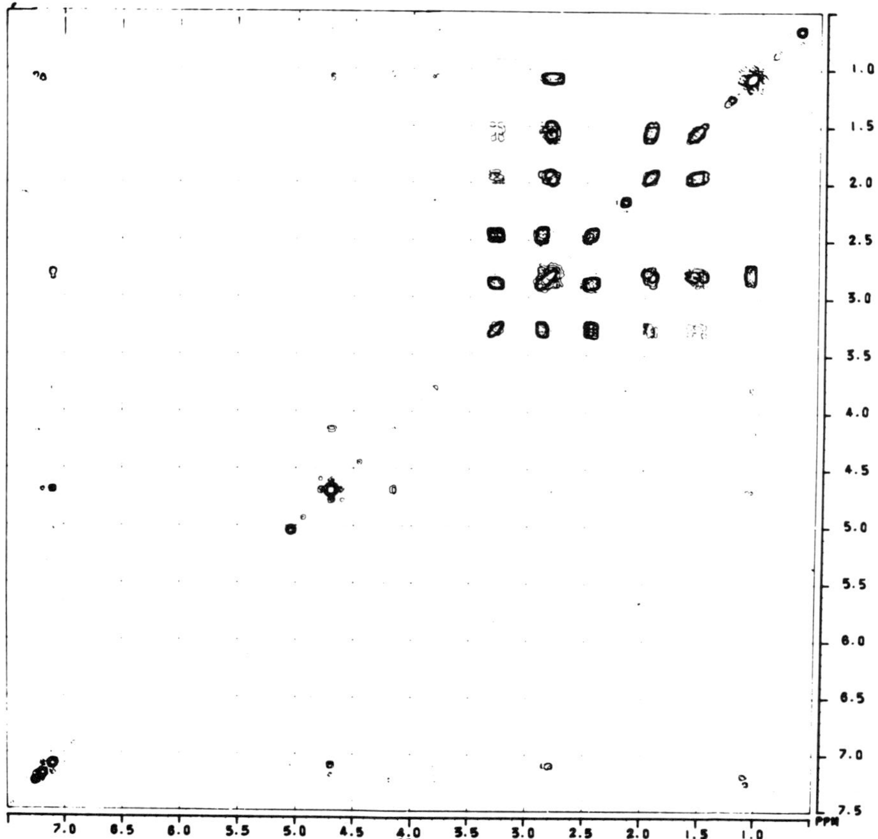

Figure 4. Homonuclear Chemical Shift Correlated (COSY-45) 2D NMR Spectrum of Oxamniquine.

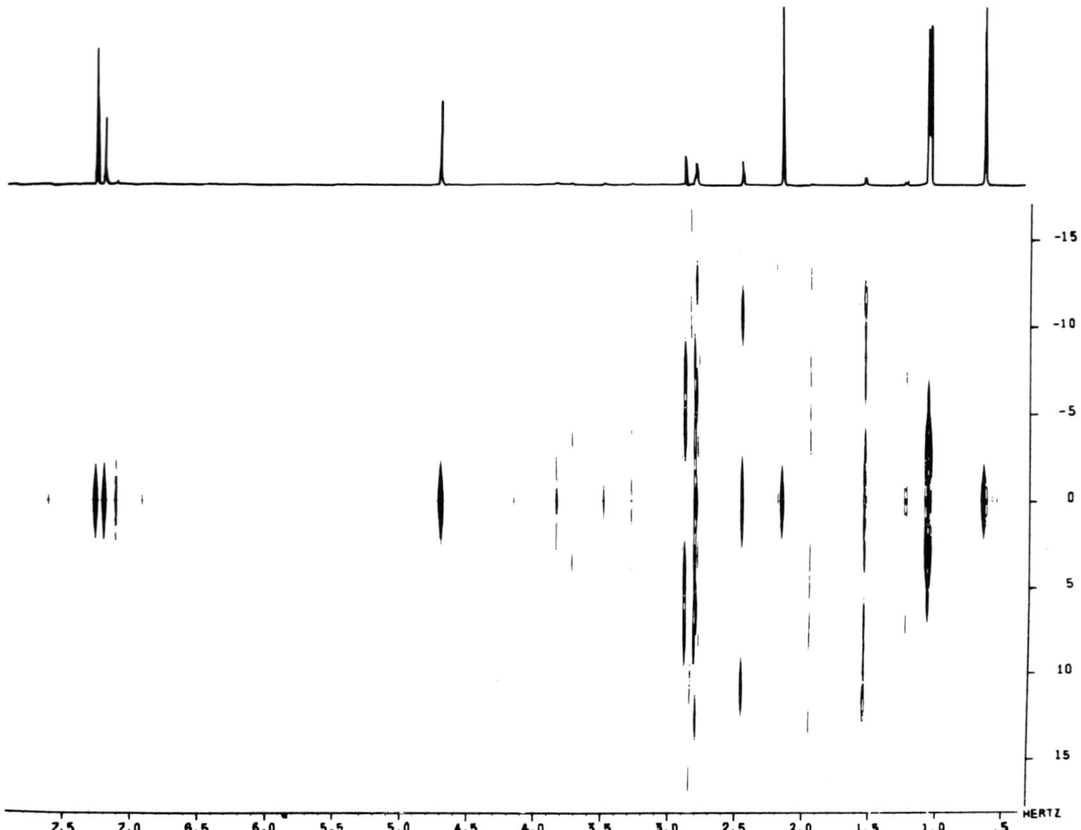

Figure 5. Homonuclear 2D J-resolved NMR Spectrum of Oxamniquine.

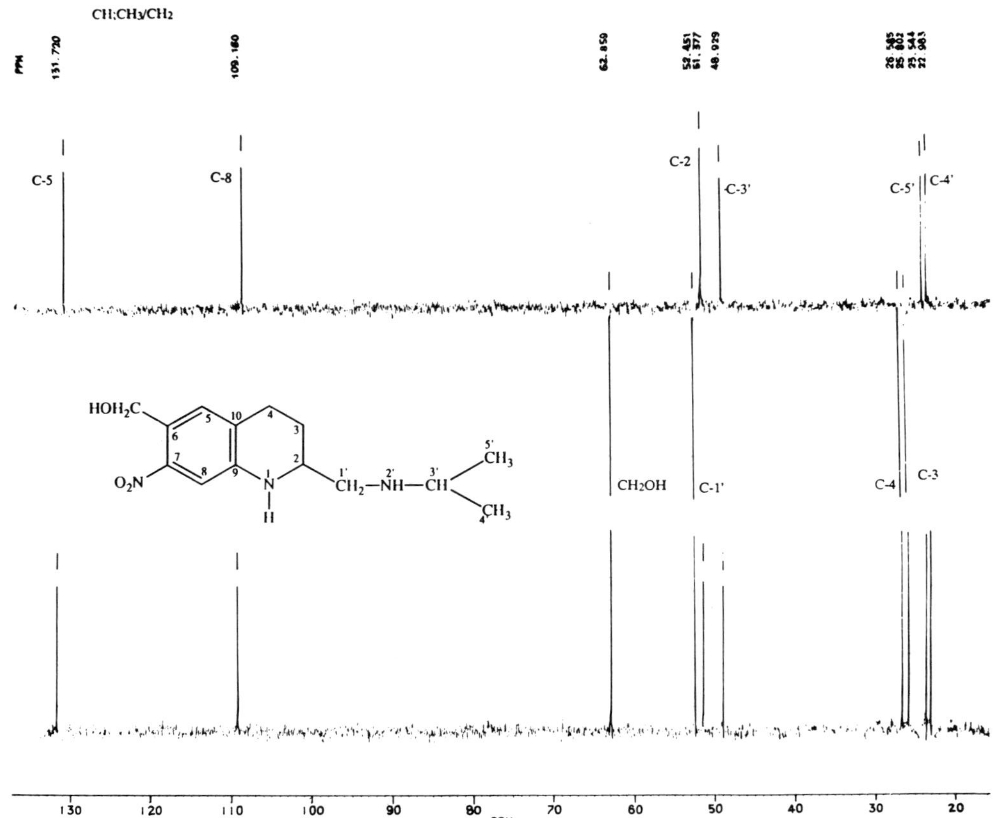

Figure 6a. 75 MHz ^{13}C-NMR (DEPT) Spectrum of Oxamniquine ($CDCl_3$).

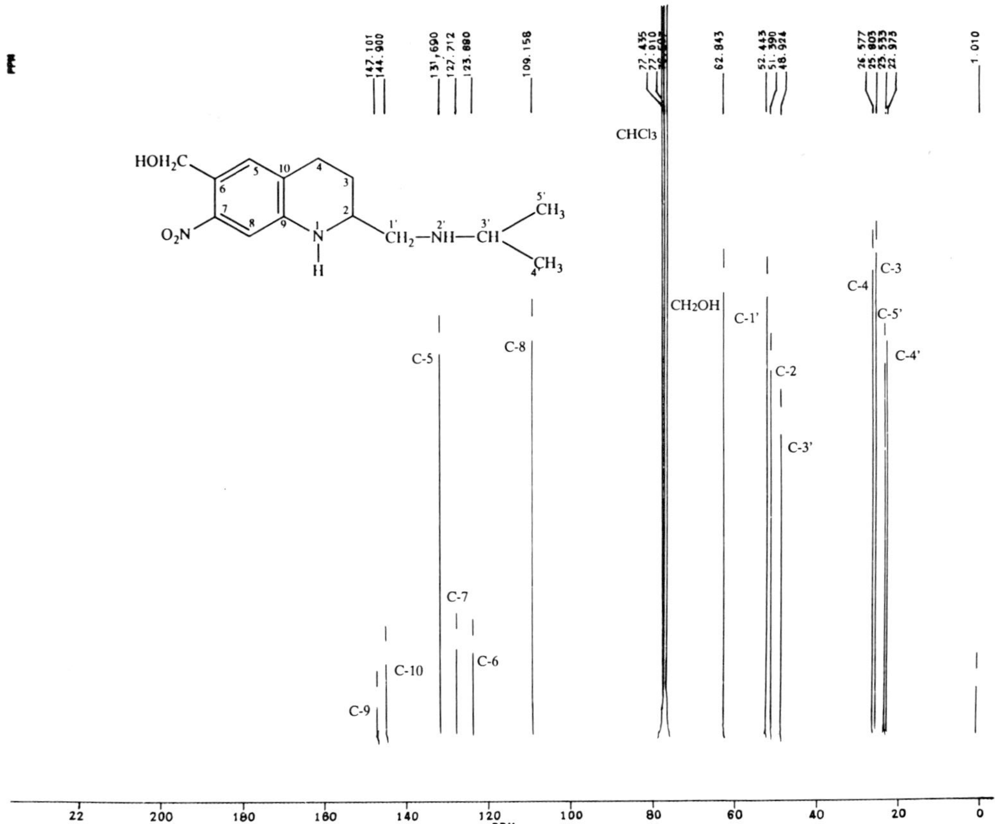

Figure 6b. 75 MHz ^{13}C-NMR Off Resonance Decoupled Spectrum of Oxamniquine ($CDCl_3$).

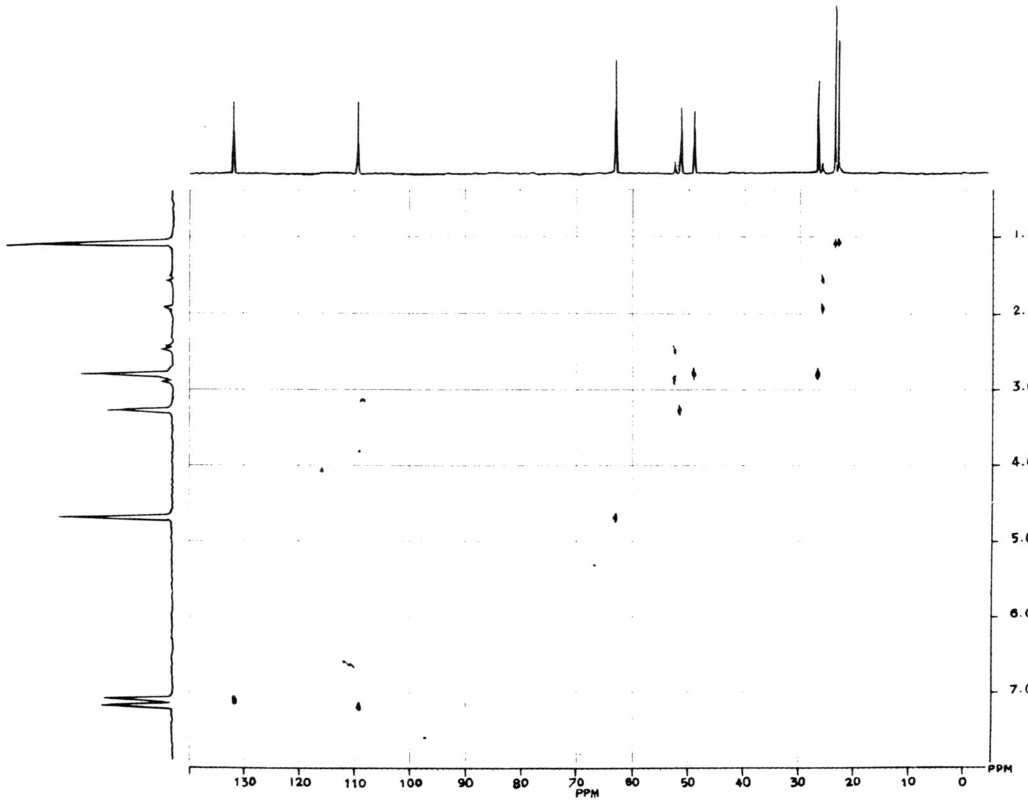

Figure 7. Heteronuclear Chemical Shift Correlated (Heterocosy) 2D NMR Spectrum of Oxamniquine.

"Table III (Continued)"

^{1}H-NMR Chemical shift ppm	Proton	Coupling constant	Hz	^{13}C-NMR Chemical shift ppm	Carbon
2.87	1a	$J_{2,1'b}$	3.65	109.15	8
3.28	2	$J_{2,3a}$	9.69	147.10	9
4.69	C<u>H</u>$_2$OH	$J_{2,3b}$	3.65	144.90	10
5.02	OH			52.44	1'
7.10	5			48.92	3'
7.18	8			22.97*	4'
				23.53*	5'
				62.84	CH_2OH

*Values may be interchanged.

4.14 Mass Spectrum

The electron impact ionization spectrum of oxamniquine obtained at 70 eV using a solid probe insertion is shown in Figure 8. The spectrum was run on a Finnigan Mat 1125 double focusing mass spectrometer connected to a PDP 11/34 (DEC) computer system. Table IV illustrates the prominent fragments and their mass/charge (m/z) ratios.

5. QUALITATIVE TESTS

5.1 Identification (12)

5.1.1 The infrared absorption spectrum of a mineral oil dispersion of oxamniquine exhibits maxima only at the same wavelengths as that of a similar preparation of USP Oxamniquine RS.

5.1.2 Prepare a solution of 25 mg in 1 ml of warm chloroform. On a suitable thin-layer chromatographic plate coated with a 0.25 mm layer of silica gel G, apply 20 ul of this solution and 20 ul of a standard solution of USP Oxamniquine RS in the same solvent having a concentration of 25 mg per ml. Allow the spots to dry, and develop the chromatogram in a solvent system consisting of chloroform, n-hexane, isopropyl alcohol, isopropylamine (200:100:20:5) until the solvent front has moved about three-fourth of the length of the plate. Remove the plate from the developing chamber, mark the solvent front and locate the spots on the plate by viewing under short-and long-wavelength ultraviolet light: The R_f value of the principal spot obtained from the test solution corresponds to that obtained from the standard solution.

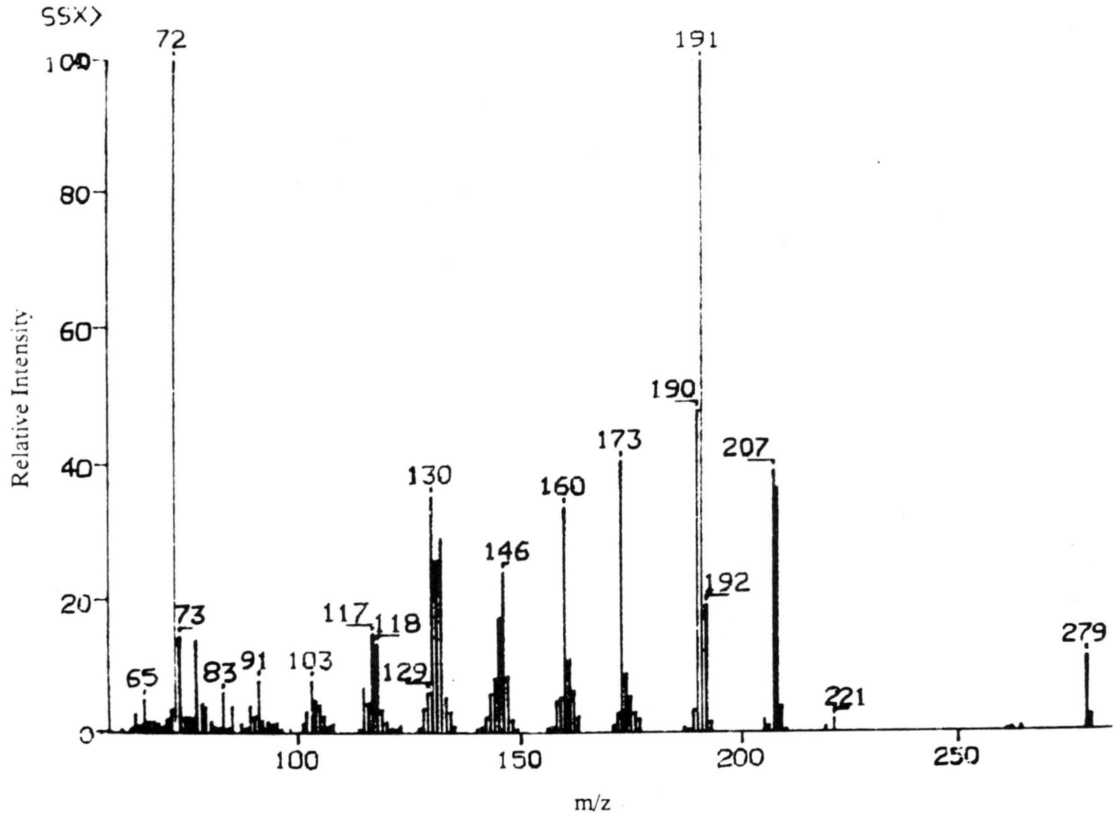

Figure 8. Electron Impact-Mass Spectrum of Oxamniquine.

Table IV

Proposed Fragmentation Pattern of Oxamniquine

m/z	Relative intensity, %	Ion
279	11.2	$M^+(C_{14}H_{21}N_3O_3)$
221	2.1	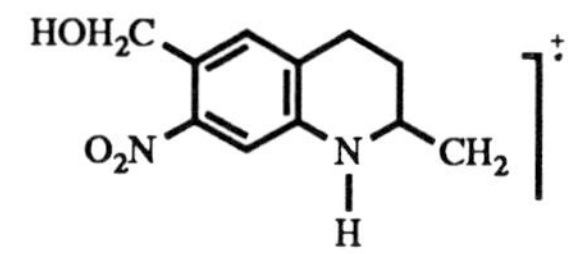
207	3.91	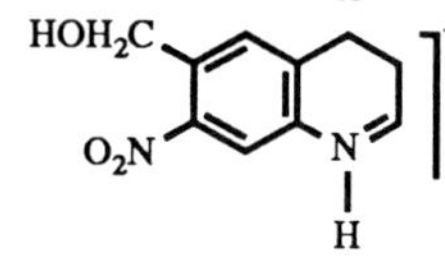
191	100	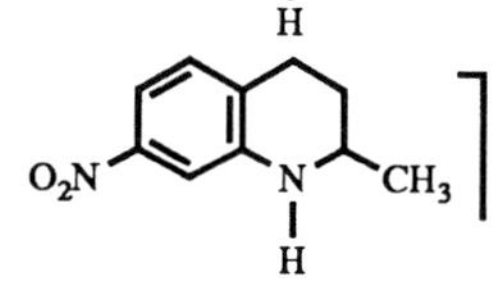
173	40.8	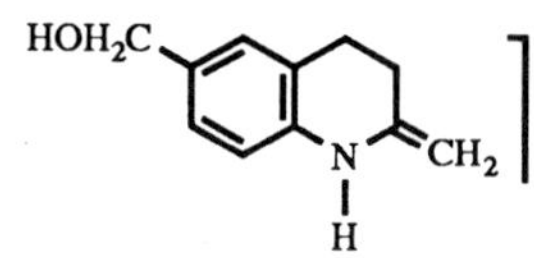
160	33.7	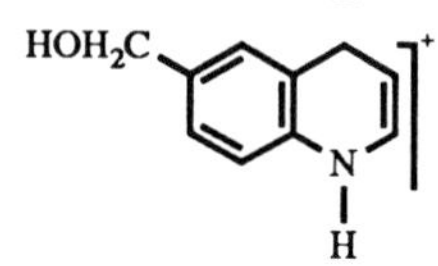
146	24.1	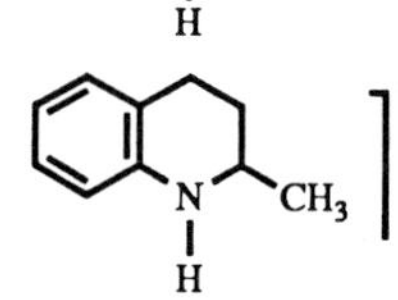
130	35.4	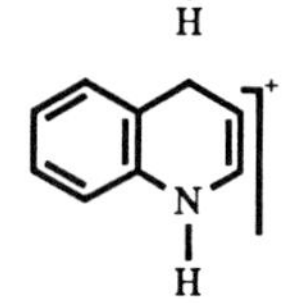
118	13.3	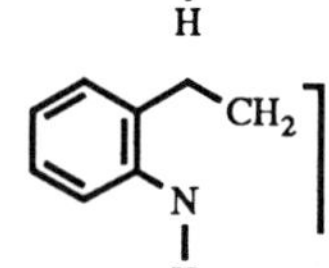
72	99	$\overset{+}{C}H_2-NH-CH(CH_3)_2$

5.2 Impurity Test for Related Compounds (12)

Prepare, by warming, a solution in chloroform containing 25.0 mg of oxamniquine per ml. Apply 20 ul of this solution to a suitable thin-layer chromatographic plate coated with a 0.25 mm layer of silica gel G. Apply to the same plate 20 ul of a methanol solution containing 175 ug per ml of USP Related Compound A RS and 20 ul of a methanol solution containing 125 ug per ml of USP Related Compound B RS. Allow the spots to dry, and develop the chromatogram in a solvent system consisting of chloroform, n-hexane, isopropyl alchol, isopropylamine (200:100:20:5) until the solvent front has moved about three-fourths of the length of the plate. Remove the plate from the developing chamber, mark the solvent front, and locate the spots on the plate by viewing under short-and long-wavelength ultraviolet light: any spots from the oxamniquine are not greater in size or intensity than the spots, occurring at the respective Rf values, produced by the standard solutions corresponding to not more than 0.5% of 1,2,3,4-tetrahydro-2-[(isopropylamino)methyl]-7-nitro-6-methyl-quinoline and not more than 0.5% of 1,2,3,4-tetrahydro-2-[(isopropylamino)methyl]-5-nitro-6-quinolinemethanol, corresponding to related compounds A and B, respectively.

6. METHODS OF ANALYSIS

6.1 Titrimetric Analysis

Korolkovas and Haraguchi (23) have described a method for the determination of oxamniquine by titration with perchloric acid in nonaqueous medium in the presence of glacial acetic acid. The end point is detected by using quinaldine red solution as indicator.

6.2 Spectrophotometric Analysis

The USP (12) assay of oxamniquine in pure form and in capsules is based on spectrophotometric determination. A quantity of oxamniquine equivalent to about 100 mg is transferred to a 200 ml volumetric flask, dissolved in methanol, made up to volume with methanol and mixed. A 5 ml aliquot of this solution is pipetted into a 500 ml volumetric flask, diluted with methanol to volume and mixed. The absorbance of the assay solution and of a standard solution containing about 5 ug per ml of USP Oxamniquine RS in methanol is measured concomitantly at the wavelength of maximum absorbance at about 251 nm against a methanol blank. The quantity, in mg, of $C_{14}H_{21}N_3O_3$ in the portion of oxamniquine taken is calculated by the formula: 20 C (A_u/A_s), in which C is the concentration, in ug per ml, of USP Oxamniquine RS in the standard solution, and A_u and A_s are the absorbances of the assay solution and the standard solution, respectively.

In the USP (12) assay of oxamniquine capsules, a quantity of the powdered material equivalent to about 250 mg of oxamniquine is transferred to a 500 ml volumetric flask and about 400 ml of methanol is added to dissolve the drug content. The solution is shaken by mechanical means for thirty minutes, diluted with methanol to volume, and filtered through filter paper, discarding the first 20 ml of the filtrate. 10 ml of the clear filtrate is transferred to a 100 ml volumetric flask, diluted with methanol to volume, and mixed. A 10 ml aliquot of this solution is pipetted into another 100 ml volumetric flask, diluted with methanol to volume, and mixed. An accurately weighed quantity of USP Oxamniquine RS is dissolved in methanol, diluted quantitatively and stepwise with methanol to obtain a standard solution having a known concentration of about 5 ug per ml. The absorbance of both solutions is determined concomitantly at the wavelength of maximum absorbance at about 251 nm using methanol as the blank. The quantity, in mg, of $C_{14}H_{21}N_3O_3$ in the portion of capsules taken is calculated by the formula: 50 C (A_u/A_s), in which C is the concentration, in ug per ml, of USP oxamniquine RS in the standard solution, and A_u and A_s are the absorbances of the solution from the capsules and the standard solution, respectively.

6.3 Chromatographic Analysis

6.3.1 Gas-liquid chromatography (GLC)

A method for the analysis of oxamniquine in serum or plasma, sensitive to 10 ng/ml, has been developed (24). Oxamniquine and a close structural analogue used as the internal standard are extracted from serum with ether. After derivatization with N,O- bis (trimethylsilyl) acetamide, oxamniquine is determined as its trimethylsilyl ether derivative by GLC using an electron-capture detector. The method has been applied to determine serum concentration profiles of different dosage forms of oxamniquine.

6.3.2 High-performance liquid chromatography (HPLC)

Jun and Radwan (25) have reported a sensitive and reliable HPLC assay procedure for the determination of oxamniquine in plasma or urine.

Separation of the drug is achieved on a reversed-phase Ph column using UV detection at 254 nm. The eluting solvent is a mixture of 0.05M acetate buffer pH 5 and acetonitrile (3:7). The average recovery of oxamniquine from three or more replicate drug plasma samples of different concentration (0.125-4.00 ug/ml) is 95.5% and its coefficient of variation is 4.17%. The detection limit in plasma is about 20 ng/ml. The method has been employed to measure the plasma level versus time profile of oxamniquine following a single bolus i.v. dose of 16 mg/kg to a dog.

7. ACTIVITY, METABOLISM AND PHARMACOKINETICS

7.1 Activity

In adult *S. mansoni,* incubated for 1 hour *in vitro* with oxamniquine and then returned into the mesenteric veins of permissive animal hosts, survival of schistosomes has been assessed after 3-4 weeks by portal perfusion. The drug proved effective and may exert its schistosomicidal activity by inhibiting nucleic acid synthesis in the parasite (26). It has been suggested that oxamniquine becomes tightly bound to the DNA of adult *S. mansoni* after 1 hour incubation. Apparently the schistosomicidal activity of oxamniquine results from the formation of a covalent bond between the drug and the parasite DNA (11).

Oxamniquine (200 mg/kg, i.m.) given to mice infested with *S. mansoni* does not appear to affect the process of cercaria-schistosomulum transformation but kills effectively most of the larvae embedded in the skin before reaching the lung (27). Single oral doses of 25-1000 mg/kg of the drug show activity against *S. mansoni* infestation in mice, but the ED causing 50% inhibition is 6.2-15.7 and 8.2-10.7 times higher in 2 resistant strains than in 4 reference strains of *S. mansoni* (28).

The chemoprophylactic activity of oxamniquine has been evaluated in mice experimentally infected with *S. mansoni.* The drug was administered as a single dose corresponding to about 1/5 of LD_{50} and the mice were found to be free of schistosomes after the treatment (29). Oxamniquine (0.4 mg/ml) has been found to kill 50 and 10% respectively of the male and female worms *in vitro* after 24 hours; a 0.6 mg/ml dose is fully schistosomicidal after 12 hours. After 6-18 hours of incubation with the drug, the ornithine-δ-transaminase activity of worms *in vitro* falls by 27-30%. At 200 mg/kg (i.m.) of the drug the enzyme activity of worms of schistosome-infected mice is decreased by 87% after 11 days. Thus the inhibition of enzyme activity by oxamniquine may be related to its schistosomicidal action (30).

7.2 Metabolism

Oxamniquine has been shown to possess higher schistosomicidal activity than that of its parent methyl compound, 2-N-isopropyl-aminomethyl-6-methyl-7-nitro-1,2,3,4-tetrahydroquinoline, which is converted to the hydroxymethyl derivative, oxamniquine, in several animal species (31). These findings, in conjunction with evidence of high *in vitro* activity of oxamniquine (32), indicates that oxamniquine *per se* is the active compound in mice and monkeys.

The metabolism of oxamniquine has been studied in various animal species and in man. The drug is converted to two acidic metabolites, the major one arising from oxidation of the 6-hydroxymethyl group to a carboxyl group and the other by oxidation of the side chain to give the 2-carboxylic acid. Both metabolites are found in appreciable quantities only in the urine of mouse, rabbit, hamster and dog (33). Conversion of oxamniquine to the -6-carboxylic acid appears to take place in the gut wall in the dog (34). Oxamniquine is not N-dealkylated by liver microsomal preparations (35).

The uptake and retention of ^{14}C-labeled oxamniquine by *S. mansoni* in the mouse host after a single oral or intramuscular dose (50 mg/kg) has been studied (36). Six days after the drug administration, at the time of an hepatic shift, more drug related material is present in male worms than in female worms. Examination of worms recovered from mice 4 hours after treatment shows that the metabolities of oxamniquine constitute 70-90% of the drug-related material present in the worms.

Oxamniquine (50 mg/kg) elevates β-glucuronidase activity in liver, spleen, kidney and bladder tissue homogenates of both the control and *S. mansoni* infected mice. The effect of oxamniquine on liver β-glucuronidase is more prolonged than the infection itself. The elevation of β-glucuronidase is more pronounced in infected and oxamniquine treated mice than in either the treated or infected controls (37). Oxamniquine may exert its effect on liver β-glucuronidase activity, at least in part, by direct action on the enzyme (38).

7.3 Pharmacokinetics

Oxamniquine is readily absorbed following oral ingestion, and a peak concentration in plasma occurs within about 3 hours. The presence of food significantly delays absorption and limits the concentration achieved in plasma during the first several hours after administration. Urinary excretion is the major route of elimination in man. Only a small portion of the drug is excreted unchanged, upto 70% appears as a single acid metabolite being present in the plasma at concentrations more than tenfold greater than those of oxamniquine. The metabolite is predominantly excreted in the first 12 hours and is devoid of schistosomicidal activity (6,33,34).

8. TOXICITY

Oxamniquine displays low acute toxicity in mice, hamsters and rabbits. In rats, however, it is much more toxic, the females being 8-10 times more sensitive than the males; single doses show an idiosyncratic sex-linked hepatic necrosis and bile duct proliferation. In chronic studies in mice (18 months) and hamsters (19 months) with intermittant dosage regimens relevant to likely usage in man, no evidence of carcinogenicity potential has been observed at dose levels upto 150 mg/kg. Oxamniquine shows no material toxicity or teratogenicity in mice and rabbits, and only slight embryotoxicity after high oral doses (39). The LD_{50} in mice and rabbits are greater than 2000 mg/kg and 1000 mg/kg, i.m., and 1300 mg/kg and 800 mg/kg orally, respectively (4,14).

In clinical use, oxamniquine is generally well tolerated with side-effects limited to a mild transient dizziness and the drug should be used with caution in patients with a history of epilepsy (1,13). No significant toxic effects on liver, kidney or heart have been observed in patients treated with single doses of 12.5-15 mg/kg of oxamniquine (40).

ACKNOWLEDGEMENTS

The authors are thankful to the United States Pharmacopeial Convention, Inc., and Pfizer Central Research, Sandwich, Kent, U.K. for donating samples of oxamniquine.

REFERENCES

1. Foster, R. (1973). Trans. R. Soc. Trop. Med. Hyg. 67, 674.
2. Richards, H.C. (1985). Chem. Br. 21, 1001.
3. Richards, H.C. and Foster, R. (1969). Nature (London), 222, 581.
4. Foster, R. (1973). Rev. Inst. Med. Trop. Sao Paulo 15, 1.
5. Webster, L.T., Jr. (1985). In "Goodman and Gilman's The Pharmacological Basis of Therapeutics", 7th Edition (A.G. Gilman,

L. S. Goodman, T.W. Rall and F. Murad, eds.), p. 1016, MacMillan Publishing Co., New York.

6. Kaye, B. (1984). "Oxamniquine: Mechanism, Pharmacokinetics and Mode of Action", WHO Scientific Working Group on the Biochemistry and Chemotherapy of Schistosomiasis, p. 1, WHO, Geneva.
7. Jewsbury, J.M. (1977). Cent. Afr. J. Med. 23, 18.
8. Woolhouse, N.M. (1980). In "Toxicology in the Tropics", 1st Int. Symp. 1979 (R.L. Smith and E.A. Bababunmi, eds.), p. 194, Taylor and Francis, London.
9. Richards, H.C. (1982). In "Chronicles of Drug Discovery" (J.S. Bindra and D. Lednicer, eds.), Vol. 1, p. 257, Wiley, New York.
10. Richards, H.C. (1985). In "Medicinal Chemistry: The Role of Organic Synthesis in Drug Research" (B.J. Price and S.M. Roberts, eds.), p. 271, Academic Press, New York.
11. Pica-Mattoccia, L., Cioli,D. and Archer, S. (1987). Biochem. Soc. Trans. 15, 70.
12. "The United States Pharmacopeia" (1990). 22nd Revision "The National Formulary" 17th Edition, p. 980, United States Pharmacopeial Convention, Inc., Rockville, Md.
13. "Martindale, The Extra Pharmacopoeia" (1989). 29th Edition (J.E.F. Reynolds, ed.), p. 61, The Pharmaceutical Press, London.
14. "The Merck Index" (1983). 10th Edition (M. Windholz, ed.), p. 992, Merck and Co., Rahway, N.J.
15. Richards, H.C. (1968). S. African Patent 6, 803, 636; C.A. (1969), 71, 30369k.
16. Richards, H.C. (1974). U.S. Patent 3,821, 228; C.A. (1974), 81, 120498n.
17. Baxter, C.A.R. and Richards, H.C. (1971). J. Med. Chem. 41, 1033.
18. Baxter, C.A.R. and Richards, H.C. (1972). J. Med. Chem. 15, 341.
19. Korolkovas, A., Itaya, M. and Coelho, Z.M.P. (1977). Rev. Farm. Bioquim. Univ. Sao Paulo 15, 27; C.A. (1979), 91, 102026j.
20. "Dictionary of Organic Compounds" (1979). 5th Edition (J. Buckingham, ed.), Vol. 4, p. 4427, Chapman and Hall, New York.
21. Kuhnert-Brandstatter, M. and Vollenklee, R. (1987). Sci. Pharm. 55, 27.
22. Kofitsekpo, W.M. (1980). Drugs Exp. Clin. Res. 6, 421.
23. Korolkovas, A. and Haraguchi, T. (1980). Rev. Farm. Bioquim. Univ. Sao Paulo 16 , 32; C.A. (1981), 94, 7846e.
24. Woolhouse, N.M. and Wood, P.R. (1977). J. Pharm. Sci. 66, 429.

25. Jun, H.W. and Radwan, M.A. (1985). Anal. Lett. 18, 1345.
26. Pica-Mattoccia, L. and Cioli, D. (1985). Am. J. Trop. Med. Hyg. 34, 112.
27. Oliveira, M.A., Pellegrino, J., Pereira, L.H. and Valadares, T.E. (1976). Rev. Inst. Med. Trop. Sao Paulo 18, 298; C.A. (1977), 86, 115034g.
28. Yeang, F.S.W., Marshall, I. and Huggins, M. (1987). Ann. Trop. Med. Parasitol. 81, 337.
29. Pellegrino, J., Pereira, L.H. and Mello, R.T. (1977). Rev. Inst. Med. Trop. Sao Paulo 19, 43; C.A. (1977), 87, 145753d.
30. Goldberg, M., Gold, D., Flescher, E. and Lengy, J. (1980). Biochem. Pharmacol. 29, 838.
31. Kaye, R. and Woolhouse, N.M. (1972). Xenobiotica 2, 169.
32. Woolhouse, N.M. and Kaye, B. (1977). Parasitology 75, 111.
33. Kaye, B. and Woolhouse, N.M. (1976). Ann. Trop. Med. Parasitol. 70, 323.
34. Kaye, B. and Roberts, D.W. (1980). Xenobiotica 10, 97.
35. Woolhouse, N.M., Kaye, B. and Parke, D.V. (1979). Xenobiotica 9, 323.
36. Woolhouse, N.M. and Kaye, B. (1977). Res. Commun. Psychol., Psychiatry Behav. 75, Pt. 1, 111.
37. El-Zoghby, S.M., Ebied, S.A., Kholy, S.A., Saad, A.A., Tawab, G.A.A., Hammouda, N.A. and El-Gohary, Y.A. (1980). Biochem. Pharmacol. 29, 429.
38. El-Sewedy, S.M., El-Melegy, S., Akhnoukh, S., Salam, N. and Rafaa, A.A. (1980). Biochem. Pharmacol. 29, 2925.
39. Chvedoff, M., Faccini, J.M., Gregory, M.H., Hull, R.M., Monro, A.M., Perraud, J., Quinton, R.M. and Reinert, H.H. (1984). Drug Dev. Res. 4, 229.
40. Huggins, D. (1978). Rev. Bras. Clin. Ter. 7, 97; C.A. (1978), 89, 71110j.

ANALYTICAL PROFILE OF PHENOLPHTHALEIN

FAHAD J. AL-SHAMMARY

MOHAMMAD SALEEM MIAN*

NEELOFUR ABDUL AZIZ MIAN

Clinical Laboratory Sciences Department, College of Applied Medical Sciences, King Saud University, P.O. Box 10219, Riyadh-11433, Saudi Arabia.

*Pharmaceutical Chemistry Department
College of Pharmacy, King Saud University
P.O. Box 2457, Riyadh-11451, Saudi Arabia.

PHENOLPHTHALEIN

CONTENTS

1. Introduction

The laxative action of Phenolphthalein was discovered by accident when, about 60 years ago, the Hungarian government decreed that phenolphthalein should be added to wines to indicate contamination with basic adjuvants. The wine thus treated were then found to have cathartic effects.

Phenolphthalein has become one of the most successful synthetic laxatives, both because it is easily manufactured from phthalic anhydride and phenol, and because in contrast to many other laxatives it does not damage the kidney. Pharmaceutical preparations containing phenolphthalein are in the market under countless names. The question as to whether phenolphthalein acts locally on the smooth intestinal muscle or by way of a nervous path has not yet been answered.

In conclusion phenolphthalein is the most widely used laxative because it is relatively cheap, uniform in action, non-toxic, stable, tasteless, non-habit forming and does not cause kidney damage (1).

2. Description

2.1. Nomenclature

2.1.1. Chemical Names

(a) 3.3.Bis(4-hydroxyphenyl)-1-(3H)-Isobenzofuranone (2).

(b) α (p-hydroxyphenyl)- α(4-oxo-2.5-cyclohexadiene-1-ylidine)-0-toluic acid (2).

(c) 3.3-bis(p-hydroxyphenyl)phthalide (2,3,4,5,6).

(d) 1(3H)-Isobenzenefuranone,3.3-bis(4-hydroxyphenyl) (5).

(e) α αdi-(4-hydroxyphenyl)phthalide (7).

2.1.2. Generic Names

Bom-Bon: Chocolax; Darmol; Dihydroxyphthalophenone; Ex-lax: Evac-U Gen: Euc-chessina, Evac-Q-Kwik Tablet; Evac-Q-Tabs; Fenoltaleina: Fructines Vichy: Fructine Vichy;

Laxante Yer;Laxatone; Laxen Busto; Laxin; lilo; Neopru-nex;neo-purghes; phenolax;prulet; purganol; purgante; purgante el Aleman; Purgante Orravan; Purgestol.

2.2. Formulae

2.2.1. Empirical (2)

$C_{20}\ H_{14}\ O_4$

2.2.2. Structural

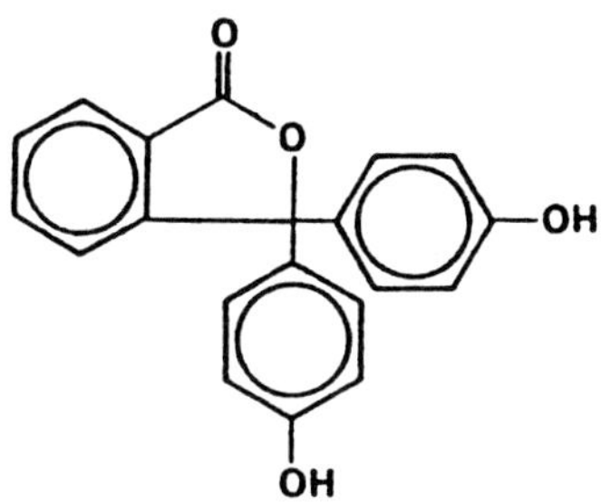

2.2.3. CAS (Chemical Abstract Service Registry Number)

77-09-8

2.2.4. Optical Density (8)

1.277^{32}_{4}

2.2.5. Wiswesser-Line Notation (8)

T56 BVO DHJ D R DQ and DR DQ

2.3. Molecular Weight (2)

318.31

2.4. Elemental Composition (2)

C 75.6% H 4.43% O 20.10%

2.5. Appearance, Color, Odour and Taste

White or yellowish white. Minute, triclinic crystals, often twinned (2) white or yellowish white crystalline or amorphous powder; odourless and tasteless(3,6) yellow phenolphthalein is said to be about 3 times stronger, on a weight basis, than is white phenolphthalein, but this has not been adequately supported by clinical studies (9).

3. Physical Properties

3.1. Melting Range (2)

258° - 262°C

3.2. Solubility

Almost insoluble in water, soluble in alcohol (95%) and in ether (6) slightly soluble in chloroform(2), soluble in dilute solution of alkali hydroxides and hot solutions of alkali carbonates forming a red solution (3). Phenolphthalein in solution can exist in different forms, which are in equillibrium, representing the various forms by the classical formulas presented in **Fig.(A)**. At a pH lesser than 8 practically all of the Phenolphthalein is present in the lacton form I (10).

3.3. pH (2)

Colorless to pH 8.5; pink to deep red above pH 9.

3.4. Dissociation Constant

The drug has a Pk_a (Dissociation constant) 9.7(25°C) (4,9).

3.5. Partition Coefficient (4)

Log p(octanol/pH 7.4), 2.4

3.6. Loss on drying (6)

When dried to constant weight at 105°C, loses not more than 1.0% on its weight.

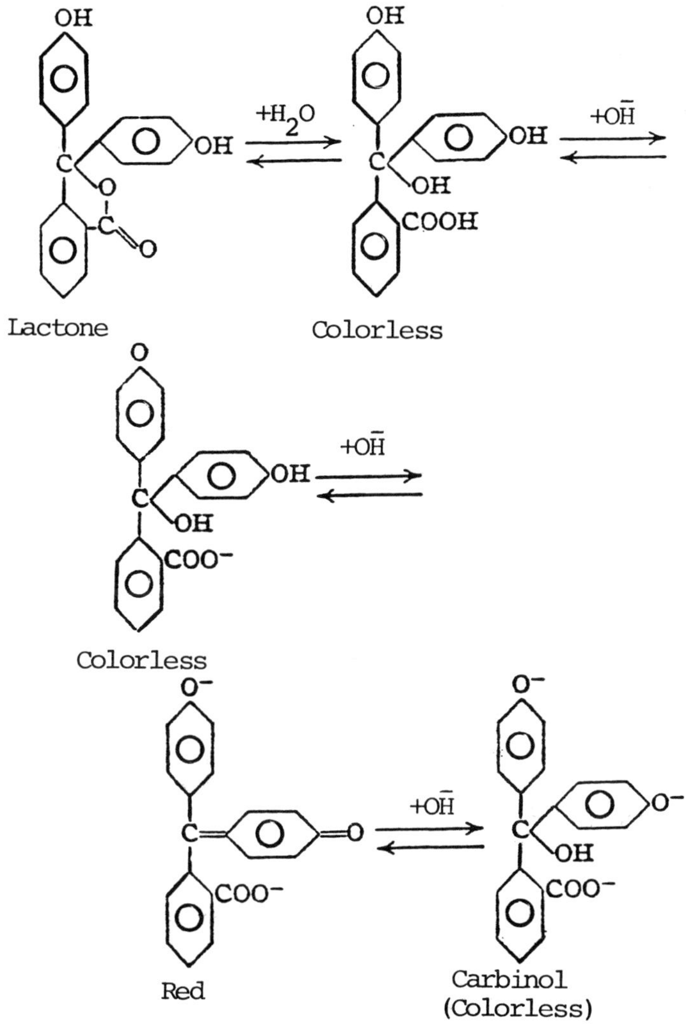

Fig. A: Various forms of Phenolphthalein in Solution.

3.7. Residue on ignition (5)

Not more than 0.1%

3.8. Moisture Content (11)

Not more than 1% determined by drying at 105°C

3.9. Hygroscopicity (11)

It absorbs significant amounts of moisture at 25°C at relative humidities up to about 90%.

3.10. Toxicity

Phenolphthalein is usually non-toxic even in relatively large doses, although fatalities have occured in young children after ingestion of 0.6-1.8g (4).

3.11. X-Ray Powder Differaction (12)

The x-ray differaction pattern of Phenolphthalein was determined using Philips full automated x-ray differaction spectrogoniometer equipped with PW 1730/10 generator. Radiation was provided by copper target (Cu annode 2000 W) and high intensity x-ray tube operated at 40KV and 35MA. Divergence slit and the receiving slit wire 1° and 0.1° respectively. The unit was equipped with philips PM 8210 printing recorder and digital printer. The x-ray differaction pattern of Phenolphthalein is presented in **fig.(1)**. The values of scattering angle 2θ interplanner distance dA° and relative intensity I/Io are shown in **table (1).**

3.12. Spectral Properties

3.12.1. Ultraviolet Spectrum (UV) (12)

The UV spectrum of Phenolphthalein in ethanol (3mg%) was scanned from 200 to 360 nm using LKB 4054 UV/vis spectrophotometer **(Fig. 2)**. It exhibited the following uv data **(Table 2)**.

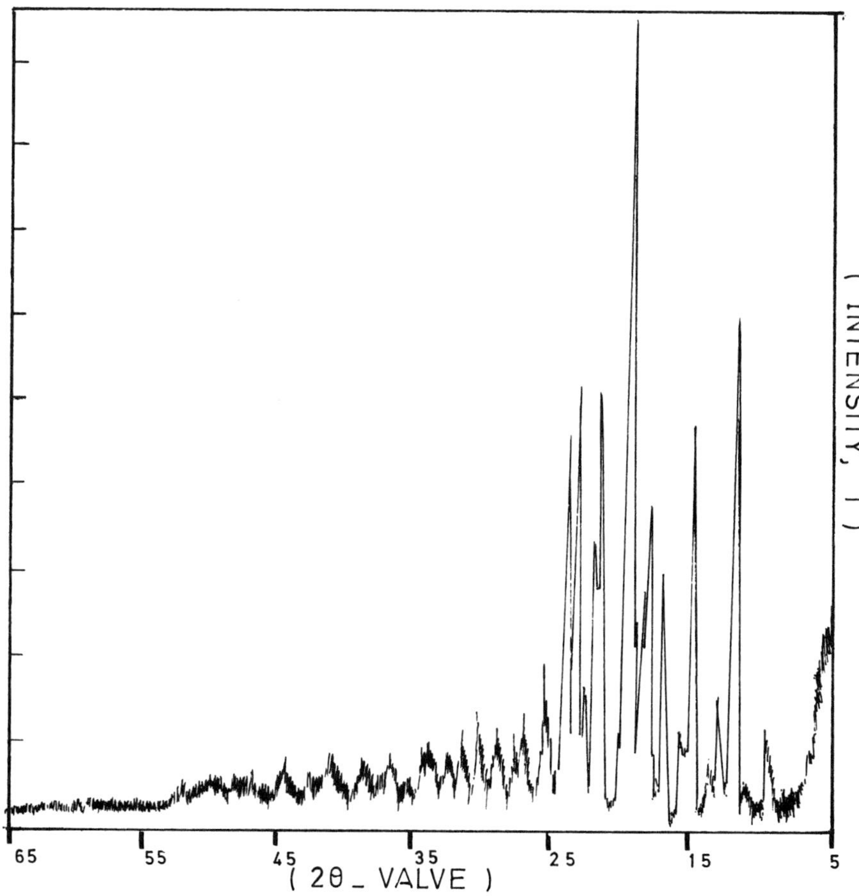

Fig (1) CHARACTERISTIC PRINCIPAL LINES OF X-RAY POWDER DIFFRACTION OF PHENOLPHTHALEIN.

TABLE (I). Characteristic lines of the X-ray powder diffraction of Phenolphthalein.

2θ	dÅ	I/Io %	2θ	dÅ	I/Io %
7.785	11.3563	6.044	30.505	2.9303	14.579
9.308	9.5010	14.276	31.546	2.8360	12.608
9.919	8.9173	5.112	32.533	2.7522	10.116
11.013	8.0339	7.777	33.792	2.6525	9.921
12.148	7.2853	62.954	34.321	2.6128	11.026
12.943	6.8397	17.742	35.377	2.5371	6.629
13.549	6.5349	10.030	36.614	2.4543	8.318
15.125	5.8578	49.003	37.308	2.4102	7.365
15.667	5.6560	13.734	37.733	2.3840	6.109
17.093	5.1873	31.693	38.776	2.3223	9.012
18.226	4.8674	40.445	39.277	2.2938	4.419
18.571	4.7777	29.440	39.960	2.2561	5.480
19.259	4.6085	25.216	40.281	2.2389	6.065
19.805	4.4828	100	40.957	2.2035	8.427
20.013	4.4366	59.769	42.557	2.1243	7.213
21.858	4.0661	54.483	43.718	2.0705	5.870
22.132	4.0164	35.831	44.179	2.0500	7.755
22.883	3.8862	19.074	46.565	1.9503	5.979
23.510	3.7840	54.332	47.281	1.9225	6.065
24.217	3.6750	48.678	47.934	1.8978	6.109
25.306	3.5193	17.200	48.995	1.8591	6.130
25.768	3.4573	22.118	50.073	1.8216	6.000
27.112	3.3889	16.096	51.976	1.7593	5.372
27.880	3.2000	12.846	53.794	1.7041	3.184
29.000	3.0789	12.239	59.610	1.5510	3.336
30.248	2.9547	2.816			

2θ = scattering angle.

dÅ = interplanner distance.

I/Io % = relative intensity based on highest as 100.

TABLE (2) UV Characteristics of Phenolphthalein.

λ_{max}	ε (molar absorptivity) $cm^{-1}gm\ mol\ L^{-1}$
205	27261.146
229	14692.144
276	2006.369

Other reported uv spectral data for Phenolphthalein in methanol as λ_{max} 277nm (A^1_1 = 150a) (4)

3.12.2. Infrared Spectrum (IR) (12)

The IR spectrum of Phenolphthalein as KBr disc was recorded on a Perkin Elmer 580B Infrared spectrometer to which an Infrared data station is attached (Fig. 3).

The structural assignments have been correlated with various frequencies (Table 3).

TABLE 3. IR Characteristics of Phenolphthalein.

Frequency cm^{-1}	Assignment
3230	-OH group
2940	-CH group
1720, 1610	C=O group
1520, 1470, 1430	-CH
1240, 1180	-O- linkage
760, 730	adjacent aromatic hydrogens

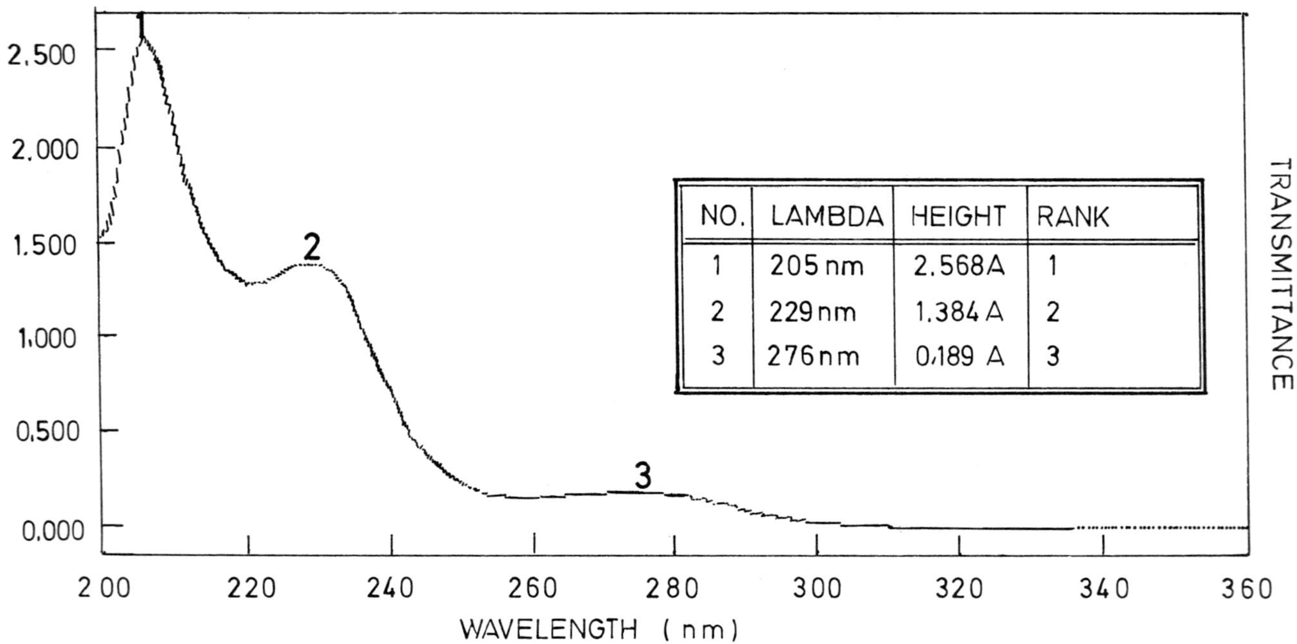

NO.	LAMBDA	HEIGHT	RANK
1	205 nm	2.568A	1
2	229 nm	1.384 A	2
3	276 nm	0.189 A	3

Fig (2) UV SPECTRUM OF PHENOLPHTHALEIN

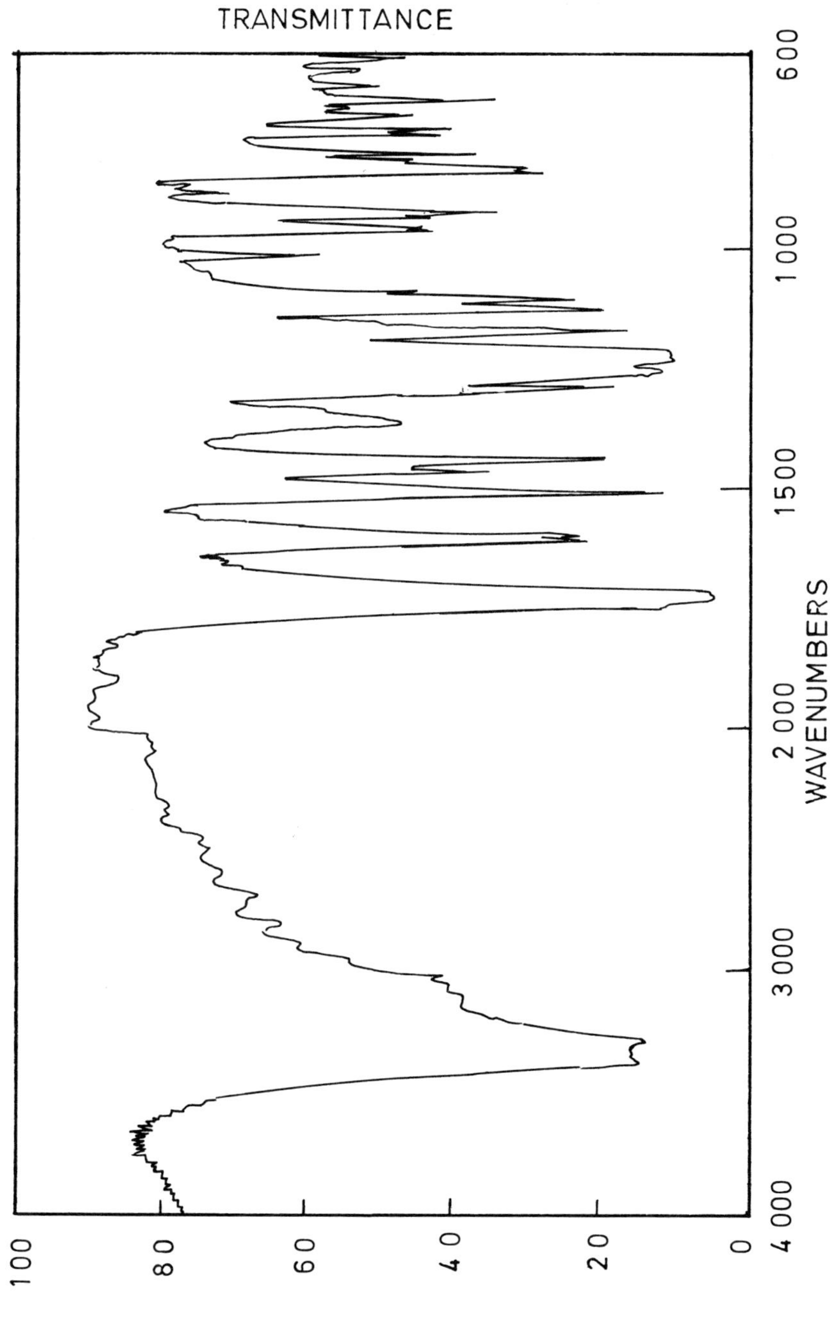

Fig (3) I.R. SPECTRUM OF PHENOLPHTHALEIN

3.12.3. Nuclear Magnetic Resonance Spectra (12)

3.12.3.1. Proton Spectra (PMR)

The PMR spectra of Phenolphthalein in DMSO-d_6 was recorded on a varian T-80A, 80MH$_z$ NMR spectrometer using TMS as internal reference and is shown in **Fig. (4)**. The interpretation of the spectrum is given below.

Chemical Shift (PPM)	No. of Protons	Assignment
9.75	2	OH's
7.88 to 6.81	12	Aromatic

3.12.3.2. ^{13}C-NMR Spectra

The ^{13}C-NMR noise decoupled and off resonance spectrum of Phenolphthalein in DMSO using TMS as an internal reference is recorded on Jeol FX-100 FT NMR spectrometer and are presented in **Fig.(5) and Fig.(6)**.

3.12.4. Mass Spectrum

The mass spectrum (12) of Phenolphthalein is presented in **Fig. (7)**. This was obtained by electron impact ionization on a Finnigen-Mat 1020 GC/Mass spectrometer by direct inlet probe at 270°C. The electron energy was 70ev. The spectrum scanned to mass 340amu. The base peak is 43.2 with a relative intensity 100%.

Secondary-ion mass spectrum of Phenolphthalein has also been reported by Kloeppel, K.D. et al (13), in which secondary ions were produced with use of silver targets having chemically roughened surfaces with a capacity of or about 6mm^{3}cm^{-1}. Samples were applied to the target either as solution or as suspensions in glycerol and exposed in vacuo to the primary-ion beam (mainly Ar^+). The resulting secondary ionization was less destructive than chemical ionization and yielding more fragment ions than did field ionization.

4. Synthesis

Scheme 1: (14)

Phenolphthalein is prepared by the condensation of

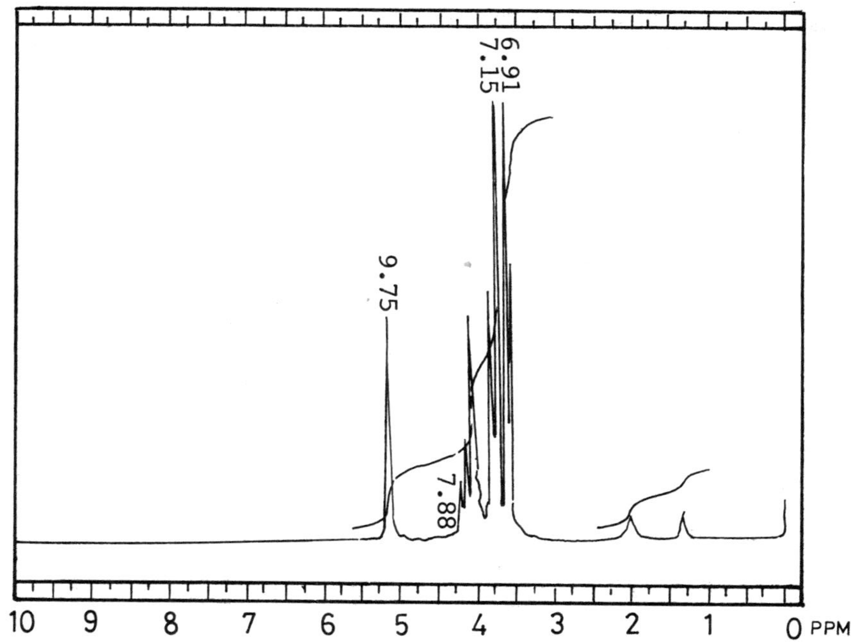

Fig 4 : PMR SPECTRUM OF PHENOLPHTHALEIN IN DMSO_d_6

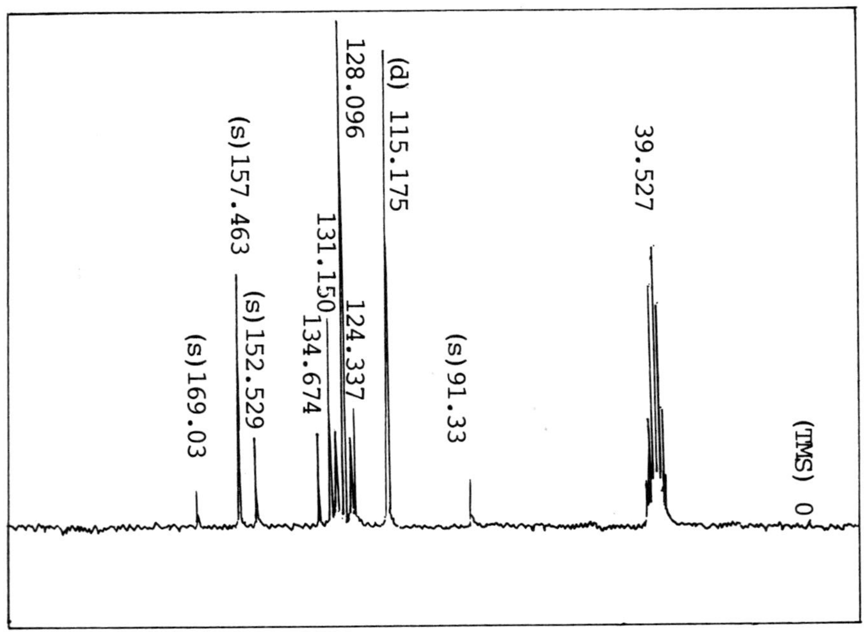

Fig(5) ^{13}C_NMR NOISE DECOUPLED SPECTRUM OF PHENOLPHTHALEIN

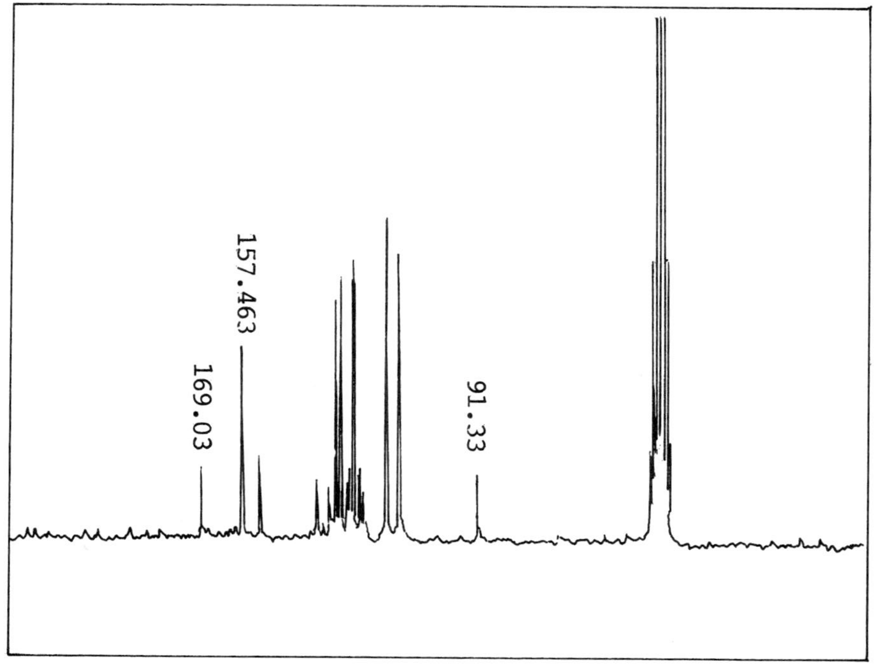

Fig (6) ^{13}C_NMR OFF RESONANCE SPECTRUM OF PHENOLPHTHALEIN

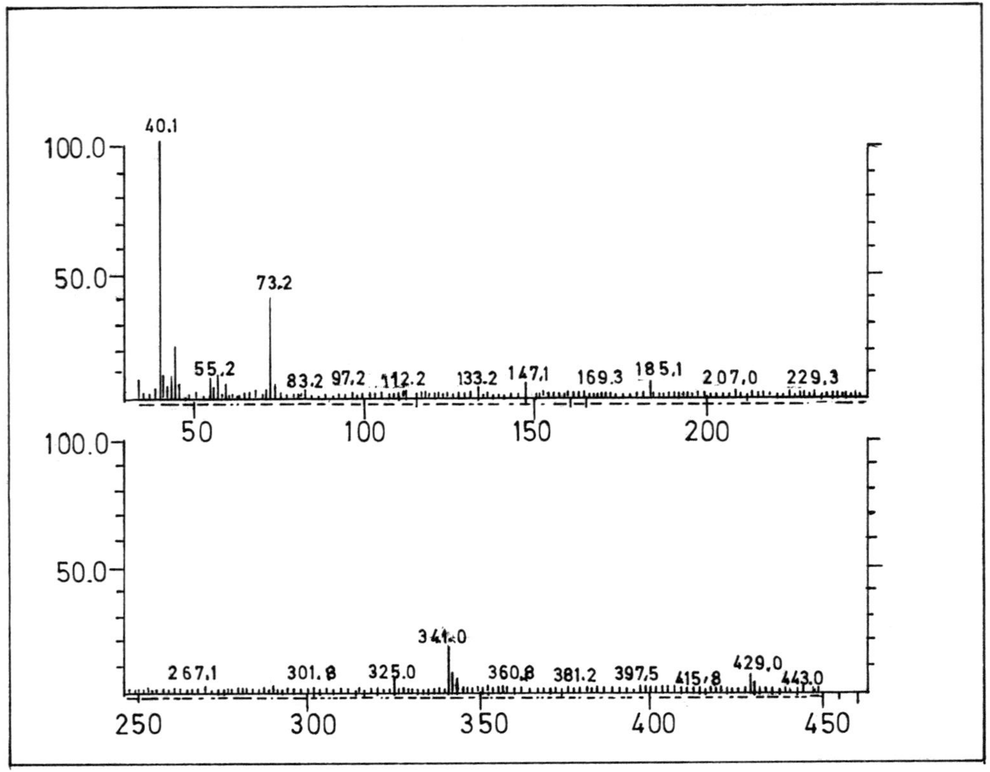

Fig 7 : MASS SPECTRUM OF PHENOLPHTHALEIN

phthalic anhydride with phenol.

A mixture of 10 parts of phenol, 5 parts of phthalic anhydride, and 4 part of sulfuric acid is heated at 120°C for 10 to 12 hours. The product is extracted with boiling water, then the residue dissolved in dilute sodium hydroxide solution filtered and precipetated with acid.

Scheme 1:

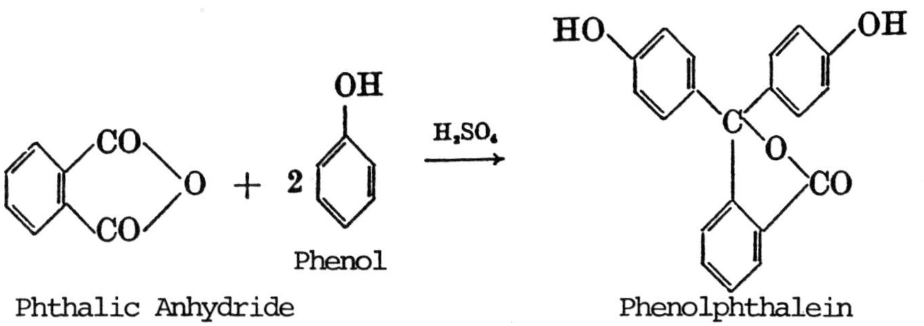

Scheme 2: (15)

Phenol condenses with pthalic anhydride in the presence of concentrated sulfuric acid or anhydrous zinc chloride to yeild the colourless phenolphthalein as the main product. When dilute caustic alkali is added to an alcoholic solution of phenolphthalein, an intense red colouration is produced. The alkali opens the lactone ring in phenolphthalein and forms a salt at one phenolic group. The reaction may be represented in steps, with the formation of a hypothetical unstable intermediate that changes to a coloured ion. The color is probably due to resonance which places the negative charge on either of the two equivalent oxygen atoms. With excess of concentrated caustic alkali, the first red colour disappears; this is due to the production of the carbinol and attendant salt formation, rendering resonance impossible. The various reactions are presented in scheme II.

Scheme 2:

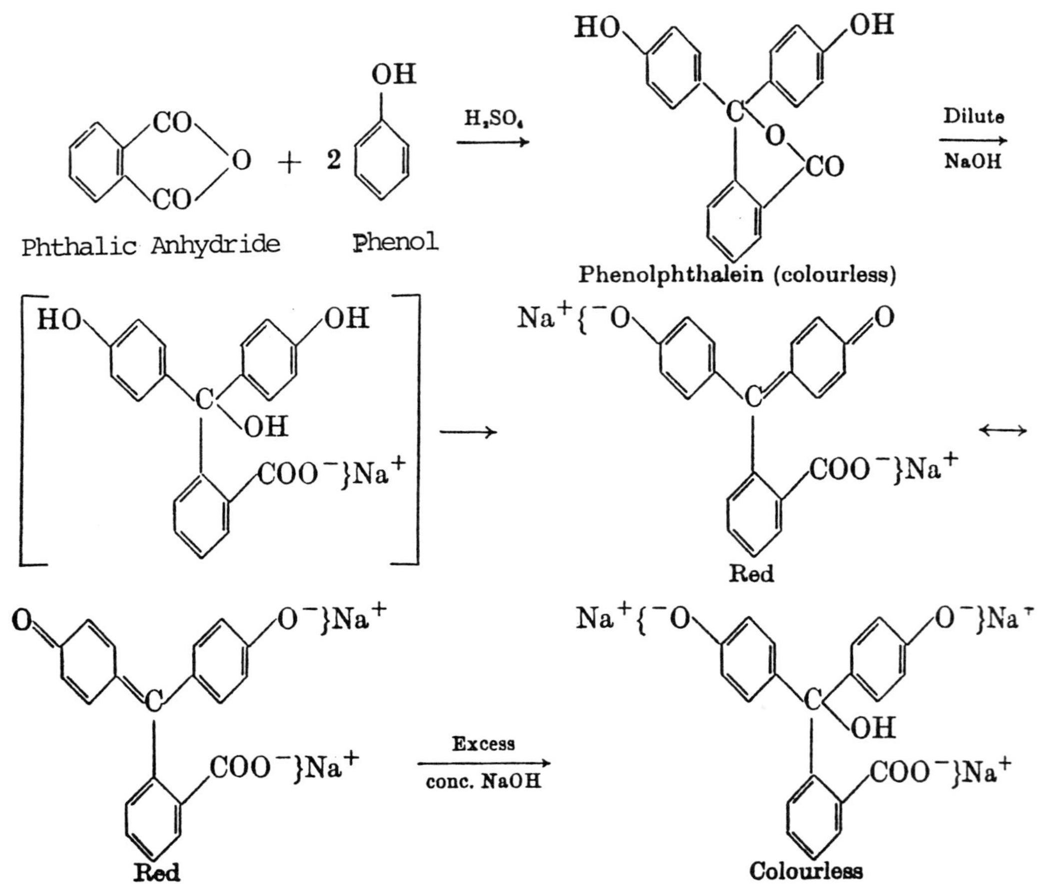

CO
CO
O
+ 2
OH
Phthalic Anhydride
Phenol
H_2SO_4
HO
OH
C
O
CO
Phenolphthalein (colourless)
Dilute
NaOH
HO
OH
C
OH
$COO^-\}Na^+$
$Na^+\{^-O$
O
C
$COO^-\}Na^+$
Red
O
C
$O^-\}Na^+$
$COO^-\}Na^+$
Red
Excess
conc. NaOH
$Na^+\{^-O$
$O^-\}Na^+$
C
OH
$COO^-\}Na^+$
Colourless

5. Metabolism and Pharmacokinetics

5.1. Absorption and Elimination

Phenolphthalein is poorly absorbed after oral administration (up to about 15%) and partially excreted in the bile as the active glucuronide, and in the urine, if the urine is alkaline the excreted compound will impart a red colour. The greater part of the ingested material is eliminated in the faeces in both free and conjugated forms (4).

Phenolphthalein is an irritant purgative which is usually taken at night to act in the morning. Some of it may be absorbed and excreted partially in the bile, so that purgative effects may continue for several days. It is excreted chiefly in the faeces, but some may be excreted by the kidneys, imparting a red colour to alkaline urine. Insignificant amounts are excreted in the milk (9).

5.2. Uses

Phenolphthalein is a diphenylmethane stimulant laxative. It usually has an effect within 6-8 hours but because of enterohepatic circulation, its action may continue for several days. It is usually administered in pills or tablets, or it may also be given as an emulsion with liquid paraffin. It is usually given in the dose of 30-200 mg taken at bedtime; doses of 270 mg daily should not be exceeded. Yellow phenolphthalein, an impure form has been claimed to be more active than white phenolphthalein (3).

The usual oral laxative dosage of phenolphthalein is 30-270mg daily for adults and children 12 years of age and older, 15-30 mg daily for children 2-5 years of age, given as a single dose or divided doses (9).

It is also used as an indicator in volumetric analysis (14).

A 1% alcoholic solution as an indicator in titrations of mineral and organic acids and most alkalies. Not suitable for ammonia. Very sensitive to CO_2, and in estimating carbonates the liquid must be boiled. Borax can be titrated with phenolphthalein as an indicator only when glycerol is present, because the color gradually fades away as the acid is added. Useable with few alkaloids (2).

5.3. Adverse Effects

Excessive bowl activity, usually diarrhea or abdominal discomfort, nausea, vomiting, cramps, weakness, dizziness, palpitations, sweating and fainting may follow the administration of an evacuant. Diarrhea may lead to fluid and electrolyte deficits.

Accidental or intentional overdosage may lead to severe diarrhea and excessive loss of body fluids and electrolytes, muscle weakness and tremor, gastrointestinal disturbances and fainting (16).

Allergic reactions usually as skin rashes or erruptions have occured with phenolphthalein. Cardiac and respiratory distress, and albuminuria and haematuria have also been reported. Abdominal discomfort such as cramps or colic may occasionally occur. Prolonged use of overdosage can result in diarrhea with excessive loss of water and electrolytes, particularly potassium, there is also the possibility of developing an atonic non-functioning colon. Phenolphthalein may cause pink or red discoloration of the urine or faeces (3).

Dermatitrs (fixed drug eruptions, pruritus, burning, vesiculation, and residual pigmentation) may occur in hypersensitive patients. Fatal anaphylactic reactions have been reported, but an absolute causal relationship with phenolphthalein has not been established. Non-thrombocytopenic purpura has been reported occasionally (3).

Osteomalacia in a 51-year old woman was attributed to depletion of body calcium as a result of diarrhea due to long-term phenolphthalein ingestion (17).

Tablets containing about 1.8g of phenolphthalein were taken by a 3-year old child. Despite gastric lavage, the patient developed pulmonary oedema and became comatose. Death occured after 13 hours and postmortem examination revealed cerebral and pulmonary oedema and phenol in the gastric juice (18).

Peritonial dialysis for 24 hours and the administration of isoprenaline was successful treatment for the hypotension, severe acidosis, pulmonary oedema, and oliguria which occured in a 35-year old man who had taken phenolphthalein 2g as a chocolate laxative preparation (19).

A report of accute pancreatitis occuring in a 34-year old man who had inadvertently ingested 2g phenolphthalein . The patient had been taking phenolphthalein for 2 years for the treatment of chronic constipation. There was complete recovery and no sequelae from the pancreatitis (20).

5.4. Precautions

Phenolphthalein should not be given to patients who have previously shown sensitivity to the drug (3).

The mixture should not be taken within 30 minutes of mealtime. It should not be used for longer than a few days wtihout medical advice. It should be supplied in wide-mouthed bottles (9).

A high-fluid intake and a low residue diet are essential for a successful colon preparation. Recommended liquids are black coffee, plain tea, strained fruit juice, soft drinks or water and NO MILK or CREAM (16).

The drug should not be used when abdominal pain, nausea or vomiting are present. Frequent and continued use may cause dependence upon laxatives. Rectal bleeding or failure to respond may indicate a serious underlying condition and require a physician's consultation. As phenolphthalein is also excreted in human milk, therefore, caution should be exercised when drug is administered to a nursing woman.

6. Methods of Analysis

6.1. Identification Methods

a. It is readily dissolved by solutions of alkali hydroxides and by hot solutions of alkali carbonates yeilding red liquids. These solutions are decolorized by the addition of an excess of acid or by high concentrations of alkali hydroxides (5).

b. It melts at about 260° (11).

c. The UV absorption of phenolphthalein in methyl alcohol is maxima at 276nm (E1%, 1cm = 145) (9).

d. Dissolve 0.1g in 100 ml of (95%) alcohol dilute 5 ml to 50 ml with alcohol (95%) and evaporate 5 ml of the resulting solution to dryness on a water bath. Dissolve the residue in sufficient aminoacetate buffer solution pH 11.3 to produce 100 ml, mix, and immediately measure the extinction of a 1cm layer at the maximum at about 555nm. Calculate the content of $C_{20}H_{14}O_4$, taking 1055 as the value of A(1%, 1cm) at the maximum at about 555nm (6).

6.2. Titrimetric Method

(a) U.S.P. XXI describes the method for the estimation of phenolphthalein. Take about 100mg of accurately weighed phenolphthalein in 250 ml iodine flask and add 5 ml of dimethylformamide, and agitate carefully untill the phenolphthalein has dissolved completely. Add 25 ml of 0.2M Sodium carbonate, 25 ml of 0.2M Sodium bicarbonate, and 50.0 ml of 0.1N iodine vs, and swirl the mixture gently for 1 minute. Immediately add 10 ml of chloroform and 5 ml of 4 N sulfuric acid, and titrate the excess iodine with 0.1 N sodium thiosulfate vs, adding 3 ml of Starch TS as the end point is approached. Each ml of 0.1 N iodine is equivalent to 3.979 mg of $C_{20}H_{14}O_4$ (5).

(b) S. Doernberg et al determined phenolphthalein in chocolate preparations. A sample of chocolate containing 80-100 mg of phenolphthelein extracted with 3 x 5ml portions of CCl_4 (to remove fat) and then 3 x 5 ml portions of acetone to extract phenolphthalein. The acetone is removed on water bath, and 25 ml of (98%) pyridine and 5 ml of anhydrous ethylendiamine are added and then 1% thymol blue indicator in pyridine. The solution is titrated slowly to a colour yellow to yellow-green (this is the blank) with a solution of 1.25 g of sodium in 100 ml of ethanolamine diluted to 500 ml with pyridine. The titration is then continued potentiometrically. The second titration figure x 0.0150 = wt. of phenolphthalein (21).

(c) An improved semimicro method for the estimation of phenolphthalein was described by D. Subrahmanyam et al. About 20 mg of drug is dissolved in 10 ml of 0.1M Na_2CO_3 by heating on a water bath at 90°C. The solution is cooled and transferred to a 100 ml calibrated flask. Then 10 ml of 0.1M $NaHCO_3$ is added followed by 10 ml of iodine solution (12.69g of I, 18g of KI, add H_2O to one litre). The solution is made up to 100 ml with H_2O. To a 50 ml portion are added 5 ml of chloroform and 5 ml of H_2SO_4. The liberated iodine is titrated against 0.2N $Na_2S_2O_3$. The standared error of the method is $\pm$0.07% (22).

(d) A.I. Gengrinovich, et al. suggested the quantitative method for the determination of phenolphthalein. Dissolve 0.05-1.5g of sample in hot 10% Na_2CO_3 solution, cool transfer to a glass-stoppered flask, add 15-50cc 0.1 N iodine, followed, after thorough shaking by 12cc of 2 N HCl. Add 10cc of ether and 1.2cc starch solution and titrate with 0.1 N $Na_2S_2O_3$. When ICl is used 10cc of 10% KI must be added before acidification. To prepare ICl, transfer 3.5 gm KIO_3, 5.5 gm KI, 40cc HCl (37%), and 40cc H_2O to a glass-stoppered flask and shake the mixture until iodine has completely dissolved. Add 15cc $CHCl_3$ and decolorized the $CHCl_3$ layer by adding dropwise 1% KIO_3 solution. Decant aqeous layer and transfer to a 1000-cc measuring flask and dilute to the mark. Determine the titer by removing by 25cc, adding 10cc, 10% KI, and titrating with $Na_2S_2O_3$ (23).

(e) The method suggested by Zoltai, et al, is that Phenolphthalein is titrated as a weak acid, with NaOH solution, in aqeous 30% ethanol medium; conductometric or oscillometric indication of the end point is used. The inflection on the titration curve correspond to titration of the two phenolic protons. The method has been used to determine phenolphthalein in drug mixtures without interference from, e.g. sucrose, fennel oil, or constituents of senna or belladonna leaves (24).

(f) Phenolphthalein in antispasm tablets is estimated by Sankar Sakti Bandyophyay (25).

(g) Phenolphthalein in mineral oil emulsions was determined by Alexander et al. The emulsion is broken by the addition of an acidified solution of NaCl, any glycerin or other solvent capable of absorbing iodine is removed, mixture made alkaline with NaOH and treated with an excess of standered iodine solution. After iodination is complete, the excess iodine is determined (26).

(h) Phenolphthalein can be determined by measuring the excess $KBrO_3$ used to oxidize it. Dissolve 0.1g of sample in 15 ml of 0.1 N NaOH, add 8-10 ml 0.5 N $KBrO_3$, 5 ml 10% KBr, and while shaking rapidly, add 25 ml acid mixture(20 ml 80% ACOH and 5 ml conc. HCl); after shaking one minute add 1gm KI and titrate the liberated iodine with 0.1 N $Na_2S_2O_3$. One ml of 0.1 N $KBrO_3$ = 0.003977gm phenolphthalein. Blank was also run in parallel (27).

In mineral oil emulsions phenolphthalein is estimated iodimetrically (28, 29, 30) in an alkaline medium by adding an excess of the standard iodine solution and titrating back the residual iodine with standard thiosulphate. Instead of sodium thiosulphate, hydrizine solution is also used as a titrant (31, 32). The end point is detected by using starch or by applying the dead-stop amperometric technique. Gravimetrically, Phenolphthalein is determined in mineral oil emulsions (33,34) in tablets (35,36) and in perparations containg chocolate (37).

6.3. Colourimetric Method

(a) A colourimetric procedure for the determination of phenolphthalein based upon the condensation with 4-Aminophenazone in prescence of an alkaline oxidizing agent and measuring the absorbance at 460nm, is proposed. The procedure determines 0.08-0.6 mg/25 ml with mean recovery of 99.54 ± 0.64%. The method is applied to the determination of phenolphthalein in pharmaceutical preparations (38).

(b) **For the authentic sample:** This method is based upon nitrosation of phenolphthalein with sodium nitrite and glacial acetic then the O-nitroso compound is determined by the formation of the cobaltic chelate compound and measuring the absorbance at 388nm. The colourimetric analysis of authentic phenolphthalein described by El-sayed L et al (39) is by preparing a stock standard solution of phenolphthalein (1mg/1ml) in ethanol. In a

100 ml-volumetric flask dilute 20 ml of the stock standard solution with 0.001N NaOH solution to the mark. Measure accurately 5 ml of prepared solution in 25 ml volumetric flask and add 5 ml of solvent solution (20% ethanol in 0.001 N NaOH).

Add 0.5 ml glacial acetic acid followed by 1 ml saturated solution of sodium nitrite and 2 ml of a 5% cobalt nitrite. Heat the mixture for 15-20 minutes on a boiling waterbath or until the blank turns pale pink. Then, cool to room temperature. Dilute the reaction mixture to volume with water. Measure the coloured solution at 388nm against reagent blank.

For Tablets: The method applied for the determina tion of phenolphthalein in tablets is described as: Weigh and powder 20 tablets of the sample. Extract an accurately weighed amount of the powdered tablets, equivalent to one tablet of phenolphthalein, three times each with 10 ml of ethanol. Filter the extract and wash the residue with ethanol. Transfer the combined filterate and washings to a 50 ml volumetric flask and complete the volume with ethanol, accurately measure 5 ml of the last obtained solution into 50 ml volumetric flask and complete to volume with 0.001 N sodium hydroxide. Transfer an aliquot of the final solution to a 25 ml volumetric flask and complete the procedure for the development as given for authentic sample.

(c) The colorimetric methods used for its analysis are based on the appearance of the red colour in alkaline medium at pH 10.4-11 and measurement of the colour at 550nm within 15 minutes (40,41). Phenolphthalein was determined in biological fluids (42) by hydrolysing with 8 N hydrochloric acid, then adding sodium hydroxide and glycine of pH 10.6, and measuring the colour at 550nm.

A condensation reaction with 4-aminophenozone in the presence of an alkaline oxidising agent was used by (39) for the determination of phenolphthalein in the pure form and in tablets. The British pharmacopoeia 1980 (43) estimated phenolphthalein by measurement of its colour at pH 11.3 in an amino acetate buffer at 555nm.

(d) Estimation of Phenolphthalein in anthelmintics is based upon the principal as the Phenolphthalein was detected by the typical red coloration with alkali and was estimated colorimetrically at pH 10.4-11.0 after extraction with acetone distilling off the solvent, dissolving in 0.1 N Na_2CO_3 solution and adjusting the pH (44).

Kover, K.A also reported the method for the analysis of Phenolphthalein (45).

6.4. Spectrophotometric

(a) Phenolphthalein in biological fluids is determined spectrophotometrically by Morris et al (46). Samples are deproteinised with acetone, the supernatant liquid is evaporated and phenolphthalein in an ethanolic solution of the residue is determined by spectrophotometry at 550nm after adjusting the pH to 12 with glycine-NaOH solution. Conjugated phenolphthalein is determined similarly after enzymic hydrolysis.

(b) Suzuki et al (47) assayed phenolphthalein in alkaline media. 50 mgs of drug is dissolved in 100 ml of ethanol and the solution is diluted with ethanol to a concentration of 50 to 500 ug in 5 ml. A 5 ml aliquot of this solution is evaporated to dryness on a water bath. The residue is dissolved in 0.1 M carbonate buffer of pH 10.6, to give 50 ml of solution. and the absorbance is measured immediately at 555nm and at 20°C against the buffer solution. Use of carbonate buffer gave better results than did the use of 0.1 M glycine or 0.1 M phosphate buffer of pH 11.3.

(c) In another method (48) phenolphthalein is determined by using sulfuric acid. To the sample solution containing 5 to 25 ug of phenolphthalein is added 0.5 ml of H_2SO_4-ethanol(3:1) with cooling in ice water and the volume is adjusted to 5 ml. The absorbance of the solution is measured at 506nm against the H_2SO_4-ethanol. The colour is stable for four hours at room temperature.

(d) Phenolphthalein is also determined spectrophotometrically (49).

(e) The drug is also estimated in the Hung. Formulae Nomales, four compounded Phenolphthalein preparations are official (3 powders and 1 liquid paraffin emulsion). For the assay weigh a sample containing about 40 mg of phenolphthalein. Extract with MeOH. and add 0.4-0.5 g white clay (bolus alba), shake and filter into a 50 ml measuring flask. Make up to volume. Evaporate 5.00-10.00 ml of the filterate to dryness on a water bath. Dissolve the remainder in 4.0ml MeOH. add 10.0ml 0.1N NaOH. and determine the extinction with a pulfrich photometer using the filter No. S 55. Calculate the amount of Phenolphthalein from a standard plot of the extinction as the function of Phenolphthalein concentration. (40-120r in 4.0 ml MeOH = 10.0 ml 0.1 N NaOH). Extinction measurement should be made within 15 minutes after mixing the MeOH extinction with the NaOH. The producibility of the process is within 4% ± 2%) (50).

6.5 Polarographic Method

(a) Polarographic determination (51) of phenolphthalein in certain pharmaceuticals is based on the extraction of phenolphthalein with 96% ethanol and the solution is diluted with water to an ethanol concentration of 25%. For polarography, the solution is made acidic (3.8 N HCl). The evaluation is carried out by means of a calibration curve. Drug is extracted from liquid paraffin containing emulsions with ethanol, the determination of phenolphthalein in chocolate tablets is not possible.

(b) M. Traiter et al (52) determined phenolphthalein polarographically by dissolving the sample 40 mgs in 100 ml of 96% ethanol and dilute an aliquot (25 ml) to 100 ml with 4 N.HCl. Bubble an aliquot of this solution for 5 mins. with H_2 and record the polarographic waves of phenolphthalein.

(c) Phenolphthalein is determined polarographically in pharmaceuticals with glacial acetic acid as solvent by G. Dusinsky et al. (53) Drug is dissolved in glacial acetic acid and sodium acetate. In 12.5% acetic acid and 0.7 M sodium acetate. In 12.5% acetic acid and 0.75 M sodium acetate the height of the wave of phenolphthalein $E_{1/2}$ = -0.69V vs the S.C.E.) is linearly proportional to concentration in the range 10^{-4} to 7.5×10^{-4} M.

(d) **For Authentic Samples:** This investigation (54) deals with the polarographic behaviour of phenolphthalein in aqueous solution containing different supporting electrolytes. The cathodic irreversible diffusion controlled wave, obtained in 0.1 N hydrochloric acid is used for its determination. By preparing a stock standard solution of phenolphthalein (1mg/ml) in ethanol. Transfer different aliquots of the standard stock to a 25 ml volumetric flask and dilute to the mark with 0.1N HCl. In each case, the final concentration of ethanol, has to amount to 20%. Before polarographic measurements are performed the dissolved oxygen must be removed by bubbling a stream of pure nitrogen for about three minutes.

For Tablets: Weigh and pulverize 20 tablets. Extract an accurately weighed amount of the powder equivalent to one tablet, three times with 10 ml ethanol. Filter the extract and wash the residue with the ethanol. Transfer this residue into 100 ml volumetric flask and make up to the volume with ethanol. Transfer an aliquot of this solution to a 25 ml volumetric flask. dilute with 0.1 N HCl and then proceed as given for the authentic sample The polarographic behaviour of phenolphthalein has been investigated (10) under conditions of varying pH, solvent and temperature. The reversibility and reaction mechanism of phenolphthalein at the droping mercury electrode with a discontinously alternating applied potential is described (55).

Polarographic behaviour of Phenolphthalein (10)

1. The polarographic behavior of phenolphthalein has been investigated under widely varying conditions. In 25% ethanol phenolphthalein gives a one-step reduction wave in a pH range between 0 and 9. In this range the half-wave potential changes according to $\pi 1/2$ (vs. S.C.E.) = 0.54 - 0.046 pH. The red form of phenolphthalein yields two waves of equal height, each wave giving a well-defined diffusion current. The sum of the two diffusion currents is equal to that of the single wave observed at lower pH. The polarographic reduction of phenolphthalein to phenolphthalin is accompanied by a transfer of two electrons.

2. In the presence of a phosphate buffer of a pH of 7 the diffusion current decreases abnormally when the ethanol concentration is increased. In 60% ethanol the

"apparent" diffusion current becomes extremely small at 25°. It is concluded that the lactone form of phenolphthalein is not reduced, but its hydrated form is. At alcohol concentrations of 25% or less the rate of hydration at the surface of the electrode is so large that a normal diffusion current is observed. At high alcohol concentrations the "apparent diffusion current" becomes entirely rate- and not diffusion-controlled. Experiments carried out at various temperatures substantiate this interpretation.

3. In 25% ethanol in the presence of a buffer of pH 7 the diffusion current remains constant in a potential range between -1.0 and -1.2 volt; it decreases slightly between -1.2 and -1.4 volt and then drops suddenly to attain a small minimum value at a potential of about -1.6 volt. Experimental evidence has been given substantiating the interpretation that the occurence of the minimum is to be attributed to a decreased rate of transformation of the lactone form into the reducible hydrated form at the surface of the dropping electrode.

6.6. Chromatographic Methods

6.6.1. Paper chromatography

Phenolphthalein is identified (56) by using the procedure consisting in impregnating a suitable paper with Na_2CO_3 solution and using a mixture of ethanol and H_2O as solvent in the descending method; fourteen hours are required for the solvent to travel 45cm. The drug for analysis is either directly extracted with methanol or first hydrolysed with a mixture of methanol and HCl. The diluted with H_2O and extracted with ether.

6.6.2. Gas Chromatography

Urine containing drug was treated with β-D-glucuronidase and adjusted to pH 7.5, then the laxative were extracted into ethyl ether with use of an Extrelut Column. The trimethylsilyl derivatives were prepared and analysed by g.l.c.-m.s.(70ev). A salinised U-shaped column (180cm x 3mm) packed with 3.8% of SE-30 on chromosorb W AW-DMCS HP (80-100 mesh) was used operated at 280°, and the mass spectrometer was operated in the electron impact mode, with multiple-ion monitoring. The detection limits were 0.03 for the derivative of Phenolphthalein (57).

6.6.3. Thin Layer Chromatography

A summary of some of the TLC method for the analysis of phenolphthalein are given in the **table (4)**.

6.6.4. High Performance liquid Chromatography (HPLC)

Aitzemueller et al (64) determined phenolphthalein by HPLC. The constituents of the tablets are simultaneously determined on a column (25cm x 4.6mm) of Li Chrosorb Si 60 (5 um), with $CHCL_3$-ethanol (19:1) as mobile phase and detection at 254nm, with 4-hydroxyacetophenone as internal standard.

6.7. Thermal Behaviour (DSC) (12)

The differencial scanning calorimetry (DSC) thermal curve for phenolphthalein was obtained **Fig. (8)** on a perkin-Elmer DSC-2C differencial calorimeter. Nitrogen was used as the purge gas. The scan was performed at the rate of 10°C min^{-1} from 230 to 290°C. The DSC curve revealed an endothermic melting peak Max. 258.8°C.

7. Acknowledgements

The authors are highly thankful to Liberty S. Matibag and Mr. Babikir Awad Mustafa, College of Applied Medical Sciences, King Saud University for their secretarial and technical assistance respectively in preparing the manuscript.

8. Reference

1. Burger, A. (Editor), "Medicinal Chemistry" 3rd Edition Wiley-Interscience, New York, (1970).

2. "The Merck Index", 11th ed. p.1150 (1989).

3. "Martindale" The Extra Pharmacopeia" 29th Ed., p.1102, The Pharmaceutical Press, London, (1989).

4. "Clark's Isolation and Identification of Drugs" 2nd Ed. The Pharmaceutical Press, London (1986).

5. "The United States Pharmacopeia" 16th Ed. United States

DSC

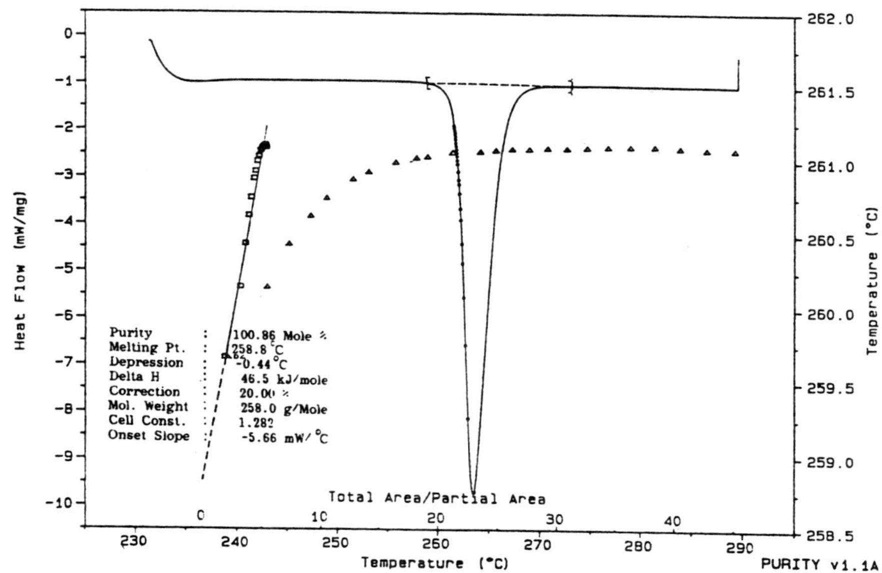

Fig 8 : THE DIFFERANTIAL SCANNING CALORIMETRIC (DSC) THERMAL CURVE OF PHENOLPHTHALEIN

TABLE (4). SUMMARY OF TLC CONDITIONS USED FOR THE DETERMINATION OF PHENOLPHTHELEIN

SUPPORT	SOLVENT SYS.	DETECTION	EXT. SOLVENT	RF.	REF.
Kiesel gel HF_{254}	$CHCl_3CH_3OH$	UV	-	-	58
0.25nm layer of silica gel G	Benzene:ethyl-acetate: acetic acid (75:24:1)	sprayed with 5% alcoholic KOH	-	0.48	59
Silica gel GF	UV 254nm	-	-	-	60
Silica gel	$CHCl_3$.ethyl ether(4:1)	-	-	-	61
Silica	$CHCl_3$-butyl-acetate (9:2)	-	-	-	62
Silica gel G	ethyl acetate:methanol: H_2O (10:2:1)	-	sprayed with 3% NN. dimethyl-p-phenylenediamine hydrocholoride solution in methanol:H_2O (1:1)	-	63

Pharmacopeial Convention, Inc., 12601 Twinbrook Partway, Rockville, Md. 20852, p. 820 (1985).

6. "British Pharmacopeia" Her Majesty's Stationary Office, London, p. 430 (1988).
7. "British Pharmacopeia" Her Majesty's Stationary Office, London, p. 360 (1973).

8. Atlas of Spectral Data and Physical Constants for Organic Compounds. J.C. Grasselli and W.M. Ritchy, 2nd Ed., p. 96 'CRC Press Inc., Cleaveland, Ohio (1975).

9. "Drug Information 90" p. 1639. American Society of Hospital Pharmacists.

10. I.M. Kolthoff and D.J. Lehmicke, J. Amer. Chem. Soc. 70 p. 1879-1885 (1948).

11. "The Pharmaceutical Codex" p. 688 11th Ed., The Pharmaceutical Press, London (1979).
12. Fahad J. Al-Shammary, Neelofur Abdul Aziz Mian. Unpublished data 1991.

13. Kloeppel, K.D.; Weyer, K.; von Buenau G. Fachz. Lab., 28(4) 258-262, 264-265 (1984).

14. "Remington's Pharmaceutical Sciences" 13th Ed. p. 879 Mack Publishing Company, Easton, Pennsylvania (1965).

15. A Text book of Practical Organic Chemistry by Arthur I. Vogel 3rd Ed. p. 984. Longman Group Limited, London (1972).

16. "Physician's Desk Reference" by Edward R. Branhart Medical Economics Company, Inc., U.S.A. 42nd Ed. (1988).

17. B., Frame; et al., Archs intern Med., 128, 794 (1971).

18. L., Sarcinelli; et al., Proc. Eur. Soc. Stud. Drug 11, 261 (1970).

19. N. Buchanan; et al., S. Afr. Med. J. 50, 1060 (1976)

20. A.L. Lambrianides and R.D. Rosin, Post Grad. Med. J., 60, 491 (1984).

21. S. Doernberg, M. Hubacher and I. Lysyj J. Amer. Pharm. Asso., Sci. Ed., 43, 418-420 (1954).

22. D. Subrahmanyam and M. Srinivasan J. Sci. Ind. Res., B, India, 15(1) 30-33 (1956).

23. A.I. Gengrinovich and I. Mansurkhanova Aptechnoe Delo. 3(6) 9-12 (1954).

24. Zoltai, Emma; and Szabo Zsuzsa. Acta Pharm. Hung., 51(5) 204-207 (1981).

25. Sankar Sakti Bandyopadhyay, J. Instn. Chem. India, 40(2) 57-59 (1968).

26. Alexander T. Warren, John E. Logun, and Raymond L. Thatcher. J. Am. Pharm. Assoc. 39, 10-11, (1950).

27. Popou S.F., Med. Prom. S.S.S.R., 6, 25; (1949) Chem. Abstr.; 45, 500d; (1951).

28. "The National Formulary" 14th Ed.; published by American Pharmaceutical Association, p. 563; (1975).

29. Teresa M., Fuentes Paz. Anales Fac. Farm. y Biuoquim, 2, 693; (1951); Chem. Abstr.; 48, 11006a (1954).

30. Menezes, F.; Rev Quim e Farm. Rio de Janero 16 317; (1951); Cnem. Abstr. 48 2983h; (1954).

31. Panwar, K.S., Raot S.P.; Gaur, J.N.; Anal. Chim. Acta. 25 218; (1961).

32. Panwar. K.S., Mathur, N.K.; Ras, S.P. Anal.Chim. Acta. 24: 541; (1961).

33. Bickford, C.F.; Schoetzow, R.E., J. Amm. Pharm. Assoc., 25 1128; (1936).

34. O'Keefe, M.F.; J. Assoc. Off. Agr. Chem. 25, 843; (1942).

35. Thomas, G.; J. Pharm. Belg.; 21 361; (1939).

36. Bendyopadyay, A.: Sakti. S. J. Inst. Chem (India) 40 57; (1968). Chem. Abstr.; 70 6582h; (1969).

37. Harold. C.; Ser. Chem. Pharm. 1: (1952).

38. El-Sayed, L.; Amer, M.M.; Bebawy, L.; Egypt J. Pharm. Sci.; 30(1-4) 221-228 (1989).

39. El-Sayed. L.; Ellaithy. M.M.; Bebawy, L. Il Farmaco. Ed. Pr. 40(8) 273-282 (1985).

40. Wahbi. A.M.; Abdine. H.; Korany. M.; Abdel-Hay. M.H.; J. Assoc. Off. Anal. Chem., 61, 1113; (1978).

41. Ljungberg, S., Pharm. Acta Helv., 25 351; (1950).

42. Gustalson. J.H.; Benet, L.Z., J. Pharm. Pharmacol., 26 937 (1974).

43. The British Pharmacopeia, H.M. Stationary Office, London, p. 342 (1980).

44. A. Burgin, Hunder; Jahre Schweiz Apoth. Ver 1843-1943, 342-6. (1943).

45. Kovar, K.A. Arch. Pharm. Berl., 307(4), Germany, 264-272 (1974).

46. Morris, Sarah M.; and Powell, D.W. Anal. Biochem., 95(2) 465-471 (1979).

47. Suzuki, Masao; Kanbe, Kunio; Asabe, Yoshihiro; and Takitani, Shoji. Bunseki Kagaku 27(8), 525-527 (1978).

48. Suzuki, Masao; Kanbe, Kunio; Asabe, Yoshihiro, and Takitani,Shoji. Bunseki Kagaku, 27(8), 495-499 (1978).

49. A. Fischer. Pharmazie 25(3) 155-156 (1970).

50. Emma Zolotay-Mataolesy Gyogyszereszet 10(9). 344-346 (1966).

51. J. Blazek and Z. Stejskal; Ceskosl. Farmac., 5(6), 344-345 (1956).

52. M. Traiter, Ceskosl. Farm., 10(7), 352-354 (1961).

53. G. Dusinsky and Z. Bruntova Ceskosl Farmae 5(6), 340-343 (1956).

54. M.M. Ellaithy; L. El-Sayed; L. Bebawy; M.M. Amer Il Farmaco. Ed. Pr. 41 (1986).

55. Suzuki, M.: J. Electrochem Soc. Japan, 22 220: (1954).

56. H. Auterhoff and N. Maiss Dtsch. Apoth Ztg. 98(26), 659 (1958).

57. Kok, Robert M.: and Faber, Dick B. J. Chromatogr., 222(3): Biomed. Appl. 11(3) 389-398 (1981).

58. Rousselet, R.; and Paris, R.R. J. Pharm. Belg. 26(5) 568-576 (1971).

59. D.D. Datta and D. Ghosh Indian J. Pharm. 29(6) 179 (1967).

60. R.R. Bhandari and H. Walker J. Chromat. 45(2) 324-325 (1969).

61. Tatsutoyo Fuwa, Teruko Kido and Hiroko Tanaka Yakuzaigaku 24 123-127 (1964).

62. Akimova, T.G. ; Syanava, E.M.; and Savvin, S.B. Zh. Anal. Khim. 37(6) 999-1007 (1982).

63. N.D. Gyanchandani; M. Yamamoto and I.C. Nigam. J. Pharm. Sci. 58(2) 197-200 (1969).

64. Aitzetmueller, K. HRC CC J. High Resolut. Chromatogr. Chromatogr. Commun. 1(4), 220 (1978).

POLYTHIAZIDE

NEGENDRA VARA PRASAD, T.*
VENKATA RAO, E.
SARVESWARA SASTRY, B.

Department of Pharmaceutical Sciences
Andhra University
Waltair – 530 003, India.

SURYA PRAKASA SASTRY, C.

Food & Drugs Laboratories
School of Chemistry
Andhra University
Waltair – 530 003, India.

Leonard J. Kostek
Pfizer Inc. Central Research
Groton, Connecticut 06340

* Present Address: TTK Pharma
8 Old Trunk Road
Madras 600-043
India

CONTENTS

2.3 Formula, Molecular Weight, Structure, C.A.S. Registry No.

$C_{11}H_{13}ClF_3N_3O_4S_3$ 439.90 (5,6)

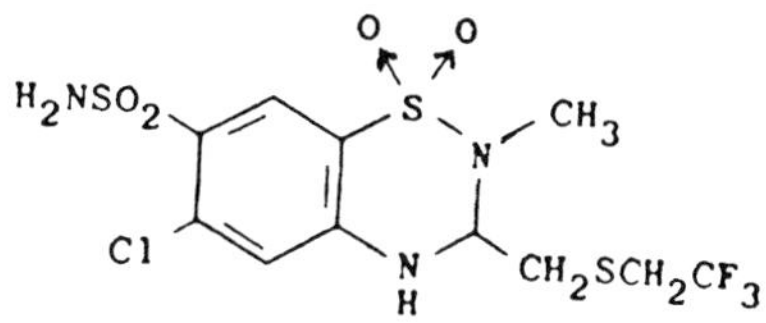

C.A.S. Registry No. 346 – 18 – 9

2.4 Other Names (8)

Nephril	U.K., Australia, India
Drinsil	Germany, South Africa
Renese	Belgium, Canada, Denmark, France, Netherlands, Spain, Sweden, U.S.A., Switzerland, Norway.
Minizide	U.S.A.

2.5 Dosage Forms

Polythiazide (1 mg)
Polythiazide (1 mg) + Reserpine (0.25 mg)
Polythiazide + Prazosin (Minizide)

3. PHYSICAL PROPERTIES

3.1 Appearance, Color, Odor

A white, almost white crystalline powder with an alliaceous odor. (7, 8, 12, 13)

3.2 Melting Point

The following melting points were reported in literature.

209	(7, 10)
202.5	(5, 6, 8)
207 – 217 with decomposition	(4, 9, 14)
202 – 202.5	(17)
202 – 203, 207 – 208	(28)

1. INTRODUCTION

Polythiazide is an orally effective non-mercurial diuretic and antihypertensive drug supplied as a tablet alone or in combination with reserpine (1). Polythiazide exerts an equivalent increase in urinary sodium excretion at about onetenth the dose of hydrochlorothiazide.

2. DESCRIPTION

2.1 Chemical Names

Polythiazide (PT) is the nonproprietary name of 2H-1,2,4-Benzothiadiazine-7-sulphonamide, 6-chloro-3,4-dihydro-2-methyl-3-[[(2,2,2-trifluoroethyl) thio]methyl]-,1,1-dioxide(3, 4, 14) or

6-chloro-3,4-dihydro-2-methyl-3-[[(2,2,2-trifluoroethyl)thio]methyl]-2H-1,2,4-benzothiadiazine-7-sulfonamide-1,1-dioxide(2-6, 12-15) or

2-methyl-3-(β,β,β,-trifluoroethylthiomethyl)-6-chloro-7-sulfamyl-3,4-dihydro-1,2, 4-benzothiadiazine-1,1,-dioxide (5, 6) or

6-chloro-3,4-dihydro-2-methyl-3-(2,2,2-trifluoroethylthiomethyl)-2H-1,2, 4-benzothiadiazine-7-sulphonamide-1,1-dioxide (7-9) or

2-methyl-3,4-dihydro-3(2,2,2-trifluoroethylthiomethyl)-6-chloro-7-sulphamyl-1,2, 4-benzothiadiazine-1,1-dioxide (11) or

6-chloro-3,4-dihydro-2-methyl-3-(2,2,2-trifluoroethylthiomethyl)-1,2,4-benzothiadiazine-7-sulphonamide-1,1-dioxide (12, 13) or

6-chloro-3,4-dihydro-2-methyl-7-sulphamoyl-3-(2,2,2-trifluoroethylthiomethyl)-2H-benzo-1,2,4-thiadiazine-1,1-dioxide (5, 6, 7, 9, 10).

2.2 COMPENDIA

Polythiazide is official in BP, USP and NF (12-15). Other compendia, including analytical and pharmaceutical information, are given in The Pharmaceutical Codex (7), in Kirk-Othmer, Encyclopedia of Chemical Technology (26) and in Doctor's Desk Reference (16).

3.3 Solubility

Practically insoluble in water and chloroform, soluble in methanol, acetone and 1:40 in alcohol (7, 8). Soluble in aqueous solution made alkaline with hydroxide of alkali metals. Rate of decomposition increases with increase of pH. (5, 6)

3.4 Ionization Constants

Various pKa values reported in literature are listed below, which shows that polythiazide is a monoprotic drug exhibiting only one pKa value.

pKa	9.10	(18)
	9.82 ± 0.04	(19)
	9.10 ± 0.10	(20)
	11.0	(21)
	9.58	(22)

3.5 Storage

Polythiazide is to be stored in air tight containers, protected from light. (8)

3.6 Ultraviolet Spectrum

The details of Ultraviolet Spectra in various media are listed below.

Extinction co-efficients $\left[E \ ^{1\%}_{1cm}\right]$ were shown in parenthesis.

Solvent	*Principal Bands (nm)*	
MeOH	226, 268 (500), 317	(13, 14)
EtOH (95%)	270 (468), 231, 318	(7, 63)
0.1N NaOH	226, 264, 307	(23)
Distilled Water made to alkaline to pH 10.0 with 2N ammonia	315 (80), 264, (460)	(10)
Distilled Water made acidic to pH 2.0 with H_2SO_4	270 (550), 310 (135)	(10)

3.7 Infrared Spectrum

The Infrared Spectrum of Polythiazide in KBr disc was presented in The Pharmaceutical Codex, BP (7,13). The spectrum shows principal peaks at the following wave numbers (cm^{-1}).

1071, 1120, 1162, 1326, 1345, 1597 (7)
1120, 1162, 1240, 1500, 1597, 1071 (9)
3360, 3280, 1350, 1330, 1160, 1125 (23)
1310, 705, 1610, 1575, 1500, 1460, 1415, 1425.

An infrared absorption spectrum of polythiazide was obtained as a KBr pellet using a Perkin-Elmer model 283 infrared spectrophotometer. The infrared spectrum of polythiazide is shown in Figure (1) on page 5.

3.8 Mass Spectrum

The direct probe electron impact (EI) mass spectrum of polythiazide as shown in Figure (2) on page 6 was obtained on a VG 70/250S mass spectrometer using 70 ev electron energy and a probe temperature of 250 degrees C. The mass spectrum of polythiazide Figure (2) on page 6 shows a base molecular peak at 439 m/z. Other prominent peak are observed at m/z = 310, 299 and 206. These fragmentation species lead to a breakdown pattern as suggested in Figure (3) on page 7. These data are consistent with the fragmentation pattern suggested by Clarke.[9]

3.9 NMR Spectra – Proton and Carbon-13 (25)

The proton and Carbon-13 magnetic resonance spectra of polythiazide are shown in Figures (4) on page 8 and (5) on page 9 respectively. The sample was run in deuterated dimethylsulfoxide and the spectra were obtained on a Brucker model 300 MHz instrument.

3.10 X-ray Crystallography (24)

X-ray Crystallographic data of polythiazide, chlorothiazide, hydrochlorothiazide have been determined from Weissenberg method by Dupout and Didelberg.

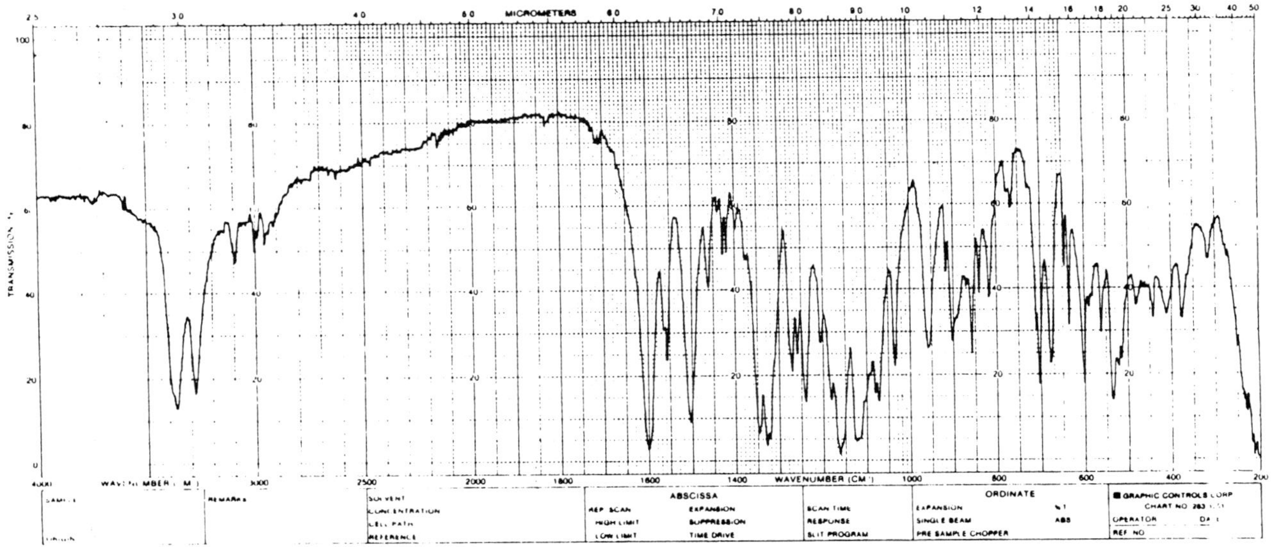

Figure (1) Infrared Absorption Spectrum of Polythiazide in a KBr Disc

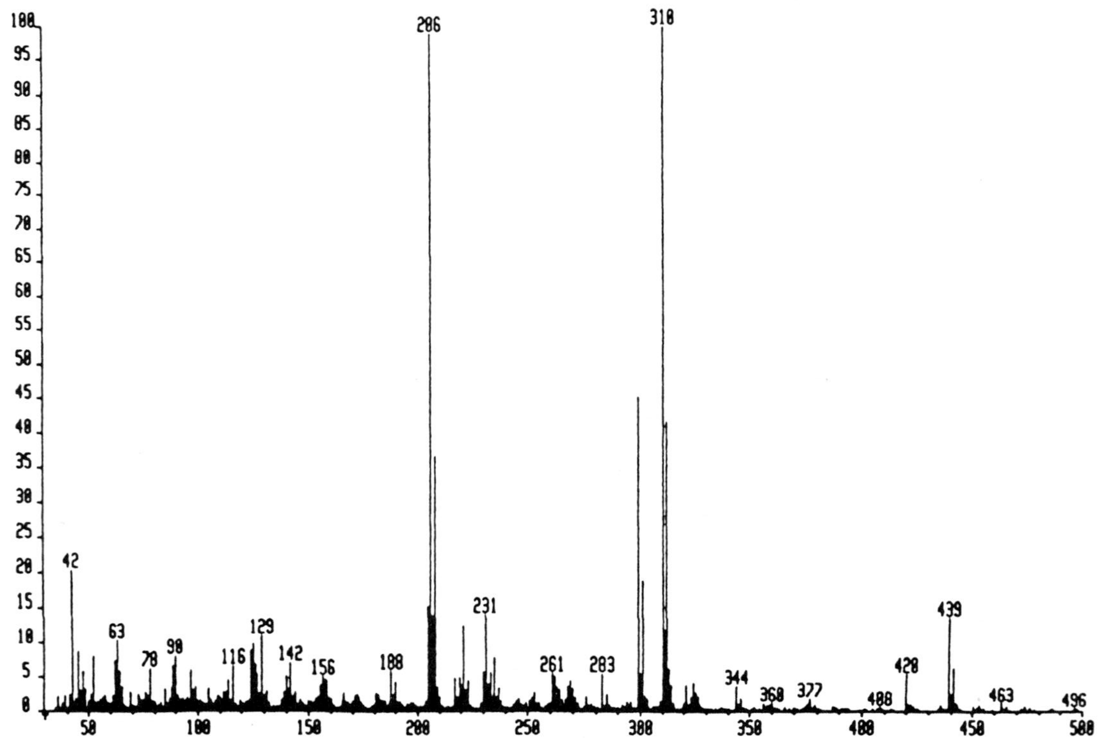

Figure (2) Electron Impact (EI) Mass Spectrum of Polythiazide

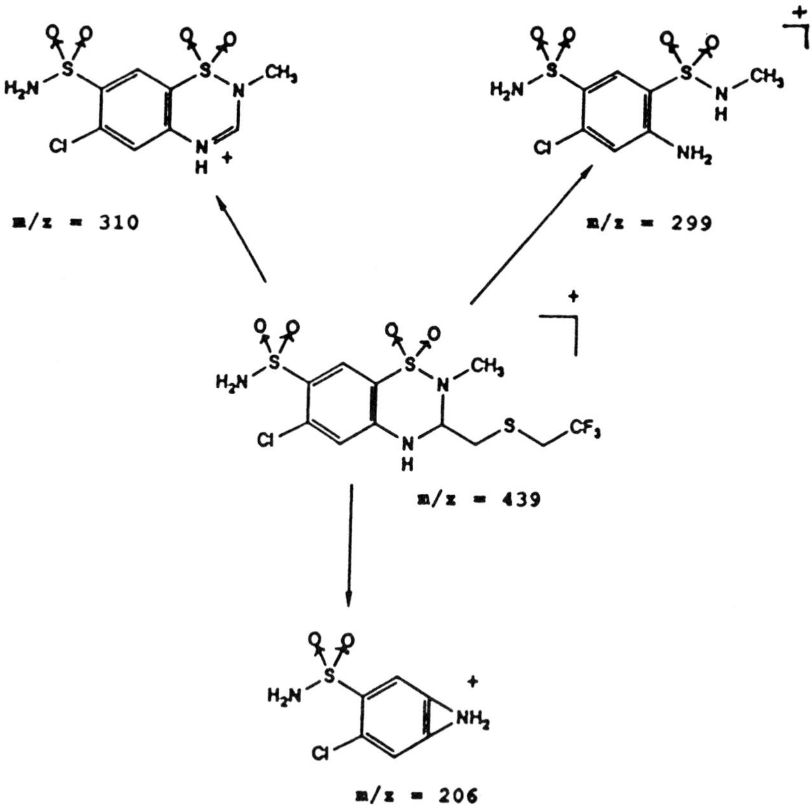

Figure (3) Mass Spectral Fragmentation Pattern for Polythiazide

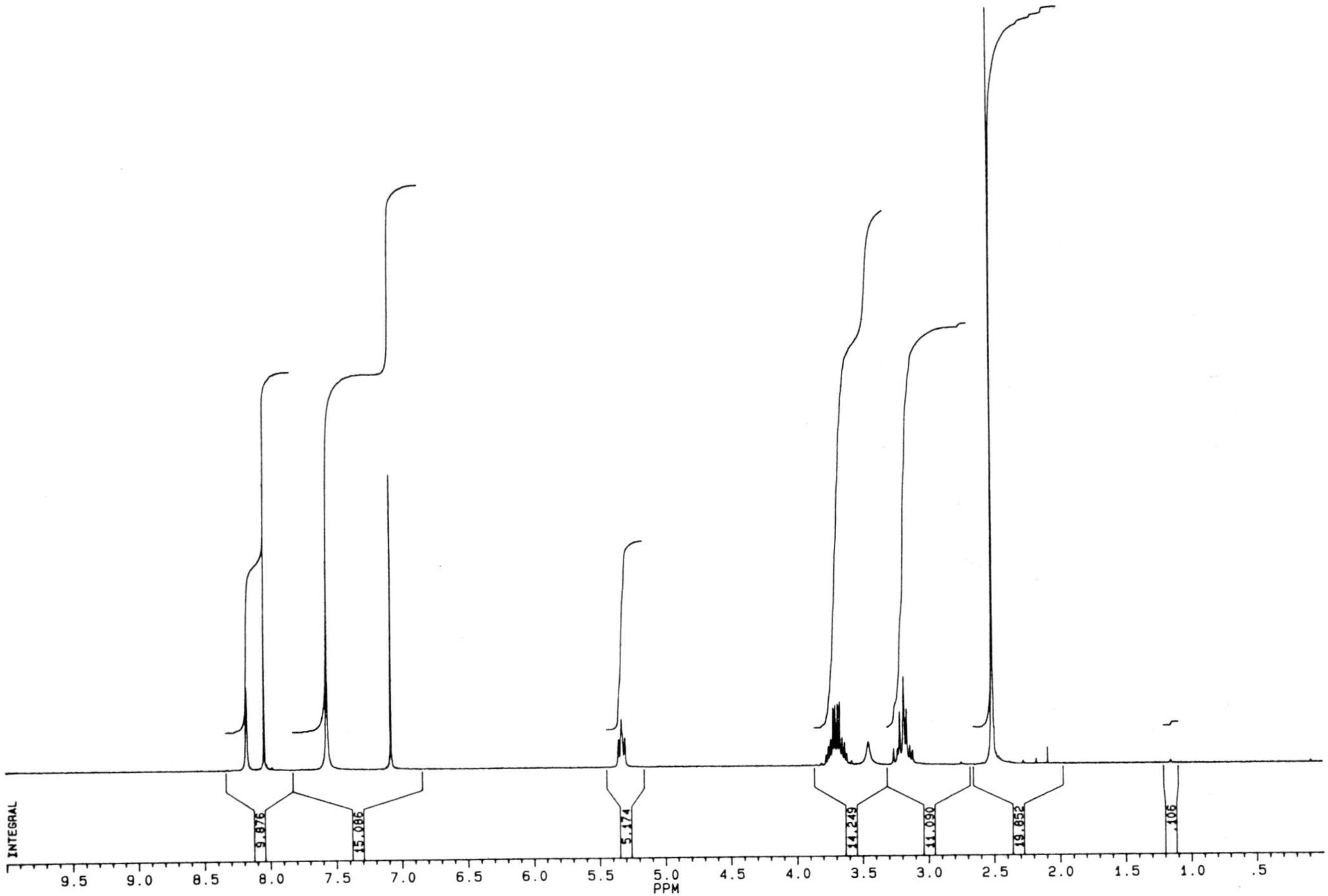

Figure (4) Proton NMR Spectrum of Polythiazide

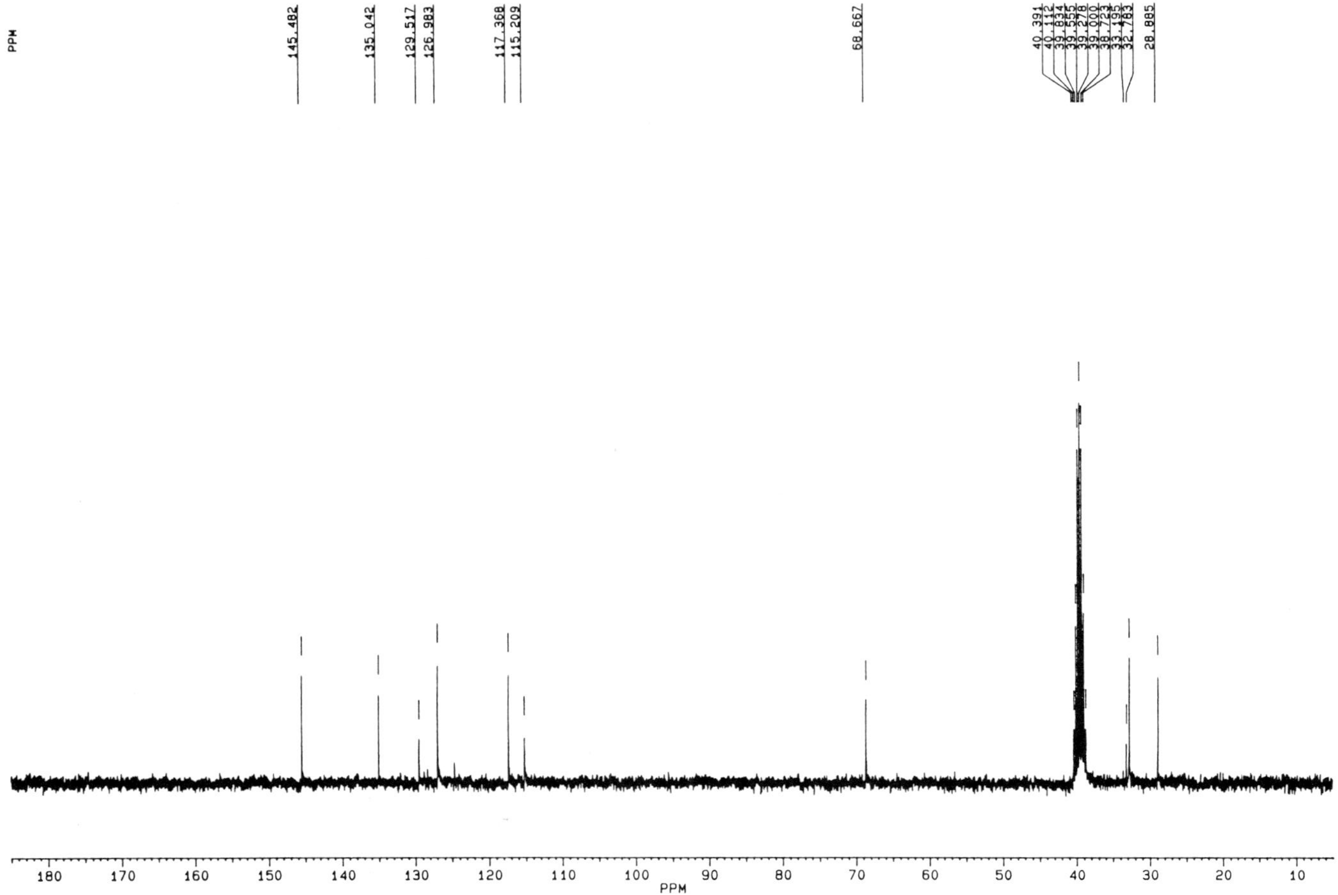

Figure (5) Carbon-13 NMR Spectrum of Polythiazide

3.11 Crystal Properties (23)

The operating parameters and X-ray diffraction data of polythiazide are given below in Table I:

Instrument	:	PW 1730 X-ray Diffractometer
Target	:	CU
Energy	:	KV : 35; mA : 20
Detector	:	Proportional
HV	:	2 x 796 volts
Filter	:	Ni
P.H.S. Window	:	LL : 22% ; W : 48, 4%
Gain	:	64
Time Constant	:	2
Chart Speed	:	1 cm/min.
20 scanning speed	:	1°/min.
Full scale	:	4×10^3

TABLE I

20	*d*	*Relative intensity* *I/Io*	*20*	*d*	*Relative intensity* *I/Io*
11.45	7.7215	5	30.50	2.9284	2
11.90	7.4305	2	32.30	2.7692	8
13.35	6.6266	6	34.25	2.6158	2
14.80	5.9804	100	34.75	2.5793	4
16.05	5.5174	26	35.20	2.5474	5
20.00	4.4357	31	37.65	2.3875	1
20.30	4.3708	28	38.50	2.3363	4
21.00	4.2267	44	39.45	2.2821	2
22.0 (b)	4.0368	18	40.40	2.2307	9
22.95	3.8717	3	42.10	2.1444	2
23.7 (s)	3.7509	2	44.1 (b)	2.0517	2
24.15	3.6820	38	44.75	2.0234	4
25.90	3.4371	2	45.10	2.0085	2
26.20	3.3984	5	46.30	1.9592	2
27.35	3.2580	7	49.15	1.8520	2
28.80	3.0972	18	28.98	3.0784	17
31.00	2.8823	4	31.50	2.8376	3

4. STABILITY (27)

Stability of polythiazide alone under different conditions, based on the available information, is found to be destroyed up to 100% within 1 hr. at 35° in NaOH solution at pH 12.0.

5. SYNTHESIS (4, 17, 28-31)

Preparation of polythiazide involves high temperature chlorosulfonation of *m*-chloroaniline (I), ammonolysis of the resulting bisulfonylchloride (II) leading to the corresponding bisulfonamide (III). Condensation of III with urea yields a heterocycle (IV). Treatment of IV with CH_3I and a base results in the alkylation of the more acidic ring nitrogen (V). Basic hydrolysis of V gives the N-methylated sulphonamide (VI). Condensation of VI with trifluoroethylthio acetaldehydedimethylacetal (VII) yields polythiazide which is recrystallized from isopropanol. The schematic representation of the synthetic process for polythiazide is shown in Figure (6) on page 12.

ALTERNATE SYNTHESIS (78)

An alternate synthesis involves treating 2,4-disulfamyl-5-chloroaniline with 2,2,2-trifluoroacetaldehyde in equimolar amounts to produce 3-(2,2,2-trifluoroethylthiomethyl)-6-chloro-7-sulfamyl-3,4-dihydro-1,2,4-benzothia-diazine-1,1-dioxide. Upon treatment with sodium hydride followed by a methyl iodide addition, the mixture is added to water where a white solid material forms. The product is recrystallized from acetone-toluene.

6. METABOLISM AND PHARMACOKINETICS (7, 10)

Polythiazide has a much greater potency and duration of action than chlorothiazide. The diuretic action is initiated within about 2 hrs and may last up to 48 hrs.

Absorption: Well absorbed after oral administration, $t\frac{1}{2}$ is 1.2 hrs.

Metabolism: In dogs, hydrolytic ring fission observed to form 4-amino-3-methyl sulphamoylbenzenesulphonamide and s-trifluoroethylthioglycollic acid as metabolites.

Excretion: Polythiazide is completely absorbed in dogs after oral or intravenous administration. It is excreted completely in 5 days, 80 - 85% in urine and 15 - 20% in feces (excretion $t\frac{1}{2}$ is 25.7 hrs, plasma half-life is 26 hrs.).

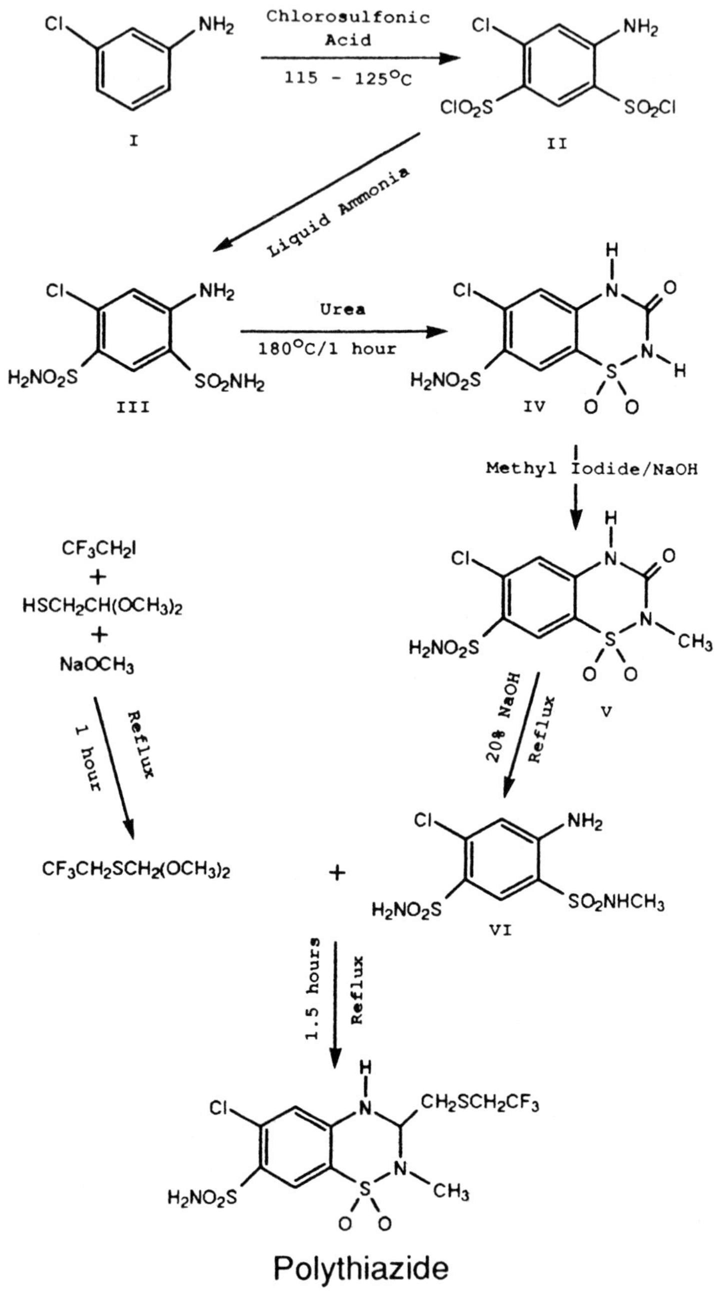

Figure (6) Synthetic Pathway for Polythiazide

7. METHODS OF ANALYSIS

7.1 Elemental Analysis (5)

Calculated % C:30.03, H:2.98, Cl:8.06, F:12.96, N:9.55, O:14.55, S:21.87.

7.2 Identification

BP and USP proposes methods of identification based on UV and IR spectra. The following are some tests of identification recorded in literature. (9)

Koppanyi-Zwikker's test	–	Violet
Libermann's test	–	Red-brown
Palladium Chloride	–	Orange
Sulphuric acid	–	Orange
PDAB or Vanillin	–	Violet (60)

7.3 Loss on Drying (12, 13)

Should not be more than 1% at 105°.

7.4 Sulphated Ash (12, 13)

Should not be more than 0.1%.

7.5 Spectrophotometric Analysis (12-15)

BP, USP or NF XIII describes UV spectrophotometric assay for polythiazide at 268 nm using methanol as the solvent.

7.6 Titrations

A general non-aqueous titration procedure was described for polythiazide along with other weak acids (34). It involves dissolving the drug in DMSO (DMF, Pyridine or $BuNH_2$ for a few compounds) and titrating with 0.1N KOH/PrOH. For the detection of end point, one of the following indicators was utilized: Metanil yellow, Azo yellow, Quinaldine red, Thymolphthaline, Phenol violet or Thymolphthaline-Tropeolin O (4:1). An iodometric method of determination was reported for polythiazide and other diuretics (35). This method involves preliminary treatment of polythiazide with potassium hexacyanoferroate in presence of H_2SO_4 and $ZnSO_4$ at 50° for 35 minutes or N-chlorosucciamide-HCl for 20 minutes followed by addition of KI. Titration of liberated Iodine was performed with sodium thiosulphate. A semi-aqueous potentiometric titration was performed for polythiazide (18) in acetone-water mixture(5:45, 10:40, 15:35 and 20:30). Titrations are performed by measuring the pH after the addition of each 0.1 ml of 0.05N NaOH.

7.7 Polarography (36)

A polarographic method using a Ag-AgCl reference electrode was reported. The half wave potential of the system was found to be −1.62 volts in tetraethylammonium bromide due to the reduction of the C-Cl bond of polythiazide.

7.8 Colorimetric Analysis

The colorimetric methods for polythiazide which were reported in literature are indicated in Table II below:

Table II

Treatment	*Reagent*	*max nm*	*ref*
Hydrolysis/ Diazotization	BM Reagent	510, 525	14, 15, 32
–do–	Pot. guiacol sulphomate	520	37
–do–	Phloroglucinol	415-20	38
Hydrolysis	PDAB/acid medium	445	39
–do–	MBTH-$NaIO_4$	550	40
–do–	PDMPD-NaOCl	540	42
----	Haematoxylin	555	43
----	Folin-Ciocalteu Reagent	760	44
Strong Acid treatment	Thymol	490	41

7.9 Chromatographic Techniques

7.9.1 Column Chromatography (45)

A method was reported for the separation and quantification of chlorothiazide, methyclothiazide and polythiazide using basic celite column. Chlorothiazide was eluted from a K_2HPO_4-celite column with acetic acid – ether mixture and extracted from organic phase into 0.2N HCl for the spectrophotometric assay, whereas polythiazide and methyclothiazide were eluted from 0.1M $NaHCO_3$-celite column with $CHCl_3$ and assayed spectrophotometrically.

7.9.2 Paper Chromatography

Pilsbury and Jackson (61) identified polythiazide (R_f 0.88) on Whatmann No. 1 paper after developing with acetone for 6½ hrs. and using UV (254 nm) detection or alkaline sodium naphthoquinone spray which gave orange spots within 10-15 minutes. Adam and Lapiere (48) reported the use of two paper chromatography methods for separation of polythiazide from other thiazides. One of these methods involves the use of Whatmann No. 1 paper and Pentanol – 12N aq. NH_3-H_2O (4:1:3). The other method involves the use of Whatmann No. 1 paper saturated with formamide and $CHCl_3$ – Pentanol (4:1) which is saturated with formamide for elution. Detection was done in both cases by UV (254 nm).

7.9.3 High Performance with Thin-Layer Chromatography (62)

A high performance with thin-layer chromatography method for the identification of polythiazide and 10 other diuretics was reported by Uchino *et al*. The thin-layer method uses a Wako gel FM plate and iso PrOH – NH_4OH (11:1) as an eluent. The HPLC method used a Nucleosil 5 C_{18} column.

7.9.4 Thin Layer Chromatography

The analyses of diuretics by thin-layer and paper chromatography were reviewed by Karel Macek (46). The reported TLC procedures for the identification of polythiazide are tabulated and presented in Table III. These procedures include both quantitative and qualitative separations.

Table III. TLC methods for the identification of polythiazide

Drugs separated	*Stationary phase*	*Solvent systems*	*Detection*	*Ref.*
PT, 8 other diuretics	Silica gel 60F254	MeOH-strong NH_3 (100:1.5) R_f 0.8 EtOAc:BZ (4:1) R_f 0.48 EtOAc-BZ (100:1.5) R_f 0.66	UV 254 nm hydrolysis and Azocoupling with BM reagent	7
PT	–do–	Toluene-Acetone (3:2)	UV 254 nm	13
PT	Silica gel	$CHCl_3$-MeOH (9:1) R_f (0.45 – 0.55)	UV, diphenylcarbazone	14
PT	–do–	–do–	UV 268 nm	15
PT, TCM, HCT	–do–	EtOAc-BZ (8:2)	UV 254, 1% PDAB containing HCl	18
PT	Silica AR 7 GF	CH_2Cl_2-MeOH (90:2) Rf 0.15	UV 268 nm	47
Saluritic thiazides including PT	Alumina G Silica gel G	EtOAc (R_f 0.1) EtOAc-BZ (4:1) R_f 0.85	UV 250 nm	48
–do–	Silica gel G	EtOAc-H_2O (100:1.5) R_f 0.58	– – –	50

Table III (continued)

Drugs separated	*Stationary phase*	*Solvent systems*	*Detection*	*Ref.*
PT, 8 other diuretics	Silica gel	EtOAc-$CHCl_3$-Heptane (1:1:1) R_f 0.85	B-M reagent UV $HgNO_3$, SNP/NaOH Van-urk reagent	49
PT, related subs.	–do–	Acetone	UV 254 nm	51
PT, 6 other thiazides	Kiesel gel	Toluene-Xylene-Dioxan-isoPrOH-25% aq NH_3 (1:1:3:3:1) R_f 0.84, PrOH-EtOAc-H_2O-aq NH_3 (5:1:3:1) R_f 0.89, EtOAc-BZ (3:1)	10% aq. $CuSO_4$-20% aq. NH_3 (5:1)	52
–do–	Silica gel	EtOAc-BZ (8:2)	Modified Folin reagent FMIR	53
Pt, five thiazides	–do–	Acetone-BZ-H_2O (3:3:1, 4:3:1, 5:3:2)	5% $BrCl_4$ NaF/EtOH	54
Pt, RS, sixteen diuretics	–do–	Methyl Ethyl Ketone-n-hexane (1:1, 2:1, 3:2) R_f 0.46, 0.67, 0.70 $CHCl_3$-Acetone-Tri Ethanolamine (50:50:1.5) R_f 0.65	Dragendroff's reagent diphenyl carbazone, Anisaldehyde, $KMnO_4$, BM reagent alkaline PDAB Mercurous Nitrate Chloroplatonic acid	55

Table III (continued)

Drugs separated	*Stationary phase*	*Solvent systems*	*Detection*	*Ref.*
Pt Degrn. products	–do–	$CHCl_3$-MeOH (9:1) R_f 0.45 - 0.55	UV H_2SO_4 spray	56
PT, 360 other S-contg. drugs	Silica Gel	Several systems	$Na_2Fe(CN)_5NO$ and $K_3Fe(CN)_6$, Ehrich's reagents	57
PT, CT, HCT	TLC, GC	----	BM reagent	58
Drugs contg. amino, imido, acetylinic, ethylenic, carboxylic purine groups		EtOAc-MeOH-NH_4OH (85:10:5) MeOH-NH_4OH (100:1.5) Hexane-$CHCl_3$-Me_2CO (2:3:1)	15% $FeCl_3$/2% HCl-15% aq. KI (1:1)	59
PT	----	EtOAc-BZ (8:2) R_f 0.55	Acidified PDAB or Vanillin	60
PT 14 diuretics, 7 antihypoglycemics	Silica gel G	$CHCl_3$-MeOH (9:1)	Quinones in DMF, DMSO or dioxan	77

7.9.5 Gas Chromatography

Andrey and Moffat (9, 63) analyzed polythiazide using a glass column containing 2–5% –SE 30 on Chromosorb G (80/100 mesh) at 238°C using Nitrogen (45 ml/mm) as the carrier gas. RI is 2380 using even number n-alkanes as a reference compound.

Hobbs and Twomey (11) estimated polythiazide in plasma via librated volatile trifluoroethylthioacetaldehyde. A glass column containing 5% Carbo wax 20M on Chromosorb G DMCS (60/80 mesh) at 130°C with Argon-methane (9:1) (70 ml/min) and ^{63}Ni electron capture detector were used.

Fragerlund *et al.* (25) estimated polythiazide after converting it to its tetramethyl derivative using FID or ^{63}Ni ECD detectors and nitrogen as a carrier gas.

7.9.6 High Performance Liquid Chromatography

The reported High Performance Liquid Chromatographic methods for the determination of polythiazide in dosage forms, biological fluids or in the presence of other therapeutically active substances have been compiled in Table IV.

Table IV. High performance liquid chromatographic methods for polythiazide

Compounds separated	*Column*	*Mobile Phase*	*Detector*	*Ref.*
PT/tablets	Spherisorb ODS	MeOH – 0.1M KH_2PO_4 at pH 3.5 (45:55)	UV 268 nm	13
PT, TCM, CYT, PFZ	Hitachi gel #3010	60% MeOH – 40% Na_2HPO_4-H_3PO_4 pH 3–10	UV 280 nm	22
PT, RS, Vanillin	μ Bondapak pH/corasil	MeOH–H_2O (35:65)	UV 254 nm	64, 65
PT, Vanillin	Partisil 10-ODS	MeCN–H_2O (98:2)	UV 254 nm	66
Thiazides in urine	Reverse phase	----	----	67
PT, Prazosin in plasma	μ Bondapak CN	$CHCl_3$–MeOH (97:3)	UV 264 nm	68
PT, RS, Vanillin	ODS column	MeOH–H_2O–AcOH (55:44:1)	UV 254 nm	69
PT, Prazosin	μ Bondapak CN	MeCN–Sod. Phosphate buffer pH 3 (7:13)	UV 268 nm	70
PT 8 diuretics	RP Monomeric column	30% aq.MeOH or 30% aq. MeOH + pH 4.0 acetate buffer	variable UV or fluorescence detector	71

Table IV (continued)

Compounds separated	*Column*	*Mobile Phase*	*Detector*	*Ref.*
PT, 560 compds.	C_{18}–SIL X–10	----	UV 220 nm	73
Relative retention data of 570 compds. including PT	RP 18	MeCN–pH 2.30 Phosphate buffer (156:344)	UV 220 nm	72
Thiazides, Pot. sparing loop diuretics	----	MeCN–H_2O (40:6)	UV	74
PT	ODS–Hypersil	MeCN–H_2O contg. 10 ml AcOH/ltr. (30:70)	UV	75

8. OTHER TECHNIQUES (24, 76)

Techniques such as radio immuno assay, lasers etc., have been used although less frequently for the determination of polythiazide.

9. ABBREVIATIONS USED

PT: Polythiazide, CT: Chlorothiazide, HCT: Hydrochlorothiazide, TCM: Trichloromethiazide, MCT: Methyclothiazide, CYT: Cyclothiazide, PFZ: Penfluazide, RS: Reserpine, MBTH: 3-methyl benzothiazolinonehydrazoneHCl, PDAB: p-dimethyl aminobenzaldehyde, PDMPD: p-N,N-dimethylaminophenylene diamine HCl, SNP: Sodium Nitro prusside.

10. ACKNOWLEDGEMENTS

The authors are grateful to the Quality Assurance Manager, Pfizer Limited for the generous reference sample of polythiazide. Thanks are also due to Dr. G. Ramana Rao and Dr. S. S. N. Murthy, Quality Control Labs, IDPL, Hydrabad for their encouragements and technical help.

11. REFERENCES

1. 'Physician's Desk Reference' (1969) 30th Ed., P2169. Mediean Economic Co., New Jersey.

2. Daniels, T. C. and Jorgensen, E. C. (1979) in Wilson and Giswold's Textbook of Organic, Medicinal and Pharmaceutical Chemistry, Doerge, R. F. (Ed) 18th Ed., P515, J. B. Lippincot Co., Philadelphia.

3. Index Guide (1985) Chemical Abstracts.

4. Swinyard, E. A. (1980) in 'Remington's Pharmaceutical Sciences', Arthur Osol, (Ed) 16th Ed., P881, Mack Publishing Co., Easton, Pa. 18042, U.S.A.

5. The Merck Index (1983) 10th Ed., P1095.

6. The Merck Index (1976) 9th Ed., P986.

7. The Pharmaceutical Codex (1979) 11th Ed., P717, 1029.

8. Martindale – The Extra Pharmacopoeia (1989) 29th Ed., P999, Reynolds, J. E. F. (Ed), The Pharmaceutical Press, London.

9. Isolation and Identification of Drugs (1986) 10th Ed., P913, Clarke, E. G. C. (Ed), The Pharmaceutical Press, London.

10. Isolation and Identification of Drugs (1969) 9th Ed., P506, Clarke, E. G. C. (Ed), The Pharmaceutical Press, London.

11. Hobbs, D. C., Twomey, T. M., *Clin. Pharmacol. Ther.* (1978) **23**, 241.

12. B. P. (1980), P357, 808.

13. B. P. (1988), P451, 988, S99.

14. USP XXI/NF XVI (1985), P850-2.

15. NF XIII (1970), P564-6.

16. Doctor's Desk Reference (1986-87), P555-7, R. Lakshmipathy (Ed), Professional Publications Ltd., Madurai, India.

17. 'Pharmaceutical Manufacturing Encyclopedia' Chemical Technology Review No. 124 (1979), P517, Marshall, S. (Ed), Noyes Data Corporation, N.J., U.S.A.

18. Hennig, U. G., Moskalyk, R. E., Chatten, L. G., Chan, S. F., *J. Pharm. Sci.* (1981), **70**, 317-9.

19. Khan, A. S., Cantwell, F. F., *Talanta* (1986), **33**, 119.

20. Hennig, U. G. (1979), M.Sc. Thesis Univ. of Alberta, Canada (through ref. 19).

21. Schriebine, A., Schreiber, E. C., Yu.M., Wiseman, E. H., P.S.E.B.M. (1962) (through ref. 19).

22. Yamazaki, M., Suzuka, T., Ito, Y., Itoh, S., Kitamura, M., Ohashi, K., Takeda, Y., Kameda, A., Orita, Y., Nakahama, H., *Chem. Pharm. Bull.* (1984), **32**, 2380.

23. Prasad, T. N. V. (1986), unpublished data.

24. Dupout, L., Didelberg, O., *Acta Crystallographica Sect. B,* (1970), **26**, 1884-5.

25. Fragerlund, C., Hartvig, P., Lindstrom, B., *J. Chromatogr.* (1979), **168**, 107-16.

26. Kirk-Othmer Encyclopedia of Chemical Tech. (1979), **8**, 1-33, 3rd Ed. Wiley Inter. Science, New York.

27. Matsushima, K., Kiyota, K., Iryo (1969), **23**, 1561-3.

28. McManus, J. M., U.S. Patent 3,009,911, Nov. 21, 1961 assigned to Chas. Pfizer & Co., Inc.

29. Benedict, R. G., Stodola, F. H., U.S. Patent 2,771,397, Nov. 20, 1956 assigned to U.S. Secretary, Agriculture.

30. Petty, M. A., U.S. Patent 2,595,605, May 1952 assigned to American Cynamed.

31. Close, W. J., Leo, R. S., Leonard, E. B., James, H. S., Maynette, V., *J. Amer. Chem. Soc.* (1960), 82, 1132-5.

32. Pinson, R. Jr., Schreiber, E. C., Wiseman, E. H., Hanmgartner, D., *J. Med. Pharm. Chem.* (1962), **5**, 491-503 (CA. 57;7842C).

33. Chas. Pfizer & Co., British Patent 910,145, Nov. 14, 1962 (CA. 58;3448G).

34. Schnekinburger, J., Quade-Henkel, M., Dtsch. Apoth-Z tg (1984), **124**, 1167-70 (CA. 101;177587b).

35. Shukla, I.C., Ahmed, S., Singh, D., Srivastava, M. K., Ind. *J. Pharm. Sci.* (1983), **45**, 249.

36. Grantes, P., Barat, J., *Ann. Pharm. Fr.* (1967), **25**, 447.

37. Magalhaes, J. F., Piros, M. G., Rev. Farm. Biquim. Univ., Sao Paulo (1971), **8**, 273.

38. Sastry, C. S. P., Prasad, T. N. V., Aruna, M., Suryanarayana, M. V., Venkata, Rao E., *Eastern Pharmacist* (1988), **3 1(2)**, 125.

39. Namigohar, F., Khorrami, J., Soltani, A., *J. Pharm. Belg.* (1977), **32**, 162 (CA. 87;90791S).

40. Sastry, C. S. P., Prasad, T. N. V., Sastry, B. S., Rao, E. V., *The Analyst* (1988), **113**, 255-8.

41. Perez, J. B., Danhier, A. C., Brieva, J. A., *An. Real. Acad. Farm.* (1967), **33(2)**, 233 (CA. 68;98681Z).

42. Sastry, C. S. P., Suryanarayana, M. V., Tipirneni, A. S. R. P., *Talanta* (1989), **36**, 491.

43. Suryanarayana, M. V., Ph.D. thesis (1989), Andhra University, India.

44. Sastry, C. S. P., Suryanarayana, M. V., Prasad, T. N. V., Tipirneni, A. S. R. P., Indian Drugs (in press).

45. Fazzari, F. R., *J. Assoc. Off. Anal. Chem.* (1974), **56**, 677-80.

46. Pharmaceutical Application of Thin Layer Chromatography and Paper Chromatography, Karel Macek (Ed), (1972), P241-4, Elsevier.

47. Flinn, P. E., Smith, T. G., *J. Assoc. Off. Anal. Chem.* (1974), **57**, 82-4.

48. Adam, R., Lapiere, C. L., *J. Pharm. Belg.* (1964), **19**, 78-89 (AA. 12;4190).

49. Gecgil, S., *Eczalik Bul.* (1965), **7 (8-10)**, 100-8 (CA. 64;14032d).

50. DuChene, M., Lapiere, C. L., *J. Pharm. Belg.* (1965), **20**, 275.

51. Margasinski, Z., Danielak, R., Rafalowska, *Acta Pol. Pharm.* (1966), **23**, 35-7 (AA. 14;3545).

52. Guven, K. C., Cobanlar, S., *Eczacilic Bult.* (1967), **9**, 98-103 (CA. 67;102839f).

53. Smith, P. J., Hermann, T. S., *Anal Biochem.* (1968), **22**, 134-44.

54. Ren., T. W., Ming, C. T., *Chemistry Taipei* (1970), **1-2**, 25-7.

55. Stohs, S. J., Scratchley, G. A., *J. Chromatogr.* (1975), **114**, 329.

56. Wessinger, W., Auterhoff, H., Dtsch.Apoth-Ztg (1979), **119(35)**, 1377 (CA. 91;199020p).

57. Frank H., Heinisch, G., Rank, W., Sci. Pharm. (1979), **47(2)**, 135-42 (CA. 92;11291w).

58. Merriken, R. A., *Aviat. Space Environ. Med.* (1980), **51**, 996-8.

59. Fiorese, F., Vermeulen, G., Turcotte, C., Toxicol Aspects. (1980), **9**, 431-5, Kovatsis, A. V. J. M. (Ed), (Int. Congr. Eur. Assoc. Poison Control Cent.).

60. Thakur, A. B., Dayal, S., *J. Pharm. Sci.* (1982), **71**, 1422.

61. Pilsbury, V. B., Jackson, J. V., *J. Pharm. Pharmacol.* (1966), **18**, 713.

62. Uchino, K., Yamamura, Y., Saitoh, Y., Isozaki, S., Tamura, Z., Nakagawa, F., Sekine, K., Kojima, I., Yakugaku Zasshi (1984), **104**, 1101 (CA. 102;41073u).

63. Andrey, R. E., Moffat, A. C., *J. Chromatogr.* (1981), **220**, 195.

64. Cohen, D. M., Munnelly, K. P., *J. Pharm. Sci.* (1976), **65**, 1413.

65. Moskalyk, R. E., Lockck, R. A., Chatten, L. G., Veltman, A. M., Nielech, M. F., *J. Pharm. Sci.* (1975), **64**, 1406.

66. Wong, C. K., Tsan, D. V. J., Cohen, D. M., Munnelly, K. P., *J. Pharm. Sci.* (1977), **66**, 736.

67. Tisdall, P. A., Moyer, T. P., Anhalt, J. P., *Clin. Chem.* (1980), **26**, 702 (CA. 93;36651x).

68. Dokladalova, J., Coco, S. J., Lemke, P. R., Quercia, G. T., Korst, J. J., *J. Chromatogr.* (1981), **224**, 33 (CA. 95;35150n).

69. Belchmann, W. J., *J. Liq. Chromatogr.* (1986), **9**, 1463-78 (CA. 105;49166j).

70. Belchmann, W. J., *J. Liq. Chromatogr.* (1986), **9**, 1033 (CA. 105;12233c).

71. Shah, V. P., Lee, J., Prasad, V. K., *Anal. Lett.* (1982), **15B**, 529-36.

72. Daldrup, T., Susanto, F., Michalke, P., *Fr. Z. Anal. Chem.* (1981), **308**, 413 (CA. 96;29498).

73. Daldrup, T., Michalke, P., Boehme, W., *Chromatogr. Newsl.* (1982), **10**, 1-7 (CA. 98;84433c).

74. DeCroo, F., Vanden, B. W., DeMoerloose, P., *J. Chromatogr.* (1985), **325**, 395 (CA. 103;27411s).

75. Gill, R. Unpublished data through ref. 9.

76. Orita, Y., Ando, A., Musuhara, H., Mataga, N., Urakabe, S., Arzneim-Forsch (1977), **27**, 940.

77. Agarwal, S. P., Nwaiwu, J., *J. Chromatogr.* (1986), **351**, 383-7.

78. Buck, C. J., McManus, J. M., U.S. Patent 3,133,060, May 12, 1964 assigned to Chas. Pfizer & Co., Inc.

TERAZOSIN

Zui L. Chang and John F. Bauer

Abbott Laboratories
North Chicago, Illinois 60064

Contents

1. Description

1.1 Nomenclature

1.11 Chemical Name

(a) 2-[(4-(Tetrahydro-2-furanyl) carbonyl)-1-piperazinyl]-6,7-dimethoxy-4-quinazolinamine, monohydrochloride, dihydrate

(b) 1-(4-Amino-6,7-dimethoxy-2-quinazolinyl)-4-[(tetrahydro-2-furanyl)carbonyl)]piperazine

(c) 2-[(Tetrahydro-2-furoyl)-1-piperazinyl]-4-amino-6,7-dimethoxyquinazoline

(d) CAS: 63590-64-7

1.12 Generic Name

Terazosin
Terazosin Hydrochloride
Terazosin Hydrochloride Dihydrate

1.13 Trade Name

Hytrin (Abbott)
Heitrin (Abbott)
Hytrinex (Abbott)
Vasocard (Abbott)
Vasomet (Mitsubishi Kasei)

1.2 Formulas and Molecular Weight

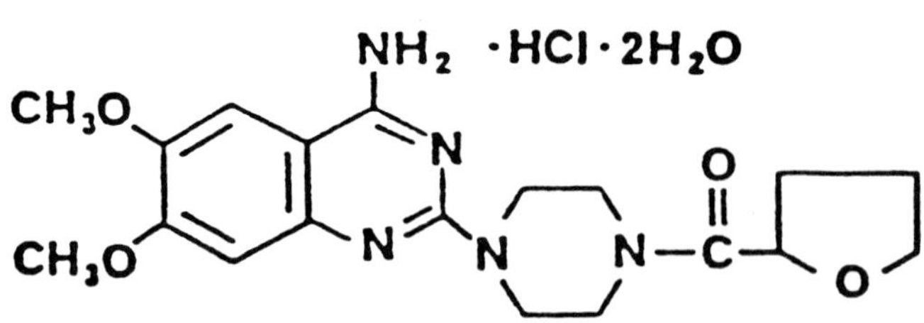

$C_{19}H_{25}N_5O_4 \cdot HCl \cdot 2H_2O$ M.W. 459.93

1.3 Appearance, Color, and Odor

Terazosin is a white fine crystalline powder with no discernible odor.

2. Physical Properties

2.1 Infrared Spectrum

The infrared spectrum of terazosin is presented in Figure 1. The spectrum was measured in the solid state as a potassium bromide dispersion. The following bands (cm^{-1}) have been assigned for Figure 1 (1).

(a) Between 3600 and 2500 cm^{-1}

A broad complex absorption due to superimposing N-H stretching vibrations of the protonated amine, and to the OH stretching vibrations of the water of crystallization.

(b) 1640 cm^{-1}

A band due to C=O stretching vibration of the tertiary amide.

(c) 1620 cm^{-1}

A band due to N-H bending vibrations of the protonated amine.

(d) Between 1585 and 1522 cm^{-1}

Bands due to skeletal stretching modes of the quinazoline nucleus.

2.2 Raman Spectrum

The Raman spectrum of terazosin as shown in Figure 2 was obtained in the solid state on a Cary Model 83 Spectrophotometer. The following bands (cm^{-1}) have been assigned for Figure 2 (2).

Figure 1 - Infrared Spectrum of Terazosin

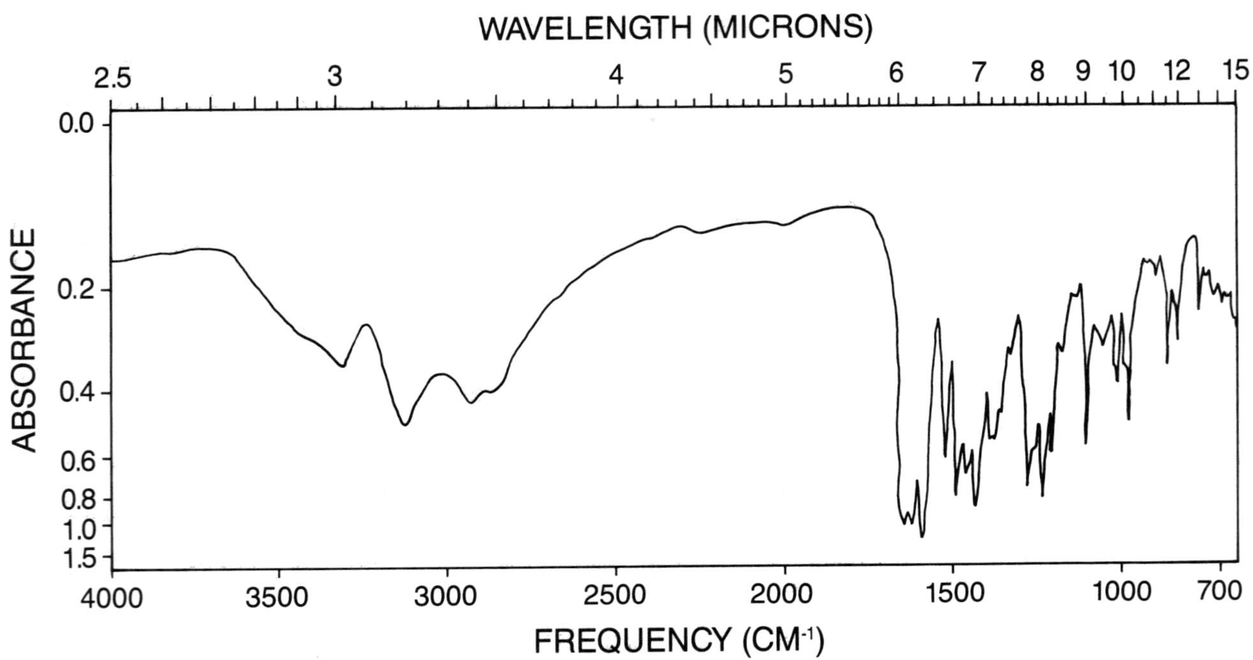

Figure 2 - Raman Spectrum of Terazosin

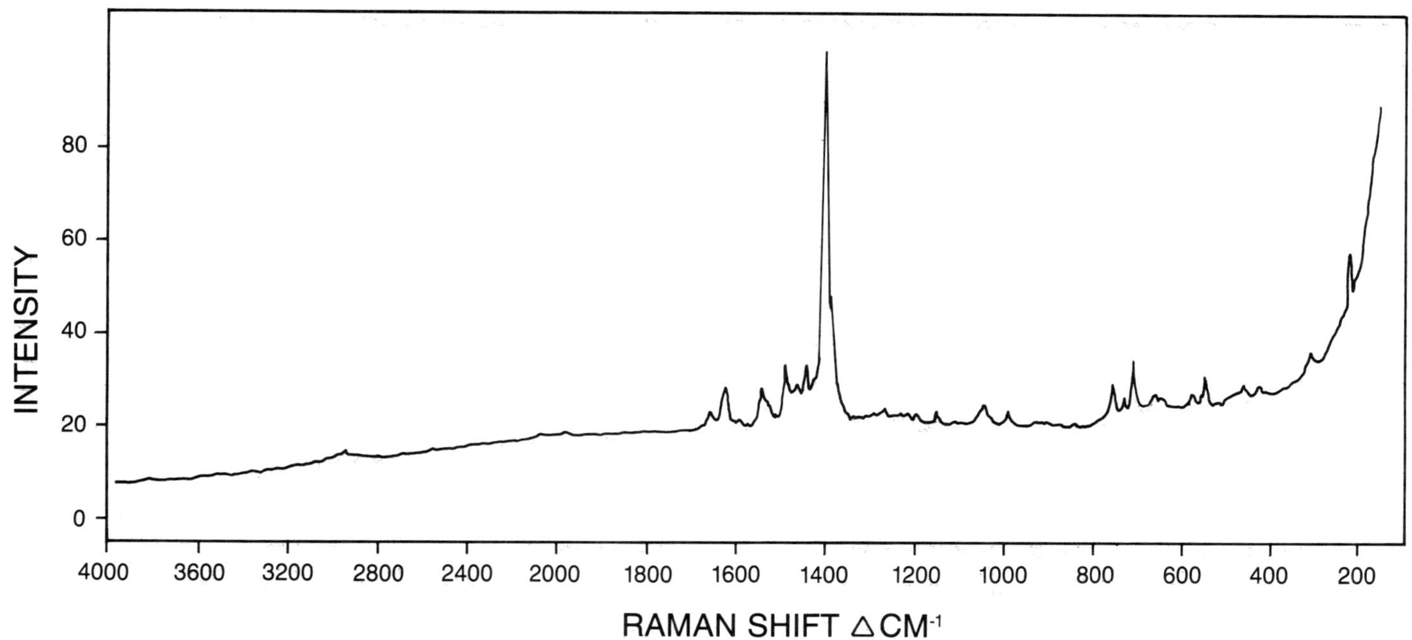

(a) 1629 cm^{-1}

A band due to skeletal stretching vibrations of the benzene ring.

(b) 1406 cm^{-1}

A very strong band probably due to ring vibrations of the quinazoline nucleus.

2.3 Proton Nuclear Magnetic Resonance Spectrum (PMR)

The proton magnetic resonance spectrum of terazosin as shown in Figure 3 was obtained on a General Electric GN 300 Spectrometer with a 1% w/v solution in a solvent of deuterated dimethyl sulfoxide. The spectral peak assignments (3) are presented in Table I.

Structure

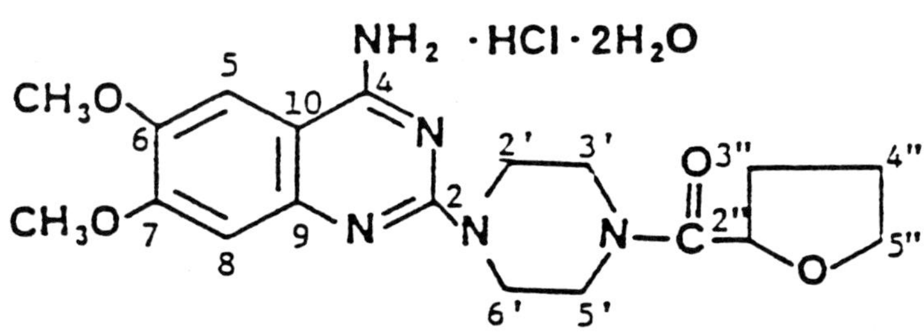

Figure 3 - Proton Nuclear Magnetic Resonance Spectrum of Terazosin

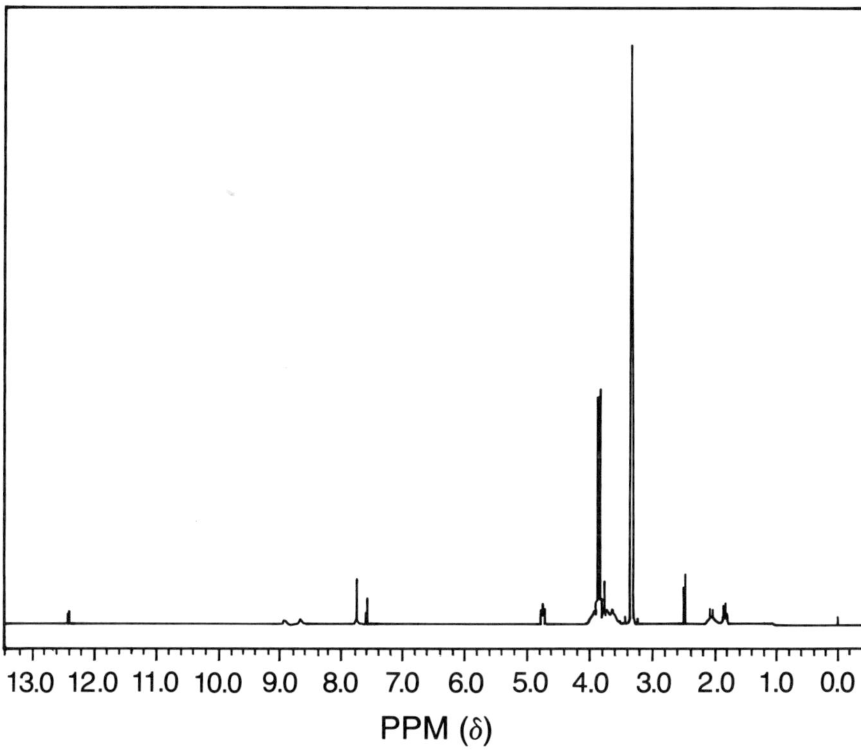

Table I.

Spectral Assignments for Terazosin

Proton Assignment	Chemical Shift (ppm)	Multiplicity
5	7.76	singlet
8	7.60	singlet
2′	4.05-3.5	complex
3′		
5′		
6′		
2"	4.75	doublet of doublets
		$3J_{HH}$ = 5.5 Hz
		$3J_{HH}$ = 7.3 Hz
3"	2.05	complex
4"	1.85	complex
5"	3.78	complex
		$3J_{HH}$ = 6.7 Hz
		$3J_{HH}$ = 4.3 Hz
7-OCH_3	3.88	singlet
6-OCH_3	3.85	singlet
NH_2	8.93	broad
	8.66	broad
HCl	12.41	broad singlet

Chemical shifts are reported from internal TMS (0.0 ppm).

2.4 Carbon-13 Nuclear Magnetic Resonance Spectrum (CMR)

The carbon-13 nuclear magnetic resonance spectrum of terazosin as shown in Figure 4 was obtained on a General Electric GN 300 Spectrometer with a 1% solution in a solvent of deuterated dimethyl sulfoxide. The spectral peak assignments (3) are presented in Table II.

Figure 4 - Carbon-13 Nuclear Magnetic Resonance Spectrum of Terazosin

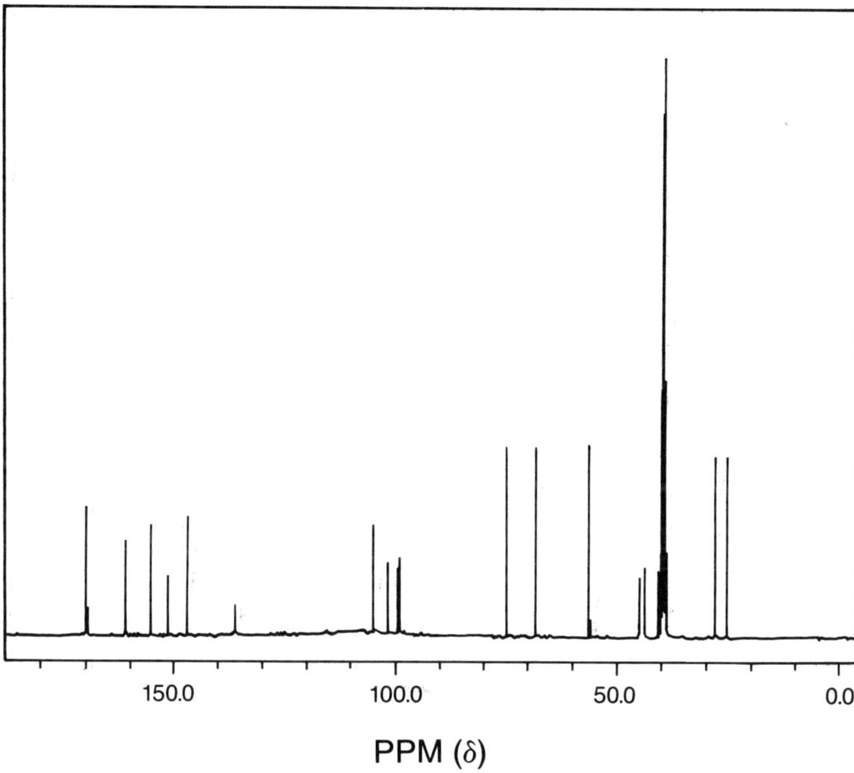

Table II

CMR Spectral Assignments for Terazosin

Carbon Assignment	Chemical Shift ppm	Attached Protons by DEPT*
2	151.3	0
4	161.2	0
5	105.0	1
6	146.7	0
7	155.1	0
8	99.2	1
9	136.2	0
10	101.7	0
2′	44.8	2
3′	44.2	2
5′	43.9	2
6′	40.7	2
2"	74.9	1
3"	27.9	2
4"	25.2	2
5"	68.2	2
O=C	169.7	-
7-OCH_3	56.0	3
6-OCH_3	56.3	3

The DMSO-d6 solvent peak septet is set at 39.5 ppm as a reference point.

* DEPT is distortional enchancement polarization and transfer.

2.5 Mass Spectrum

The mass spectrum of terazosin as shown in Figure 5 was obtained using a A.E.I. MS-902 Mass Spectrometer with an ionization electron beam energy of 70 eV. High resolution data were compiled and tabulated with the aid of a PDP-11 computer.

The mass spectrum assignments of the prominent ions and subsequent fragments are shown in Table III and Figure 6 (4).

Figure 5 - Mass Spectrum of Terazosin

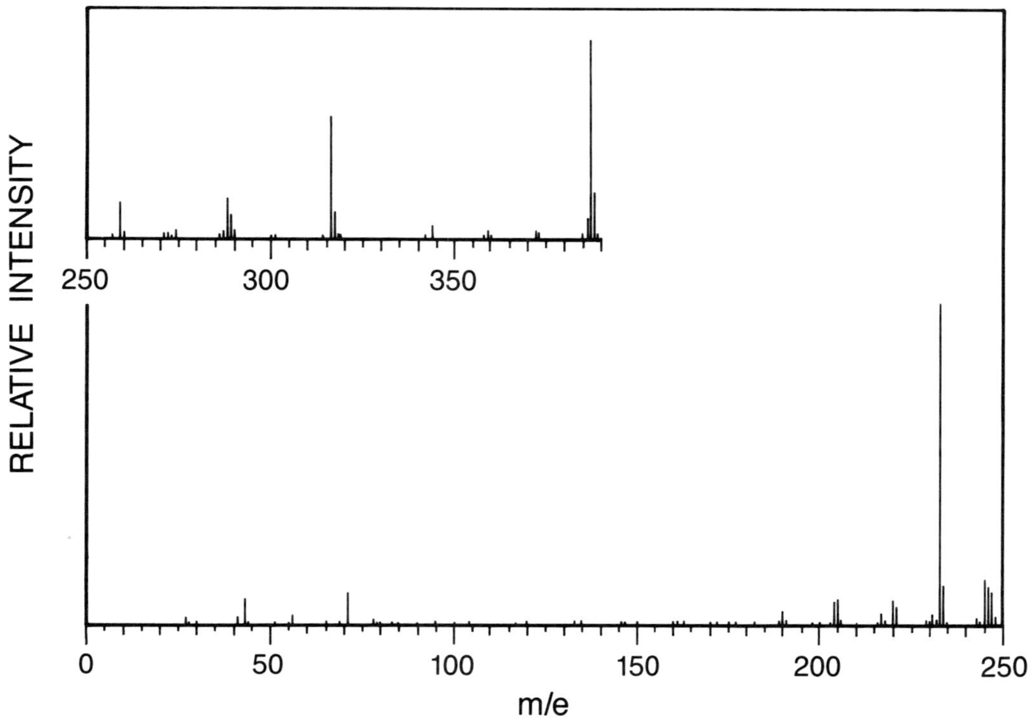

Figure 6 - Fragmentation Pathway of Terazosin

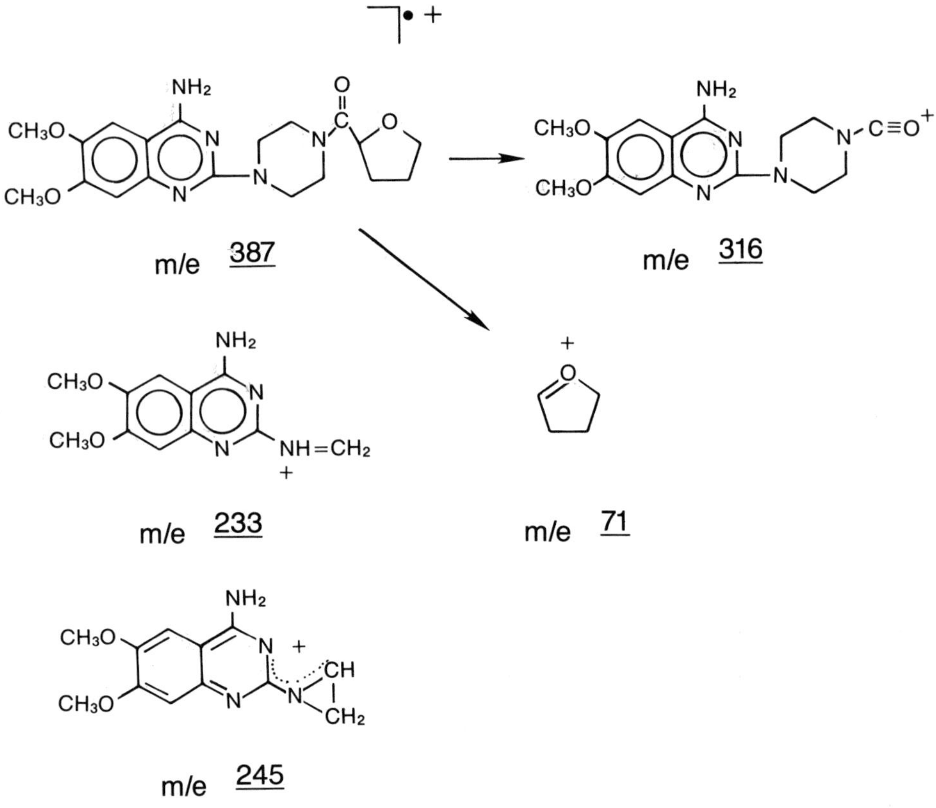

Table III

High Resolution Mass Spectrum of Terazosin

Measured Mass	Calculated Mass	Formula
387.1900	387.1906	$C_{19}H_{25}N_5O_4$
316.1391	316.1409	$C_{15}H_{18}N_5O_3$
245.1029	245.1038	$C_{12}H_{13}N_4O_2$
233.1031	233.1039	$C_{11}H_{13}N_4O_2$
71.0493	71.0497	C_4H_7O

2.6 Ultraviolet Spectrum (UV)

When a 0.005% solution of terazosin in water was scanned from 400 to 200 nm, three maxima at 212 nm (a=65.7), 245 nm (a=127.5) and 330 nm (a=24.0) were observed (Figure 7).

2.7 Fluorescence Spectrum

A 0.0009% solution of terazosin in water exhibits fluorescence when excited with ultraviolet light. The emission spectra of terazosin in water, given in Figure 8 show an emission maximum at 390 nm (5). These spectra were obtained using excitation wavelengths of 257 nm and 345 nm which correspond to the maxima of the absorption spectra.

2.8 Solubility

The following solubility data have been determined for terazosin at room temperature:

Solvent	Solubility, mg/mL
Methanol	33.7
Water	29.7
95% Ethanol	4.1
0.1 N Aqueous HCl	3.8
Chloroform	1.2
Acetone	0.01
Hexane	Practically insoluble

Figure 7 - Ultraviolet Absorption Spectrum of Terazosin

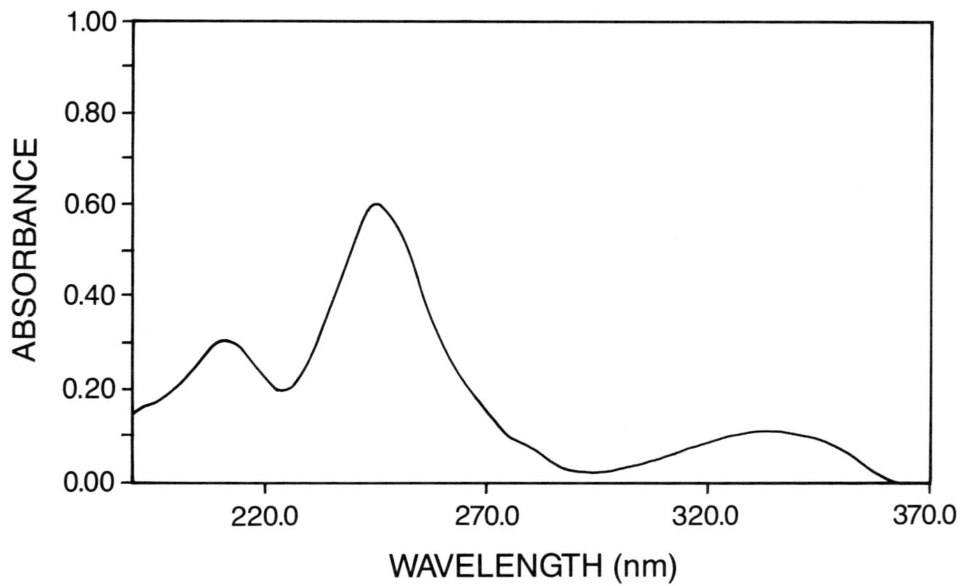

Figure 8 - Emission Spectra of Terazosin
Excitation Wavelengths at 257nm and 345nm

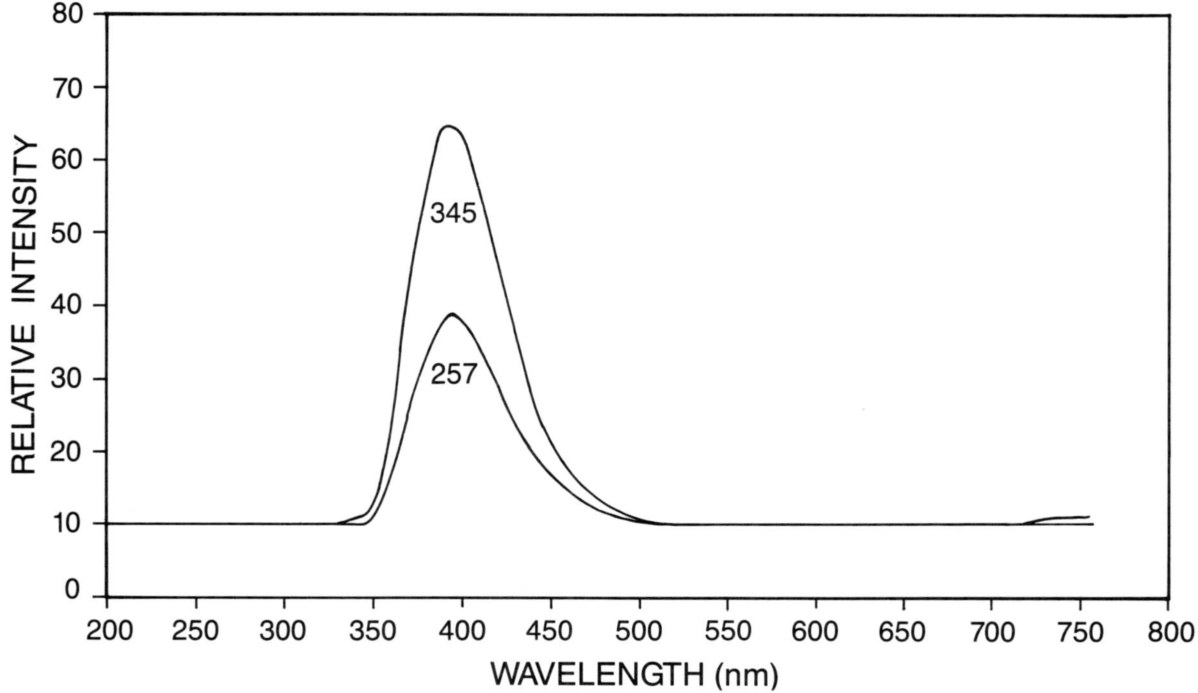

2.9 X-Ray Powder Diffraction

The x-ray powder diffraction pattern was determined by visual observation of a film obtained with a 143.2 mm Dehye-Scherrer Powder Camera (Table IV). An Enraf-Nonius Difractis 601 Generator; 38 KV and 18 ma with nickel filtered copper radiation, $\lambda = 1.5418$, was employed (6).

Table IV

X-Ray Powder Diffraction Pattern of Terazosin d-Spacings and Intensities

d(Å)*	I/I_0**	d(Å)*	I/I_0**
9.9	10	3.71	100
8.7	10	3.65	5
8.0	5	3.45	10
7.6	50	3.11	5
6.9	2	3.05	2
5.6	8	2.95	5
5.35	5	2.84	5
5.2	5	2.79	5
4.95	10	2.68	5
4.75	30	2.58	2
4.45	5	2.47B	5
4.30	10	2.12	2
4.15	2	1.94	2
4.0	10	1.91	2
3.91	20	1.79	2

B = Broad line.

$$*d = \frac{n\lambda}{2 \text{ Sin A}} = \text{interplanar distance}$$

**I/I_0 = Relative intensity (based on the highest intensity of 100).

2.10 Melting Range

Terazosin melts in the range of 272.6° and 274°C (USPXXI, P. 1250, Class Ia).

2.11 Differential Thermal Analysis

The differential scanning calorimetry (DSC) curve (7) of terazosin obtained on a DuPont Model 910 Differential Scanning Calorimeter and Model 990 Differential Thermal Analyzer at a scan rate of 5°C/minute exhibits a broad endotherm at 131°C which is due to loss of water of hydration. A sharp endothermic peak at 272°C is indicative of the melting point of terazosin (Figure 9).

The thermogravimetric analysis curve (7) of terazosin obtained on a DuPont Model 951 Thermal Gravimetric Analyzer at a scan rate of 5°C/minute exhibits a weight loss corresponding to the first endotherm on the DSC. The loss represents 8.0% and is due to loss of water of hydration. Some indications of sublimation and/or decomposition are apparent above 200°C (Figure 10).

2.12 pKa

Titration of terazosin with 0.1 N aqueous NaOH using water as the sample solvent gave a pKa value of 7.1 (proton gained).

2.13 Sublimation

Terazosin showed some sublimation at 242-244°C. This is also observed by thermogravimetric analysis.

2.14 Hygroscopicity

Terazosin (dihydrate) showed negligible hygroscopicity. Samples of terazosin were subjected to 50% and 70% relative humidities, and less than 0.1% moisture absorption was observed for both conditions.

2.15 Optical Rotation

A 1% water solution of terazosin exhibited no optical activity when measured on a Perkin-Elmer Model 141 Polarimeter at 589 nm in a 1 dm cell at 25°C. Although the compound has an asymmetric center, it is isolated as the racemic mixture.

Figure 9 - Differential Scanning Calorimeter Analysis Curve of Terazosin

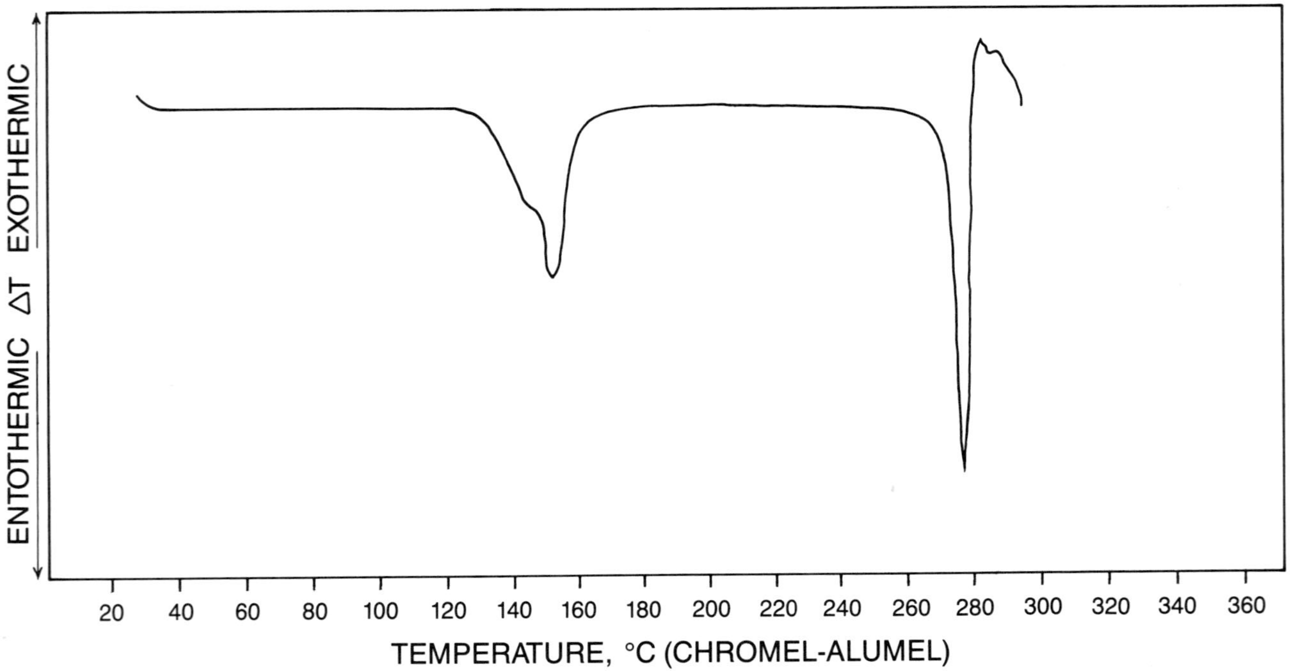

Figure 10 - Thermogravimetric Analysis Curve of Terazosin

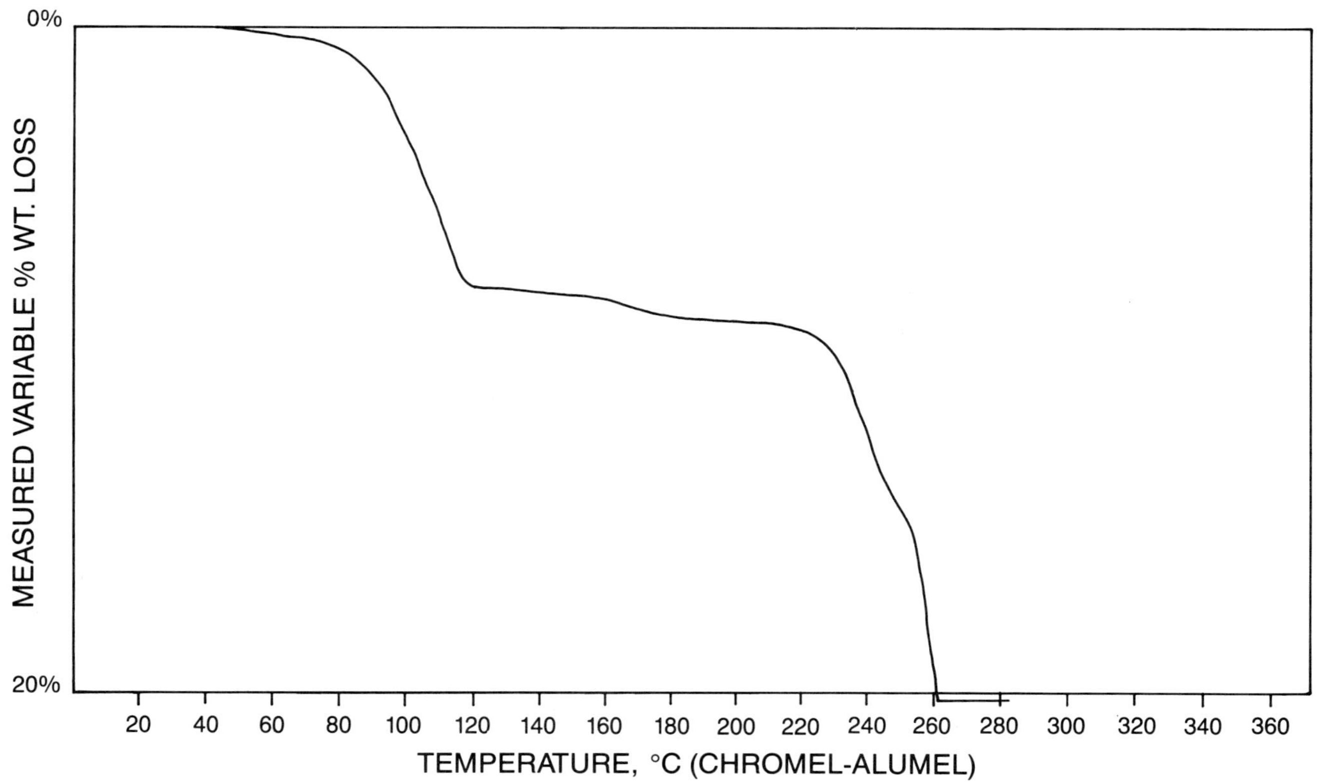

3. Synthesis

Terazosin was first prepared by M. Winn et al. (8) and the process was modified by R. Roteman (9).

Terazosin may be synthesized by the reaction scheme shown in Figure 11 (10).

4. Stability

In the solid state, terazosin is relatively stable under high-stress thermal and photochemical degradative conditions. It is also relatively stable in water at room temperature and at elevated temperature for a short period. However, the compound shows degradation in weak aqueous acid at room temperature and rapid degradation in weak aqueous acid and weak aqueous base at elevated temperatures.

The major degradation product observed by thin-layer chromatography and high performance liquid chromatography in neutral, weak acid, or weak base is 2-piperazinyl-6,7-dimethoxy-4-aminoquinazoline dihydrochloride (Figure 12, Compound II). The amide linkage is broken by hydrolysis to form Compound II and III. Compound III could further be degraded to Compounds IV by decarboxylation in water, acid or base with the presence of heat (Figure 12) (11).

5. Method of Analysis

5.1 Elemental Analysis

A typical elemental analysis of a sample of terazosin is presented in Table V.

Table V

Elemental Analysis of Terazosin
(as the dihydrate; the sample was analyzed without drying)

Element	% Theory	% Found
Carbon	49.62	49.66
Hydrogen	6.57	6.80
Nitrogen	15.23	15.32
Oxygen	20.87	20.96
Chlorine	7.71	7.68

Figure 11 - Synthetic Pathway of Terazosin

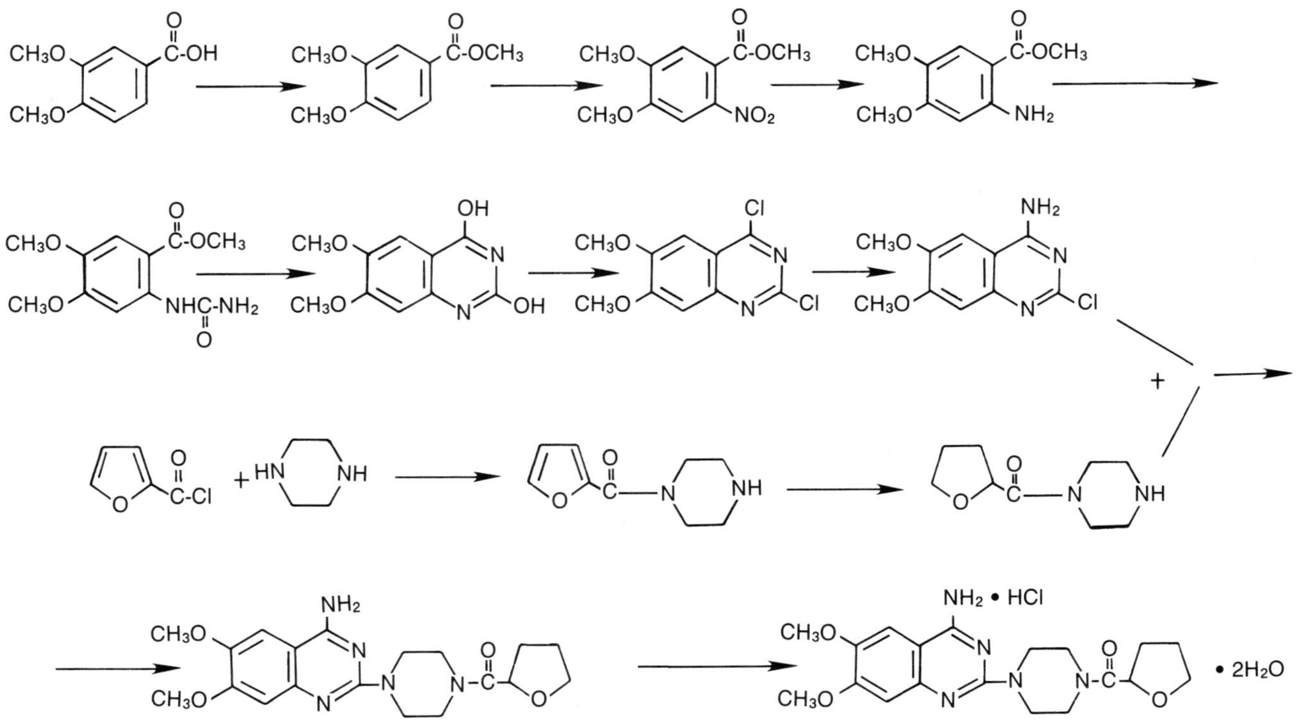

Figure 12 - Degradation Pathway of Terazosin

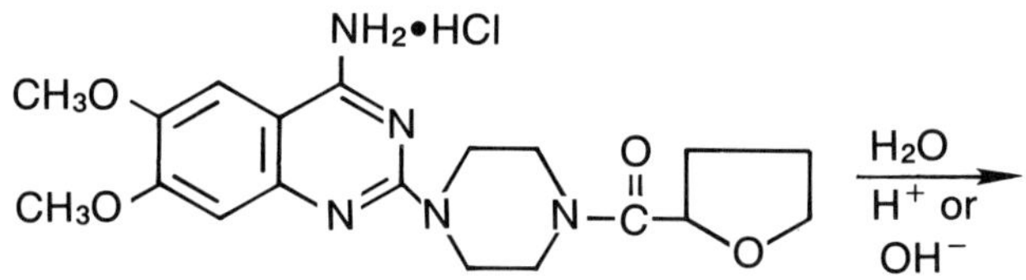

Terazosin (I)

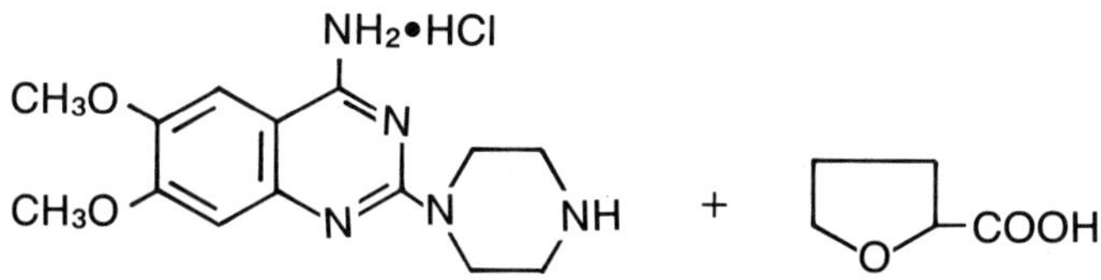

2-Piperazinyl-6,7-dimethoxy-4-aminoquinazoline monohydrochloride (or dihydrochloride)

(II)

Tetrahydro-2-furoic acid

(III)

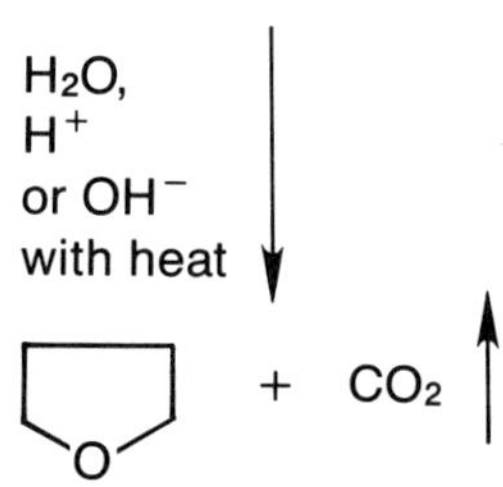

Tetrahydrofuran

(IV)

5.2 Chromatographic Analysis

5.21 Thin-Layer Chromatography

Several thin-layer chromatographic systems have been investigated to evaluate the purity of terazosin. The systems giving the best separation of terazosin, its impurities and degradation product are listed in Table VI.

Table VI

Thin-Layer Chromatographic Systems for Terazosin

System No.	Solvent System	Adsorbent	Visualization
1.	Chloroform:Toluene: Methanol (9:1:6)	Silica Gel 60 F_{254}	Shortwave Ultraviolet light and starch-potassium iodide spray.
2.	Methylene Chloride: Ethyl Acetate: Acetic Acid (19:1:0.2)	Silica Gel 60 F_{254}	Ceric sulfate-ammonium molybdate spray

(a) System 1.

2-Piperazinyl-6,7-dimethoxy-4-aminoquinazoline (R_f 0.05) and 2-chloro-4-amino-6,7-dimethoxy quinazoline (R_f 0.74) can be separated from terazosin (R_f 0.60) and detected by shortwave ultraviolet light. 2-Tetrahydrofuroyl piperazine (R_f 0.10) can be detected by starch-potassium iodide reagent.

(b) System 2.

This system is used to detect a possible degradation product, tetrahydro-2-furoic acid which is difficult to differentiate and observe in System 1. Tetrahydro-2-duroic acid has an R_f at 0.11 in System 2 and can be detected by ceric sulfate-ammonium molybdate spray. Terazosin has an R_f at origin.

5.22 High Performance Liquid Chromatography

The following HPLC methods were developed for quantitative determination of terazosin and of its by-products:

System 1

Mobile phase: Water:Acetonitrile:Methanol (68:22:10) containing 0.02 M citrate buffer.

Column: 30 cm x 1/4" (o.d.) x 4 mm (i.d.) packed with μ-Bondapak™ C_{18}, 10μ (Waters Associates)

System 2

Mobile phase: Water:Acetonitrile:Isopropyl alcohol (72:7:2) containing 0.02 M citrate buffer.

Column: 30 cm x 1/4" (o.d.) x 4 mm (i.d.) packed with Zorbax Rx_x C_8, 5μ (MAC MOD Analytical)

Flow Rate: 2.0 mL/minute

Detector: 254 nm

5.23 Chiral Separation by High Performance Liquid Chromatography

A direct enantioselective separation of the enantiomers of terazosin was developed using the following conditions (12):

Mobile phase: Acetonitrile:50 mM potassium phosphate buffer, pH 7.4 (6:94)

Column: Chiral-AGP, 100 x 4.0 mm (i.d.), 5μ (Chrom Tech, Huddinge, Sweden)

Column Temperature: 5°C

Flow Rate: 0.9 mL/minute

Detector: 254 nm, 0.02 AUFS

Injection Amount: 5 μL (0.1 mg/mL water)

5.3 Titrimetry

5.31 Perchloric Acid Titration

Terazosin exhibits basic properties. It can be potentiometrically titrated with standardized 0.1 N perchloric acid using a modified glass-calomel electrode system, in which 0.1 N lithium perchlorate in acetic acid has been substituted for potassium chloride, employing glacial acetic acid as the sample solvent and in the presence of mercuric acetate.

5.32 Silver Nitrate Titration

The chloride ion of terazosin can be titrated potentiometrically with standardized 0.1 N silver nitrate solution using a glass-silver electrode pair and employing water as the sample solvent.

6. Analysis of Pharmaceutical Formulations

Terazosin tablet formulations can be analyzed by a high performance liquid chromatography procedure using 3,4-dimethoxybenzoic acid or m-nitroaniline as the internal standard. The method uses a μ-Bondapak C_{18}, 10μ column and 0.1 M, pH 4.0 citrate buffer : acetonitrile: methanol (16:3:1) or 0.005 M, pH 4.0 acetate buffer: acetonitrile: methanol (7:2:1).

7. Drug Metabolism and Pharmacokinetics

The absorption, distribution and excretion of terazosin were studied in rats after oral (males and females) and I.V. (males) administration of 1 mg/kg dose of ^{14}C-terazosin hydrochloride (13). After oral administration of ^{14}C-terazosin, the blood concentration of radioactivity reached a peak at 3 hr. and declined thereafter with a half-life of 2.3 hr. in both males and females. After I.V. administration of ^{14}C-terazosin hydrochloride, the blood concentration of radioactivity declined biphasically with half-lives of 2.0 hr. and 20 hr. in males.

After oral administration of ^{14}C-terazosin to male rats, the plasma concentration of radioactivity reached a peak at 3 hr.; unchanged terazosin accounted for 88-100% of the radioactivity. After oral administration of ^{14}C-terazosin to male rats, high concentrations of radioactivity accumulated in the digestive tract, liver, kidneys, urinary bladder, adrenals and pituitary gland, but a low concentration of radioactivity accumulated in the brain. From 81.9 to 83.2% of the orally administered radioactivity was excreted in the feces and from 13.0 to 14.9% in the urine in the first 120 hrs. On the other hand, 69.4% of the I.V. administered radioactivity was excreted in the feces and 26.2% in the urine. The plasma protein binding of radioactivity was 68.7% at 3 hr. and 65.4% at 6 hr. after oral administration of ^{14}C-terazosin to male rats.

The metabolism of terazosin hydrochloride was studied in rats after oral administration of ^{14}C-terazosin hydrochloride, 1 mg/kg (14). Unchanged terazosin and the following five metabolites were identified in the bile and urine: 4-amino-2-[4-(4-carboxy-2-

hydroxybutanoyl)-1-piperazinyl]-6,7-dimethoxyquinazoline (HG-PAD), 4-amino-2-[4-(tetrahydro-2-furoyl)-1-piperazinyl]-6-hydroxy-7-methoxyquinazoline (6H-TRZ), 4-amino-2-[4-(tetrahydro-2-furoyl)-1-piperazinyl]-7-hydroxy-6-methoxyquinazoline (7H-TRZ), 4-amino-2-piperazinyl-6,7-dimethoxyquinazoline (PAD) and 2,4-diamino-6,7-dimethoxyquinazoline (DD).

The major radioactive component excreted in the urine was unchanged terazosin (11.2% of dose). The main metabolites in the urine were HG-PAD (1.1% of dose) and the conjugates of 6H-TRZ (1.1% of dose). In the bile, unchanged terazosin (0.4% of dose) was a minor radioactive component and the major metabolite was HG-PAD (10.7% of dose).

The main metabolic pathway of terazosin hydrochloride in rats involved the formation of lactone by oxidation of the tetrahydrofuran ring of terazosin followed by the formation of HG-PAD by hydrolysis of the lactone (Figure 13) (14).

The placental transfer and excretion into milk of terazosin hydrochloride in rats was reported by Shibata, et.al. (15).

The metabolism of terazosin hydrochloride in humans was reported by Uchide, et.al. (16). The following metabolites together with unchanged terazosin were found in the urine: terazosin glucuronide, 4-amino-2-[4-(4-carboxy-2-hydroxybutanoyl)-1-piperazinyl]-6,7-dimethoxyquinazoline (HG-PAD), 4-amino-2-[4-(tetrahydro-2-furoyl)-1-piperazinyl]-6-hydroxy-7-methoxyquinazoline (6H-TRZ), 4-amino-2-piperazinyl-6,7-dimethoxyquinazoline (PAD) and 2,4-diamino-6,7-dimethoxyquinazoline (DD). Free terazosin excreted in the urine in 24 hr. accounted for 12.9% of the administered dose and the glucuronide for 4.6%. PAD, HG-PAD, DD, and 6H-TRZ accounted for 3.6%, 3.1%, 0.8% and 0.3% of the dose, respectively.

The effects of terazosin on blood coagulation, fibrinolysis systems, immunological function and enzymes were studied in animals and humans by Horii, et.al. (17). Kyncl, et.al. evaluated the antihypertensive activity, pharmacokinetics and

Figure 13 - Proposed Metabolic Pathways of Terazosin Hydrochloride in Rats

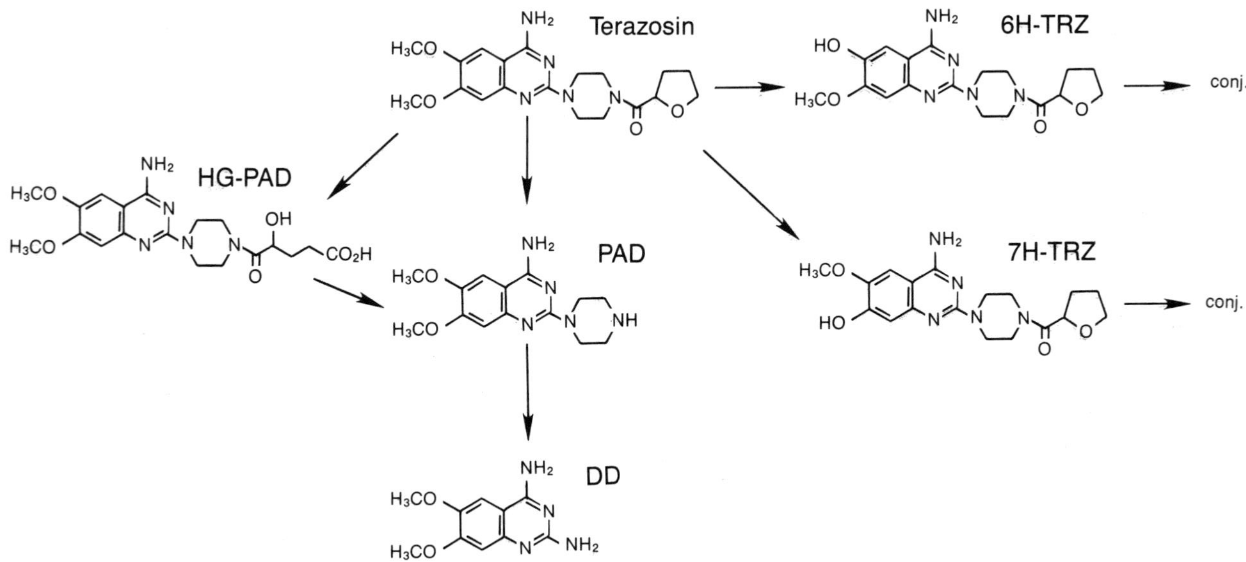

metabolism of orally administered terazosin in various animal models and in man (18).

Terazosin kinetics are linear after oral and intravenous doses (19). Terazosin (0.1 to 3.0 mg/kg) lowered blood pressure without increasing heart rate when given orally to spontaneously hypertensive rats. When administered intravenously to dogs, terazosin lowered blood pressure primarily by decreasing peripheral vascular resistance. The nature of the alpha-blocking properties of terazosin was evaluated in vitro using both radioligand binding studies and functional tests in rabbit aorta and pulmonary artery. These studies demonstrate that terazosin is highly selective for alphasub 1 receptors (20).

Effect of age on pharmacokinetics of terazosin and blood pressure responses to terazosin were reported by McNeil, et.al. (21,22).

The pharmacokinetics of terazosin have been assessed in human volunteers, hypertensive patients, a limited number of elderly volunteers, and a small number of patients with congestive heart failure. Terazosin undergoes extensive hepatic metabolism, and the major route of elimination is via the biliary tract. Small amounts of terazosin are excreted in the urine. Plasma and renal clearances are 80 and 10 mL per minute, respectively. The mean beta-phase half-life is approximately 12 hours. The pharmacokinetics of terazosin were not influenced by age, congestive heart failure, or hypertension (23).

Renal insufficiency was not shown to affect the pharmacokinetics of terazosin in fifteen patients receiving oral terazosin (1 mg once daily) for two weeks. The plasma half-life of terazosin in patients with normal renal function was 10.0 hours, compared with 8.4 hours in patients with moderate renal insufficiency and 9.8 hours in the group with severe renal insufficiency. There was also no apparent relationship between renal insufficiency and the maximum change in blood pressure or pulse rate. Renal excretion was found to play a minor role in the elimination of terazosin, and this explains the lack of a relationship between renal insufficiency and the pharmacodynamics of terazosin (24).

Sennello, et.al. evaluated thirty healthy male and female volunteers between the age of 23 and 75 years who received 1-mg oral and intravenous doses of terazosin. For both routes of administration, the only pharmacokinetic variables significantly correlated with age were terminal elimination rate constant and the area under the plasma concentration-time curve (AUC). However, the differences in half-life and AUC between the youngest and oldest subjects were modest and not of practical clinical significance. There was no evidence of an enhanced pharmacologic of toxic effect in older subjects (25).

A review of pharmacodynamic and pharmacokinetic properties, and therapeutic efficacy in essential hypertension was given by Titmarsh and Monk (26).

8. Determination of Terazosin in Biological Fluids

Spectrophotofluorimetry (27) and high performance liquid chromatography (HPLC) with UV (28) or fluorescence (29-38) detection methods have been used for the determination of terazosin in biological fluids.

The HPLC methods have the required sensitivity for studying the pharmacokinetics of either compound following single intravenous or oral dosage. However, in general these methods require a relatively large sample size (0.1-10 mL) and involve lengthy extraction procedures which either incorporate back-extraction (28,29,35,36) or require evaporation of the organic extract and reconstitution prior to analysis (28,30,32,33,35,37,38). Holt et.al. (39) developed a method which is based on the direct extraction of a small volume (50 μL) of sample into an organic solvent at alkaline pH followed by HPLC analysis with fluorimetric detection. This method has been used for the determination of terazosin in human plasma and urine. Another rapid and sensitive analysis of terazosin in plasma, peritoneal dialysis solution, and urine using high-performance liquid chromatography with fluorescence detection was developed by Patterson (40).

9. References

1. J. Fornnarino and W. Washburn, Abbott Laboratories, Personal Communication.

2. W. Washburn, Abbott Laboratories, Personal Communication.

3. M. Cirovic, R. Stanaszek, and D. Whittern, Abbott Laboratories, Personal Communication.

4. S. Mueller, Abbott Laboratories, Personal Communication.

5. D. Pacenti, Abbott Laboratories, Personal Communication.

6. J. Quick, Abbott Laboratories, Personal Communication.

7. J. Fornnarino, Abbott Laboratories, Personal Communication.

8. M. Winn et.al., Assignor to Abbott Laboratories, Ger. Patent 2,646,186 and U.S. Patent 4,026,894 (1977).

9. R. Roteman, Assignor to Abbott Laboratories, Ger. Patent 2,831,112 (1979) and U.S. Patent 4,251,532 (1981).

10. R. Roteman, Abbott Laboratories, Personal Communication.

11. E. Earl Royals, Advanced Organic Chemistry, (1964), P. 108.

12. M. Fitzgerald, Abbott Laboratories, Personal Communication.

13. K. Shibata, R. Igusa, K. Inoue, H. Mukouyama, J. Nakajima, A. Fujino, and M. Uchide, Oyo Yakuri, 33(5), 775-83 (1987).

14. K. Shibata, R. Igusa, K. Inoue, H. Mukouyama, J. Nakajima, A. Fujino, T. Sekiya and M. Uchide, Oyo Yakuri, 33(5), 765-74 (1987).

15. K. Shibata, R. Igusa, K. Inoue, H. Mukouyama, Y. Hozumi, A. Fujino, and M. Uchide, Oyo Yakuri, 33(5), 785-91 (1987).

16. M. Uchide, Y. Takamatsu, K. Ohnuma, T. Sekiya and A Iwamoto, Yakuri to Chiryo, 15(4), 1513-24 (1987).

17. D. Horii, M. Ohkubo, S. Fukuda, Y. Jindo, Y. Arai, T. Kanayama and S. Kadowaki, Iyakuhin Kenkyu, 18(2), 230(9) (1987).

18. J. Kyncl, R. Sonders, D. Sperzel, M. Winn and J. Seely, New Cardiovasc, Drugs 1, 18 (1986).

19. S.E. Patterson, Clin. Pharmacol. Ther., 38(4), 423-427 (1985).

20. J.J. Kyncl, Am. J. Med., 80(5B), 12-19 (1986).

21. J.J. McNeil, O.H. Drummer, E.L. Conway, B.S. Workman and W.J. Louis, J. Cariovasc. Pharmacol., 10(2), 168-75 (1987).

22. G.S. Stokes, J. Cardiovasc. Pharmacol. 12/Suppl. 8, S109-S115 (1988).

23. R.C. Sonders, Am. J. Med. 80(5B), 20-4 (1986).

24. P. Jungers, D. Ganeval, N. Pertuiset and P. Chauveau, Am. J. Med., 80(5B), 94-99 (1986).

25. L.T. Sennello, R.C. Sonders, H.N. Glassman, D.C. Jordan, R.R. Luther and K.G. Tolman, Clin. Ther. 10(5), 600-607 (1988).

26. S. Titmarsh and J.P. Monk, Drugs, 33(5), 461-77, (1987).

27. A.J. Wood, P. Bolli and F.O. Simpson, Br. J. Clin. Pharmacol., 3, 199 (1976).

28. S.E. Patterson, J. Chromatogr., 311, 206 (1984).

29. Y.G. Yee, P.C. Rubin and P. Meffin, J. Chromatogr., 172, 313 (1979).

30. P.A. Reece, J. Chromatogr., 221, 188 (1980).

31. E.T. Lin, R.A. Baughman and L.Z. Benet, J. Chromatogr., 183, 367 (1980).

32. P. Larochelle, P. Du Souich, P. Hamet, P. Larocque and J. Armstrong, Hypertension, 4, 93 (1982).

33. T.M. Twomey and D.C. Hobbs, J. Pharm. Sci., 67, 1468 (1978).

34. P.C. Rubin, J. Brunton and P. Meredith, J. Chromatogr., 221, 193 (1980).

35. M.K. Dynon, B. Jarrot, O. Drummer and W.J. Louis, Clin. Pharmacokinet., 5, 583 (1980).

36. V.K. Piotrovskii, V.G. Belolipetskaya, A.R. El'Man and V.I. Metelitsa, J. Chromatogr., 278, 469 (1983).

37. P. Jaillon, P. Rubin, Y.G. Yee, R. Ball, R. Kates, D. Harrison and T. Blaschke, Clin. Pharmacol. Ther., 25, 790 (1979).

38. J. Dokladalova, S.J. Coco, P.R. Lemke, G.T. Quercia and J.J. Korst, J. Chromatogr., 224, 33 (1981).

39. R.K. Bhamra, R.J. Flanagan and D.W. Holt, J. Chromatogr., 380, 216-221 (1986).

40. Patterson, Sean E., J. Chromatogr., 311(1), 206-12 (1984).

ZIDOVUDINE

Manohar L. Sethi

Department of Medicinal Chemistry
College of Pharmacy and Pharmacal Sciences
Howard University
Washington, D.C. 20059

1. Description

1.1 Name

Compound, S or BW A509U or Azidothymidine (AZT) is the early name (1). Zidovudine is the generic name. 3'-Azido-3'-deoxythymidine or 3'-azido-2',3'-dideoxythymidine is the chemical name. The structure of zidovudine is given below:

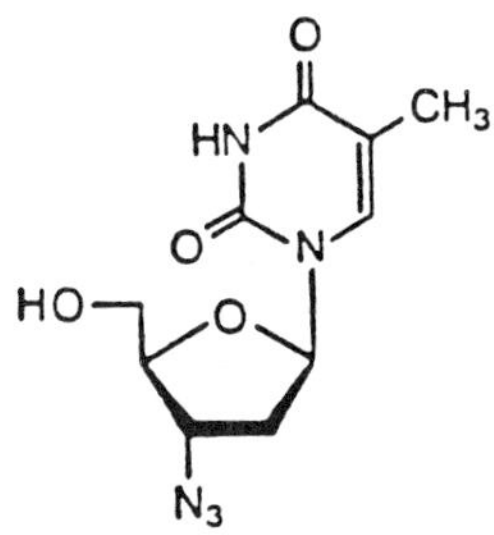

Zidovudine
(3'-azido-3'-deoxythymidine, or AZT)

Molecular weight: 267.24
Molecular formula: $C_{10}H_{13}N_5O_4$

1.2 Isomeric Forms, Structure and Activity

Zidovudine is a thymidine analogue in which the 3-hydroxyl (OH) group is replaced by an azido ($-N_3$) group. With B-D glucopyranuronosyl at 5-hydroxyl group, it forms a nucleoside which exists in two isomeric forms, unphosphorylated inactive threo isomer and triphosphate active erythro isomer. Phosphorylation occurs at the 5-hydroxyl group. The structure of two isomeric forms of Zidovudine is given below:

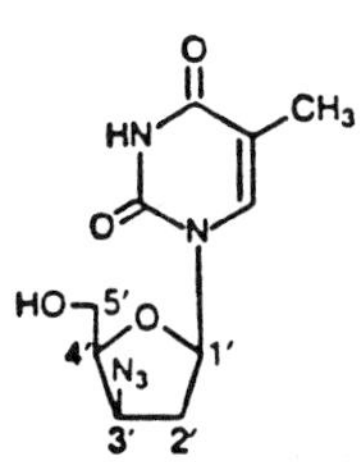

Threo-zidovundine

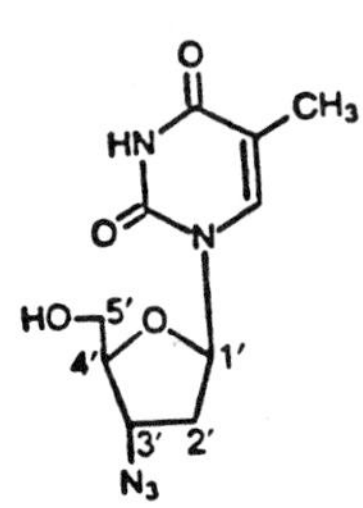

Erythro-zidovudine

1.3 Color, Odor and Appearance

Zidovudine is a white to off-white, odorless, and crystalline solid.

1.4 Brand Name (Prescription Drug) and Manufacturer's Name (2)

Retrovir is the only USA FDA approved prescription drug of zidovudine manufactured by Burroughs Wellcome Co., Research Triangle Park, NC 27709.

1.5 History

Zidovudine was synthesized in 1964 by Horwitz et al. (3). In 1974, Ostertag, W. et al. reported that zidovudine inhibited C-type murine retrovirus replication *in vitro* (4). In 1985, Mitsuya et al. (1) initiated clinical application of zidovudine and reported its inhibitory effect against HIV-1 (Human Immunodeficiency Virus type-1), the etiologic agent of AIDS (Acquired Immunodeficiency Syndrome) and AIDS-related Complex (ARC) patients. Within two years, zidovudine emerged as the only prescription product in the United States for adult AIDS patients suffering from a history of cytologically confirmed *Pneumocystis carinii* pneumonia (PCP) and/or an absolute decrease in the number of peripheral blood helper/inducer lymphocytes (< 200 mm^3).

2. Synthesis and Physical Properties

Nucleosides can be synthesized by the methods reported in the literature (5-18). These methods describe the synthesis of zidovudine (azidothymidine) or its intermediates. For example, methyl 5-O-tert-butyldiphenylsilyl-2-deoxy-α,β-D *threo* pentofuranoside, **1** is prepared from D-xylose as a divergent intermediate for the synthesis of azidothymidine (9). An alternative method of synthesizing azidothymidine has been reported in which a suitable derivative of thymine is coupled with a protected furanoside of 3-azido-2,3-dideoxy-D-erythropentose, **2** (13-16). Synthesis of azidothymidine has been utilized in the synthetic sequence leading to nucleotides, 3′-amino-3′-deoxythymidine 5′-phosphate, 3 and 5′-amino-5′-deoxythymidine 3′-phosphate, 4 (7).

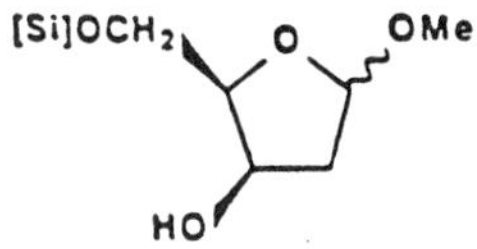

1

[Si] = Ph$_2$^{t}BuSi

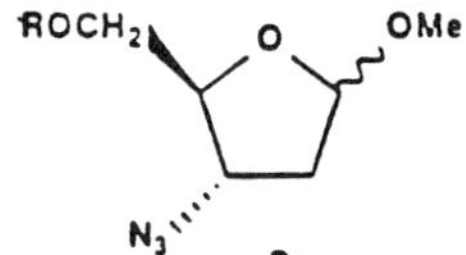

2

R= p-Me-C_6H_4CO

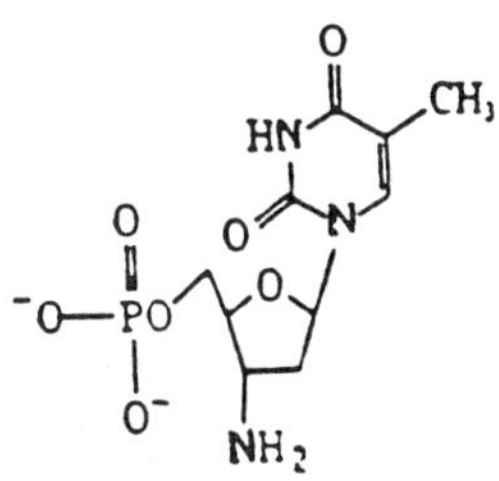

3

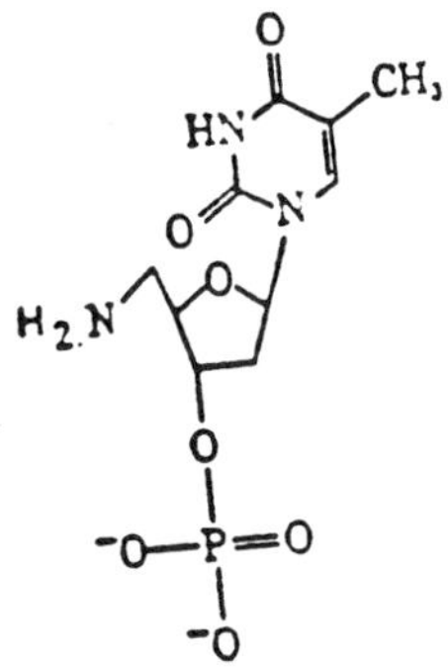

4

Zidovudine (3′-Azido-3′-deoxythymidine, 8) was first prepared by mesylation of 5 (1′-(2′-Deoxy-5′-O-trityl-B-D-lyxosyl) thymine) to 6 (sulfonate of 5) which was treated with lithium azide in N,N-dimethylformamide (DMF) at 100° under an atmosphere of nitrogen for 3 hours (3). The reaction mixture was poured with stirring in ice water and the off-white solid product was dried in air. The crude product 7 showing a prominant azide peak (λ max 4.76 μ) was dissolved in chloroform and converted to hydrochloride salt by mixing with a chloroform solution containing hydrogen chloride. The reaction mixture was neutralized with sodium bicarbonate and the 3′-Azido-3′-deoxythymidine product 8 was purified and dried m. p. 106-112°C (dried product m.p. 119-121°C). Solubility in water is 25/mg ml at 25° C.

λ_{max}^{KBr} 4.76 μ (azide)

$\lambda_{max\ min}^{H_2O}$ 266.5 mμ (e 11,650), 234.5 (2610)

$[\alpha]_D^{25}$ + 99 (c 0.5, water);

Anal. Calcd. for $C_{10}H_{13}O_5N_4$:

C, 44.94, H, 4.9; N, 26.21
Found C, 44,98; H, 4.74; N, 26.22

The essential steps of synthesis of zidovudine (3) are shown below.

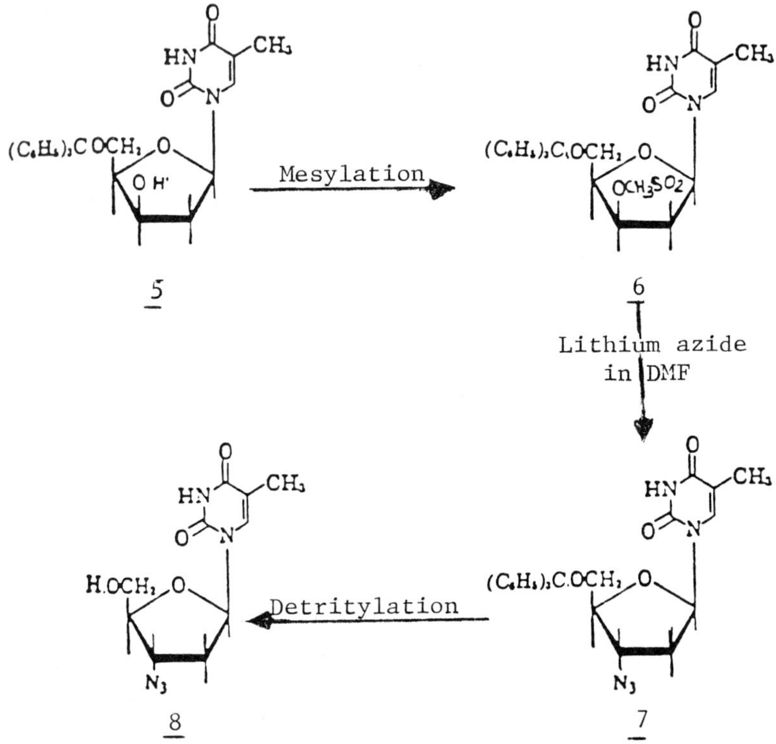

Several 3′-deoxythymidine analogues with an heterocyclic five-membered ring in the 3′-erythro position have been synthesized. Most of the compounds were inactive except the 3′-pyrrol-1-yl derivative which indicated marginal antiviral activity against HIV (19). Sythesis of chemically reactive analogues of zidovudine are carried out and tested against HIV (19-23). Some inactive acyclic derivatives of zidovudine have been synthesized by Michael addition which provide a versatile new route to nucleoside analogues (24). Some other compounds related to zidovudine, such as 1′,2′-seco-2′,3′-dideoxycytidine, -guanosine, -adeonosine and -insosine have been synthesized. These compounds are reported to be potential anti-HIV agents (25). Synthesis and anti-HIV activity of several fluorinated nucleoside analogues related to azidothymidine have been reported (23-30).

A series of prodrugs of zidovudine have been synthesized in order to improve uptake by the HIV-1 infected cells and to increase the plasma half-life of zidovudine. The prodrug moieties include (a) morpholine and n-phenylpiperazine-1-acetic acid (b) 1,4-dihydro -1-methyl-3 nicotinic acid (c) retinoic acid and (d) certain amino acids and (e) dihydropyridine derivatives. The prodrugs are generally less cytotoxic as compared to zidovudine and are relatively stable toward plasma esterases with t 1/2 ranging from 10 to 240 minutes (31-35). Synthesis and antiretroviral activity of phospholipid analogues of zidovudine and other antiviral nucleosides have been reported (35,36) recently. Derivatives of zidovudine active against the HIV virus are under application for European patent (37).

3. <u>NMR and X-ray Crystallography</u>

^{1}H and ^{13}C NMR study of zidovudine has been undertaken. Modified Karplus relations have been used to obtain molecular structure for the indirect constants. Table 1 shows proton chemical shifts and indirect coupling constants in zidovudine (38). Results from molecular confirmational studies (38-41) differ from those obtained in solid state by X-Ray diffraction as shown in Figures 1 and 2 (42). Crystalline structure and possible functional role of azido group in zidovudine has been studied (43,44). An independent publication of X-ray crystallography of zidovudine has been reported by other workers (45,46).

4. <u>Method of Analysis and Chromatography</u>

A simple and rapid HPLC assay of zidovudine in plasma has been undertaken using reversed phase (RP 8) column with 1% acetate buffer pH 5-methanol (82/18, V/V) as the solvent and detection at 265 nm. Dectection limits in plasma and urine were 20 and 200 ng/ml, respectively (47-50). A HPLC of zidovudine in plasma and urine sample is shown in Figures 3 and 4, respectively (47).

Table 1 - Proton chemical shifts and indirect coupling constants in AZT

Chemical shifts (ppm)			Coupling constants* (Hz)		
Protons	In DMSO	In D_2O	Protons	In DMSO	In D_2O
1'	6.075	6.064	1'2'	6.4	a
2'	2.359	2.357	1'2"	6.7	a
2"	2.257	2.357	1'3'	0.5	0.5
3'	4.387	4.211	2'2"	- 13.7	- 13.9
4'	3.800	3.866	2'3'	7.2	a
5'	3.624[b]	3.715[b]	2"3'	5.2	a
5"	3.582[b]	3.643[b]	3'4'	4.9	5.5
6'	5.218	c	4'5'	4.1	3.5
5(methyl)	1.764	1.738	4'5"	3.9	4.6
6	7.665	7.494	5'5"	- 12.4	- 12.6
			5'6'	5.7	c
			5"6'	5.5	c

* $J_{C6-1'} = 3.9$ Hz and $J_{C2-1'} = 2.3$ Hz.

a 2' and 2" are equivalent and only the sums $J_{1'2'} + J_{1'2''} = 13.0$ Hz and $J_{2'3'} + J_{2''3'} = 13.0$ Hz are obtained.

b by convention 5' is at a lower field than 5".

c In D_2O, the hydroxy proton (6') is not observed.

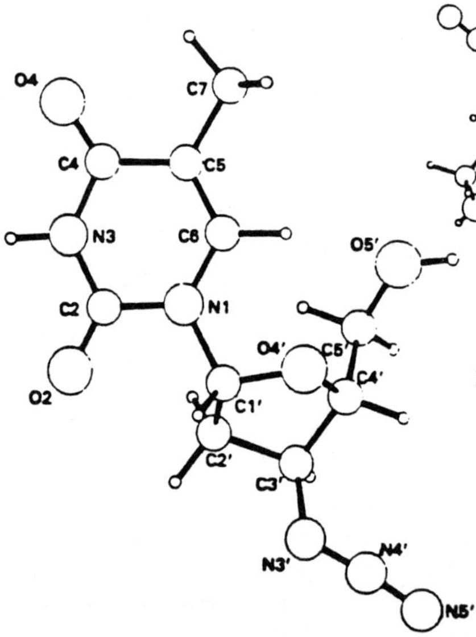

Fig. 1. View of the molecule giving atomic numbering.

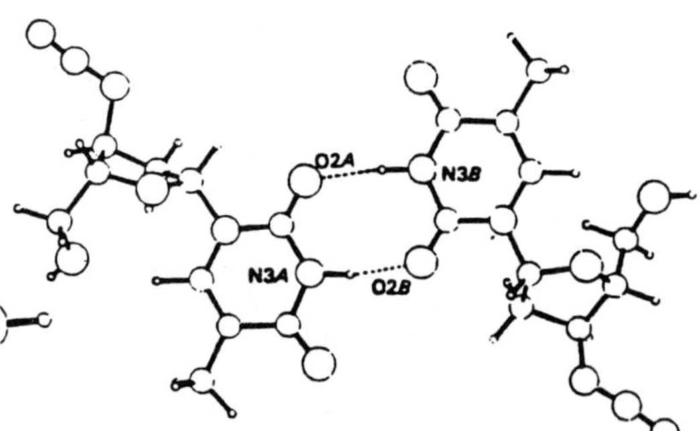

Fig. 2. The asymmetric unit showing the base pairing.

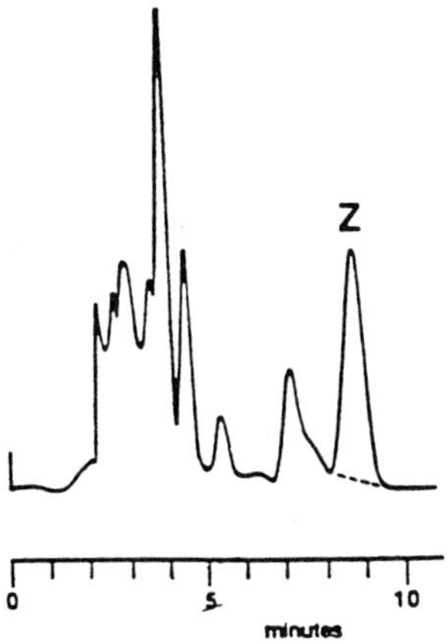

Fig. 3. HPLC chromatogram of a urine sample spiked with zidovudine (Z) (10 μg/ml). The broken line presents the blank chromatogram. For chromatographic conditions see text.

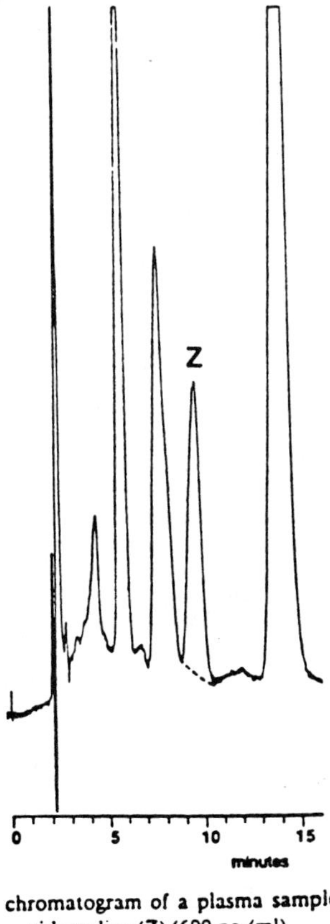

Fig. 4. HPLC chromatogram of a plasma sample spiked with zidovudine (Z) (600 ng/ml).

Analysis of zidovudine levels in serum, milk and tissues by isocratic HPLC has been reported (51). A preparative-scale HPLC separation and purification of zidovudine phosphate has been reported recently (52). A colorimetric method of assay has been developed for anti-HIV agents including zidovudine (53,54). Simultaneous quantification of zidovudine and its metabolites in serum and urine by HPLC using a column-switching technique has been reported (55,56). This technique describes a direct switching method which reduces the management of samples to a minimum and hence shortens the assay time considerably. Figure 5 shows a HPLC of azido-thymidine (AZT) and its 5-0-glucuronide (GAZT) metabolite (55).

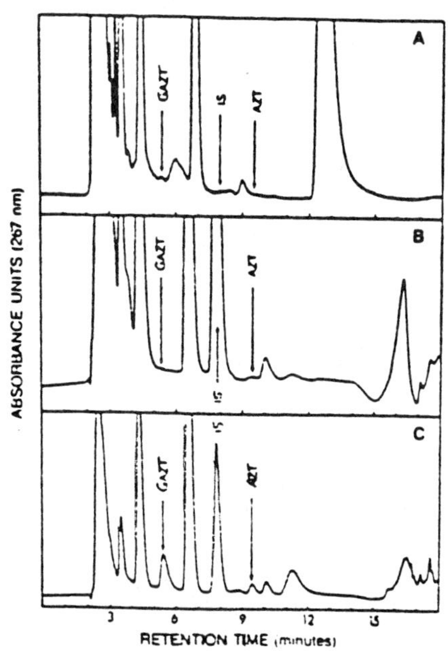

Fig 5. Representative chromatographic profiles obtained from the analysis of extracted human sera. Panel A, normal serum pooled from five donors (0-0.01 a.u.f.s.); panel B, serum obtained from an AIDS patient prior to receiving oral AZT (0-0.01 a.u.f.s.); panel C, serum obtained 2.5 h after therapy, containing 1.53 μM GAZT and 0.45 μM AZT (0-0.03 a.u.f.s.). Retention times for GAZT, I.S. and AZT are indicated.

Current analytical techniques for zidovudine include radioimmunoassay (57,58) which are comparable to HPLC method (59). Determination of zidovudine concentration in serum by enzyme-linked immunosorbent assay and by time-resolved fluoroimmunoassay has recently been performed (60,61). Under this method, two solid-phase, nonradioactive immunoassays were developed and evaluated for determining zidovudine concentration in serum. These methods have been compared with HPLC and RIA (62-65). Other methods for the analysis of zidovudine include cytometric/fluorometric assay for the evaluation of the anti-HIV activity of antiviral compounds in MT-4 cells (66,67).

The paper and thin layer chromatography of nucleoside derivative including zidovudine has been reported (9,18). The following solvent systems have been used.

Paper

Solvent systems,
- A, 1-Butanol-acetic acid - water (5:2:3, V/V).
- B, 2-Propanol-water-conc. NH_4OH (7:2:1, V/V)
- C, 2-Propanol-aqueous 1% ammonium sulfate (5:2, V/V).
- D, 2-Ethanol-aqueous 1% ammonium acetate (5:2, V/V).

TLC

Solvent systems,
- A, Diethyl ether
- B, Ethyl acetate-ethanol-tetrahydrofuran (1:1:1, V/V)
- C, Ethyl acetate
- D, Chloroform-1% ammonium hydroxide in methanol (3:2, V/V),
- E, 2-Propanol-water-concentrated ammonium hydroxide (7:2:1, V/V).

The spots were visualized with UV light and compounds containing a trityl group were detected by spraying aqueous 5% perchloric acid and heating at 80°C for 5 minutes.

5. Pharmacology (68-72)

5.1 Antiviral Activity

Zidovudine is an inhibitor of the *in vitro* replication of some retroviruses including HIV (also known as Human T cell lymphotropic Virus-type III, HTLV-III; Lymphadenopathy-associated-virus, LAV; AIDS-related Virus, ARV). Retroviruses contain reverse trancriptase, RT (RNA-directed DNA polymerase), an essential enzyme for the life cycle of the retrovirus. Unlike other viruses, retrovirus RT catalyzes the conversion of genetic RNA to DNA and in a series of steps completes its life cycle. Retroviruses affect the cellular immune system (T lymphocyte and B lymphocytes) (73). Both lymphocytes originate from bone marrow but T lymphocytes mature in the thymus and produce long-lived immunity. T lymphocytes exist in different subsets, (i) The helper/inducer subset called T4 (CD4) responsible for delayed hypersensitivity reactions (ii) cytotoxic T lymphocytes specific antigens and (iii) suppressor T lymphocytes suppressing T4 and B lymphocytes. Cytotoxic and suppressor T cells are called T8 lymphocytes. Zidovudine possesses antiretroviral activities both *in vitro* and *in vivo*.

Zidovudine possesses *in vitro* antiretroviral activity in animal and human cells. It has effects on viral replication, enzymes and proteins as given below:

a. Simian T cell Leukemia Virus (STLV) of macaque cells is inhibited at a concentration of 0.4, 1.9 and 19 μmol/L, (74).

b. Friend virus particles of mouse cells are inhibited at a concentration of 0.25 μmol/L (75).

c. Kirsten murine sarcoma virus of mouse and rat cells is inhibited at a concentration of 10 μmol/L and 100 μmol/L, respectively (76).

d. Infectious anemia virus of equine epidermis cells is inhibited at a concentration of 30 µmol/L (76).

e. Antigen expression of H9, MT2 and MT4 cell lines (Human T lymphocyte lines) is inhibited at a concentration of 1-2 µmol/L (77,78).

f. Cytopathogenecity protection of TM3, ATH8, MT2 and MT4 (Human T lymphocyte lines) is at a concentration of 0.006-10 µmol/L (79).

g. Reverse transcriptase (retroviral enzyme) activity and replication of PBM (peripheral blood mononuclear) cells, H9 and jurkat cells (Human T lymphocyte line) is inhibited at a concentration of 0.002-0.5 µmol/L (80-84).

h. p24 antigen (HIV protein) production in tetanus toxoid stimulated by T4 lymphocytes is inhibited (85,86).

5.1.2 <u>Other Retro and Non-retroviral Activity</u>

Zidovudine is inactive against cytomegalovirus, vaccinia, herpes simplex, vasicular stomatitis, varicella zoster, adenovirus, cornavirus, influenza A virus, respiratory syncytial measles, rhinovirus, bovine rotavirus and yellow fever virus infections (87). Zidovudine inhibits Epstein-Barr Virus (EBV) (Ic 50, 0.14 µmol/L) and hepatitis B Virus (Ic 50, 0.2 µmol/L) (88). Zidovudine is active <u>in vivo</u> against Rauscher murine leukemia virus which encouraged initiation of zidovudine for medical application (89).

5.1.3 <u>Synergistic Activity of Zidovudine in Combination with Antiviral or Other Agents</u>

Additive <u>in vitro</u> activity is reported when the <u>following</u> antiviral or other agents are combined with zidovudine:

a. Acyclovir (88)
b. Foscarnet (90)
c. Castanospermine (91)
d. Dideoxycytidine (92)
e. α, β and γ-Interferons (93)
f. Dextran sulfate (94)
g. Glycyrrhizin (95)
h. Granulocyte-macrophage colony stimulating factor (96)
i. Ampligen (90)

5.1.4 <u>In vivo Activity</u> (88-72)

a. Zidovudine protects kittens from FeLV (Feline Leukemia Virus) in a dose-related fashion in pre- and post-treatment (97,98).

b. Zidovudine prolongs the life of MuLV (Murine Leukemia Virus)-infected mice in pre- and post-infection (99).

c. Zidovudine protects mice from MSV (Mouse Sarcoma Virus)-infected tumor formation and increases survival of rats with higher doses (100).

5.1.5 <u>Effect on Host Cells and Tissues</u>

Zidovudine has effect on the following host cells and tissues:

a. Colony formation of granulocyte-monocyte colony forming units and erythroid burstforming units (human cell lines) are inhibited by zidovudine, Ic50 0.9 and 2.4 µmol/L, respectively (101).

b. Growth rate of jurkat cells (Human T cell lymphocyte line) is inhibited by zidovudine (80-84).

c. Slight impairment of antigen -induced T lymphocyte responses at 1 µmol/L, of T lymphocyte activation at 50 µmol/L, and of peripheral blood mononuclear cell activation at 10 μmol/L concentration of zidovudine is obtained (1).

d. Zidovudine at a 50 μmol/L concentration partially inhibits immunoglobulin production by B lymphocytes. However, activity and cytopathogenic effects of HIV are strongly inhibited by 1-10 μmol/L zidovudine concentration (1).

e. Zidovudine protects kittens from FeLV and mice from MuLV infection (102). It also produces a dose-related reversible macrocytic anemia in cynomolgus monkeys.

5.2 Mechanism of Action

Zidovudine is converted to monophosphate, diphosphate and triphosphate derivatives by the host cell thymidine kinase, thymidylate kinase and other cellular enzymes (nucleoside diphosphate kinases), respectively. Zidovudine triphosphate is the active form which competes with thymidine 5-triphosphate for binding to the HIV RT. The HIV by mistake incorporates zidovudine triphosphate in viral replication. Since the 3'-hydroxyl group in zidovudine is replaced by an azide group, subsequent 5'-3'phosphodiester linkages cannot be formed. Consequently, zidovudine acts as a chain terminator in the DNA synthesis resulting in inhibition of HIV replication (103-107).

5.3 Antibacterial Activity

Zidovudine has effect on different types of bacteria as given below:

a. Zidovudine inhibits strongly Gram-negative bacteria such as *Salmonella typhimurium*, *Shigella flexneri*, *Escherichia. coli*, *Enterobacter aerogens*, *Vibrio anguillarum* and *Klebsiella pneumoniae*. Moderate activity was observed against *Yersinia enterocolitica* and *Vibrio cholerae* (108).

b. Zidovudine is active against Gram-positive bacteria, such as *Staphylococcus aureus*, *Streptococcus pyogenes*, *Pseudomonas aeruginosa*, *Mycobacterium intracellulare* and *Mycobacterium tuberculosis* (108).

6. Therapeutic Categoration

6.1 Dosage Forms (2)

Zidovudine is available in:
100 mg capsule
50 mg/5 ml Pediatric syrup
10 mg/ml Intravenous infusion

6.2 Administration and Dosage

The recommended oral starting dose in adults is two 100 mg capsule or four teaspoonfuls (20 ml) syrup given every 4 hours around the clock. With the appearance of toxic symptoms, daily dose can be reduced to 100 mg every 4 hours. Intravenous infusion is administered 1 to 2 mg/kg infused over 1 hour six times daily.

6.3 Clinical Application

6.3.1 Effect on Immune Cells

Clinically, zidovudine was recommended by the FDA for AIDS or ARC patients with PCP and T lymphocytes(CD4 cells) count below $200/mm^3$ in the peripheral blood. Recently, zidovudine is recommended for use in the HIV-infected individuals with CD4 (T4 helper/inducer) lymphocyte counts below $500/mm^3$ (109). Zidovudine intravenous and oral treatment protected T4 lymphocytes infected by HIV and improved T4:T8 lymphocyte ratio. However, these benefits are transient and T4 lymphocyte counts fell to pretreatment levels after an average of 24 and 50 weeks of treatment in AIDS and related ARC patients, respectively (110). Zidovudine appears to increase platelet counts after 4 to 12 weeks in 34 AIDS patients (111). The platelet counts of 21 out of 23 thrombocytopenic patients are normalized (112). Zidovudine also increases natural killer cell activity. Zidovudine reduces P24 antigen in serum or CSF and increases T4 cell counts during early treatment. The discontinuation of therapy lead to virus and P24 antigen production to the pretreatment levels and decreases T4 lymphocyte counts (111,113).

6.3.2 Effect on Opportunistic and Specific Infections

Preliminary data indicate that zidovudine-treated AIDS or ARC adults and children develop fewer new opportunistic infections and improve immunological and neurological conditions (110,114). The usual opportunistic infections are PCP, a typical mycobacterial infection, toxoplasmosis, oral candidiasis, nocardiosis, bacterial and viral infection, Kaposi's sarcoma and neurologic diseases (115). HIV infected adults and children with dementia or peripheral neuropathy improve with zidovudine therapy (116). HIV-infected patients with Kaposi's sarcoma without opportunistic infections and low T4 lymphocyte count have limited benefit with zidovudine therapy (110,117). However, zidovudine is highly beneficial in some HIV-associated psoriasis (118).

The FDA approved zidovudine in the treatment of young children with advanced symptoms of AIDS in November 1989. The use of zidovudine in children with AIDS indicates significant specific neurologic impairment consisting of lymphadenopathy, encephalopathy and myelopathy. A marked increase in total lymphocyte and CD4 cell count is observed. Although P24 serum antigen detected in the children disappears temporarily during intravenous use of zidovudine, it reappears progressively during oral treatment. The hematologic toxicity is similar to the adult AIDS patients (114).

6.3.3 Early Therapeutic Use

Zidovudine in early treatment results in:

a. Increased T4 (helper/inducer) lymphocyte counts (119).

b. Improved T4:T8 (helper/inducer: cytotoxic/suppressor) lymphocyte ratios and total lymphocyte count (120).

c. Improved chronic opportunistic infections (109).

d. Improved AIDS symptoms, such as peripheral neuropathy, nightsweats, weight loss and temperature (109).

e. Decreased P24 antigen in serum and CSF (121,122).

f. Delayed progression of HIV disease (123).

g. Decreased drug-induced toxicity in the subjects with mildly symptomatic HIV disease and count less than 500 CD4 T lymphocytes/mm^3 (123).

6.3.4 Long-term Therapeutic Use

Zidovudine administration in AIDS and ARC patients on long-term basis causes:

a. Overall prolonged survival of the patient (124).

b. Bone marrow depression leading to hematologic toxicity including granulocytopenia and severe anemia requiring transfusion.

6.3.5 Future Prospects of Zidovudine

Zidovudine has shown remarkable results in the beginning of AIDS epidemic crises. But potential bone marrow toxicity has hampered the use of the drug. Therefore, future prospects of zidovudine include systematic study of:

a. Combination therapy of zidovudine with other antiviral and or immunomodulators.

b. Effect of zidovudine in children with AIDS or ARC symptoms.

c. HIV isolated resistance to zidovudine.

d. Effect of zidovudine in pregnant women with AIDS and ARC symptoms.

e. Transferance of HIV through placenta to the child.

f. Use of zidovudine in the early course of HIV infection or asymptomatic HIV infected patients indicating HIV-positive serum.

g. HIV-infection to health professionals caring for AIDS patients.

6.4 Combination Therapy

The aims of combining zidovudine with other agents are:

a. Prophylaxis of PCP when zidovudine is combined with trimethoprim/dapsone, pyrimethamine or sulfadoxine/pyrimethamine (125).

b. Prevention of blindness in HIV-infected patients with cytomegalovirus retinitis when zidovudine is combined with ganciclovir (126).

c. Clinical improvements, such as weight gain, Karnofsky scores, increased lymphocyte counts and substantial reduction in serum P24 antigen when zidovundine is combined with acyclovir (127,128).

d. Regression of Kaposi's sarcoma lesions when zidovudine is combined with interferons (129,130).

6.5 Toxicology and Adverse Effects

Zidovudine-treated patients revealed serious hematological abnormalities due to toxic effects on bone marrow. This included granulocytopenia anemias requiring transfusion or erythropoietin (131). Some adverse effects of zidovudine are severe headache, rash, myalgia, nausea, vomiting, nail pigmentation, hepatotoxicity and neurologic symptoms (132,133). In rats and mice, median lethal dose of zidovudine is reported greater that 750 mg/kg intravenously and greater than 3000 mg/kg orally (110). In oral toxicity studies, rats exhibited slight to mild macrocytic anemia in 3-6 months. In monkeys, zidovudine causes reversible dose-related macrocytic anemia when 35, 100 or 3,000 mg/kg zidovudine is given intravenously for 3-6 months. No teratogenecity in rats or rabbits is observed (102). Zidovudine is weakly mutagenic at concentrations of 1,000 to 5,000 ug/ml in mammalian cells but is negative in bacterial mutagenecity assay. An overview of the preclinical toxicity of zidovudine is reported by Kenneth M. Ayers (102).

6.6 Drug Interactions

No detailed studies of zidovudine interaction with other drugs are reported. However, acetaminophen is associated with an increased frequency of neutropenia during drug therapy (115). Since both acetaminophen and zidovudine are metabolized by glucuronidation, it is possible that other agents, such as indomethacin, sulfadoxine and sulfamethoxazole metabolized by the same method are contraindicated with concurrent use of zidovudine.

6.7 Monitoring Drug in Patient (134)

Complete blood test should be performed before and during zidovudine therapy. A low or decreasing reticulocyte count with decreasing hemoglobin indicates anemia. The following tests are important to monitor drug therapy every two weeks for the first 3 months.

a. CD4 (T4 helper/inducer) counts

b. T4:T8 lymphocyte ratio

c. P24 antigen

d. Differential platelet count

e. Total leukocyte count

f. Hemoglobin

g. Hematocrit

h. B_2 Microglobulin

i. IgA or immune complex levels

7. Pharmacokinetics

7.1 Absorption and Bioavailability

Zidovudine is completely and rapidly absorbed through the gastrointestinal tract and bioavailability is approximately 65% (135). Peak plasma concentration occur in HIV-infected patients with oral dose 100, 200, and 250 mg to 1.27 + 0.64, 2.35 + 1.75 and 3.44 + 1.8 μmol/L, respectively (115). Intravenous zidovudine doses of 1, 2.5 and 5 mg/kg infused over 1 hour resulted in peak plasma concentrations ranging from 1.5 to 2.5, 4 to 6 and 6 to 10 μmol/L, respectively (136-141). Figures 6 and 7 show mean plasma concentration-time profile of zidovudine following intravenous and oral administration (139). Enhancement of the <u>in vitro</u> skin permeability of azidothymidine via iontophoresis and chemical enhancer is studied (142).

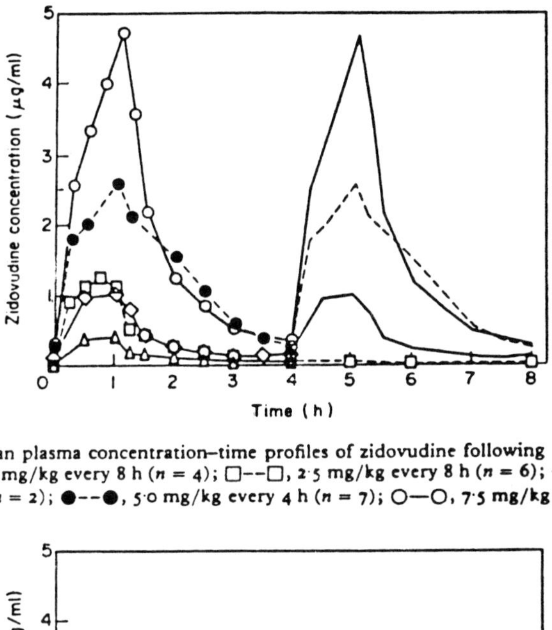

Fig 6. Mean plasma concentration–time profiles of zidovudine following IV administration. △—△, 1·0 mg/kg every 8 h (n = 4); □--□, 2·5 mg/kg every 8 h (n = 6); ◇—◇, 2·5 mg/kg every 4 h (n = 2); ●--●, 5·0 mg/kg every 4 h (n = 7); ○—○, 7·5 mg/kg every 4 h (n = 3).

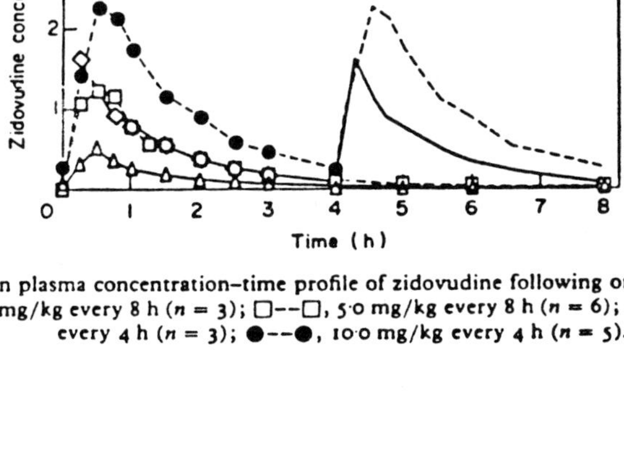

Fig. 7. Mean plasma concentration–time profile of zidovudine following oral administration. △—△, 2·0 mg/kg every 8 h (n = 3); □--□, 5·0 mg/kg every 8 h (n = 6); ◇—◇, 5·0 mg/kg every 4 h (n = 3); ●--●, 10·0 mg/kg every 4 h (n = 5).

7.2 Volume Distribution and Protein Binding

The volume distribution at steady-state is: 1 L/Kg in dog (135) and 1.4 L/Kg in AIDS patients(139,141, 143-145). In children volume distribution of zidovudine is 56 L/mm2 (141). Zidovudine is approximately 20% bound to plasma proteins in rhesus monkey (146-148), 15% bound to dog plasma (149) and 34-38% to human plasma (150).

7.3 Penetration into Central Nervous System (CNS), Maternal Fetal Transfer and Semen

The extent of penetration of zidovudine in CNS and brain depends upon route and time after administration. In cerebrospinal fluid (CSF), concentration of zidovudine reached a maximum 2 hours after starting a 1 hour intravenous infusion of drug. The peak concentrations are much lower than those achieved in plasma (151). The CSF:plasma ratio ranged from 0.15 to 1.35 which contributes to the anti-HIV effect of the drug on CNS (152,153). Zidovudine crossed the placenta in mice, non-human primates and humans (154-157). However, drug concentration in the fetus was not higher than in the mother. In human, zidovudine has been shown to concentrate in the semen of HIV-infected patients with semen ratio ranging from 1.3 to 20.4 at 45 to 75 minutes and AUC (Area under the Plasma Concentration-time Curve) ratio 3 after oral administration of 200 mg drug on a 4-hourly schedule (150).

7.4 Metabolism and Excretion

Zidovudine in monkeys and AIDS patients is converted to inactive 3′-azido-3′deoxy-5-0-β-D-glucopyranosyl thymidine which is excreted in the urine. Approximately 17-23% metabolite and 60-72% unchanged zidovudine is recovered from the urine after intravenous or oral administration of drug (151). The mean half-life of the drug in HIV-infected patients is approximately 1 hour for both intravenous and oral administration of drug. The plasma clearance is 1.3 L/Kg (145). Metabolism of zidovudine is influenced by enzyme present in rat and human liver microsomes (158,159).

7.5 Pharmacokinetics in AIDS or ARC Children

In children less than 4 years old, oral dose of zidovudine (80 or 120 mg/mm^2) results in peak plasma concentration of 2.96 and 4.43 µmol/L, respectively after half hour administration (160). CSF: plasma zidovudine concentration ratios are 0.80 and 0.58 after intravenous and oral doses, respectively. Total clearance of zidovudine is 730 $ml/min/m^2$ in children with symptomatic HIV infection. The mean steady-state volume distribution of zidovudine is 56 L/m^2. Approximately 81% of zidovudine is recovered in the urine as unchanged drug and metabolite (161-162). Children with asymptomatic HIV-infection from 14 months to 12 years indicate slightly different pharmacokinetic data. An initial single dose infusion of zidovudine (80 mg/m2 over 1 hour) was given, followed 6 hours later by a continuous zidovudine infusion. The mean steady-state plasma zidovudine concentrations were 1.9, 2.8, 3.1 and 4.5 µmol/L after infusions of 0.5, 0.9, 1.4 and 1.8 mg/kg/h, respectively. The ratio of CSF: plasma zidovudine concentration was 0.24 at steady-state level.

7.6 Effect of Probenecid on Zidovudine Pharmacokinetics

Alteration of zidovudine pharmacokinetics by probenecid in patients with AIDS or ARC was reported (163). The mean zidovudine plasma levels were significantly higher than in its absence after concurrent administration of probenecid. This resulted in a two-fold increase in the mean AUC, a corresponding decline in the apparent total clearance and a prolongation of the mean half-life. The urinary excretion ratio of the 5-glucuronide to zidovudine and a decline of the metabolite in the renal clearance was markedly reduced (163-167).

Acknowledgement

The author is thankful to the Drug Information Department of Burroughs Wellcome Company for providing part of the literature search on zidovudine.

References

1. Mitsuya, H., Weinhold, K.J. et al. Proc. Natl. Acad. Sci. USA 82 7096 (1985).

2. PDR, 45th Ed., Medical Economics Company, 1990, pp. 788-794.

3. Horwitz, J.P., Chau, J. and Noel, M. J. Org. Chem. 29, 2076 (1964).

4. Ostertag, W., Roseler, G. et al. Proc. Natl. Acad. Sci. USA 71, 4980 (1974).

5. Horwitz, J.P., Urbanski, J.A. and Chua, J. J. Org. Chem. 27, 3301 (1962).

6. Horwitz, J.P., Chau, J. and Urbanski, J.A. J. Org. Chem. 28, 942 (1963).

7. Glinski, R.P., Khan, M.S. et al. J. Org. Chem. 38, 4299 (1973).

8. Baker, B.R., Schaub, R.E. and Williams, J.H. J. Am. Chem. Soc. 77, 7 (1955).

9. Fleet, G.W., Son, J.C. and Derome, A.E. Tetrahedron 44, 625 (1988).

10. Lin, T-S., and Mancini, W.R. J. Med. Chem. 26, 544 (1983).

11. Lin-T-S., Gao, Y-S and Mancini, W. R. J. Med. Chem. 26, 1691 (1983).

12. Lin, T.S. and Prusoff, W.H. J. Med. Chem. 21, 109 (1978).

13. Dyatkina, N. B., Krayevsky, A.A. and Azhayev, A.V., Synthesis 410 (1985).

14. Dyatkina, N.B. and Azhayev, A.V. Synthesis 961 (1984).

15. Dyatkina, N.B., Krayevsky, A.A. and Azhayev, A.V. Bioorg Khim. 12, 1048 (1986).

16. Fleet, G.W.J. and Son, J.C. Tetrahedron Lett. 28, 3615 (1987).

17. Vince, R. and Hua, M. J. Med. Chem. 33, 17 (1990).

18. Hill, J.A. and Freeman, G.A. J. Label Compounds Radiopharm. 25, 277 (1988).

19. Wigerinck, P., Aerschot, A.V. et al. J. Med. Chem. 33, 868 (1990).

20. Murata, M. and Achiwa, K. Chem. Pharm. Bull. (Tokyo) 38, 836 (1990).

21. Warshaw, J.A., Watanabe, K.A. et al J. Med. Chem. 33, 1663 (1990).

22. Schreiber, S.L. and Ikemoto, N. Tetrahedron Lett. 29, 3211 (1988).

23. Lin T.S., Chen, M.S. et al. J. Med. Chem. 30, 440 (1987).

24. Scheiner, P., Geer, A. et al. J. Med. Chem. 32, 73 (1989).

25. Abushanab, E. and Sarma, M.S. J. Med. Chem. 32, 76 (1989).

26. Sochacka, E., Nawrot, B. et al. J. Med. Chem. 33, 1995 (1990).

27. Watanabe, K.A., Harada, K. et al. J. Med. Chem. 33, 2145 (1990).

28. Sterzycki, R.Z., Ghazzouli, I. et al. J. Med. Chem. 33, 2150 (1990).

29. Mettes, E., Lehmann, C. et al. Biochem. Biophys. Res. Commun. 153, 825 (1988).

30. Koshida, R., Cox, S. et al. Antimicrob. Agents Chemother. 33, 2083 (1989).

31. Aggarwal, S.K., Gogu, S. R. et al. J. Med. Chem. 33, 1505 (1990).

32. Chu, C. K., Bhadti, V. S. et al. J. Med. Chem. 33, 2188 (1990).

33. Kawaguchi, T., Ishikawa, K. et al. J. Pharm. Sci. 79, 531 (1990).

34. Gogu, S. R., Aggarwal, S.K. et al. Biochem. Biophys. Res. Commun. 15, 1591 (1989).

35. Hostetler, K.Y. et al. J. Biol. Chem. 265, 6112 (1990).

36. Devine, K.G., McGuigan, C. et al. AIDS 4, 371 (1990).

37. Bethune, D., Pierre, M. et al. Chemical Abstract 113, EP 357495, 1990.

38. Swapna, G.V.T., Jagannadh, B. et al. Biochem. Biophys. Res. Commun. 164, 1086 (1989).

39. Herzyk, P., Beveridge, A. and Neidle, S. Biochem. Biophys. Res. Commun. 145, 1356 (1987).

40. Parthasarthy R. and Kim, H. Biochem. Biophys. Res. Commun. 152, 351 (1988).

41. Wilson, C.C. and Tollin, P. Nucleosides Nucleotides 6, 643 (1987).

42. Dyer, I., Low, J.N. et al. Acta Crystallogr. C 44, 767 (1988).

43. Camerman, A., Mastropaolo, D. and Camerman, N. Proc. Natl. Acad. Sci, USA 84, 8239 (1987).

44. Biswas, G., Das, M.C. and Banerjee, A. Indian J. Phys. 63 A, 134 (1989).

45. Birnbaum, G.I., Giziewicz, J. et al. Can. J. Chem. 65, 2135 (987).

46. Gurskaya, G.V., Tsapkina, E. N. et al. Dokl. Akad Nauk SSSR, 291, 854 (1986).

47. Underberg, W.J.M., Underberg-Chitoe, U.K. et al. I.J. Pharmaceut. 50, 175 (1989).

48. Unadkat, J.D., Crosby, S.S. et al. J. Chromatogr. 430, 420 (1988).

49. Mazzei, M., Balbi, A. et al. Farmaco 45(S6), 737 (1990).

50. Kupferschmidt, R. and Schmid, R.W. Chin. Chem. 35, 1313 (1989).

51. Ruprechts, R.M., Sharpe, A.H. et al. J. Chromatogr. 528, 371, (1990).

52. Taurcotte, J.G., Pivarnik, P.E. et al J. Chromatogr. 499, 55 (1990).

53. Schwartz, O., Henin, Y. et al AIDS Res. Hum. Retroviruses 4, 441 (1988).

54. Pauwels, R. Balzarini, J. et al. J. Virol. Methods 20, 309 (1988).

55. Good, S.S., Reynolds, D.J. and De Miranda, P. J. Chromatogr. Biomed. Appl. 431, 123 (1988).

56. Lacroix, C., Hoang, T.P. et al J. Chromatogr. Biomed. Appl. 525, 240 (1990).

57. Henry, K. Chinnock, B.J. et al. JAMA 259, 3023 (1988).

58. Quinn, R.P., Orban, B. et al. J. Immunoassay 10, 177 (1989).

59. Mathes, L.E., Muschik, G. et al. J. Chromatogr. 432, 346 (1988).

60. Tadepalli, S.M., Quinn, R. P. et al. J. AIDS 3, 19 (1990).

61. Puckett, L.D., Orf, J.W. et al. Fourth Intern. Conf. on AIDS, Stockholm, June 12-16, 1988 Vol. 2. Stockholm: Swedish Ministry of Health and Social Affairs, 1988: abstract no. 177.

62. Klecker, R.W. et al. Clin. Pharmacol. Ther. 41, 407 (1987).

63. Hedaya, M.A. and Sawchuk, R.J. Clin. Chem. 34, 1565 (1988).

64 Tadepalli, S.M. and Puckett, L. et al. Clin. Chem. 36, 897 (1990).

65. Batra, K., Dosouza, M. et al. Intl. Conf. on AIDS, Montreal, Quebec, Canada, June 4-9, 1990, Abstract no. 345 p. 279.

66. Schols, D., Pauwels, R. et al. J. Immunol. Methods 114, 27 (1988).

67. Granich, G.G., Eleland, M. R. et al. Antimicrob. Agents Chemother. 33, 1275 (1989).

68. Langtry, H.D. and Campoli-Richards, D.M. Drugs 37, 408 (1989).

69. Schinazi, R.F., Sommadossi, J.P. et al. Antimicrob. Agents Chemother. 34, 1061 (1990).

70. Hartmann, H., Vogt, M. E. et al. AIDS Res. Hum. Retroviruses 4, 457 (1988).

71. Perno, C.F., Yarchoan, R. et al. J. Exp. Med. 168, 111 (1988).

72. Olsen, J.C., Furman, P. et al. J. Virol. 61, 2800 (1987).

73. Fauci, A.S. and Lane, H.C. The Acquired immunodeficiency Syndrome (AIDS) in Braunwald, et al. (Eds.) Harrison's Principles of Intern. Medicine, 11th ed. New York: McGraw-Hill, 1987, pp. 1392-1396.

74. Tsai, C-C, Follis, K.E. et al. Proc. Soc. Exp. Biol. 185, 240 (1987).

75. Krieg, C.J., Ostertag, W. et al. Exper. Cell Res. 116, 21 (1978).

76. Dahlberg, J.E., Mitsuya, H. et al. Proc. Natl. Acad. Sci. USA 84, 2469 (1987).

77. Nakashima, H., Tochikura, T. et al. Virology 159, 169 (1987).

78. Hamamoto, Y., Nakashima, H. et al. Antimicrob. Agents Chemother., 31, 907 (1987).

79. Pauwels, R., De Clerq, E. et al. J. Virology Methods 16, 171 (1987).

80. Smith, M.S., Brian, E. L. et al. J Virol. 61, 3769 (1987).

81. Fridland, A. Connelly, M.C. and Ashmun, R. Mol. Pharmacol. 37, 665 (1990).

82. Eriksson, B.F., Chu, C.K. and Schinazi, R.J. Antimicrob. Agents Chemother. 33, 1729 (1989)

83. Baba, M., Pauwels, R. et al. Biochem. Biophys. Res. Commun. 142, 128 (1987).

84. Furman, P.A. Fyfe, J.A. et al. Proc. Natl. Acad. Sci. USA 83, 8333 (1986).

85. Lyerly, H.K., Cohen, O.J. and Weinhold, K.J. AIDS Res. Hum. Retroviruses 3, 87 (1987).

86. Nakashima, H.K., Matsui, T. et al. Antimicrob. Agents Chemother. 30, 933 (1986).

87. Nordenfelt, E. Lofgren, B. et al. J. Med. Virol. 22, 231 (1987).

88. Lin, J-C., Zhang, Z-X et al. Antimicrob. Agents Chemother. 32, 265 (1988).

89. De Clerq, E., Balzarini, J. et al. Biochem. Pharmacol. 29, 1849 (1980).

90. Schinazi, R.R., Erikasson, B. F. H. et al. Antiviral Res. 9, 119 (1988).

91. Johnson, V.A., Walker, B.D. et al. IV Intern. Conf. AIDS, Stockholm, June 12-16, 1988, Abstract no. 3618.

92. Tornevik, Y and Eriksson, S. IV Intern. Conf. on AIDS, Stockholm, June 12-16, 1988 Abstract no. 3150.

93. Williams, G.J. Antiviral Res. 9, 138 (1988).

94. Ueno, R. and Kuno S. Lancet I, 1379 (1987).

95. Ito, M., Nakashima, H. et al. Antiviral Res. 7, 127 (1987).

96. Hammer, S.M. and Gillis, J.M. Antimicrob. Agents Chemother. 31, 1046 (1987).

97. Tavares, L., Roneker, C. et al. Cancer Res. 47, 3190 (1987).

98. Hardy, W.D., Zuckerman, E. E. et al. J. Cell Biochem. S 10A, 207 (1986),

99. Ruprecht, R.N., O'Brien, L.G. et al. Nature 323, 467 (1986).

100. Baba, M., Pauwels, R. et al. III Intern. Conf. on AIDS, Washington, D.C. June 1-5, 1987, Abstract no. T 4.2, p 54.

101. Sommadossi, J-P, and Carlisle, R. Antimicrob. Agents Chemother. 31, 452 (1987).

102. Ayers, K.M. Am. J. Med. 85, 186 (1988).

103. Mitsuya, H., Yarchoan, R. et al. J. Am. Acad. Dermatol. 22, 1282 (1990).

104. Olsen, J.C., Furman, P. et al J. Virol. 61, 2800 (1987).

105. Tarusova, N.B., Khorlin, A.A. et al. Mol. Biol. (Mosk) 23, 1716 (1989).

106. Komig, H., Behr, E. et al. Antimicrob. Agents Chemother. 33, 2109 (1989).

107. Kedar, P.S., Abbotts, J. et al. Biochem. 29, 3603 (1990).

108. Elwell, L.P., Ferone, R. et al. Antimicrob. Agents Chemother. 31, 274 (1987).

109. Volberding, P.A., Lagakos, S.W. et al. N. Engl. J. Med. 322, 941 (1990).

110. Fischl, M.A., Richman, D.D. et al. N. Engl. J. Med. 317, 185 (1987).

111. Spear, J., Steinberg, J. et al. 27th Intersci, Conf. on Antimicrob. Agents Chemother., October 4-7, 1987, New York, Abstract no. 384, p. 163.

112. Finaud, M., Gallais, M. et al. IV Intern Conf. on AIDS, Stockholm, June 12-16, 1988, Abstract no. 3636.

113. De Wolf, F., Lange, J. M. A. et al. Lancet 1, 373 (1988).

114. Blanche, S., Caniglia, M. et al. Am. J. Med. 85, 203 (1988).

115. Morse, G. D., Lechner, J. L. et al. DICP, The Ann. Pharmacother. 24, 754 (1990).

116. Pizzo, P. A., Eddy, J. et al. N. Engl. J. Med. 319, 889 (1988).

117. Lane, H. C., Falloon, J. et al. Ann. Intern. Med. 111, 41 (1989).

118. Ruzicka, T., Froschl, M. et al. Lancet 2, 1469 (1987).

119. Cotton, P. JAMA 263, 1605 (1990).

120. Fleury, H. J. A. , Morere, G. et al. Lancet, 1, 880 (1988).

121. Brew, B.J., Rosenblum, M. and Price, R.W. J. Neuroimmunol. 20, 133 (1988).

122. Diquet, B., Simon, F. et al. V Intern. Conf. on AIDS, Montreal, Quebec, Canada, June 4-9, 1989, Abstract no. 343, p. 279.

123. Fischl, M. A., Richman, D.D. et al. Ann. Intern. Med. 112, 727 (1990).

124. Creagh-Kirs, T., Doi, P. et al. JAMA 260, 3009 (1988).

125. Flanigan, T., Shapiro, E. et al. N. Engl. J. Med. 317, 1155 (1987).

126. Henry, K., Cantrill, H. et al. JAMA 257, 3066 (1987).

127. Surbone, A., Yarchoan, R. et al. Ann. Intern Med. 108, 534 (1988).

128. Mitsuya, H., Matsukura, M. and Broder, S. Rapid in vitro systems for assessing activity of agents against HTLV-III/LAV In Broder S. (ed.) AIDS: modern concepts and therapeutic challenges, New York: Marcel Dekker, 1987 pp.303-333.

129. Orholm, M. Pedersen, C. et al. IV Intern Conf. on AIDS, Stockholm, June 12-16, 1988, Abstract no. 3626.

130. Hartshorn, K.L., Vogt, M. W. et al. Antimicrob. Agents Chemother. 31, 168 (1987).

131. Richman, D.D., Fischl, M. A. et al. N. Engl. J. Med. 317, 192 (1987).

132. Dubin, G. and Braffman, M.N. Ann. Intern. Med. 110, 85 (1989).

133. Loke, R.H., Murray-Lyon, I.M. and Carter, G.D. Brit. Med. J. 300, 163 (1990).

134. Camsonne, R., Bazin C. et al. Therapie 44, 417 (1989).

135. Collins, J.M. and Unadkat., J.D. Clin Pharmacokinet. 17, 1 (1989).

136. Drew, R.H. et al. Antimicrob. Agents Chemother. 33, 180 (1989).

137. Morse, G.D., Olson, J. et al. Antiviral Res. 11, 57 (1989).

138. Boudinot, F.D., Schinazi, P.C. et al. AIDS Res. Hum. Retroviruses 6, 219 (1990).

139. Cload, P. A. J. Infect. 18(S1), 15 (1989).

140. Unadkat, J.D., Wang, J.P. et al. Pharm. Res. 6, 734 (1989).

141. Blum, M. R., Liao, S.H.T. et al. Am. J. Med. 85 (S 2A), 189 (1988).

142. Wearley, L and Chien, Y. M. Pharm. Res. 7, 34 (1989).

143. De Miranda, P., Burnette, T.C. and Good S. S. Drug Metab. Dispos. 18, 315 (1990).

144. Tartaglione, T.A., Holeman, E. et al. J. AIDS 3, 32 (1990).

145. Balzarini, J., Herdewijn, P. and De Clerq. E. J. Biol. Chem. 264, 6127 (1989).

146. Collens, J.M., Klecker, R.W. et al. J. Pharmacol. Exp. Ther. 245, 466 (1988).

147. De Miranda, P., Burnette, T.C. and Good S. S. 27th Intersci. Conf. Antimicrob. Agents Chemother., New York, October 4-7, 1987, Abstract no. 378, p. 162.

148. Good. S.S., Koble, C. S. et al Drug Metab. Dispos. 18, 321 (1990).

149. Krasny, H.C. and Good, S.S. Fed. Proc. 45, 207 (1986).

150. Henry, K., Chinnock, B.J. et al. JAMA 259, 3032 (1988).

151. Klecker, R.W., Collins, J.M. et al. Clin. Pharmacol. Ther. 41, 407 (1987).

152. McCloskey, S.E., Rege, A.B. et al. Clin Res. 36, 21 A (1988).

153. Torrence, P.F., Kinjo. et al. FEBS-Lett 234, 135 (1988).

154. Gillet, J-Y, Bougain, A. et al. J. Gynecol. Obstet. Biol. Reprod. (Paris) 19, 177 (1990).

155. Liebes L., Mendoza, S. et al. J. Infec. Dis. 161, 203 (1990).

156. Doshi, K.J. Gallo, J.M. et al. Drug Metab. Dispos. 17, 590 (1989).

157. Gillet, J.Y., Garrafo, R. et al. Lancet 2, 269 (1989).

158. Cretton, E.M., Waterhous, D.V. et al. Drug Metab. Dispos. 18, 369 (1990).

159. Haumont, M.,Magdalou, J. et al. Arch. Biochem. Biophys. 281, 264 (1990).

160. McKinney, R.E. et al. J. Pediatr. 116, 640 (1990).

161. Ballis, F., Pizzo, P. et al. Proc. Am. Soc. Clin. Oncol. 7, 1 (1988).

162. Ballis, F.M., Pizzo, P.A. et al. J. Pediatr. 114, 880 (1989).

163. De Miranda, P., Good, S.S. et al. Clin. Pharmacol. Ther. 46, 494 (1989).

164. Steffe, E.M., King, J.H. et al. J. AIDS 3, 691 (1990).

165. Hedaya, M.A. and Sawchuk, R. J. J. Pharm. Sci. 78, 716 (1989).

166. Kornhauser, D.M., Petty, B.G. et al. Lancet 2, 473 (1989).

167. Hedaya, M.A., Elmquist, W.F. and Sawchuk, R. J. Pharm. Res. 7, 411 (1990).

CUMULATIVE INDEX

Bold numerals refer to volume numbers